Space-Time Integration in Geography and GIScience

Mei-Po Kwan • Douglas Richardson
Donggen Wang • Chenghu Zhou
Editors

Space-Time Integration in Geography and GIScience

Research Frontiers in the US and China

Springer

Editors
Mei-Po Kwan
Department of Geography
and Geographic Information Science
University of Illinois at
Urbana-Champaign, Champaign
IL, USA

Department of Human Geography
and Spatial Planning
Utrecht University, Utrecht
The Netherlands

Donggen Wang
Department of Geography
Hong Kong Baptist University
Hong Kong, China

Douglas Richardson
Association of American Geographers
Washington, DC, USA

Chenghu Zhou
Institute of Geographical Sciences
and Natural Resources
Chinese Academy of Sciences
Beijing, China

ISBN 978-94-017-7905-0 ISBN 978-94-017-9205-9 (eBook)
DOI 10.1007/978-94-017-9205-9
Springer Dordrecht Heidelberg New York London

Printed on acid-free paper

Springer is part of Springer Science+Business Media (www.springer.com)

Acknowledgments

The editors would like to acknowledge and thank the following for their support of this book: the US National Science Foundation, the Chinese Academy of Science, the US National Institutes of Health, the National Natural Science Foundation of China, the Geographical Society of China, the Association of American Geographers, and our editors at Springer Publishing. We would also like to thank Miranda Lecea for her editorial production work on the book and Xiaoyi Ma, who coordinated the translation of the book to Mandarin Chinese.

Contents

Contributors

Yanwei Chai Department of Urban and Economic Geography, College of Urban and Environmental Sciences, Peking University, Beijing, China

Yongwan Chun School of Economic, Political and Policy Sciences, University of Texas at Dallas, Richardson, TX, USA

Yingying Duan State Key Laboratory of Resources and Environmental Information System, Institute of Geographical Sciences and Natural Resources Research, Chinese Academy of Sciences, Beijing, People's Republic of China

Michael R. Evans Department of Computer Science and Engineering, University of Minnesota, Minneapolis, MN, USA

Zhixiang Fang State Key Laboratory for Information Engineering in Surveying, Mapping, and Remote Sensing, Wuhan University, Wuhan, People's Republic of China

Betsy George Department of Computer Science and Engineering, University of Minnesota, Minneapolis, MN, USA

Peng Gong Department of Environmental Science, Policy, and Management, University of California, Berkeley, CA, USA

Center for Earth System Science, Tsinghua University, Beijing, China

Michael F. Goodchild Department of Geography, University of California, Santa Barbara, CA, USA

Daniel A. Griffith School of Economic, Political and Policy Sciences, University of Texas at Dallas, Richardson, TX, USA

Zhipeng Gui Center for Intelligent Spatial Computing, George Mason University, Fairfax, VA, USA

Viswanath Gunturi Department of Computer Science and Engineering, University of Minnesota, Minneapolis, MN, USA

Qunying Huang Center for Intelligent Spatial Computing, George Mason University, Fairfax, VA, USA

Yunfeng Jiang Center for Intelligent Spatial Computing, George Mason University, Fairfax, VA, USA

Mei-Po Kwan Department of Geography and Geographic Information Science, University of Illinois at Urbana-Champaign, Champaign, IL, USA

Department of Human Geography and Spatial Planning, Utrecht University, Utrecht, The Netherlands

Harry F. Lee Department of Geography and International Centre of China Development Studies, University of Hong Kong, Hong Kong, Hong Kong

Qingquan Li Shenzhen Key Laboratory of Spatial Smart Sensing and Services, College of Civil Engineering, Shenzhen University, Shenzhen, People's Republic of China

Rongrong Li Institute of Space and Earth Information Science, The Chinese University of Hong Kong, Hong Kong, Hong Kong

Zhenlong Li Center for Intelligent Spatial Computing, George Mason University, Fairfax, VA, USA

Hui Lin Institute of Space and Earth Information Science, The Chinese University of Hong Kong, Hong Kong, Hong Kong

Desheng Liu Department of Geography, The Ohio State University, Columbus, OH, USA

Kai Liu Center for Intelligent Spatial Computing, George Mason University, Fairfax, VA, USA

Zhilin Liu School of Public Policy and Management, Tsinghua University, Beijing, China

Peter Lostritto Center for Intelligent Spatial Computing, George Mason University, Fairfax, VA, USA

Feng Lu State Key Laboratory of Resources and Environmental Information System, Institute of Geographical Sciences and Natural Resources Research, Chinese Academy of Sciences, Beijing, People's Republic of China

Harvey J. Miller Department of Geography, The Ohio State University, Columbus, OH, USA

Pradeep Mohan Department of Computer Science and Engineering, University of Minnesota, Twin Cities, Minneapolis, MN, USA

Atsushi Nara Geography, San Diego State University, San Diego, CA, USA

Zhixin Qi Department of Urban Planning and Design, The University of Hong Kong, Hong Kong, Hong Kong

Douglas Richardson Association of American Geographers, Washington, DC, USA

Shih-Lung Shaw Department of Geography, University of Tennessee, Knoxville, TN, USA

Shashi Shekhar Department of Computer Science and Engineering, University of Minnesota, Twin Cities, Minneapolis, MN, USA

Kee-long Billy So Division of Humanities, Hong Kong University of Science and Technology, Clear Water Bay, Kowloon, Hong Kong

Ying Song Department of Geography, The Ohio State University, Columbus, OH, USA

Kathleen Stewart Department of Geography, The University of Iowa, Iowa City, IA, USA

Bingxia Sun Department of Geography, Hong Kong Baptist University, Kowloon Tong, Kowloon, Hong Kong, Hong Kong

Guibo Sun Institute of Space and Earth Information Science, The Chinese University of Hong Kong, Hong Kong, Hong Kong

Min Sun Center for Intelligent Spatial Computing, George Mason University, Fairfax, VA, USA

Ka-chai Tam Department of History, Hong Kong Baptist University, Kowloon Tong, Kowloon, Hong Kong

Tana, Ph.D. Department of Urban and Economic Geography, College of Urban and Environmental Sciences, Peking University, Beijing, China

Yu-Shiuan Tsai Department of Geography, National Taiwan University, Taipei, Taiwan

Wei Tu Shenzhen Key Laboratory of Spatial Smart Sensing and Services, College of Civil Engineering, Shenzhen University, Shenzhen, People's Republic of China

Jaclyn Vogliano NextEra Energy Resources, LLC, Juno Beach, FL, USA

Donggen Wang Department of Geography, Hong Kong Baptist University, Kowloon Tong, Kowloon, Hong Kong, Hong Kong

Wei Wang Department of Geography, The University of Iowa, Iowa City, IA, USA

Tzai-Hung Wen Department of Geography, National Taiwan University, Taipei, Taiwan

Jizhe Xia Center for Intelligent Spatial Computing, George Mason University, Fairfax, VA, USA

Chen Xu Center for Intelligent Spatial Computing, George Mason University, Fairfax, VA, USA

Chaowei Yang Center for Intelligent Spatial Computing, George Mason University, Fairfax, VA, USA

KwangSoo Yang Department of Computer Science and Engineering, University of Minnesota, Minneapolis, MN, USA

Anthony Gar-On Yeh Department of Urban Planning and Design, The University of Hong Kong, Hong Kong, Hong Kong

Manzhu Yu Center for Intelligent Spatial Computing, George Mason University, Fairfax, VA, USA

May Yuan Geospatial Information Science, University of Texas at Dallas, Richardson, TX, USA

David D. Zhang Department of Geography and International Centre of China Development Studies, University of Hong Kong, Hong Kong, Hong Kong

Yan Zhang Institute of Beijing Studies, Beijing Union University, Beijing, China

Chenghu Zhou Institute of Geographical Sciences and Natural Resources, Chinese Academy of Sciences, Beijing, China

Nanying Zhou Center for Intelligent Spatial Computing, George Mason University, Fairfax, VA, USA

Xun Zhou Department of Computer Science and Engineering, University of Minnesota, Twin Cities, Minneapolis, MN, USA

Chapter 1
Introduction

Mei-Po Kwan, Douglas Richardson, Donggen Wang, and Chenghu Zhou

This book grew out of two recent initiatives. The first was the development of long-term collaboration between the Association of American Geographers (AAG) and the Geographical Society of China (GSC). This activity resulted in a formal AAG-GSC memorandum of understanding, addressing ten areas of mutual interest and coordination. These included research, publications, holding of joint meetings, educational seminars, translations of key geographical articles and books, and editorial collaboration on the Wiley-AAG *International Encyclopedia of Geography*.

The second major initiative leading to the development of this book was the special symposium entitled "Space-Time Integration in Geography and GIScience" organized by the editors and others and held within the AAG's 2011 Annual Meeting at Seattle. This symposium met with overwhelming response with five

M.-P. Kwan (✉)
Department of Geography and Geographic Information Science, University of Illinois at Urbana-Champaign, Champaign, IL 61820, USA

Department of Human Geography and Spatial Planning, Utrecht University, Utrecht, The Netherlands
e-mail: mpk654@gmail.com

D. Richardson
Association of American Geographers, 1710 16th Street NW, Washington, DC 20009, USA
e-mail: drichardson@aag.org

D. Wang
Department of Geography, Hong Kong Baptist University, Hong Kong, Hong Kong
e-mail: dgwang@hkbu.edu.hk

C. Zhou
Institute of Geographical Sciences and Natural Resources, Chinese Academy of Sciences, 11A Datun Road, Chaoyang District, Beijing 100101, China
e-mail: zhouch@lreis.ac.cn

M.-P. Kwan et al. (eds.), *Space-Time Integration in Geography and GIScience*, DOI 10.1007/978-94-017-9205-9_1

concurrent sessions over three days including 300 presentations. These addressed cutting edge space-time GIScience research challenges from theoretical foundations to research in topical areas ranging from transportations and urban systems to climate change and historical GIS. This symposium saw strong participation and attendance by researchers from China.

This book seeks to enhance research on space-time integration and bring together two major clusters of research activity in this rapidly growing field. Juxtaposing work side by side, from the U.S. and China, it sets the stage for new collaborative/joint research progress, uniting the strengths of each country in synergistic ways. In part this book also is a first step in developing "shared language" regarding complex concepts related to space-time integration research. Chinese and U.S. terminology and approaches to space and time concepts and their integration in GIScience research draw from distinct deeply established cultural, historical, religious, and scientific traditions and foundations. That is part of what, for us, makes this book so exciting, and innovative. It represents a mechanism and venue for helping to bridge ontological and philosophical divides and to create common terminology and conversations that help translate, combine, delineate, and make understandable these varying fundamental approaches to ideas of such core concepts as space and time and how they interact in GIScience.

Clearly, both cultures and countries have much to gain, intellectually and scientifically, and also pragmatically, if we can effect greater communication and understanding at all levels, including scientifically on the topics such as those addressed by this book.

Research in the field of space-time integration is of course by no means limited to the U.S. and China, and the papers that follow acknowledge precedents and ongoing work from around the world, including Europe, Japan, Canada, and elsewhere. However, the increasingly dynamic scholarly and research interaction between the U.S. and China is particularly notable in these fields, witnessed by a proliferation of conferences, publications, workshops, student and faculty exchange, and joint research programs. In fact, many U.S. researchers are now based in China for extended periods of time and many leading Chinese researchers are now based at U.S. universities and research labs. The intellectual exchange between the U.S. and China is rapidly growing and the collaborations in geography and GIScience are increasingly open and capable of producing exciting advances.

The authorship of this book also illustrates the already intertwined nature of research among U.S. and Chinese scholars. Many of the following chapters are jointly authored by U.S. and Chinese scientists. Authors include Chinese researchers temporarily or permanently based in the U.S., and U.S. researchers who have spent significant time working at Chinese universities and research centers.

Thus, we use the term "U.S.-based researchers" and "China-based researchers" instead of American or Chinese researchers to reflect these existing collaborations. The sampling of research from each country published here was selected with future collaborations in mind.

1.1 Synopsis of Chapters Led by U.S.-Based Authors

The eight chapters by U.S.-based researchers focus on various topics on space-time integration in GIScience research. They address issues such as the processing and analysis of spatiotemporal big data, mining spatiotemporal data for knowledge discovery, and modeling dynamic exposure to air pollution.

The opening chapter by Goodchild and Gong examines the distinctive and evolving tradition of science and geographic research in the U.S. and China. The chapter discusses issues ranging from scientific methods to data collection and sharing to the cultural and political contexts of the U.S. and China. It suggests that space-time research in U.S. and Chinese science will continue to converge in the future.

The chapter by Kwan explores three notions in geographic research that will benefit greatly through the lenses of time and human mobility: racial segregation, environmental exposure, and accessibility. It argues for the need to expand our analytical focus from static residential spaces to other relevant places and times in people's everyday lives. It seeks to conceptualize and understand people's spatiotemporal experiences based on their everyday mobility, since these complex experiences cannot be fully understood by just looking at where people live.

Evans et al. discuss spatiotemporal networks as time-aggregated graphs useful for reasoning, analysis and algorithm design in various social applications. The chapter presents a spatiotemporal network model called time-aggregated graph that represents network properties as a time series. The model reduces the temporal replication needed in other models while retaining spatial network information. Griffith and Chun implement an extension of the eigenvector spatial filtering principle to eigenvector space-time filtering to analyze the space-time structure underlying the Census mail response rates (MRRs) in the U.S. for three decennial censuses. The results provide helpful insights into various census data collection issues that are useful for future census data collection.

Kwan, Liu, and Vogliano present a method for improving air pollution exposure assessment through integrating the spatiotemporal dynamics of pollution concentrations and detailed space-time activity patterns of individuals. The chapter differentiates two sources of exposure to air pollution: (a) static exposure evaluated at an individual's residential location; and (b) dynamic exposure assessed based on the individual's locations at different times of the day. The method was implemented to examine individual exposure to particulate matter PM_{10} in Columbus, Ohio (U.S.).

Song and Miller focus on the internal structure of the space-time prism. The chapter discusses and illustrates two properties of prism interior: the probability to visit each location within the prism interior and the velocity profile associated with each location. Two examples are provided to demonstrate the benefits of applying these two properties: a modified utility-based accessibility benefit measure and a new measure for the potential environmental costs of accessibility.

Wang and Stewart develop a method for automatically representing the spatiotemporal information and tracking moving entities over space-time to show the spatial extent and trends of the dynamics of hazard-related events. The method automatically processes event semantics to add meaning to the information mapped from documents. The chapter also discusses methods for combining natural language processing techniques and ontologies to improve the extraction of semantics of hazard-related events sourced from web text. Using their recent projects as examples, the chapter by Yang et al. discusses how cutting-edge computing technologies can be utilized to handle and process spatiotemporal big data. The technologies discussed in the chapter include: (1) grid and cloud computing; (2) sensor web, Earth observations, and simulation models; (3) standard-based systems; and (4) multidimensional visual analytics.

Yuan and Nara provide an overview of space-time trajectory analysis and discusses the challenges of scaling it up to spatiotemporal big data. The chapter provides a conceptual framework and methodology for performing space-time analysis with a massive number of space-time tracks and illustrates its usefulness for extracting complex activity patterns with several examples.

The chapter by Zhou, Shekhar, and Mohan seeks to provide an interdisciplinary perspective on methods for mining spatiotemporal change patterns and promotes cross-fertilization of ideas among researchers of diverse backgrounds. The chapter analyzes these taxonomies and patterns from a space-time perspective to compare related concepts from different disciplines.

In the concluding chapter, Richardson addresses the origins and emergence of real-time space-time integration in GIScence and geography, and develops a working research agenda for moving this dynamic field forward. He also explores the concept of transformational research in geography and GIScience, and raises key questions about how innovation and discovery in GIScience might be better fostered by examining the actual practices of GIScience research and innovation, and the social structures and contexts which constrain or enable our science.

1.2 Synopsis of Chapters Led by China-Based Authors

The contributions by China-based scholars cover various spatiotemporal issues at scales spanning global to micro levels. They include methodological developments and a review of the state-of-the-art research and applications in different fields. Many of these chapters also provide case studies in China. They reflect not only the progress in space-time research in China, but also examine prominent environment, social, and urban issues in China.

Lee and Zhang explore the association between climate change and human crises in pre-industrial societies. Addressing a spatiotemporal issue at the global and geological time scale, this study seeks to establish potential linkages between climate changes and population checks, population collapses, and socio-political chaos in China, Europe, and other countries/regions in the Northern Hemisphere during the pre-industrial era. Challenging the Malthusian explanation for human misery, the study argues that deteriorating climate, which led to reduced land carrying capacity, was the ultimate cause of human crises in pre-industrial societies. Addressing a local issue of land use changes, Yeh and Qi explore the potentials of using Polarimetric synthetic aperture radar (PolSAR) images in short-interval monitoring of land use and land cover changes. The study involves the use of a time series of RADARSAT-2 PolSAR images to monitor change in land development at its early stages.

Tam and So present a study of historical GIS. It introduces a method of transforming the verbal and qualitative information collected by traditional local gazetteers in China into quantitative data that can be visualized in a modern historical GIS. The method is illustrated with a case study of drawing the township boundaries of the Songjiang prefecture circa the Ming-Qing periods of China. The method can be applied to build a GIS infrastructure for China's historical geography at a spatial scale lower than the prefectural level presently used by most of the country's historical GIS. The chapter by Wen and Tsai presents a methodology on modeling spatiotemporal diffusion of epidemics. The methodology is illustrated by a case study on a dengue epidemic in Kaohsiung, Taiwan. The results showed that this method can be used to detect the possible origin of an epidemic and to differentiate patterns of spatial diffusion.

Five chapters focus on the application of the space-time approach to urban and transportation research. The chapter by Chai et al. provides an overview of individual-based spatiotemporal behavior studies in China. It addresses issues concerning methodologies, data collection and empirical applications. Applications in urban transportation, tourism, and urban planning in China are discussed. It argues that the individual-based space-time behavioral approach has contributed to improving our understanding of the implications of the dramatic spatial restructuring in urban China for individuals' daily life, social equity and environmental issues. Lin et al. provide a comprehensive review of the theoretical considerations and measurement issues concerning the influence of built environment on walking behavior in China. They explore the use of positioning technologies such as global positioning system (GPS) to collect activity-travel behavior information and to derive activity types and trip purposes from the trajectory data.

Wang and Sun propose a data mining approach for detecting activity type and trip purpose from passive GPS trajectory data. The authors argue that the proposed approach has self-learning and self-adaptive capabilities and needs neither prior knowledge nor artificial interference in the process of searching for the optimal solutions. The chapter by Fang et al. presents a time-dependent prism-based approach to identifying critical links and opportunities for scheduling joint activities of multiple individuals. A scenario involving joint activities among four persons is designed and implemented to demonstrate the feasibility of this approach. Duan and Lu present a study of employing a complex network-based method to explore the robustness of city road networks at three different granularities: road segment-based, stroke-based, and community-based. They argue that the community-based method is a suitable method for evaluating the robustness of city road networks.

The topics addressed in this book open many new research pathways forward. The editors and leading researchers in geography and GIScience represented in the following chapters welcome ongoing discussion and interaction with the readers of this book, and with space-time researchers from around the world. It has been a great personal pleasure for us, as editors, to engage in this joint process, and we are convinced that these initial efforts will yield significant advances in research on geographical and geospatial concepts and applications in the future.

Chapter 2
Perspectives on Space and Time in US and Chinese Science

Michael F. Goodchild and Peng Gong

2.1 Introduction

In principle, scientific knowledge is universal, and the methods of science similarly pay no attention to national boundaries, languages, or cultures. In practice, of course, these assertions are far from true, and great differences exist in how science is conceived and practiced around the world. Nowhere is this more true than in comparing scientific practice in the US and in China. The emergence of China onto the international scientific stage is comparatively recent, following the years of the Cultural Revolution, the reopening of universities, and the very rapid growth of the Chinese economy of the past three decades.

These issues penetrate science at all levels and in all domains, and nowhere more than in geography. The very definition of geographic science demands international consistency and collaboration, because geography is at once global and transcending of boundaries. While human systems are to some extent defined by language, culture, and boundaries, environmental systems are not, and lack of free exchange of environmental data is clearly an impediment to our understanding of how the Earth's environmental systems work.

This chapter focuses on these issues as they affect research that is framed in space and time. The authors represent the US (Goodchild) and China (Gong),

M.F. Goodchild (✉)
Department of Geography, University of California, Santa Barbara, CA 93106-4060, USA
e-mail: good@geog.ucsb.edu

P. Gong
Department of Environmental Science, Policy, and Management, University of California, Berkeley, CA, USA

Center for Earth System Science, Tsinghua University, Beijing, China
e-mail: penggong@berkeley.edu

M.-P. Kwan et al. (eds.), *Space-Time Integration in Geography and GIScience*, DOI 10.1007/978-94-017-9205-9_2

and while their interests as geographers span both the social and the environmental, one has worked in recent years more on the social side (Goodchild), and one on the environmental side (Gong). Thus the chapter presents two distinct perspectives, and ranges in content from the abstract issues of scientific method to the practical issues of data collection and sharing.

The next two sections present the US and Chinese perspectives. These are then followed by a brief conclusion.

2.2 A US Perspective

2.2.1 Methodological Perspectives

Science has long given greatest importance to discoveries that apply everywhere, at all times. In mathematics, physics, and chemistry, for example, it would be absurd to announce a discovery that was only valid on certain days of the week, in certain countries or places. Newton's laws of motion, Einstein's principles of special and general relativity, and Mendeleev's periodic table are all believed to be entirely independent of space and time.

The social and environmental sciences also inherited this perspective, though there has been almost continuous debate about its merits throughout their history. In economics, for example, a strong predilection for theory emerged in a discipline that showed little interest in how the principles of economic behavior might vary through time, and between different areas of the Earth's surface. The expression of economic principles might vary, of course, because of varying conditions, but the principles themselves should remain constant. In biology, Darwin's observations of finches in the Galapagos Islands led to a unified and universal principle of natural selection, rather than to a science based on detailed description of spatial or temporal variations.

In the United States, the discipline of geography has long struggled with these issues. The *nomothetic* argument holds that geography should be concerned with discovering principles that apply everywhere and at all times in the domain of geography, that is, the surface and near-surface of the Earth (see, for example, Bunge 1966). In the late 1950s it appeared that such principles might indeed be found, and great interest was expressed in apparent empirical regularities such as Zipf's rank-size rule and Horton's laws of stream number. Arguments grounded in economics led to the regularities predicted by Christaller and Lösch for patterns of settlement, and by von Thünen for patterns of land use. The spatial interaction model predicted a regular decline of interaction with distance. But many of these regularities turned out to be no more than the most likely outcome of random processes (Goodchild 1992), and the regularities predicted by Christaller and others turned out to be far too occluded by noise to be observable in any but the weakest forms (Goodchild 1969).

Many US geographers were unconvinced by the nomothetic arguments, preferring a discipline that focused on documenting the unique properties of places, a perspective termed *idiographic*. The nomothetic perspective lost further ground when the promised empirical regularities failed to materialize, when many found its version of *explanation* to be fundamentally weak, and when others saw it as detached from the world, rather than engaged with improving the world. By the late 1970s the nomothetic perspective had lost almost all of the momentum it had gained in the previous decade.

Maps have always been an important tool for explicit display of space for research, both inside and outside the discipline of geography. Biologists have long made maps of habitat and the distribution of species; sociologists have used maps to record varying human conditions in urban environments; and political scientists have used maps to study the dynamics of elections. In the 1960s it became possible for the first time to store the contents of maps in computers, and to use them for various purposes: editing the contents of maps prior to printing, making accurate measurements of properties such as area and distance from maps, and performing the detailed mathematical calculations needed to project the curved Earth onto flat paper. By the 1970s these and other applications had coalesced into the concept of a geographic information system (GIS), an integrated computer application for capturing, editing, storing, manipulating, and analyzing the contents of maps. The first commercially viable applications appeared in the 1980s, and GIS was well on the way to becoming the multi-billion-dollar industry that it is today.

In the four decades since the initial development of GIS, its widespread adoption has led to a profound reassessment of the nomothetic/idiographic debate, of the nature of geography as a discipline, and of the significance of space and time in US science. Much of the work in this area has occurred within the field of geographic information science (GIScience), which can be defined as the scientific study of the issues surrounding geographic information and its associated technologies.

First, there are two concepts that are now recognized as universal properties of geographic information – spatial dependence and spatial heterogeneity – and several additional candidates (Anselin 1989; Goodchild 2004). Spatial dependence is the foundation principle of the disciplines of spatial statistics and geostatistics, and in its most conceptual form amounts to the observation that nearby things are more similar than distant things, a statement often attributed to Tobler (Sui 2004). Spatial heterogeneity implies that the Earth's surface exhibits uncontrolled variation, and that no place can therefore be representative of the whole. It has led to recent interest in *place-based* techniques such as geographically weighted regression (Fotheringham et al. 2002) that propose general principles, but allow the specific parameters of those principles to vary spatially and temporally over the geographic domain.

In a GIS the database captures and represents the properties of locations and thus expresses the idiographic perspective, or Varenius's concept of *special geography* (Warntz 1989). The software performs functions on the data, including analysis and modeling, applying the same functions uniformly, and thus can be seen as expressing the nomothetic perspective, or Varenius's concept of *general geography*. In this sense GIS straddles the nomothetic/idiographic debate, unifying them in a single structure.

Some of the earliest US applications of GIS were in resources management, especially in the forestry industry, where it is used to manage and plan the growth and exploitation of timber. GIS is now extensively used in planning, landscape architecture, and other disciplines that adopt an *interventionist* stance, applying science to the design, modification, and hopefully improvement of the geographic domain. Recently the term *geodesign* has come to be associated with this perspective (Steinitz 2012), which is exclusively neither idiographic nor nomothetic, but closer to the activist ideals of the 1970s critics of scientific geography.

GIS has also done much to increase the scientific community's interest in space and time, by making it much easier to study phenomena within a spatiotemporal framework (Liang et al. 2010). In epidemiology, for example, it is now much easier to make maps of rates of disease, to identify places where rates are anomalous, and to access other information about those places and their surroundings. Maps have been used to link rates of certain cancers to exposure to certain carcinogens, for example. A map (and a GIS) is now much more likely to be seen as an essential tool in any of the disciplines dealing with phenomena distributed over the surface of the Earth. Yet substantial skepticism remains. Some years ago one of the authors was engaged in conversation by a well-known US economist, who argued that the explanatory power of location, in the form of latitude and longitude, was no greater than that provided by *any* pair of socially significant variables, such as income and level of education. But this argument misses an essential point – that it is not latitude and longitude per se that explain (although causal arguments can sometimes be made, especially for latitude). Instead properties derived from latitude and longitude, such as distance, are more explanatory, and latitude and longitude provide the essential links needed to establish context, and to show how phenomena can be explained by what is near in space and time. Clearly similar explanatory arguments cannot be made for what is near *in income and education*.

2.2.2 New Data Sources and New Questions

The original motivation for GIS was the map, and the need to perform various functions on its contents. But the contents of a map are often highly synthesized and compressed versions of original observations. A topographic map, for example, is a synthesis of perhaps millions of observations of ground elevation and the positions of identified features. A printed map is also very selective in its contents, focusing on the more stable and persistent phenomena of the Earth's surface and avoiding the transitory, in order to be valid for as long as possible after printing (Goodchild et al. 2007).

Technical developments over the past few decades, many of them in the US, have brought profound changes to the nature of maps and the availability of geographic information. Instead of being exclusively the preserve of experts, maps can now be made by anyone using readily available data and software, at very little cost. Maps can be made of real-time traffic conditions, the locations of aircraft, or the ground-based view from the current location of a smart-phone user. Time has new importance, and is now increasingly integrated into GIS and geographic information generally. Yet it has proven difficult to escape the power of the map metaphor (Goodchild 1988) and its constraining influence on research and development. In the case of the automobile it took many years before the metaphor of the horse and carriage (the *horseless carriage*) could be dropped and the full potential of the new technology realized, and this same pattern has clearly influenced the integration of time into GIS.

In the US, researchers now have access to a vast flood of geographic information, a significant part of what is variously referred to as the *exaflood* or *Big Data*. The locations of aircraft, buses, fleet vehicles, and ships are now continuously monitored and in many cases made available through the Internet. Mobile phones are continuously tracked when turned on, with positional uncertainties as small as a few meters. Credit and debit cards produce spatiotemporal records whenever they are used, though these will typically be available only to banks and law-enforcement agencies. Imaging sensors on satellites create terabytes of geographic information daily, as do sensors on aircraft and unmanned drones. Ground-based sensors monitor traffic conditions at points on road networks, and companies such as Google collect ground-based images from moving vehicles and even bicycles.

While these new data are abundant, major issues exist in linking and synthesizing them into the forms needed for research and applications. Traditional research sources, such as the US Census, are collected subject to rigid procedures and norms, compiled using replicable processes, and documented with detailed metadata. As a result it has been possible for vast amounts of research to be conducted on them over the years. But the new data sources often are incomplete, with no rigorous sampling design, and little in the way of metadata or quality assurance. Elwood et al. (2013) have argued that this new world of Big Data will be dominated by the need for synthesis and *post-hoc* quality control, and will require the development of a range of new techniques.

These arguments are especially apposite in the case of geographic information developed and contributed by citizens, as distinct from qualified experts. So-called *volunteered geographic information* (VGI; Goodchild 2007) or *crowd-sourced* geographic information is exemplified by projects such as Open Street Map and Wikimapia, and now constitutes a significant source of knowledge about the geographic world that is comparable in its value to that produced by the traditional mapping agencies. Many companies now rely on VGI to update their digital street maps, and mapping agencies such as the US Geological Survey make use of citizen input as well. Goodchild and Li (2012) have reviewed the available strategies for quality assurance, and the growing literature on VGI quality has often shown accuracies at least as good as those of traditional sources, especially with respect to timeliness.

It was argued earlier that maps traditionally emphasized stable phenomena such as buildings, streets, and topography over transitory phenomena such as traffic conditions. Map-making was expensive, requiring the fielding of crews of experts. But VGI has the potential to provide timely updates and observations of phenomena in near-real time because it takes advantage of a dense distribution of citizens rather than a sparse distribution of experts. As a result VGI has become increasingly important in the creation and distribution of time-critical information about disasters such as wildfires (Goodchild and Glennon 2010). Interesting results have been obtained from the monitoring of twitter traffic, either by relying on locations embedded in messages or by using the georeferences that are associated with a small fraction of tweets.

2.3 A Chinese Perspective

2.3.1 Traditional Chinese Science

Why did modern science, with its emphasis on experiment and logical positivism not originated in China? This has puzzled many. The early maturity of the Chinese culture with its orientation towards statesmanship is part of the reason (Gong 2012a). The development of traditional Chinese science valued practicality and relied heavily on the experiences of others. The knowledge acquired from traditional Chinese science was often spread through private and mostly oral imparting from a master to his limited number of apprentices. For example, knowledge in traditional Chinese medical practice was usually handed down secretly in the family from generation to generation. Compared to the widespread publication of knowledge in western societies, knowledge from traditional Chinese science could be easily lost. The pattern of knowledge loss and re-invention in the long Chinese history has been facilitated by a culture that promotes obedience in all aspects of daily life to rules made up by ancestors. Patenting knowledge through private impartment is an important characteristic in traditional Chinese science and has had a strong influence.

The 1637 completion of the book *Tian Gong Kai Wu* ("making things like in the heaven with human technology"), written by Yingxing Song (1368–1644), was a highlight of traditional Chinese science, occurring in the late Ming Dynasty (Song 1637). It included detailed descriptions of traditional Chinese practical technologies of crop growing, silkworm growing, harvesting tools, weaving, ceramics, mining, and metallurgy. Some of the techniques and wooden instruments described in this book for wheat and rice growing and mining are still in use today. However, few abstract principles of physics and chemistry, that can be applied everywhere and at all times, have been put forward in traditional Chinese science. At around the same time, Guangqi Xu (1562–1633), mentored by foreign missionaries, became a pioneer modern scientist in China through his introduction of western geometry, arithmetic, surveying, and geography. His book entitled *Nong Zheng Quan Shu* ("comprehensive topics on agricultural affairs"), published in 1639, not only integrated historical agricultural achievements in China but also included his own contributions, obtained through his experiments in Shanghai and Tianjin (Xu 1639). Xu had hoped that a revolution in Chinese thinking would have occurred in the decades after he and Matteo Ricci translated part of Euclid's geometry into Chinese, in the early 1600s. Unfortunately, due to the downfall of the Ming Dynasty in 1644 and the succession of the Qing Dynasty (1636–1911), ruled by a minority nationality, widespread dissemination of western scientific thoughts did not happen for another 300 years.

Traditional Chinese science is primarily descriptive and less quantitative than western science. Despite the work of Guangqi Xu and Matteo Ricci in the early seventeenth century, it was unable to take advantage of western modern mathematics until recently. Traditional Chinese concepts of time are imprecise, being based on the division of the day into 12 time slots. Due to ignorance of western science and technology by the government and by humanities scholars, and the vulnerable approach to knowledge inheritance, many technologies developed in earlier Chinese history were lost and can only been seen through mentions in later writings.

Geographical knowledge is often recorded in local chronicles at the national, provincial and prefecture levels, impeding the search for nomothetic rules. Even idiographic approaches cannot be carried out thoroughly due to the lack of skills in quantitative analysis and lack of support from other disciplines such as mathematics, physics, chemistry, and biology. For example, a great expeditionary work entitled *Travel Notes of Xiake Xu* was completed in 1641, initially with 2.4 million words (only 0.4 million words remaining now). Self-supporting, Xiake Xu travelled over half of the provinces in China, primarily in southern China and mostly by himself. His travel diaries are primarily descriptive. Different from local chronicles, his diaries captured a bigger picture of China and its cross-provincial geographical conditions, including climate, landforms, living habits, products, mining, and hydrological properties (Xu 1641) at various locations. An important geographical work entitled *Du Shi Fang Yu Ji Yao* ("highlight of landmarks in historical works") was completed by Zuyu Gu in 1692, with 2.8 million words. In this book, Gu (1692) compiled from over 800 kinds of literature on geographical knowledge related to land resources, military fortresses, political evolution, and relative distances of every county in the late seventeenth century.

Modern earth science with a focus on location began in the late nineteenth century in China, when overseas Chinese scholars returned to China and began conducting field studies for mineral and coal exploration. Modern geological maps were being produced at the site and regional levels. For example, the first geological map made by a Chinese is the Geological Map of Zhili Province (now part of Hebei Province and Tianjin Municipality), developed by Rongguang Kuang in 1910. This map contains geological layers, drainage, cities, and railroad information. In the early 1900s, a handful of Chinese earth scientists returned from overseas established earth science programs in universities and research organizations. Modern geography and meteorology had been established by Dr. Kezheng Zhu in China in the 1920s. Station-based geographical data collection with instruments such as weather stations, runoff gauge stations, and water-level gauges also began in the early twentieth century. However, due to continuous national and international wars in the late 1800s and the first half of the 1900s, adding new geographical knowledge had been sporadic in space and time.

Large-scale field investigations of China's natural environment took place after 1949 in support of the rapid development of the agricultural and industrial sectors, and geographical science achieved significant progress in the 1950s and 1960s. As a result, a large number of maps at scales ranging from 1:4 million to 1:1 million were produced for the entirety of mainland China. These include maps of geological structure and mineralogy, soil, and physical geography. Large-scale topographic maps at 1:10,000–1:50,000 were being produced for the eastern part of China. In the meantime, relevant institutions developed rapidly in various levels of government. By the 1970s, in addition to agricultural bureaus established in the 1950s and 1960s, every county had a weather station, a hydrological bureau, and a forest bureau, part of whose responsibilities were collecting weather, hydrological, and timber data, respectively. Networks of weather observation, hydrological measurements, and timber inventory were rapidly developed in support of weather forecasting, flood and drought control, and planned timber production. In 1986, China began its second soil inventory which led to the production of soil maps at three scales: 1:50,000 at the county level, 1:500,000 scale at the provincial level and 1:1 million at the national level. However, due to poor archiving, many of the raw data of soil surveys were lost. In the same year, China began to monitor environmental quality in 113 selected cities under the administration of a newly established governmental agency: the State Environmental Protection Agency.

Mineral, hydrological, weather, soil, forest, environmental, and many more different types of data are collected directly under the coordination of governmental agencies at various levels. With multiple regional research institutes in geography and botany, the Chinese Academy of Sciences has completed several types of maps at the national level since the 1980s. These include China's 1:4 million vegetation map (Hou 1979) and the 1:1 million land use map in the 1980s (Wu 1990).

Geographic information systems gained popularity in the 1980s in the western world, and Chinese scholars quickly followed the trend. A state key lab was established in the Chinese Academy of Sciences on resources and environmental information systems by Shupeng Chen as early as 1989. GIS has been quickly included in university curricula. Now almost all geography programs in China offer GIS courses and many universities have established GIS departments. A large number of students specialized in GIS are employed in government agencies at various levels and in the industrial sector. This has effectively facilitated the digitizing of location-related data. Quick adoption of western concepts by Chinese scholars is also exemplified by the organization of the first International Symposium on Digital Earth in 1999 following the 1998 speech of Al Gore. In the following 9 years, the Chinese Academy of Sciences funded the International Society for Digital Earth in 2006 and the *International Journal of Digital Earth* in 2008 (Goodchild et al. 2012).

Since the 1990s, more detailed maps at the national scale have been made with digital remote sensing. Land-use and land-cover maps at 30 m resolution have been made every 5 years since 1995 (Liu et al. 2002). Wetland maps over the entirety of China have been made for 1978 (80 m resolution), 1990 (30 m), 2000 (30 m), and 2008 (20 m) (Gong et al. 2010; Niu et al. 2012). Urban expansion maps at 30 m resolution for all cities in China have been made for 1990, 2000, and 2010 (Wang et al. 2012). In addition, the 1:1 million vegetation map was compiled by Zhang (2007) and the 1:1 million landform map by Zhou (2009). In the meantime, most paper maps have been digitized and stored in various databases.

However, the temporal aspect of geographical data has lagged behind. This is reflected in a recent atlas that compiled population census data and environmental data in order to track changes from 1949 and 2009 (Gong and Liu 2010). Although China has conducted six population censuses since 1952, the finest mapping unit is at the county level, and population migration data can only be recovered at the provincial level. This has substantially restricted their use in city planning, not to mention location-based service analysis. Environmental data have a short history since 1986 and are limited to only 113 cities. This makes it impossible to have a comprehensive view of China's environmental change, not to mention forming widely applicable environmental policies. Recently atmospheric-quality analysis by third parties using satellite remote sensing has been demonstrated to be a viable solution (Zhang et al. 2012).

The situation with respect to temporal data is quickly improving thanks to the large number of satellites for environmental observation that are scheduled to be launched by the Chinese government (Kramer and Cracknell 2008), as well as to an intensification of ground-based observation networks. For example, in the 1990s the Chinese Academy of Sciences began to establish a network of ecological monitoring stations (the Chinese Ecosystem Research network) (Sun 2006; Yu et al. 2004). It has since grown to 46 stations in various ecosystems that provide data supporting research for thousands of researchers. Monitoring networks empowered by Internet and wireless sensor-network technologies serving different functions are being widely established by governmental agencies. For example, in order to prevent and control infectious diseases similar to the 2003 Severe Acute Respiratory Syndrome (SARS) from causing large-scale damage to public health, China developed in 2004 the Infectious Disease and Emergent Public Health Event Web-based Direct Reporting System linking all hospitals at township level and above. This system has been in operation for 9 years and has collected detailed information on 39 infectious diseases for the entirety of China on a daily basis. Also, many hydrological gauge stations are based on wireless sensor-network technology.

In summary, traditional Chinese science is empirical, descriptive, and handed down often privately through family inheritance or imparted from masters to apprentices. These characteristics are different from western science, which is experimental, quantitative, and analytic, and in which the spread of knowledge is rapid through publishing and open speech, that is later transformed into university lectures, seminars, and workshops. This restricted approach to the spread of knowledge made it hard for new ideas and new technologies to be quickly transmitted in ancient China. Geographical knowledge in space and time has been acquired only since the early twentieth century, particularly after 1949, following the widespread introduction of modern science. The collection of spatial and temporal data has been dramatically accelerated since the turn of the twenty-first century, supported by modern information technologies.

2.3.2 Challenges and Opportunities

Although significant progress has been made in spatiotemporal data collection and handling in China, some unique challenges exist. The most significant one is data sharing. From the above introduction it can be seen that with the exception of data created by the Chinese Academy of Sciences, most of the data products at the national level have been developed by governmental agencies. Access to small amounts of data at limited locations may be possible, but it is extremely difficult to obtain data for the entirety of China. For example, survey data such as topographic data at scales greater than 1:1 million are regarded as secret. Although hydrological data are not regarded as secret they are rarely accessible by the research community. Even among governmental agencies data are not sharable. For example, both the State Meteorology Administration (SMA) and the State Hydrology Bureau (SHB) collect precipitation and temperature data, and the number of weather stations operated by the latter is an order more than that operated by the former. But they do not share the data. Neither the SMA nor the research community has the opportunity to use the weather data from SHB to intensify the observational network and improve weather prediction. The second challenge is the lack of quality assurance to the data owned by the governmental agencies. For example, forest inventory data are collected through reports from provincial forest bureaus, who obtain their data from reports from the counties. Manipulation of data cannot be well tracked and quantified. Thus, the credibility of governmental data is uncertain. The third challenge, as discussed earlier, is the lack of creativity in making use of spatial and temporal data.

Chinese scientists currently face the best opportunities in history. They are now relatively well funded even compared with their colleagues in the western world. More creativity is naturally expected as the outcome. While sharing geographical data in China is a great bottleneck to the scientific community, additional efforts must be made to change the Chinese bureaucracy to improve data sharing among governmental agencies and the general public. The lack of quality assurance of the governmental data can be alleviated by using direct mapping methods, as in mapping wetlands and atmospheric quality using widely available remotely sensed data (Gong 2012b; Niu et al. 2012; Zhang et al. 2012). With the same group of people, the same set of tools, and the same sources of data, the whole China can be mapped more consistently under proper quality control. It is expected that through interdisciplinary collaboration and improved quantitative skills, Chinese scientists will bring more creativity in the analysis and effective use of spatial and temporal data (Xu et al. 2010). Lastly, China has a rich archive of local chronicles that contain geographic information through time. This unique source has been explored by some scholars of historical climate (Zhu 1972; Zhang 2004; Ge 2010). More coordinated extraction of spatiotemporal information from such data sources is necessary.

2.4 Conclusion

Tools for spatiotemporal analysis and modeling are currently in a state of flux, a result of rapid innovation in data collection and techniques of analysis, as well as in greater demand for real-time decisions and better planning. Yet as we have shown in this chapter, great differences still exist in the cultural and political context within which these tools must be used. While discussions aimed at improving data sharing and exchange are under way at several levels, the cultural and linguistic differences that have created the formidable impediments we see today will take much longer to address. We would be wise, also, to reflect on Chinese history and the political earthquakes that in the past have halted fragile progress. Nevertheless there are reasons to be optimistic, especially given the very rapid growth in scientific communication and exchange of personnel that has occurred between the US and China in the past decade. Organizations such as CPGIS (the International Association of Chinese Professionals in Geographic Information Sciences) are doing much to foster exchange, through cross-appointments, joint conferences, and publications. China's economic awakening of the past three decades is now reflected in a steady scientific awakening, and to creativity and innovation on a new and massive scale.

References

Anselin, L. (1989). *What is special about spatial data? Alternative perspectives on spatial data analysis*. Technical Paper 89-4. Santa Barbara: National Center for Geographic Information and Analysis.

Bunge, W. (1966). *Theoretical geography*. Lund: Gleerup.

Elwood, S., Goodchild, M. F., & Sui, D. Z. (2013). Prospects for VGI research and the emerging fourth paradigm. In D. Z. Sui, S. Elwood, & M. F. Goodchild (Eds.), *Crowdsourcing geographic knowledge: Volunteered Geographic Information (VGI) in theory and practice* (pp. 361–375). New York: Springer.

Fotheringham, A. S., Brunsdon, C., & Charlton, M. (2002). *Geographically weighted regression: The analysis of spatially varying relationships*. Hoboken: Wiley.

Ge, Q. S. (2010). *Climate change in China's history*. Beijing: Science Press (in Chinese).

Gong, P. (2012a). Cultural history holds back Chinese research. *Nature, 481*, 411.

Gong, P. (2012b). Remote sensing of environmental changes over China, a review. *Chinese Science Bulletin, 57*(22), 2793–2801.

Gong, P., & Liu, Y. (Eds.). (2010). *The atlas of population and environmental change of China (1949–2009)*. Beijing: Science Press (in Chinese).

Gong, P., Niu, Z. G., Cheng, X., et al. (2010). China's wetland change (1990–2000) determined by remote sensing. *Science in China: Earth Sciences, 53*(7), 1036–1042.

Goodchild, M. F. (1969). Local structure in the town populations of Iowa. *Geographical Analysis, 1*, 404.

Goodchild, M. F. (1988). Stepping over the line: Technological constraints and the new cartography. *American Cartographer, 15*, 311–319.

Goodchild, M. F. (1992). Analysis. In R. F. Abler, M. G. Marcus, & J. M. Olson (Eds.), *Geography's inner worlds: Pervasive themes in contemporary American geography* (pp. 138–162). New Brunswick: Rutgers University Press.
Goodchild, M. F. (2004). The validity and usefulness of laws in geographic information science and geography. *Annals of the Association of American Geographers, 94*(2), 300–303.
Goodchild, M. F. (2007). Citizens as sensors: The world of volunteered geography. *GeoJournal, 69*(4), 211–221.
Goodchild, M. F., & Glennon, J. A. (2010). Crowdsourcing geographic information for disaster response: A research frontier. *International Journal of Digital Earth, 3*(3), 231–241.
Goodchild, M. F., & Li, L. (2012). Assuring the quality of volunteered geographic information. *Spatial Statistics, 1*, 110–120.
Goodchild, M. F., Fu, P., & Rich, P. (2007). Sharing geographic information: An assessment of the geospatial one-stop. *Annals of the Association of American Geographers, 97*(2), 249–265.
Goodchild, M. F., Guo, H., Annoni, A., Bian, L., de Bie, K., Campbell, F., Craglia, M., Ehlers, M., van Genderen, J., Jackson, D., Lewis, A. J., Pesaresi, M., Remetey-Fülöpp, G., Simpson, R., Skidmore, A., Wang, C., & Woodgate, P. (2012). Next-generation digital earth. *Proceedings of the National Academy of Sciences, 109*(28), 11088–11094.
Gu, Z. Y. (1692). *Du Shi Fang Yu Ji Yao*. Reprinted by ZhongHua Book Company in 2005, Beijing (in Chinese).
Hou, X. Y. (1979). *The vegetation map of China (1:4,000,000)*. Beijing: Map Press (in Chinese).
Kramer, H. J., & Cracknell, A. P. (2008). An overview of small satellites in remote sensing. *International Journal of Remote Sensing, 29*(15), 4285–4337.
Liang, L., Xu, B., Chen, Y. L., Liu, Y., Cao, W. C., Fang, L. Q., Feng, L. M., Goodchild, M. F., & Gong, P. (2010). Combining spatial-temporal and phylogenetic analysis approaches for improved understanding on global H5N1 transmission. *PloS One, 5*(10), e13575.
Liu, J. Y., Liu, M. L., Deng, X. Z., et al. (2002). The land-use and land-cover change database and its relative studies in China. *Journal of Geographical Sciences, 12*(3), 275–282 (in Chinese).
Niu, Z. G., Zhang, H. Y., Wang, X. W., et al. (2012). Mapping wetland changes in China between 1978 and 2008. *Chinese Science Bulletin, 57*(22), 2813–2823.
Song, Y. X. (1637). *Tian Gong Kai Wu*. Reprinted in by the Commercial Press in 1928, Beijing (in Chinese).
Steinitz, C. (2012). *A framework for geodesign: Changing geography by design*. Redlands: Esri Press.
Sui, D. Z. (2004). Tobler's first law of geography: A big idea for a small world? *Annals of the Association of American Geographers, 94*, 269–277.
Sun, H. L. (2006). Scientific questions and research methods in ecosystem assessment: Chinese ecosystem research network provides supports to ecosystem assessment. *Resources Science, 28*(4), 2–3 (in Chinese).
Wang, L., Li, C. C., Ying, Q., et al. (2012). China's urban expansion from 1990 to 2010 determined with satellite remote sensing. *Chinese Science Bulletin, 57*(22), 2802–2812.
Warntz, W. (1989). Newton, the Newtonians, and the Geographia Generalis Varenii. *Annals of the Association of American Geographers, 79*(2), 165–191.
Wu, C. J. (1990). *1:1000000 China's land use map*. Beijing: Science Press (in Chinese).
Xu, G. Q. (1639). *Nong Zheng Quan Shu*. Reprinted by YueLu Publishing House in 2002. Changsha, China (in Chinese).
Xu, X. K. (1641). *Travel notes of Xiake Xu*. Reprinted by HuaXia Publishing House in 2006. Beijing (in Chinese).
Xu, G. H., Ju, H. B., He, B., et al. (2010, August 1). A perspective on the development of China's earth sciences in the 21st century. *Science and Technology Daily* (in Chinese).
Yu, G. R., Zhang, L. M., Sun, X. M., et al. (2004). Research progress of terrestrial ecosystem carbon flux in Asia. *Science in China Series D: Earth Sciences, 34*(S2), 15–29.
Zhang, D. E. (2004). *General collection of 3000 years of meteorological records in China*. Nanjing: Fenghuang Press (in Chinese).

Zhang, X. S. (2007). *1:1000000 vegetation map in China*. Beijing: Geological Publishing House (in Chinese).

Zhang, Q., Geng, G. N., Wang, S. W., et al. (2012). Satellite remote sensing of changes in NOx emissions over China during 1996–2010. *Chinese Science Bulletin, 57*(22), 2857–2864.

Zhou, C. H. (2009). *Geomorphologic atlas of the People's Republic of China (1:1,000,000)*. Beijing: Science Press (in Chinese).

Zhu, K. Z. (1972). A preliminary study on the climatic fluctuations during the last 5000 years in China. *Acta Archaeologica Sinica, 1*, 226–256 (in Chinese).

Chapter 3 Space-Time Behavior Research and Application in China

Yanwei Chai, Zhilin Liu, Yan Zhang, and Tana

3.1 Introduction

The space-time behavior approach provides a unique perspective for understanding complex relationships between human activity and urban environments in space and time (Hanson and Hanson 1993; Kwan 2002a; Miller 2004). Built largely on the theoretical framework of time geography proposed by Hägerstrand (1970), space-time behavior research has grown into an influential approach in urban geography, urban planning and transportation planning (Dijst 1999; Kwan 2002b; Timmermans et al. 2002). Recently, this approach has enjoyed even greater popularity and impact in contemporary theoretical and methodological debates, a wider variety of empirical topics, and more active engagement with planning practices, thanks to the development of GIS-based analytic tools and high-quality individual space-time activity data (Kwan 2004; Shaw and Yu 2009; Neutens et al. 2011).

The activity-based approach using disaggregate survey data has also become influential in China over the last two decades, as urban geographers and planners have tried to make sense of the dynamic interactions between individual life

Y. Chai (✉) • Tana
Department of Urban and Economic Geography, College of Urban and Environmental Sciences, Peking University, Beijing 100871, China
e-mail: chyw@pku.edu.cn; tanapku@gmail.com

Z. Liu
School of Public Policy and Management, Tsinghua University, Beijing 100084, China
e-mail: zhilinliu@tsinghua.edu.cn

Y. Zhang
Institute of Beijing Studies, Beijing Union University, Beijing 100101, China
e-mail: yanzhang@buu.edu.cn

M.-P. Kwan et al. (eds.), *Space-Time Integration in Geography and GIScience*, DOI 10.1007/978-94-017-9205-9_3

experiences on a micro scale and urban social and spatial transformations on a macro scale. The space-time behavior approach was introduced to China in this context and has subsequently been widely applied to empirical research and planning practices. In 1996, the National Natural Science Foundation of China sponsored the first urban geography research that adopted a time-geographic framework. Since then, more than 100 academic research articles have been published in Chinese and English, presenting empirical studies on coastal cities (e.g., Beijing, Tianjin, Shanghai, Shenzhen, Guangzhou, Nanjing) and inland cities (e.g., Wuhan, Urumqi, Lanzhou).

Aimed at promoting better communication and exchange among urban geographers from different contexts, this article provides a critical overview of space-time behavior research in China's urban geography. We first present a brief history of space-time behavior research in China, focusing on key methodological progress such as collection of high-quality space-time data and development of GIS-based analytic techniques. This summary is followed by a review of applications of space-time behavior approach to empirical studies and planning practices in Chinese cities. We argue that space-time behavior research offers a new perspective for understanding the complexity and diversity of human behavior patterns during the large-scale spatial and institutional transition in these cities. While learning from the west, improvement of data quality and analytic tools has facilitated GIS-based geovisualization of space-time activity patterns in Chinese cities and improved understanding of space-time behavior decision within constraints by the built environment. Nonetheless, we also recognize that while the techniques and tools used in space-time behavior research in China have caught up with that in the West, its theoretical development still lags behind.

3.2 Theoretical Introduction and Methodological Progress

3.2.1 Introduction of the Space-Time Behavior Approach to China

In Western countries, space-time behavior research had achieved a great deal of theoretical progress by the 1990s, but limitations in data quality and analytical techniques led to a period of relative stagnation in empirical studies (Kwan 2004; Shaw 2012). Nonetheless, it was during the 1990s that time geography and the space-time behavior approaches were introduced to China, where they met with a growing interest in Chinese urban geography.

To be sure, the space-time behavior research in China has benefited by learning from the fruitful theoretical progress and empirical applications made in Western geographical research. Academic journals in China have published a series of articles in the late 1990s that introduce key concepts, theoretical arguments, and empirical studies in Sweden, America and Japan (e.g., Chai and Wang 1997; Chai 1998; Chai et al. 2000), and more recently in 2009 and 2010 that further review recent progress in GIS-based time-geographical research and activity-based research (e.g., Chai and Zhao 2009; Chai et al. 2010c). In 2010, *Urban Planning International*, a Chinese journal, published a special issue on time geography co-edited by Chinese and American time geographers. This special issue included literature review articles and translated articles from western works that introduced original thoughts, theoretical framework, and methodological implementation of time geography (Chai et al. 2010a). The issue also showcased empirical studies on different urban issues in China, and discussed the prospects of the time-geography approach in urban planning practice in China (Chai et al. 2010c; Huang 2010; Tana and Chai 2010). Whereas theoretical progress has still lagged behind in Chinese space-time behavior research, more methodological progress has been made in data collection and analytical tools for space-time behavior research, which is discussed below.

3.2.2 Progress in Collection of Space-Time Data

The lack of individual-level space-time data has always been a major constraint for urban geographers in China. In the early stage, individual-level trip-based or time-use data were limited to a few government-sponsored surveys in large Chinese cities such as Beijing and Guangzhou,[1] while most data were not available or made public for academic research. Thus, while activity-diary surveys – or other types of activity surveys – were the key source of data for space-time behavior research in the West in the 1990s (e.g., Kwan 1998; Dijst and Vidakovic 2000; Shaw and Wang 2000), Chinese urban geographers also began to design and implement activity-diary surveys in Chinese cities to collect space-time data.

Funded by the National Natural Science Foundation of China (NSFC), the first series of activity-diary surveys was conducted in Dalian in 1996, Tianjin in 1997, and Shenzhen in 1998 (see Chai et al. 2002). The surveys have collected a consecutive 48-h activity diary (from 0:00 am Sunday to 11:59 pm Monday) with a temporal resolution of 1 h. A more recent survey was conducted in Beijing in 2007, which developed the survey design in the previous survey with by combining activity records with detailed travel information and by refining the temporal resolution from 1 h to 1 min (Chai et al. 2009). More surveys have been conducted in Beijing in 2008 (Lan and Feng 2010), Guangzhou in 2007 (Zhou and Deng 2010), Nanjing in 2002 (Liu 2005), and Urumqi in 2008 (Zheng et al. 2009).

[1]Beijing has large sample trip survey in 1986, 2000, and 2005. Guangzhou has trip survey in 1984 (3 % population sample) and in 1998 (ten thousands sample). The National Bureau of Statistics of China launched a national level time-use survey in 2008.

While location-based-service-facilitated data collection became popular for space-time behavior research in the Western geography after 2000, Chinese urban geographers have started catching up only in the last 5 years. GPS and GMS phone tracking technology have been integrated into activity-diary surveys to collect more accurate and real-time activity-travel information. In a pilot study conducted in two suburban neighborhoods in Beijing in 2010, 100 volunteers are recruited to participate in, each of whom is asked to carry a mobile tracking facility with built-in GPS and GMS phone chips over a week and complete an activity-diary survey through an online survey platform. Different from a traditional diary survey, the online survey platform allows respondents to interactively see the trajectory (i.e., space-time path) of their activities on a city map once respondents filled out the activity-diary information (Chai et al. 2013) (See Fig. 3.1). Respondents also can edit their activity information in case of recall error or system resolution problem. Though this pilot survey has a small sample size, it nevertheless is the first successful experiment conducted by Chinese geographers that integrates traditional diary survey method and the mobile tracking technology in space-time data collection. Furthermore, this pilot survey demonstrates the advantage of such integrated approach in the collection of long-term, high-resolution individual space-time data and the potential of scaling up this experiment into a full-scale survey with a larger sample size.

Nevertheless, scholars also recognize that most surveys have only a small sample size (often around several hundred households, or around 1,000 individuals) and that information collected is challenged with possible recall errors. Luckily,

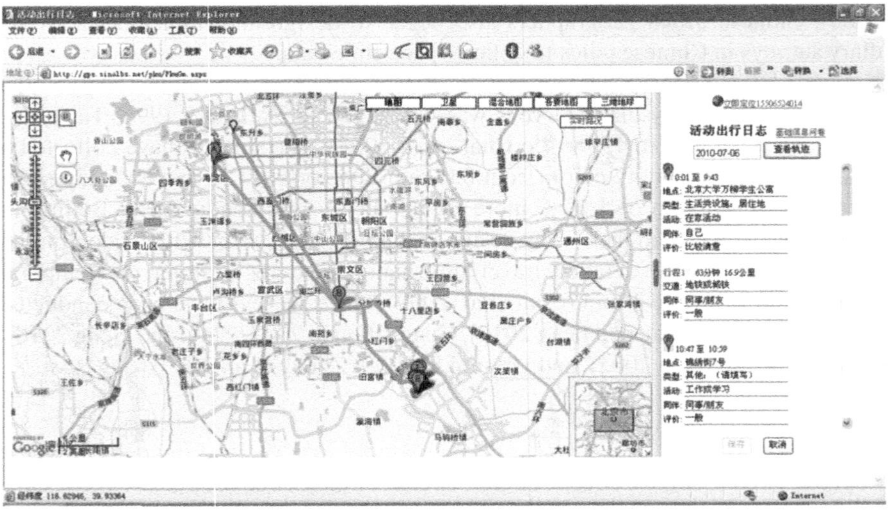

Fig. 3.1 The online survey platform in Beijing in 2010

large-sample activity-travel data, often collected by government agencies, have become more available and significantly increased the data capacity of space-time behavior research in China. For instance, scholars have used floating-car data, smart-card data, and mobile phone data in travel-behavior research, which contributed to urban research and planning practices (Li et al. 2011; Long et al. 2012; Liu et al. 2012).

3.2.3 Progress in Analytic Techniques

Another area of major development has occurred in the analytical techniques adopted in Chinese space-time behavior research. In the 1990s, empirical studies in China often adopted statistical tools to describe space-time activity patterns on an aggregate level (e.g., Chai 1996; Chai et al. 2002), while Western geographers had begun to adopt advanced statistical models and explore the geovisualization of space-time activity patterns and accessibility measurements based on space-time prisms. These advanced methods have been adopted in more recent empirical literatures in China. For instance, structural equation models (SEM), nested logit models, and other statistical tools have been used to understand how urban built environments and institutional contexts shape individual space-time behavior (e.g., Zhang and Li 2006 on shopping location choices; Cao and Chai 2007 on gender differences in time use; Zhu and Wang 2008 on consumer travel behavior; Wang and Chai 2009 on commuting behavior; Chai et al. 2010b on the spatial and temporal decision making of travel behaviors; Zhao and Chai 2010; Ma and Chai 2011 on trip chaining behaviors).

Despite slower progress until early 2000s, the disaggregate-level analysis of space-time paths has achieved major methodological breakthroughs owing to the greater number of scholarly interactions occurring between Chinese and Western geographers. Early analysis of space-time paths analysis often used a two-dimensional diagram, with the horizontal axis indicating distance from home. However, the rapid development of GIS techniques and high-resolution space-time data has enabled Chinese geographers to conduct 3D GIS-based space-time path analyses instead.

For instance, a space-time GIS toolkit named Activity Pattern Analyst (APA) has been developed to geovisualize space-time paths, thereby allowing the exploration of space-time activity patterns in a 3D environment (Chen et al. 2011, see Fig. 3.2). The APA toolkit has six basic functions for space-time pattern exploration, including space-time path generation, query, and segmentation, as well as activity query, activity density, and space-time path based clustering analysis. In addition, other computer-aided data mining methods were introduced to explore the complexity of space-time activity patterns. For instance, Li et al. (2009) have adopted the sequence alignment tool to analyze the sequencing structure of spatial and temporal

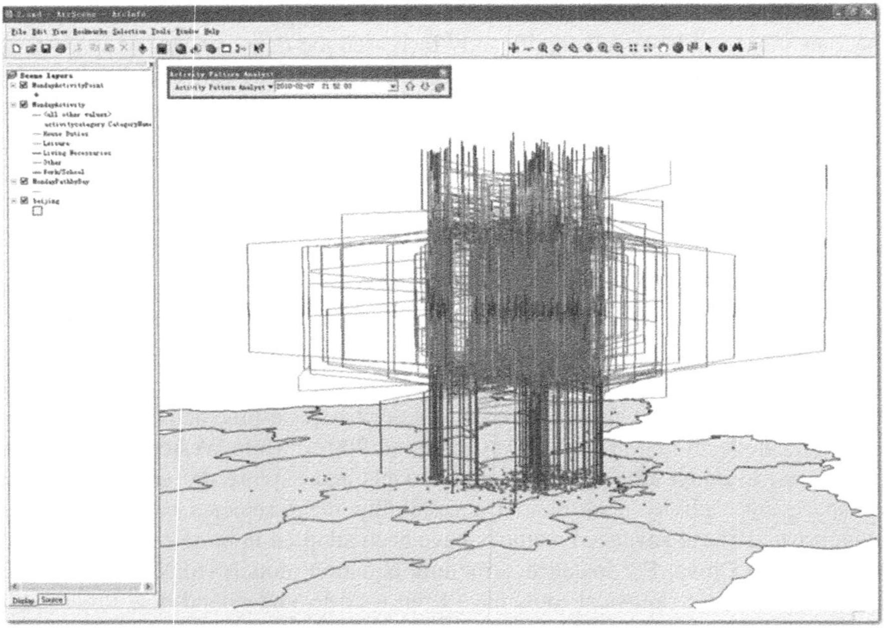

Fig. 3.2 Visualization of activity types in space and time in APA (Source: Chen et al. 2011, 399)

sequences of individual activities across 24 h. Zhang and Chai (2011) have used space-time activity density surfaces to map the spatial and temporal characteristics of the activity patterns of low-income households in Beijing.

3.3 Space-Time Behavior in Urban China: Key Themes and Empirical Findings

Inspired by more theoretical and methodological progress, empirical research also made tremendous progress to apply the space-time behavior approach to the urban context of China. Similar to that existing in Europe and the US, this literature become central to the progress of China's urban geographical research over the last 20 years. Learning from Western literatures, empirical themes have expanded from the space-time patterns of daily activities and travel behavior to more urban issues related to social equity, public health, and environmental sustainability (e.g., Lan and Feng 2010; Zhang and Chai 2011; Ma et al. 2011; Wang et al. 2012). Notably, urban geographers in China have interpreted micro-level patterns of space-time behavior within the context of macro-level institutional transformation and spatial restructuring in Chinese cities. This section reviews the key themes and empirical findings of space-time behavior research in urban China.

3.3.1 Understanding Urban Spatial Structure Through Describing Individual Space-Time Activity Patterns

Earlier empirical studies from space-time activity approach have focused on describing the spatial-temporal patterns of individual daily activities, thereby providing a micro-level interpretation of China's urban spatial structure. For instance, descriptive statistics have shown how urban residents allocate their time to different types of activities, i.e. the temporal rhythms of daily activities over the course of 24 h, and the average travel distances from home based on the purpose of an activity. For instance, Chai et al. (2002) found that the activity frequency of Chinese urban residents follows the distance decay model and that communities constitute the basic level of daily-life space, particularly for residents living in *danwei* compounds compared to suburban residents who have more extensive and wider daily-life activity space (see, Chai et al. 2002).

As to the method, the time-use structure, activity rhythms, and the space-time paths were widely used to explore the overall space-time pattern. First, the time-use structure is based on the percentages of time one spends on different types of activities over the course of 24 h. Gong et al. (2000) and Liu et al. (2000) conducted clustering analysis of time-use structure to residents in Shenzhen and found six groups of residents with distinctive time-use characteristics. Meanwhile, representative households were selected as case studies for analysis of space-time paths of household members to further illustrate the typical space-time pattern of each group (Chai et al. 2002; see Fig. 3.3).

Second, the temporal rhythm of daily activities helps capture the rhythm of activity frequencies during a day. Third, a two-dimensional chart of space-time structure was developed to illustrate an aggregate-level pattern of space-time activities of all sample residents. In the chart, the X-axis refers to the distance range from home for each category of activities, such as the 0–0.5 km, 0.5–5 km, 5–10 km, and beyond 10 km. The Y-axis refers to frequency of activities (i.e. percentage of respondents participating in an activity at certain time and within certain distance range) across 24 h (Gong et al. 2000; see Fig. 3.4).

However, the early stage of space-time behavior research in China also faced great methodological challenges. For instance, time geography as Hägerstrand proposed stresses an individual-based activity analysis, integration of the temporal and spatial dimensions, and the interpretation of individual space-time activities within multiple sources of constraints. Yet earlier empirical literature was often limited to activity patterns on an aggregate level and had insufficient reference to

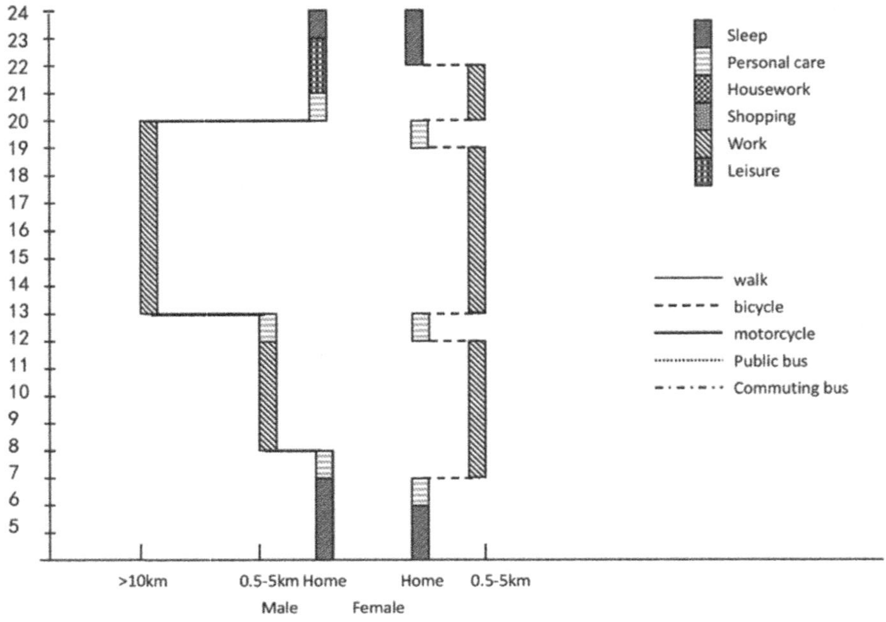

Fig. 3.3 Space-time path of typical household in Shenzhen (Source: Chai et al. 2002, 153)

constraints. In addition, key time-geographic constructs such as space-time paths and prisms were not fully applied in empirical research in this stage, partly due to limitations in GIS-based analytic tools.

3.3.2 Understanding Space-Time Behavior in the Transitional Context of China

The magnitude and complexity of urban transformation in post-reform China have attracted much scholarly attention in China and abroad. Individual residents must adapt to the constraints imposed by the changing built environment when making activity-travel decisions related to trip destination choices, trip frequency, travel modes, tour patterns, and other space-time patterns (Chai et al. 2010b). The space-time activity approach thus enables urban scholars to understand processes and consequences of China's urban transformation from the perspective of an individual's daily life experience.

One theme is the impacts of *danwei*-based urban spatial organization – as well as its dismantlement – on individual daily activity. In the socialist era, Chinese urban space was dominated by *danwei*-based compounds, which were characterized by mixed land use and high access to workplaces, services, and

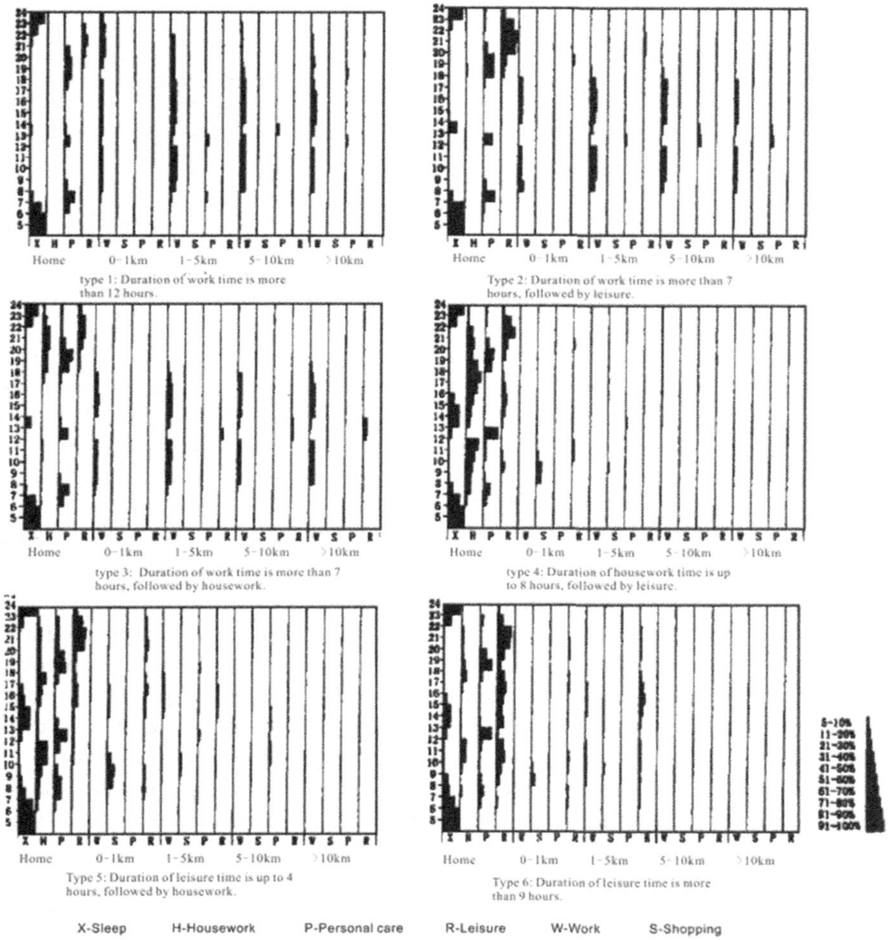

Fig. 3.4 Temporal rhythm of daily activities in Shenzhen (Source: Gong et al. 2000, 64)

facilities (Bjorklund 1986; Bray 2005). Chai's early research has concluded that in the socialist era, the daily activity space was limited to *danwei* compounds, a self-contained territorial unit in Chinese cities characterized by a mixture of workplaces, housing, and services (e.g., Chai 1996). As this *danwei*-based spatial organization was gradually dismantled due to the land and housing marketization of the 1990s, mono-functional, zoning-based land use planning began dominating Chinese cities.

Using data derived from activity diary survey in Beijing, Liu et al. (2009) and Zhang and Chai (2009) confirmed longer home-work distance and commuting time for residents in commodity housing neighborhoods compared to those in *danwei* compounds. Using a structural equation model, Wang and Chai (2009) and Wang et al. (2011b) have further tested the causal links among sources of housing

a b

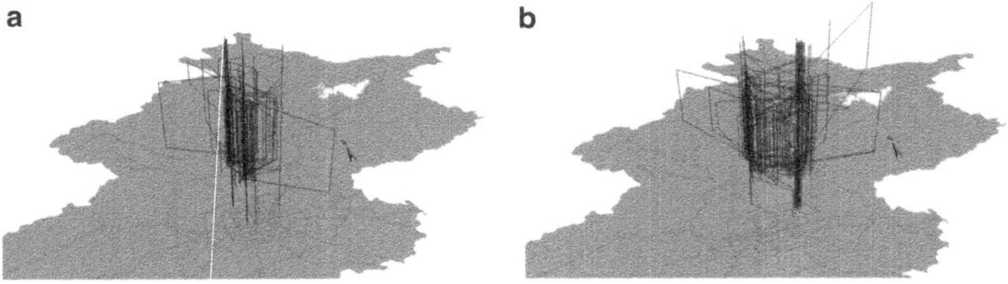

Fig. 3.5 Space-time path of residents in inner-city and suburban area (Made by APA) (**a**) *Danwei* compounds (**b**) Commercial Housing Community (Source: Zhao and Chai 2013)

(i.e., *danwei* vs. market), job accessibility, and travel behavior, which demonstrates the *danwei*'s significant effects as a source of housing on commuting, travel modes, trip rates, time spent on out-of-home activities, and other space-time patterns.

Another important theme is the impacts of suburbanization on individual space-time behavior. Zhao and Chai (2013) have found that compared to inner-city residents, those living in suburban neighborhoods have extensive space-time paths in their daily activities and longer work-activity segments along the space-time paths (see Fig. 3.5). The space-time structure of work-related activities also imposes major time-budget constraints on non-work activities and particularly leisure activities, resulting in fewer out-of-home, non-work activities during the day. By contrast, residents in *Danwei* compounds had shorter work segments on space-time paths and they tended to perform more non-work activities near the home locations, which reflected less restrictive time-budget constraint experienced by these residents.

In addition to descriptive studies of space-time activity patterns, the space-time behavior approach was adopted to investigate how individuals made decisions within the constraints of urban built environment. Other research paid more attention on impacts of urban form on non-work activities such as shopping behavior. For instance, Zhang and Chai (2009a) constructed a three-level nested logit model to investigate the influence of residential location on shopping frequency, shopping task allocation, and tour patterns. Furthermore, some other research adopted a tour-based modeling framework to examine the impact of built environment on individual's spatial-temporal decision-making (Chai et al. 2010b; Zhao and Chai 2010).

Built on the accumulation of empirical literature, Chai and colleagues proposed a conceptual framework for the urban space study based on the individual space-time behavior (Chai and Shen 2006; see Fig. 3.6). The framework aims at integrating a series of dichotomous concepts in space-time behavior research, such as space/time, choice/constraint, and aggregate/disaggregate, to better account for the relationship between the activity-travel system and the urban space become the core of this framework (Chai 2005; Zhang and Chai 2009b; Chai and Shen 2006, 2008). First, individual activity-travel behavior would be understood by exploring both revealed preferences and stated preferences within the urban spaces. Second, the framework

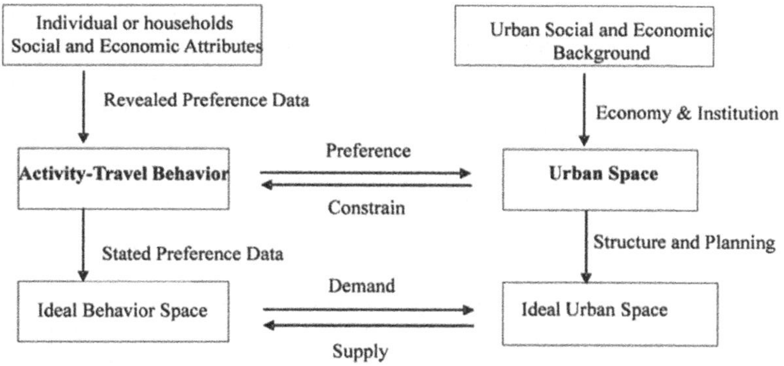

Fig. 3.6 Travel-activity based research frame of urban spatial structure (Source: Chai and Shen 2006)

also takes into account different levels of constraints that urban space (such as land use and transportation infrastructure) exerts on individual space-time behavior. Finally, such analysis provides insights for planning practices to improve urban space to better serve people's needs and demands and to promote social and environmental sustainability in Chinese cities.

Based on individual-level activity diary survey, Wang and Chai (2009), Wang et al. (2011b) used structural equation model (SEM) to test the interactions between built environment (as illustrated by housing source or types of neighborhoods, and job accessibility) and space-time behavior (measured by time spent on different out-of-home activities, car ownership and travel time) in Beijing. They found that *Danwei* as a source of housing had significant effects on commuting distance and travel modes; also that residents of different types of neighborhoods in Beijing demonstrate significant differences in car ownership, time spent for out-of-home activities, trip rate, and travel time.

3.3.3 Space-Time Behavior Research on Social Equity and Environmental Sustainability

Issues of social equity and environmental sustainability have also attracted growing attention from urban geographers in China. By learning from Western literature, urban geographers in China have quickly begun to apply the space-time activity approach to daily-activity experiences of disadvantaged populations, such as women, seniors, and the urban poor (Sun et al. 2001; Liu 2005; Lan and Feng 2010; Zhang and Chai 2011), and to environmental issues such as energy consumption and carbon emission of daily travel behaviors (e.g. Chai et al. 2011; Ma et al. 2011). These studies facilitate our understanding of tremendous and sometimes negative implications of urban transformation in Chinese cities.

One important topic concerns the space-time activity patterns of low-income residents in Chinese cities. Low-income populations tend to face greater difficulties in daily commute because of the emerging job-housing mismatch in Chinese urban spaces (Liu and Wang 2011). Zhang and Chai (2011) have found that low-income populations in Beijing "fragment" space-time activity patterns in that they have shorter work durations and irregular time allocations for work activities. Low-income residents in Beijing (Zhang and Chai 2011) and Guangzhou (Zhou and Deng 2010) have both been found to possess more constrained activity space than other income groups.

Scholars have also analyzed gender differences in space-time activity patterns in Chinese cities and compared them to that of Western cities. Among several studies, Cao and Chai (2007) have focused on time-use patterns and found that women carry more household responsibilities than men and that men spend more time on work and leisure activities. Zhang and Chai (2008) have further revealed the gender-based division of domestic labor in Chinese urban households by comparing space-time activity patterns between husbands and wives from dual-worker households in Dalian, Tianjin, and Shenzhen. They have argued that "*men are dominant in out-of-home activities, but women dominate in-home activities*" as wives had lower peaks in work activities whereas higher peaks in household activities (Zhang and Chai 2008, 1254).

In addition, urban spatial restructuring has not only changed individual activity-travel patterns but also led to substantial environmental implications. The global movement of low-carbon city in response to global climate change prompted the application of the space-time behavior approach in understanding the impacts of land use characteristics on travel-related carbon-dioxide emissions through individual travel decisions. Based on comparisons of space-time activity patterns and associated carbon emissions across different neighborhoods in Beijing, Chai et al. (2011) and Ma et al. (2011) have found that residents in mixed land-use, *danwei* compounds have lower carbon footprints due to their daily travel behavior compared to residents living in mono-functional, commodity housing communities. SEM modeling analysis has demonstrated similar results observed in Europe and the US, showing that neighborhood-based urban forma factors –land-use mix, public transit accessibility, pedestrian-friendly street design, and more – have had significant effects on reducing individual travel-induced CO_2 emissions (Ma et al. 2011; Xiao et al. 2011).

3.3.4 Application of the Space-Time Behavior Approach in Planning Practices

In addition to empirical research, Chinese scholars have tried to apply the space-time behavior approach in urban planning practices, including transportation planning and management, the smart-city master planning, tourism planning, and infrastructure site planning. The space-time behavior approach has been viewed a human-oriented approach to urban and transportation planning, as urban planning is transforming from serving the economic growth purpose to promoting livable and sustainable cities (Lu et al. 2009; Chai et al. 2010c). In particular, urban scholars and planners have begun to realize the importance of serving diverse needs of urban residents in transitional urban China. To meet this goal, the space-time behavior approach offers an effective way for understanding individuals' spatio-temporal demands in order to adjust the urban spatial planning and spatial structure to improve the quality of urban life in China.

Similar to Western cities, traffic congestion and accessibility deficiency have become primary challenges to Chinese cities during rapid urbanization. As the temporal dimension received more attention in the 1990s, the space-time behavior approach grew in popularity in transportation research, planning, and management in China (e.g. Shen and Chai 2012 on commuting flexibility; Gu et al. 2012 on traffic congestion; and Fang et al. 2012 on transportation links). Transportation research from this approach has emphasized the use of individuals as the unit of analysis, an integrated spatio-temporal analysis of activity-based travel demands, and the accounting of space-time constraints for travel decisions. Based on time-geographical framework, the accessibility-oriented planning method has been particularly useful in analyzing the spatio-temporal patterns of critical transportation links (Fang et al. 2012). The geovisualization and simulation modeling of space-time behavior have been performed to map and predict visitor flows and activity patterns (Zhu and Wang 2008; Wang et al. 2009; Huang and Ma 2011).

Similar methods have been used in other planning practices. In a case study of the Summer Palace in Beijing, Huang (2009) and Huang and Ma (2011) adopted a time-geographic approach to study the space-time patterns of tourist behaviors by combining 3D geovisualization method with statistical tools. Huang (2010) also suggested the application of the space-time behavior approach in tourists' travel route optimization and in tourists' time-sharing plan in order to achieve dynamic management of tourism attractions.

Wang and his colleagues at Tongji University have applied their studies on consumers' shopping behavior to spatial optimization of commercial streets, such as Wangfujing Street in Beijing (Wang et al. 2009, 2011a; Zhu and Wang 2007), East Nanjing Road (Zhu et al. 2006; Zhu and Wang 2008) and Xintiandi (Xu and Wang 2012) in Shanghai. Moreover, the space-time behavior simulation was used in an infrastructure site planning for the 2010 Shanghai Expo. Through a virtual survey of space-time behavior, Wang and Ma (2009) simulated the travel flows of 4,000 visitors to predict the spatial-temporal distribution of visitor flows and visitors' demands for dining and other services. Based on the simulation and prediction, suggestions of site change were proposed to improve the transportation outcomes of the Shanghai Expo infrastructure.

Finally, the space-time behavior approach has also been adopted in the smart city strategic planning for Chengde City in Hebei Province. To scale up this case, scholars from Peking University are teaming up with Beijing Transportation Research Center and Tsinghua University in the research project, "*Key Technologies of Urban*

Space-time Behavior Analysis and Application Demonstration of Smart Travel Service Platform," sponsored by China's Ministry of Science and Technology. As part of the umbrella project "*Key Technologies of Public Information Platform of Smart City Management and Application Demonstration*," this research aims at developing a smart-travel service platform in Beijing that integrates multiple sources of activity-travel data for data mining and simulation of space-time behavior, and thus facilitates smart-travel planning and management from an individual space-time behavior approach.

3.4 Conclusions and Discussion

In summary, space-time behavior research has become an influential approach in Chinese urban geography since its introduction to China in the 1990s. The progress made in the past decade has demonstrated the major potential of the space-time behavioral approach to understanding the process, dynamics, and implications of China's urban transformation from a micro-level perspective. This approach has been adopted not only in empirical research but also in planning practices to address various social and environmental issues in Chinese cities.

However, while Chinese geographers have made tremendous methodological progress in space-time behavior research, the field's theoretical progress still lags behind significantly. In particular, how can time geography and the space-time approach contribute to the theorization of current urban transformation in China that meshes with the institutional transition from a central-planned economy to a market-oriented one? Existing urban theories, while developed in the Western context of a capitalist market economy, often cannot fully explain such an unprecedented transformation that has profoundly transformed the urban landscape and individual daily-life experience. Thus, more theoretical progress is necessary for space-time behavior research on urban China so that it can contribute more directly to cutting-edge theoretical debates in the international academic community.

In addition, both Western and Chinese urban geographers face common challenges despite major technical breakthroughs. For instance, while high-quality space-time data have become more available, to what extent do more data mean better information? Although large-sample, high-resolution space-time data allow for more accurate geovisualization of space-time activity patterns, how can the interpretation of such patterns add new knowledge to the interactions between individual space-time experiences and the urban transformation process? Finally, whereas the space-time activity approach has been powerful in understanding urban spaces, how can this approach better inform urban planning practices to address critical social and environmental issues in cities? Although both Chinese and Western cities currently face major planning challenges, mutual benefit will occur based on more interaction and collaboration between scholars from both parts of the world.

Acknowledgements Research partially funded by the National Key Technology Research and Development Program of China (2012BAJ05B04).

References

Bjorklund, E. M. (1986). The DANWEI: Socio-spatial characteristics of work units in China's urban society. *Economic Geography, 62*(1), 19–29.

Bray, D. (2005). *Social space and governance in urban China: The Danwei system from origins to reform*. Stanford: Stanford University Press.

Cao, X. Y., & Chai, Y. W. (2007). Gender role-based differences in time allocation: Case study of Shenzhen, china. *Transportation Research Record: Journal of the Transportation Research Board, 2014*, 58–66.

Chai, Y. W. (1996). The internal structure of a city in Chinese arid area: A case study of Lanzhou, Gansu Province. *Chinese Journal of Arid Land Research, 9*, 169–180.

Chai, Y. W. (1998). Time-geography: Its origin, key concepts and applications. *Scientia Geographica Sinica, 18*(1), 65–72 (in Chinese).

Chai, Y. W. (2005). Methodological problems in behavioral geography study. *Areal Research and Development, 24*(2), 1–5. (in Chinese.

Chai, Y. W., & Shen, J. (2006). Travel-activity based research frame of urban spatial structure. *Human Geography* (in Chinese), *21*(5), 108–112, 54.

Chai, Y. W., & Shen, J. (2008). Activity-based approach to human spatial behavior research. *Scientia Geographica Sinica, 28*(5), 594–600 (in Chinese).

Chai, Y. W., & Wang, E. Z. (1997). Basic concepts and notation of time-geography. *Economic Geography, 17*(3), 55–61 (in Chinese).

Chai, Y. W., & Zhao, Y. (2009). Recent development in time geography. *Scientia Geographica Sinica, 29*(4), 593–600 (in Chinese).

Chai, Y. W., Li, Z. R., Liu, Z. L., & Shi, Z. H. (2000). A review of time-geography. *Human Geography, 15*(6), 54–59 (in Chinese).

Chai, Y. W., Liu, Z. L., Li, Z. R., Gong, H., Shi, Z. H., & Wu, Z. Q. (2002). *Spatio-temporal structure of Chinese cities*. Beijing: Peking University Press (in Chinese).

Chai, Y. W., Zhang, W. J., Zhang, Y., Yan, Y. N., & Zhao, Y. (2009). The production and quality management of disaggregated space-time data of individual's behaviors: A case study of activity-diary survey in Beijing. *Human Geography, 24*(6), 1–9 (In Chinese).

Chai, Y. W., Kwan, M.-P., & Shaw, S.-L. (2010a). Time geography and urban planning: An introduction. *Urban Planning International, 25*(6), 1–2 (in Chinese).

Chai, Y. W., Ma, J., & Zhang, W. J. (2010b). The residential differentiation of tour-based spatio-temporal decision-making of travel behavior in Beijing. *Geographical Research, 29*(10), 1725–1734 (in Chinese).

Chai, Y. W., Zhao, Y., & Zhang, Y. (2010c). Time geography and its application in urban planning. *Urban Planning International, 25*(6), 3–9 (in Chinese).

Chai, Y. W., Xiao, Z. P., & Liu, Z. L. (2011). Comparative analysis on CO_2 emission per household in daily travel based on spatial behavior constraints. *Scientia Geographica Sinica, 31*(7), 843–849 (in Chinese).

Chai, Y. W., Shen, Y., Ma, X. J., Zhao, Y. (2013). The collection and management of space-time data of individual behavior based on location-based technology: A case study of activity-travel survey in Beijing. *Geographical Research, 32*(3), 441–451 (in Chinese).

Chen, J., Shaw, S. L., Yu, H. B., Chai, Y. W., & Jia, Q. L. (2011). Exploratory data analysis of activity diary data: A space-time GIS approach. *Journal of Transport Geography, 19*, 394–404.

Dijst, M. (1999). Action space as planning concept in spatial planning. *Journal of Housing and the Built Environment, 14*, 163–182.

Dijst, M., & Vidakovic, V. (2000). Travel time ratio: The key factor of spatial reach. *Transportation, 27*, 179–200.

Fang, Z. X., Shaw, S.-L., Tu, W., Li, Q. Q., & Li, Y. G. (2012). Spatiotemporal analysis of critical transportation links based on time geographic concepts: A case study of critical bridges in Wuhan, China. *Journal of Transport Geography, 23*, 44–59.

Gong, H., Chai, Y. W., & Liu, Z. L. (2000). Analysis of the time-space structure of Shenzhen citizen's daily activity. *Human Geography, 15*(6), 60–66 (in Chinese).

Gu, J., Zhou, S. H., Yan, X. P., & Deng, L. F. (2012). Formation mechanism of traffic congestion in view of spatio-temporal agglomeration of residents' daily activities: A case study of Guangzhou. *Scientia Geographica Sinica, 32*(8), 921–927 (in Chinese).

Hägerstrand, T. (1970). What about people in regional science? *Regional Science Association Papers, 24*(1), 7–21.

Hanson, S., & Hanson, P. (1993). The geography of everyday life. In T. Gärling & R. G. Golledeg (Eds.), *Behavior and environment: Psychological and geographical approaches* (pp. 249–269). Amsterdam: Elsevier Science.

Huang, X. T. (2009). A study on temporal- spatial behavior pattern of tourists based on time-geography science: A case study of summer palace, Beijing. *Tourism Tribune, 24*(6), 82–87 (in Chinese).

Huang, X. T. (2010). Time geography and tourism planning. *Urban Planning International, 25*(6), 40–44 (in Chinese).

Huang, X. T., & Ma, X. J. (2011). Study on tourists rhythm of activities based on GPS data. *Tourism Tribune, 26*(12), 26–29 (in Chinese).

Kwan, M.-P. (1998). Space-time and integral measures of individual accessibility: A comparative analysis using a point-based framework. *Geographical Analysis, 30*, 370–394.

Kwan, M.-P. (2002a). Feminist visualization: Re-envisioning GIS as a method in feminist geographic research. *Annals of the Association of American Geographers, 92*, 645–661.

Kwan, M.-P. (2002b). Time, information technologies, and the geographies of everyday life. *Urban Geography, 23*(5), 471–482.

Kwan, M.-P. (2004). GIS methods in time-geographic research: Geocomputation and geovisualization of human activity patterns. *Geografiska Annaler. Series B, Human Geography, 86*(4), 267–280.

Lan, Z. M., & Feng, J. (2010). The time allocation and spatio-temporal structure of the activities of migrants in 'village in city': Surveys in five 'villages in city' in Beijing. *Geographical Research, 29*(6), 1092–1104 (in Chinese).

Li, X., Ma, X. J., Wang, C. X., Yasin, D., & Chai, Y. W. (2009). Data mining method for exploring spatio-temporal behavior sequence patterns of urban residents. *Geography and Geo-Information Science, 25*(2), 10–14 (in Chinese).

Li, Q. Q., Zhang, T., Wang, H. D., & Zeng, Z. (2011). Dynamic accessibility mapping using floating car data: A network-constrained density estimation approach. *Journal of Transport Geography, 19*, 379–393.

Liu, Y. T. (2005). *The social space of urban poverty in Transitional Urban China*. Beijing: Science Press (in Chinese).

Liu, Z. L., & Wang, M. J. (2011). Job accessibility and its impacts on commuting time of urban residents in Beijing: From a spatial mismatch perspective. *Acta Geographica Sinica, 66*(4), 457–467 (in Chinese).

Liu, Z. L., Chai, Y. W., & Gong, H. (2000). The study on citizens' leisure time use in Shenzhen city. *Human Geography, 15*(6), 73–78 (in Chinese).

Liu, Z. L., Zhang, Y., & Chai, Y. W. (2009). Home-work separation in the context of institutional and spatial transformation in urban China: Evidence from Beijing household survey data. *Urban Studies, 16*(9), 110–117 (in Chinese).

Liu, Y., Kang, C. G., Gao, S., Xiao, Y., & Tian, Y. (2012). Understanding intra-urban trip patterns from taxi trajectory data. *Journal of Geographical Systems, 14*(4), 463–483.

Long, Y., Zhang, Y., & Cui, C. Y. (2012). Identifying commuting pattern of Beijing using bus smart card data. *Acta Geographica Sinica, 67*(10), 1339–1352 (in Chinese).

Lu, H. P., Wang, J. F., & Zhang, Y. B. (2009). Models and application of transport accessibility in urban transport planning. *Journal of Tsinghua University (Science & Technology), 49*(6), 765–769 (in Chinese).

Ma, J., & Chai, Y. W. (2011). Spatio-temporal choice of individuals' shopping behavior and their differences between weekday and weekend. *Scientia Geographica Sinica, 31*(1), 29–35 (in Chinese).

Ma, J., Chai, Y. W., & Liu, Z. L. (2011). The mechanism of CO2 emissions from urban transport based on individuals' travel behavior in Beijing. *Acta Geographica Sinica, 66*(8), 1023–1032 (in Chinese).

Miller, H. J. (Ed.). (2004). Activities in space and time. In *Handbook of transportation 5: Transport geography and spatial systems*. Pergamon/Kidlington, UK: Elsevier Science.

Neutens, T., Schwanen, T., & Witlox, F. (2011). The prism of everyday life: Towards a new research agenda for time geography. *Transport Reviews, 31*(1), 25–47.

Shaw, S.-L. (2012). Guest editorial introduction: Time geography: Its past, present and future. *Journal of Transport Geography, 23*, 1–4.

Shaw, S.-L., & Wang, D. M. (2000). Handling disaggregate spatiotemporal travel data in GIS. *GeoInformation, 4*(2), 161–178.

Shaw, S.-L., & Yu, H. B. (2009). A GIS-based time-geographic approach of studying individual activities and interactions in a hybrid physical-virtual space. *Journal of Transport Geography, 17*(2), 141–149.

Shen, Y., & Chai, Y. W. (2012). Study on commuting flexibility of residents based on GPS data: A case study of suburban mega-communities in Beijing. *Acta Geographica Sinica, 67*(6), 733–744 (in Chinese).

Sun, Y., Chen, T., & Han, Y. (2001). A study on leisure behavior of the aged in Beijing. *Geographical Research, 20*(5), 537–546 (in Chinese).

Tana, & Chai, Y. W. (2010). Time geography and its enlightenments to the human-oriented community planning. *Urban Planning International, 25*(6), 36–39 (in Chinese).

Timmermans, H., Arentze, T., & Joh, C. H. (2002). Analysing space-time behaviour: New approaches to old problems. *Progress in Human Geography, 26*(2), 175–190.

Wang, D. G., & Chai, Y. W. (2009). The jobs-housing relationship and commuting in Beijing, China: The legacy of Danwei. *Journal of Transport Geography, 17*(1), 30–38 (in Chinese).

Wang, D., & Ma, L. (2009). Simulation analysis of spatial temporal distribution of visitors in Shanghai World Expo 2010. *Urban Planning Forum, 5*, 64–70 (in Chinese).

Wang, D., Zhu, W., Huang, W. S., Ma, L., Guo, J., & Nong, Y. Z. (2009). In-site visitor flow analysis and planning adjustment of the Shanghai Expo. *City Planning Review, 8*, 26–32 (in Chinese).

Wang, D., Nong, Y. Z., Zhu, W. (2011a). Consumer behavior and retail spatial structure in Wangfujing street. *City Planning Reviews* (in Chinese), *35*(7), 43–48, 60.

Wang, D. G., Chai, Y. W., & Li, F. (2011b). Built environment diversities and activity–travel behaviour variations in Beijing, China. *Journal of Transport Geography, 19*, 1173–1186.

Wang, D. G., Li, F., & Chai, Y. W. (2012). Activity spaces and sociospatial segregation in Beijing. *Urban Geography, 33*(2), 256–277.

Xiao, Z. P., Chai, Y. W., & Liu, Z. L. (2011). Quantitative distribution and related factors for household daily travel CO_2 emissions in Beijing. *Urban Studies, 18*(9), 104–112 (in Chinese).

Xu, Z., & Wang, D. (2012). Consumers' behavior and planning in commercial spaces: Shanghai Xintiandi example. *Planners, 28*(1), 23–28 (in Chinese).

Zhang, W. J., & Chai, Y. W. (2008). Theories and confirmed model of urban resident's travel demand: Considering intra-household interaction. *Acta Geographica Sinica, 63*(12), 1246–1356 (in Chinese).

Zhang, W. J., & Chai, Y. W. (2009a). The influence of residential space on household shopping tour decision-making behaviors. *Progress in Geography, 28*(3), 362–369 (in Chinese).

Zhang, W. J., & Chai, Y. W. (2009b). Urban activity-travel systems in the condition of space-time: A review of activity-based theories and models. *Urban Planning International, 24*(4), 61–68 (in Chinese).

Zhang, Y., & Chai, Y. W. (2009c). Characteristics of commuting pattern in Beijing: Based on the comparison of different urban residential areas. *Geographical Research, 28*(5), 1327–1340 (in Chinese).
Zhang, Y., & Chai, Y. W. (2011). The spatio-temporal activity pattern of the middle and the low-income residents in Beijing, China. *Scientia Geographica Sinica, 31*(9), 1056–1064 (in Chinese).
Zhang, W. Z., & Li, Y. J. (2006). Study on residents' shopping location preference and shopping location choice: A case on Xicheng and Haidian districts of Beijing. *Acta Geographica Sinica, 61*(10), 1037–1045 (in Chinese).
Zhao, Y., & Chai, Y. W. (2010). Tour-based travel decision making and related factors of urban residents. *Urban Studies, 17*(10), 96–101 (in Chinese).
Zhao, Y., & Chai, Y. (2013). Residents' activity-travel behavior variation by communities in Beijing, China. *Chinese Geographical Science, 23*(4), 492–505.
Zheng, K., Jing, K. L., Jia, L. J., Hu, J., & Wang, Y. T. (2009). A study on temporal and spatial characteristics of the shopping activities of minorities in city: A case study of the Uygur residents of Urumqi. *Yunnan Geographic Environment Research, 21*(3), 16–21 (in Chinese).
Zhou, S. H., & Deng, L. F. (2010). Spatio-temporal pattern of residents' daily activities based on T-GIS. *Acta Geographica Sinica, 65*(12), 1454–1463 (in Chinese).
Zhu, W., & Wang, D. (2007). Spatio-temporal characteristics of consumer behaviors in Wangfujing Street: An analysis based on "pedestrian grid" approach. *City Planning Review, 31*(2), 62–69.
Zhu, W., & Wang, D. (2008). Space choice behavior and multi-stop tracks of consumers in East Nanjing road. *City Planning Review, 31*(3), 33–40 (in Chinese).
Zhu, W., Wang, D., & Saito, S. (2006). Multistop shopping behavior of consumers on East Nanjing Road. *City Planning Review, 30*(2), 9–17 (in Chinese).

Chapter 4
Beyond Space (As We Knew It): Toward Temporally Integrated Geographies of Segregation, Health, and Accessibility

Mei-Po Kwan

4.1 Introduction

Geographers and scholars in cognate disciplines have long recognized the importance of time and mobility for understanding a wide range of human experiences (e.g., de Certeau 1984; Giddens 1986; Rowe and Wolch 1990; Valentine 1993; Chai et al. 2002; Lefebvre 2004; Cresswell 2006; Urry 2007). Yet many notions in geographic and social science research still tend to be conceptualized largely in static spatial terms, ignoring how our understanding of the issues we study can be greatly enriched through the lenses of time and human mobility. For instance, accessibility still tends to be understood largely in spatial terms (e.g., in terms of the distance or travel cost between facilities and the people they serve). Past research also tends to ignore various facets of time – such as rhythm, duration, and subjective experiences of time – that shape people's spatiotemporal experiences of marginalization, discrimination, and social isolation (however, see May and Thrift 2001; Valentine 2008; Dijst 2009; McQuoid and Dijst 2012; Merriman 2012; Schwanen and Kwan 2012; Schwanen et al. 2012; Rogaly and Thieme 2012; Valentine and Sadgrove 2012).

This chapter has been reprinted with the permission of the Association of American Geographers and Taylor & Francis, LLC. Originally published: Mei-Po Kwan (2013) Beyond Space (As We Knew It): Toward Temporally Integrated Geographies of Segregation, Health, and Accessibility, *Annals of the Association of American Geographers*, 103:5, 1078–1086.

M.-P. Kwan (✉)
Department of Geography and Geographic Information Science, University of Illinois at Urbana-Champaign, Champaign, IL 61820, USA

Department of Human Geography and Spatial Planning, Utrecht University, Utrecht, The Netherlands
e-mail: mpk654@gmail.com

M.-P. Kwan et al. (eds.), *Space-Time Integration in Geography and GIScience*, DOI 10.1007/978-94-017-9205-9_4

As recent research suggests, three geographic notions conceived largely in static spatial terms to date will benefit enormously from integrating time as a critical dimension (e.g., Delafontaine et al. 2011; Kwan 2012b). These three notions are racial and ethnic segregation, environmental exposure, and accessibility. They together address a wide range of social issues of interest to geographers and social scientists (e.g., social inequality, social isolation, environmental justice, environmental health, and access to social services and health care facilities). This article revisits these notions and argues that expanding our analytical focus beyond space to include time and human mobility will considerably enrich our understanding of how individuals of different social groups experience racial and ethnic segregation, exposure to environmental influences, and access to social facilities. For instance, although racial segregation is conceived as the extent to which members of one racial group live apart from those of other racial groups, people's segregation experiences are also affected by how much time they spend in areas outside of their residential neighborhoods and how much social contact they have with other racial groups there (Ellis et al. 2004; Wong and Shaw 2011). Examining segregation, environmental exposure, and accessibility in terms of residential space or location alone will only yield a partial understanding of human spatiotemporal experiences.

This article elucidates how temporally integrated geographies of segregation, environmental health, and accessibility can shed new light on many issues geographers and social scientists have been examining for decades. It suggests that new insights can be gained when commonly used spatial concepts take into account human mobility and incorporate various facets of time as integral elements. Informed by recent work on human mobility and geographies of encounter (e.g., Sheller and Urry 2006; Valentine 2008; Adey 2010), the article emphasizes that people's spatiotemporal experiences are influenced not only by where they live but also by other places they visit, when they visit these places, how much time they spend there, what they experience as they travel between these places, and who they interact with while at those places. It argues that by going beyond static residential space and examining how individuals reach relevant social opportunities or come into contact with other people or social/environmental influences as their lives unfold over space and time, our understanding of the effects of social difference on racial/ethnic segregation, health outcomes, quality of life, and many other critical social issues will be deeply enriched.

4.2 Racial and Ethnic Segregation

Geographers and social scientists have examined racial/ethnic segregation, social exclusion, and social isolation for decades (e.g., Massey and Denton 1988; Wong 1993; Darden and Kamel 2000; Schnell and Yoav 2001). Research on the segregation of different racial or ethnic groups in urban areas have been conducted largely with a focus on people's residential location (e.g., Brown and Chung 2006; Johnston et al. 2007). In these studies, segregation was examined in terms of the extent to which members of different racial or ethnic groups live apart from each other (Ellis et al. 2004). Based on the geographic clustering or concentration of different racial groups and using census data to compute summary measures of segregation, these studies examined segregation at various spatial scales based on static population distribution in the residential space (Wong and Shaw 2011).

Yet people experience segregation or social exclusion not only in their residential neighborhoods but also in other spaces as their daily lives unfold, including their workplaces and sites for social and recreational activities (Lee et al. 2008; Skans and Åslund 2010; Lee and Kwan 2011; McQuoid and Dijst 2012; Schwanen et al. 2012; Wang et al. 2012; Palmer 2013). Ignoring the time people spend outside of their residential neighborhoods and their interactions with other social groups there omits a considerable part of their everyday experiences, which may reinforce or mitigate the segregation they experience in their residential spaces and lead to erroneous conclusions about their overall segregation experiences (Wong and Shaw 2011; Jones and Pebley 2014; Kwan 2012b; Palmer et al. 2013; Krivo et al. 2013). It can also create a misleading impression of a city's racialized spaces as fixed (Kwan 2002; Ellis, Wright and Parks 2004). Considering people's daily mobility and including other spaces (e.g., school, work, or leisure spaces) in segregation studies will allow us to more accurately assess people's experience of segregation (Wong and Shaw 2011).[1]

Further, examining segregation or social isolation based on where individuals of different racial groups live ignores how the presence of others who work or undertake other activities in those residential neighborhoods influence people's segregation experience (Ellis et al. 2004; Jones and Pebley 2014). For instance, the study by Ellis et al. (2004) on the Los Angeles metropolitan area observed that segregation examined by work locations is considerably lower than by residential neighborhoods, indicating that there may be more intergroup interaction during working hours than at home. But the picture seems much more complex when other activity spaces are included, as indicated by two recent studies using the Los Angeles Family and Neighborhood Survey (L.A. FANS) dataset. Jones and Pebley (2014), for instance, found that most people experience substantial racial segregation across the range of spaces in their daily lives, not just in their residential neighborhoods (e.g., both Latinos and African Americans have activity spaces with relatively low proportion of whites). Considering a range of neighborhoods where individuals undertake their daily routines (e.g., shopping, working, seeing a

[1]However, it is important to note that living or working in less segregated environments (e.g., racially mixed residential neighborhoods, workplaces or schools) do not necessarily mean higher exposure to social advantage or more positive experience for racial minorities. As many past studies have shown, it may instead mean more intense exposure to racism and various intersectional discriminatory practices or oppressive encounters (e.g., Tatum 1987, 1997; Lim and Herrera-Sobek 2000; Li and Beckett 2006; Valentine 2007, 2010). To fully understand people's spatiotemporal experiences of marginalization, discrimination and social isolation, we need to go beyond spatial proximity of social groups to examine how oppressive power relations pervade micro spaces (and times) of everyday encounters.

doctor), Krivo et al. (2013) observed that African Americans and Latinos experience additional penalties in social isolation in where they conduct routine activities and in association with their movement in the city when compared to Whites who live in economically similar neighborhoods. These studies clearly showed that focusing only on the time people spend at home or in their residential neighborhoods can lead to misleading results.

Interestingly, urban studies scholars Atkinson and Flint (2004, 876) have provided some of the earliest arguments for expanding the focus of segregation studies to a greater consideration of "the dynamic flows of everyday life both within and outside the field of residential interaction and lived experience." They suggested that segregation needs to be considered with regard to both how people's everyday lives unfold dynamically over time (daily dynamism) and their static residential manifestations. While their study focused on how gated communities represent spaces of self-exclusion created for avoiding unwanted social contact by "elite" social groups, they argued that each of these spaces segregates its occupants from social contact with different social groups through what they called "time-space trajectories of segregation" (Atkinson and Flint 2004, 877).

Recent studies strongly corroborate this insight and the need for new perspectives on racial segregation that take time and the dynamic flows of everyday life into account. With a focus on where, when and with whom people spend their time, Lee and Kwan (2011) showed that the spatiotemporal experiences of social isolation of Korean immigrants in the U.S. can be effectively revealed through examining their daily space-time trajectories and patterns of social contacts in space-time. In another study, Wang et al. (2012) observed sociospatial segregation among residents of different types of neighborhoods in Beijing, China based on the spatiotemporal configuration of their activity spaces. Using GPS and cell phone data to examine where subjects spend time and how they move around, Palmer et al. (2013) found that for a quarter of the participants, exposures to whites when they are in and outside their residential neighborhoods are different (higher for white participants but lower for blacks and Latinos). The study concludes that time spent outside the residential neighborhood can either attenuate or intensify segregation, depending on the social group one is examining. These studies cogently illuminate the need for going beyond people's residential spaces (and times) in research on racial/ethnic segregation or social exclusion to consider how and where different social groups spend their time in their daily lives.

4.3 Environmental Exposure and Geographic Context

Geographic context is an important notion in environmental health and neighborhood effects research (e.g., Kawachi and Berkman 2003; Diez Roux and Mair 2010). It is the conceptual foundation of various methods for assessing people's exposure to contextual or environmental influences. An essential task in this kind of studies is to identify the appropriate geographic area or contextual unit for deriving exposure measures. Exposure measures in most studies to date, however, tend to ignore the critical role of time and human mobility in people's exposures to contextual or environmental influences (Kwan 2012a, b).[2] They are largely based on notions of context, neighborhood or place that conceive them as static administrative areas, often operationalized as the home census tract or block group (or other homogeneous zones constructed based on census units). These conventional notions assume that the residential neighborhood is the most relevant area affecting health outcomes or environmental exposures and neighborhood effects operate only through interactions among those who live in the same residential area.

Although these conventional exposure measures are widely used, their static conceptualizations of geographic context and ignorance of time is problematic in several respects. First, most people move around to undertake their daily activities and rarely stay at only one place (e.g., home) throughout the day. They often traverse the boundaries of several neighborhoods in the course of a day and come under the influence of many different neighborhood contexts outside of their home neighborhoods (Matthews 2008, 2011; Kwan 2009, 2012a, b). Much of the contextual or environmental influence they experience and most of the physical and social resources they utilize may be located outside or far from their home neighborhoods (Matthews et al. 2005). Residential location is thus only one of the places where people spend their time. The residential neighborhood may not include many of their daily activity locations. Recent studies that collected GPS data from participants have repeatedly shown that people spend a considerable amount of time in their daily lives outside of what has conventionally been defined as geographic context or neighborhood (e.g., Wiehe et al. 2008; Basta et al. 2010; Palmer et al. 2013).

Further, besides moving around to undertake their daily activities, people also move around over time. They may change their residence in the same city or move to another. As a result, people's exposure to environmental influences may also change considerably over time and thus studies on people's exposures to health risk factors (e.g., carcinogenic substances) also need to consider their residential history (Löytönen 1998). Contextual influences may vary over space and time in a highly complex manner. They may vary with different temporal patterns or time frames. For instance, as people move through the changing pollution field over time during the day, their exposure to traffic-related air pollution also changes (Gulliver and Briggs 2005).

Recent studies have shown that ignoring time and human mobility in environmental health and neighborhood effects research may lead to erroneous results.[3]

[2]There are exceptions to this, especially in research on individual exposures to traffic-related air pollution (e.g., Gulliver and Briggs 2005; Eleanor et al. 2010). See also Fang and Lu (2012) for a helpful review of recent studies that implemented various methods for personal real-time air pollution exposure assessment. There are also important advances in research on the role of human mobility in disease transmission (e.g., Bian et al. 2012; Wesolowski et al. 2012; Qi and Du 2013).

[3]Also see Kwan (2012b) for a discussion on the inferential errors this may cause.

For instance, Inagami et al. (2007) found that for subjects in a Los Angeles survey, greater exposure to less disadvantaged non-residential neighborhoods where subjects worked, shopped, and undertook other daily activities is associated with better self-rated health. Kwan et al. (2012b) observed that neighborhood effects based on people's residential neighborhoods tend to overestimate their actual exposure to social disadvantage for certain gender and racial groups. Both studies highlights the fact that characteristics of the non-residential neighborhoods people visit in their everyday lives may mitigate (or in other cases, exacerbate) the disadvantage they experience in their residential neighborhood. More importantly, Wiehe et al. (2013) observed using GPS data that contextual influences on adolescents' health behaviors vary by time of day, within participants' residential neighborhoods at the micro-geographic level (e.g., spending time on the front porch, street corners, or other places without adult supervision), and at various distances from home (e.g., area immediately surrounding the home versus areas farther away from home but are normally considered part of their residential neighborhood). Findings from these study clearly indicate that accurate assessment of people's exposure to contextual or environmental influences needs to be based on dynamic notions of context that take both time and human mobility into account (Kwan 2012a, b).

4.4 Accessibility: From Locational Proximity to Space-Time Feasibility

Accessibility is a widely used analytical construct in geography, urban studies, and transportation research. It helps us understand how the built environment and individual attributes affect people's access to social opportunities important to their quality of life and well-being, including jobs, social services, and health care facilities (e.g., Weber 2003; Horner 2004; Parks 2004; Casas 2007; Wang et al. 2010; Hawthorne and Kwan 2012; Shi et al. 2012; Wang 2012; Weber and Sultana 2013). However, most empirical studies to date tend to conceptualize accessibility mainly in terms of locational proximity or some closely related derivatives such as gravity-based measures. In these conceptualizations, distances or travel costs between facilities and the people they serve play an important role in determining accessibility. Although conventional accessibility measures are valuable as indicators of the relative distribution of people and the facilities that serve them (place accessibility), their ignorance of time and human mobility poses several difficulties for understanding people's experiences of access (Kwan and Weber 2003). For instance, these measures do not take into account people's need to be at certain locations at certain times of the day (e.g., chauffeuring children to or from schools and childcare providers), the amount of time they have for reaching activity locations and undertaking activities there, and facility opening hours that may render many facilities temporally unavailable and thus inaccessible (Weber and Kwan 2002, 2008; Schwanen 2007; Neutens et al. 2010; Delafontaine et al. 2011).

They also ignore the fact that people move around to undertake their daily routines and as a result may be presented with various opportunities for accessing needed material and emotional resources and improving their quality of life (Takahashi et al. 2001).

In many senses, time is a fundamental dimension that shapes people's access to and use of urban opportunities. Individuals have different space-time constraints and temporal rhythms of activities throughout the day (Kwan 1998, 1999; Dijst and Vidakovic 2000; Schwanen et al. 2008; Farber et al. 2013). These constraints and temporal rhythms may also vary considerably from day to day (Neutens et al. 2010). Facilities and services have specific temporal schedules or opening hours that render them unavailable in certain hours of the day, and individual accessibility of people with different personal and household attributes can be affected differently by changes in these opening hours (Neutens et al. 2010; Delafontaine et al. 2011). Further, not all opportunities are relevant unless the time one can spend at the activity site exceeds the threshold requires for meaningful participation in that activity (Kim and Kwan 2003). Various types of delays people encountered during their travel, such as traffic congestion or changes in transit schedules in different hours of the day, also affect the extent to which facilities can be accessed or used (Weber and Kwan 2002).

Simply put, locational proximity does not necessarily mean better access for many people. For instance, a government office is not necessarily very accessible even if it is located right next to a person's residence if the person's space-time constraints (e.g., work schedule) make it difficult to visit the office during its opening hours. Public transit is not necessarily accessible even if one lives right next to a bus stop if the bus schedule does not entail frequent service at the time it is most needed. Ignoring people's space-time constraints and the temporal schedules of facilities or services can lead to serious overestimation of the level of access people actually experience.

While only a few studies to date have compared results obtained from including and ignoring the temporal dimension in accessibility research, they provide strong evidence about the possibility of erroneous conclusions when time is ignored. For instance, Kwan (1998) compares 18 conventional accessibility measures with 12 space-time measures that take into account people's space-time constraints and the sequential unfolding of their activities over time. The study found considerable gender differences in the geographic patterns of accessibility when using space-time measures, while no such differences were observed when using conventional measures. This result means that accessibility measures that do not take time and human mobility into account may not reveal the effects of social difference (e.g., gender, race, class, age, and disability) on individual accessibility because they are not sensitive to people's space-time constraints.

More recently, Ren et al. (2014) compare the geographic patterns of demand for service generated with three conventional location-based demand measures and eight demand measures that take into account people's space-time accessibility. The study found that ignoring the temporal dimension of accessibility in demand modeling may underestimate potential demand for service in most situations and can lead to distorted demand patterns and facility location that is far from the best for clients. Studies like these have shown that ignoring time and human mobility in accessibility research can often obfuscate what people actually experience in their everyday lives and lead to erroneous conclusions. Because people with different attributes (e.g., gender, race, sexual orientation, age, and disability) face different space-time constraints, the effects of the same physical environment on accessibility, even for individuals who live at the same physical location (e.g., members of the same household), can be very different. Conceptualizing accessibility as space-time feasibility will thus have significant implications for our understanding of many important social issues.

4.5 Toward Temporally Integrated Geographies

This article argued that critical insight can be gained when commonly used spatial concepts of segregation, environmental exposure, and accessibility take time and human mobility into account. Temporally integrated human geographies have considerable potential for shedding new light on many important issues geographers and social scientists have examined for decades. The article, however, did not argue that space is no longer important. It aimed mainly to expand our analytical focus from static residential spaces to other relevant places and times in people's everyday lives: where and when people work, eat, play, shop, and socialize. Mobility is an essential element of people's spatiotemporal experiences, and these complex experiences cannot be fully understood by just looking at where people live.

While this article treated segregation, environmental exposure, and accessibility as separate notions, they are nonetheless analytically interlinked. They all focus on where and when individuals come into contact with or under the influence of other people or social/physical conditions (e.g., environmental risk factors or social opportunities) as their daily lives unfold. The places people can reach and at what time they can reach them (individual accessibility) are important determinants of their exposures to various social or environmental influences (Gulliver and Briggs 2005; Kwan 2012b). Racial or ethnic segregation not only may limit people's access to jobs and social facilities but also can expose them to higher levels of environmental risk (Chakraborty 2012; Grady and Darden 2012). There are some recent attempts to bridge these three notions and to develop new hybrid analytical constructs. For instance, Wong and Shaw (2011), Farber et al. (2012), and Palmer et al. (2013) conceptualize racial segregation as exposure to different racial groups via people's daily activity spaces. Exploring the analytical links among segregation, environmental exposure, and accessibility through some unifying notions seems a fruitful direction for the future development of temporally integrated geographies. As many social scientists are also interested in studying these three themes, interdisciplinary research along this line may have a broad impact on many disciplines beyond geography.

Moving beyond the conventional focus on static residential spaces and toward temporally integrated perspectives, however, poses many challenges. While it is now possible to collect high resolution space-time data on people's daily activities and trips using location-aware devices like global positioning systems (GPS) and mobile phones (Ahas et al. 2010; Shoval et al. 2011; Almanza et al. 2012; Rodríguez et al. 2012; Richardson et al. 2013; Wiehe et al. 2013), high quality data are still costly and time consuming to collect. Further, reliably linking the space-time data of people's movement to other relevant attributes (e.g., activity type, real-time sociogeographic context) is fraught with difficulties. There are few widely available methods for analyzing the complex relationships among human space-time trajectories, racial segregation, environmental exposure, and accessibility. Taking into account certain facets of time (e.g., people's subjective experiences of time) remains difficult. Modeling human movements and incorporating time also faces complex issues of uncertainty. Recent studies, however, have begun to address some of these difficulties. For instance, recent studies have attempted to model human mobility and travel in probabilistic terms (e.g., Ettema and Timmermans 2007; González et al. 2008; Kuijpers and Othman 2009). Qualitative approaches also provide promising alternatives for grappling with people's complex spatiotemporal experiences (e.g., Kwan and Ding 2008; Valentine and Sadgrove 2012). To fully address the challenges of temporally integrated geographies, much remains to be done in future research.

Acknowledgements I thank Martin Dijst and three anonymous reviewers for their helpful comments on earlier drafts of this article. I am also grateful to Malia Jones, Stephen Matthews and John Palmer for kindly providing me copies of their unpublished or forthcoming papers, which have helped improved this article considerably. I was supported by the following grants while writing this article: NSF BCS-1244691, NIH R01DA032371-01, NIDA R01DA028692, NIDA R01CA129771, and NSFC 41228001.

References

Adey, P. (2010). *Mobility*. New York: Routledge.

Ahas, R., Silm, S., Järv, O., Saluveer, E., & Tiru, M. (2010). Using mobile positioning data to model locations meaningful to users of mobile phones. *Journal of Urban Technology, 17*, 3–27.

Almanza, E., Jerrett, M., Dunton, G., Seto, E., & Pentz, M. A. (2012). A study of community design, greenness, and physical activity in children using satellite, GPS and accelerometer data. *Health & Place, 18*, 46–54.

Atkinson, R., & Flint, J. (2004). Fortress UK? Gated communities, the spatial revolt of the elites and time-space trajectories of segregation. *Housing Studies, 19*(6), 875–892.

Basta, L. A., Richmond, T. S., & Wiebe, D. J. (2010). Neighborhoods, daily activities, and measuring health risks experienced in urban environments. *Social Science & Medicine, 71*(11), 1943–1950.

Bian, L., Huang, Y., Mao, L., Lim, E., Lee, G., Yang, Y., Cohen, M., & Wilson, D. (2012). Modeling individual vulnerability to communicable diseases: A framework and design. *Annals of the Association of American Geographers, 102*(5), 1016–1025.

Brown, L. A., & Chung, S.-Y. (2006). Spatial segregation, segregation indices and the geographical perspective. *Population, Space and Place, 12*(2), 125–143.

Casas, I. (2007). Social exclusion and the disabled: An accessibility approach. *The Professional Geographer, 59*(4), 463–477.

Chai, Y., Liu, Z., Li, Z., Gong, H., Shi, Z., & Wu, Z. (2002). *Time-space structure of Chinese cities*. Beijing: Peking University Press (in Chinese).

Chakraborty, J. (2012). Cancer risk from exposure to hazardous air pollutants: Spatial and social inequities in Tampa Bay, Florida. *International Journal of Environmental Health Research, 22*(2), 165–183.

Cresswell, T. (2006). *On the move: Mobility in the modern western world*. New York: Routledge.

Darden, J. T., & Kamel, S. M. (2000). Black residential segregation in the city and suburbs of Detroit: Does socioeconomic status matter? *Journal of Urban Affairs, 22*(1), 1–13.

de Certeau, M. (1984). *The practice of everyday life*. Berkeley: University of California Press.

Delafontaine, M., Neutens, T., Schwanen, T., & Van de Weghe, N. (2011). The impact of opening hours on the equity of individual space-time accessibility. *Computers, Environment and Urban Systems, 35*(4), 276–288.

Diez Roux, A. V., & Mair, C. (2010). Neighborhood and health. *Annals of the New York Academy of Sciences, 1186*, 125–145.

Dijst, M. (2009). Time geographical analysis. In R. Kitchin & N. Thrift (Eds.), *International encyclopedia of human geography* (Vol. 11, pp. 266–278). Oxford: Elsevier.

Dijst, M., & Vidakovic, V. (2000). Travel time ratio: The key factor of spatial reach. *Transportation, 27*, 179–199.

Eleanor, S. C., Keller, P., Cloutier-Fisher, D., & Hystad, P. W. (2010). Gender differences in chronic exposure to traffic-related air pollution – A simulation study of working females and males. *The Professional Geographer, 62*(1), 66–83.

Ellis, M., Wright, R., & Parks, V. (2004). Work together, live apart? Geographies of racial and ethnic segregation at home and at work. *Annals of Association of American Geographers, 94*(3), 620–637.

Ettema, D., & Timmermans, H. (2007). Space-time accessibility under conditions of uncertain travel times: Theory and numerical simulations. *Geographical Analysis, 39*, 217–240.

Fang, T. B., & Lu, Y. (2012). Personal real-time air pollution exposure assessment methods promoted by information technological advances. *Annals of GIS, 18*(4), 279–288.

Farber, S., Páez, A., & Morency, C. (2012). Activity spaces and the measurement of clustering and exposure: A case study of linguistic groups in Montreal. *Environment and Planning A, 44*, 315–332.

Farber, S., Neutens, T., Miller, H. J., & Li, X. (2013). The social interaction potential of metropolitan regions: A time-geographic measurement approach using joint accessibility. *Annals of the Association of American Geographers, 103*(3).

Giddens, A. (1986). *The constitution of society: Outline of the theory of structuration*. Berkeley: University of California Press.

González, M. C., Hidalgo, C. A., & Barabási, A.-L. (2008). Understanding individual human mobility patterns. *Nature, 453*, 779–782.

Grady, S., & Darden, J. (2012). Spatial methods to study local racial residential segregation and infant health in Detroit, Michigan. *Annals of the Association of American Geographers, 102*(5), 922–931.

Gulliver, J., & Briggs, D. J. (2005). Time-space modeling of journey-time exposure to traffic-related air pollution using GIS. *Environmental Research, 97*, 10–25.

Hawthorne, T. L., & Kwan, M.-P. (2012). Using GIS and perceived distance to understand the unequal geographies of healthcare in lower-income urban neighborhoods. *The Geographical Journal, 178*(1), 18–30.

Horner, M. W. (2004). Exploring metropolitan accessibility and urban structure. *Urban Geography, 25*(3), 264–284.

Inagami, S., Cohen, D. A., & Finch, B. K. (2007). Non-residential neighborhood exposures suppress neighborhood effects on self-rated health. *Social Science & Medicine, 65*, 1779–1791.

Johnston, R., Poulsen, M., & Forrest, J. (2007). The geography of ethnic residential segregation: A comparative study of five countries. *Annals of the Association of American Geographers, 97*(4), 713–738.
Jones, M., & Pebley, A. (2014). Redefining neighborhoods using common destinations: Social characteristics of activity spaces and home census tracts compared. *Demography, 51*, 727–752.
Kawachi, I., & Berkman, L. F. (Eds.). (2003). *Neighborhoods and health*. Oxford: Oxford University Press.
Kim, H.-M., & Kwan, M.-P. (2003). Space-time accessibility measures: A geocomputational algorithm with a focus on the feasible opportunity set and possible activity duration. *Journal of Geographical Systems, 5*(1), 71–91.
Krivo, L. J., Washington, H. M., Peterson, R. D., Browning, C. R., Calder, C. A., & Kwan, M.-P. (2013). Social isolation of disadvantage and advantage: The reproduction of inequality in urban space. *Social Forces, 92*(1), 141–164.
Kuijpers, B., & Othman, W. (2009). Modeling uncertainty of moving objects on road networks via space-time prisms. *International Journal of Geographical Information Science, 23*, 1095–1117.
Kwan, M.-P. (1998). Space-time and integral measures of individual accessibility: A comparative analysis using a point-based framework. *Geographical Analysis, 30*(3), 191–216.
Kwan, M.-P. (1999). Gender and individual access to urban opportunities: A study using space-time measures. *The Professional Geographer, 51*(2), 210–227.
Kwan, M.-P. (2002). Feminist visualization: Re-envisioning GIS as a method in feminist geographic research. *Annals of the Association of American Geographers, 92*(4), 645–661.
Kwan, M.-P. (2009). From place-based to people-based exposure measures. *Social Science & Medicine, 69*(9), 1311–1313.
Kwan, M.-P. (2012a). How GIS can help address the uncertain geographic context problem in social science research. *Annals of GIS, 18*(4), 245–255.
Kwan, M.-P. (2012b). The uncertain geographic context problem. *Annals of the Association of American Geographers, 102*(5), 958–968.
Kwan, M.-P., & Ding, G. (2008). Geo-narrative: Extending geographic information systems for narrative analysis in qualitative and mixed-method research. *The Professional Geographer, 60*(4), 443–465.
Kwan, M.-P., & Weber, J. (2003). Individual accessibility revisited: Implications for geographical analysis in the twenty-first century. *Geographical Analysis, 35*(4), 341–353.
Lee, J. Y., & Kwan, M.-P. (2011). Visualization of socio-spatial isolation based on human activity patterns and social networks in space-time. *Tijdschrift voor Economische en Sociale Geografie, 102*(4), 468–485.
Lee, B. A., Reardon, S. F., Firebaugh, G., Farrell, C. R., Matthews, S. A., & O'Sullivan, D. (2008). Beyond the census tract: Patterns and determinants of racial segregation at multiple geographic scales. *American Sociological Review, 73*, 766–791.
Lefebvre, H. (2004). *Rhythm analysis: Space, time and everyday life* (S. Elden & G. Moore, Trans.). London: Continuum.
Li, G., & Beckett, G. H. (Eds.). (2006). *"Strangers" of the academy: Asian women scholars in higher education*. Sterling: Stylus.
Lim, S. G., & Herrera-Sobek, M. (Eds.). (2000). *Power, race and gender in academe: Strangers in the tower?* New York: Modern Language Association of America.
Löytönen, M. (1998). GIS, Time geography and health. In A. C. Gatrell & M. Löytönen (Eds.), *GIS and health* (pp. 97–110). London: Taylor & Francis.
Massey, D. S., & Denton, N. A. (1988). The dimensions of residential segregation. *Social Forces, 67*(2), 281–315.
Matthews, S. A. (2008). The salience of neighborhood: Some lessons from sociology. *American Journal of Preventive Medicine, 34*(3), 257–259.

Matthews, S. A. (2011). Spatial polygamy and the heterogeneity of place: Studying people and place via egocentric methods. In L. M. Burton, S. P. Kemp, M. Leung, S. A. Matthews, & D. T. Takeuchi (Eds.), *Communities, neighborhoods, and health: Expanding the boundaries of place* (pp. 35–55). New York: Springer.

Matthews, S. A., Detwiler, J., & Burton, L. (2005). Geo-ethnography: Coupling geographic information analysis techniques and ethnographic methods in urban research. *Cartographica, 40*(4), 75–90.

May, J., & Thrift, N. (2001). *Timespace: Geographies of temporalities*. London: Routledge.

McQuoid, J., & Dijst, M. (2012). Bringing emotions to time geography: The case of mobilities of poverty. *Journal of Transport Geography, 23*, 26–34.

Merriman, P. (2012). Human geography without time-space. *Transactions of the Institute of British Geographers, 37*(1), 13–27.

Neutens, T., Schwanen, T., Witlox, F., & de Maeyer, P. (2010). Evaluating the temporal organization of public service provision using space-time accessibility analysis. *Urban Geography, 31*(8), 1039–1064.

Palmer, J. (2013). *Activity-space segregation: Understanding social divisions in space and time* (Manuscript).

Palmer, J. R., Espenshade, T. J., Bartumeus, F., Chung, C. Y., Ozgencil, N. E., & Li, K. (2013). New approaches to human mobility: Using mobile phones for demographic research. *Demography, 50*(3), 1105–1128.

Parks, V. (2004). Access to work: The effects of spatial and social accessibility on unemployment for native-born black and immigrant women in Los Angeles. *Economic Geography, 80*(2), 141–172.

Qi, F., & Du, F. (2013). Tracking and visualization of space-time activities for a micro-scale flu transmission study. *International Journal of Health Geographics, 12*, 6.

Ren, F., Tong, D., & Kwan, M.-P. (2014). *Space-time measures of demand for service: Bridging location modeling and accessibility studies through a time-geographic framework*. Geografiska Annaler B, forthcoming.

Richardson, D. B., Volkow, N. D., Kwan, M.-P., Kaplan, R. M., Goodchild, M. F., & Croyle, R. T. (2013). Spatial turn in health research. *Science, 339*(6126), 1390–1392.

Rodríguez, D. A., Cho, G.-H., Evenson, D. R., Conway, T. L., Cohen, D., Ghosh-Dastidar, B., Pickrel, J. L., Veblen-Mortenson, S., & Lytle, L. A. (2012). Out and about: Association of the built environment with physical activity behaviors of adolescent females. *Health & Place, 18*, 55–62.

Rogaly, B., & Thieme, S. (2012). Experiencing space-time: The stretched lifeworlds of migrant workers in India. *Environment and Planning A, 44*(9), 2086–2100.

Rowe, S., & Wolch, J. (1990). Social networks in time and space: Homeless women in skid row, Los Angeles. *Annals of the Association of American Geographers, 80*(2), 184–204.

Schnell, I., & Yoav, B. (2001). The sociospatial isolation of agents in everyday life spaces as an aspect of segregation. *Annals of the Association of American Geographers, 91*(4), 622–636.

Schwanen, T. (2007). Gender differences in chauffeuring children among dual-earner families. *The Professional Geography, 59*(4), 447–462.

Schwanen, T., & Kwan, M.-P. (2012). Critical space-time geographies: Thinking the spatiotemporal. *Environment and Planning A, 44*(9), 2043–2048.

Schwanen, T., Kwan, M.-P., & Ren, F. (2008). How fixed is fixed? Gendered rigidity of space-time constraints and geographies of everyday activities. *Geoforum, 39*(6), 2109–2121.

Schwanen, T., van Aalst, I., Brands, J., & Timan, T. (2012). Rhythms of the night: Spatiotemporal inequalities in the nighttime economy. *Environment and Planning A, 44*(9), 2064–2085.

Sheller, M., & Urry, J. (2006). The new mobilities paradigm. *Environment and Planning A, 38*, 207–226.

Shi, X., Alford-Teaster, J., Onega, T., & Wang, D. (2012). Spatial access and local demand for major cancer care facilities in the United States. *Annals of the Association of American Geographers, 102*(5), 1125–1134.

Shoval, N., Wahl, H.-W., Auslander, G., Isaacson, M., Oswald, F., Edry, T., Landau, R., & Heinik, J. (2011). Use of the global positioning system to measure the out-of-home mobility of older adults with differing cognitive functioning. *Ageing & Society, 31*, 849–869.

Skans, O. N., & Åslund, O. (2010). Will I see you at work? Ethnic workplace segregation in Sweden 1985–2002. *Industrial and Labor Relations Review, 63*(3), 471–493.

Takahashi, L. M., Wiebe, D., & Rodriguez, R. (2001). Navigating the time–space context of HIV and AIDS. *Social Science & Medicine, 53*(7), 845–863.

Tatum, B. D. (1987). *Assimilation blues: Black families in white communities*. New York: Basic Books.

Tatum, B. D. (1997). *"Why are all the black kids sitting together in the cafeteria?" and other conversations about race*. New York: Basics Books.

Urry, J. (2007). *Mobilities*. Cambridge: Polity.

Valentine, G. (1993). Negotiating and managing multiple sexual identities: Lesbian time-space strategies. *Transactions of the Institute of British Geographers, 18*, 237–248.

Valentine, G. (2007). Theorizing and researching intersectionality: A challenge for feminist geography. *The Professional Geographer, 59*(1), 10–21.

Valentine, G. (2008). Living with difference: Reflections on geographies of encounter. *Progress in Human Geography, 32*, 321–335.

Valentine, G. (2010). Prejudice: Rethinking geographies of oppression. *Social & Cultural Geographies, 11*(6), 519–537.

Valentine, G., & Sadgrove, J. (2012). Lived difference: A narrative account of spatiotemporal processes of social differentiation. *Environment and Planning A, 44*(9), 2049–2063.

Wang, F. (2012). Measurement, optimization and impact of healthcare accessibility: A methodological review. *Annals of the Association of American Geographers, 102*(5), 1104–1112.

Wang, F., Luo, L., & McLafferty, S. (2010). Healthcare access, socioeconomic factors and late-stage cancer diagnosis: An exploratory spatial analysis and public policy implication. *International Journal of Public Policy, 5*, 237–258.

Wang, D., Li, F., & Chai, Y. (2012). Activity spaces and sociospatial segregation in Beijing. *Urban Geography, 33*(2), 256–277.

Weber, J. (2003). Individual accessibility and distance from major employment centers: An examination using space-time measures. *Journal of Geographical Systems, 5*(1), 51–70.

Weber, J., & Kwan, M.-P. (2002). Bringing time back in: A study on the influence of travel time variations and facility opening hours on individual accessibility. *The Professional Geographer, 54*, 226–240.

Weber, J., & Kwan, M.-P. (2008). Scale and accessibility: Implications for the analysis of land use-travel interaction. *Applied Geography, 28*, 110–123.

Weber, J., & Sultana, S. (2013). Why do so few minority people visit national parks? Visitation and the Accessibility of "America's best idea.". *Annals of the Association of American Geographers, 103*(3), 437–464.

Wesolowski, A., Eagle, N., Tatem, A. J., Smith, D. L., Noor, A. M., & Snow, R. W. (2012). Quantifying the impact of human mobility on malaria. *Science, 338*, 267–270.

Wiehe, S. E., Hoch, S. C., Liu, G. C., Carroll, A. E., Wilson, J. S., & Fortenberry, J. D. (2008). Adolescent travel patterns: Pilot data indicating distance from home varies by time of day and day of week. *Journal of Adolescent Health, 42*, 418–420.

Wiehe, S. E., Kwan, M.-P., Wilson, J., & Fortenberry, J. D. (2013). *Adolescent health-risk behavior and community disorder. PLOS ONE, 8*(11), e77667.

Wong, D. W. S. (1993). Spatial indices of segregation. *Urban Studies, 30*, 559–572.

Wong, D. W. S., & Shaw, S.-L. (2011). Measuring segregation: An activity space approach. *Journal of Geographical Systems, 13*(2), 127–145.

Chapter 5
The Influence of Built Environment on Walking Behavior: Measurement Issues, Theoretical Considerations, Modeling Methodologies and Chinese Empirical Studies

Hui Lin, Guibo Sun, and Rongrong Li

5.1 Introduction

The built environment provides spatial, temporal and social contexts for human behavior. They usually comprise the following elements: (1) land use patterns, including the spatial distributions of buildings and human activities; (2) transportation systems, including hard transport infrastructure and soft transit service and (3) design, including the arrangement and appearance of physical elements (Handy 2005; Saelens and Handy 2008). Certain studies of the built environment on walking behavior in Western cities are driven by planning reform movements, such as new urbanism, smart growth and transit-oriented development (TOD). By shaping the built environment, planners aim to encourage walking behavior while reducing motorized movement. A built environment is labeled as "pedestrian-oriented" if it has relatively high density, a mixture of land uses, a street network with high connectivity, human-scale streets and desirable aesthetic qualities (Cervero and Kockelman 1997). Because walking is emerging as an important form of moderate-intensity physical activity and a practical health improvement method for the general public, public health is putting great effort into researching walking behavior (Owen et al. 2004).

However, several issues require further work. (1) Measurement of the key variables: the lack of an agreed-upon definition of a "built environment" has led to various measurement approaches in empirical studies. Most studies view the relation between the built environment and human walking choices from objective or physical perspectives. The effect would obviously go unrealized without

H. Lin (✉) • G. Sun • R. Li
Institute of Space and Earth Information Science, The Chinese University of Hong Kong, Hong Kong, Hong Kong
e-mail: huilin@cuhk.edu.hk; gbsun@cuhk.edu.hk; rongrongli@cuhk.edu.hk

M.-P. Kwan et al. (eds.), *Space-Time Integration in Geography and GIScience*, DOI 10.1007/978-94-017-9205-9_5

human awareness (Kremers et al. 2006). However, perception and cognition are not physically measurable, and the objective measurements do not effectively complement their subjective counterparts. Validated and consistent measurements of the built environment, both objective and subjective, are required and will lead to more meaningful empirical studies. (2) The nature of the influence, to "forecast" or to "understand": a rarely mentioned question is "why, after all, a relationship should exist between travel behavior and built environment" (Van Acker et al. 2010). To understand the nature of the influence, we need to explore how the built environment factors into the human decision-making process (Handy 2005; Boarnet and Crane 2001). Forecast models such as the discrete choice model are useful and widely adopted in transportation research. However, the specific attributes that should be attached to choice alternatives are calibrated only from proposed built environment attributes and collected travel behavior data, and are hence not enough to truly understand walking behavior. Social psychology focuses on how people think, feel and behave, and how these thoughts, feelings and behaviors are influenced by others. The theory of planned behavior (TPB), a more structural theory, should indicate whether we should use these variables and introduce hints for understanding human behavior. The combination of a discrete choice model and the TPB holds promise. (3) The application of Western experiences to Chinese cities: a great majority of studies on this subject are based on Western cities. Conclusions drawn from developed countries' experiences may not be applicable in other contexts, especially to countries like China that are undergoing rapid economic growth and urban development (Shen 1997; Pan et al. 2009). The transferability of primary study findings and recommendations to Chinese cities remains to be tested.

This paper attempts to provide an overview of measurement issues, behavior theory underpinnings, modeling strategies concerning the influence of built environment on walking behavior, and empirical studies in a Chinese context. Following this introduction, Sect. 5.2 discusses measurement of the built environment and walking behavior. Section 5.3 introduces the theoretical basis. Section 5.4 describes our conceptual model and study design. Section 5.5 examines the specific characteristics of Chinese cities and elaborates on the study progress in China. Concluding remarks and comments on future work are given in Sect. 5.6.

5.2 Measurement Issues

Heath et al. (2006) call for a better conceptualization of the built environment to guide measurement of their components (Heath et al. 2006). Without a generic conceptualization, empirical studies could lack the appropriate guidance. Validated and consistent measurements of the built environment and walking behavior are still required (Cunningham and Michael 2004).

5.2.1 Measuring the Built Environment: Objectively or Subjectively?

In general, objective measures come from data collected in the field (Hoehner et al. 2005) or from existing land use databases available in geographic information systems (GISs) (Lin and Moudon 2010). The majority of studies show that both objective and subjective attributes are related to walking behavior (Handy et al. 2002; Lee and Moudon 2004). Subjective measures are self-reported perceptions of the environment obtained from survey questionnaires (Humpel et al. 2004). In urban planning and transportation, most empirical studies do not incorporate subjective factors. Some researchers argue that perceptions, attitudes and preferences are difficult to measure and thus cannot be taken into account (Golledge and Stimson 1997; Gärling et al. 1998). Most physical activity studies use subjective environmental measures, ostensibly because the data are obtained more economically than those of field-collected objective measures, or because detailed objective GIS land use data are not yet commonly accessible.

5.2.1.1 Objective Measures

Distance and aesthetics have the most direct influence on walking choice. In higher-density neighborhoods, land use is compact and destinations are closer, making walking more advantageous. Diversity indicates a mix of land use (Cervero and Kockelman 1997). Being equipped with more land use types that are within walking distance is favorable to increase walking rates (Handy and Clifton 2001). Density and diversity are land use indicators in conjunction with transport systems. Design refers to the aesthetic or quality of the land use and the streetscape, including the presence and attractiveness of natural sights (e.g. trees, hills), recreational facilities (e.g. public open spaces, private gardens), architectural design and pollution levels. Transit use is classified as active travel because it almost always requires walking at one or both ends of the trip (Besser and Dannenberg 2005); hence, distance to transit is an indicator. Areas well served by public transit have bus or rail stops within convenient walking distance, which could decrease automobile dependence.

Though their division is ambiguous and unsettled, these *D* variables are useful in organizing the empirical literature. However, the effects of distance and aesthetic factors may vary depending on the behavior types being studied, the competitiveness of other modal choices, the destination to be accessed and the quality of the destination to the user (Giles-Corti et al. 2005). One variable may have both positive and negative effects. For example, density can increase the capability of direct access, but lower travel speeds. Land use mixing reduces trip costs by placing destinations closer to origins. This could induce more trips and people probably travel further at the same cost, but to get more opportunities and benefits (Boarnet and Crane 2001;

van Wee 2011). An integrated indicator combining these *D* variables may increase the comparability of the studies. Accessibility has long been a central concept in urban planning. It is a measure of the spatial distribution of facilities adjusted for the desire and ability of people to overcome "spatial separation" (e.g. distance or travel time) to access a facility or activity (Handy and Niemeier 1997; McGinn et al. 2007; Hansen 1959). Accordingly, desire and ability are influenced by the importance of the trip to the user (the arrangement of that activity), attractiveness of the facility (design), location of the facility (land use) and the user's access to transport (Giles-Corti et al. 2005). Some studies adopt accessibility as an integrated indicator. For example, Kockelman (1997) uses accessibility measures by purpose and mode in both origin and destination zones; for non-work trips, the measures count sales and service jobs within 30 min walking distance (Kockelman 1997). Krizek (2003) adopts neighborhood accessibility measures that combine density, land use mix, block size and regional accessibility based on gravity (Krizek 2003).

5.2.1.2 Subjective Measures

Most studies have found little agreement between objective and perceived measures of environmental attributes (McCormack et al. 2008; Gebel et al. 2009; McGinn et al. 2007). The studies conducted to date have provided limited guidance on the relative effectiveness of subjectively measured attributes of the built environment in estimating their influence on walking behavior. It has been found that definitions of subjective measures such as accessibility to or convenience of destinations are inconsistent across studies due to different contexts and the distinct survey questionnaires applied to different types of walking. Further, the mechanism that forms individual perceptions is not explicitly instructed. Unstructured questions such as those involving the presence of destinations (e.g. shops) contribute little to the understanding of the true nature of human perception (Moudon et al. 2006). The measures employed are often the results of questions in which respondents impose their own views on the attributes. General perceptions of a residential-based hypothesis may not be enough; we need to know exactly what the respondents are referring to. These differences not only hinder comparisons across studies, but also lack instructive information for policy implications. Social psychology theories provide hints for structured subjective measures and will be illustrated later.

5.2.2 Measuring Walking Behavior: Objective or Self-Report?

There are two types of walking behavior: walking for transport and walking for leisure. While the best way to measure these behaviors is still unclear and separate measurements for the different types are still required, the importance of walking is widely recognized (Heath et al. 2006).

In transportation research, walking behavior is usually recorded by travel diary. Travel diaries ask respondents to keep a log of all trips made during a particular time period, usually 1 or 2 days. A detailed travel diary includes origins and destinations, modes of travel, durations of trips and primary activities at trip destinations. However, walking data in the diaries are usually incomplete (Handy et al. 2002) and less valid (Lee and Moudon 2004). Travel diaries typically focus on driving and public transit rather than walking, which could inevitably lead to a lack of walking data. In addition, the separation of walking for leisure from walking for transport is not necessarily straightforward (Handy 2005). Duncan and his colleagues suggests that a global positioning system (GPS), especially when used in combination with GIS, offers great promise in objectively measuring individual behavior in terms of physical and transport-related activity (Duncan et al. 2009).

In physical activity studies, self-report measures play a central role in measuring physical activity in general and walking in particular (Johnson-Kozlow and Matt 2004; Pereira et al. 1997). They are economical and allow large populations to be assessed quickly and easily (Tudor-Locke et al. 2004). Assessments are variously quantified as sessions and duration per week, number of miles per week, number of blocks walked and as walking more or less with reference to others of the same age. Many studies investigate biases and response errors in self-reported data on physical activity (Sallis and Saelens 2000). A growing number of studies are utilizing objective measures of walking behavior in conjunction with self-reporting. A common method is to utilize a device such as a pedometer or accelerometer, which detect steps and distances travelled throughout the day (Tudor-Locke et al. 2004). However, the high cost of the devices limits their usage in studies with large sample sizes. In addition, the devices alone cannot identify different built environments.

5.2.3 Specificity and Matching

Barriers are formed by spatially matching sufficiently detailed data between the built environment and walking behavior (Cervero and Kockelman 1997). Survey data are usually collected at the census collector district level (Duncan et al. 2010), leading inevitably to a lack of built environment attributes related to walking. Taking pedestrian networks as examples, street networks are too coarse to trace the paths chosen by pedestrians. A true pedestrian network should incorporate formal and informal paths, including sidewalks, laneways, pedestrian bridges and park paths that are informal but used frequently for transit. The missing pedestrian paths in the street network database are likely those that can greatly increase the connectivity of separate locations in the real world. Most accessibility studies use only street networks in their analyses, which may result in inadequacies in the description and prediction of walking travel and induce arguments about the reliability of the analysis result (Chin et al. 2008).

Introducing greater specificity to models that seek to explain the impact of the environment on behavior would greatly improve the predictability of the developed model (Giles-Corti et al. 2005). Further, correspondence between the boundaries within which environmental data are collected and the environments to which the perceptions refer is of great importance (Smith et al. 2010). For example, a researcher's definition of a neighborhood offering area-level data may be substantially different from that of a questionnaire respondent's. The lack of specificity also weakens the correlation between respondent perceptions and objective environment measures.

5.3 Theoretical Basis for Examining the Influence

The built environment's influence on walking behavior varies with walking types. Travel demand provides a theoretical foundation for the influence mechanism on walking for transport. Travel demand depends on the balance between the "utility" of the activity and the "travel cost" of reaching the destination. Because walking for leisure has health, exercise and relaxation utilities, it is included in the utility framework. It also fits in the broader daily activity context, as it offers the choice to decide where, when and how long (in the built environment) this leisure activity will be conducted. Activity patterns are thus expected to give a more complete picture of an individual's walking behavior. Discrete choice models and the TPB are useful in forecasting and understanding the travel choice that forms the skeleton of activity patterns.

5.3.1 Travel Demand Derived from Activity Patterns

The original contributions of activity pattern studies can be traced to the works of Hägerstrand (1970) and Chapin (1974). Hägerstrand (1970) focuses on three kinds of constraints restricting human movement in a spatial-temporal framework, including (1) capability constraints, (2) coupling constraints and (3) authority constraints. Capability constraints refer to limitations imposed by physiological necessities such as sleeping, eating and personal care. Coupling constraints define where, when and how long an individual can interact with other individuals. Authority constraints limit access to either space locations or time locations (e.g. a bank's business hours) (Hägerstrand 1970). The effects of the built environment can be expressed in these three constraint forms. Chapin (1974) describes a motivational framework in which activity patterns result from the interaction between individual propensities and perceived opportunities to engage in activity. He argues that individuals perform activities to meet their basic needs, which in turn determines their propensity to engage in activities (Chapin 1974). By taking opportunities into account, Chapin could assess the effect of the built environment on activity patterns (Ettema and Timmermans 1997). Hägerstrand (1970) and Chapin (1974) provide a theoretical basis for the study of human daily activity patterns, from which travel demand (including trip generation and modal choice) can be derived.

Crane (1996, 2001) explicitly incorporates the built environment into the aforementioned travel demand theory based on the utility maximization principle of microeconomics (Boarnet and Crane 2001; Crane 1996). It has been reasoned that when land use variables exert an influence on travel behavior, such an influence is seen in the effect on the relative trip cost (e.g. speed and distance) of the available modes. To include the built environment in an explicitly activity-based framework, Matt and his colleagues illustrates how the built environment can influence trade-offs between utility and cost (Maat et al. 2005). While compact built environments may reduce an individual's travel time and ability to obtain the same amount of activity benefits, the timesaving benefits of compact designs may also increase trip generations. Whether saving travel time results in less travel, longer trips to obtain extra utility or allocation of time to other activities remains in question. Maat points out that an individual's aim is not primarily to minimize travel costs, but to maximize utility within space and time constraints.

Walking for leisure can also gain theoretical support from activity-pattern studies. A better understanding of how individuals incorporate leisure walking into their daily activities allows for the development of effective policy interventions to facilitate more walking. Furthermore, because recreational activities comprise a substantial share of individuals' non-work activities, studies of participation and time use in recreational activity episodes contribute to activity-based travel demand modeling. There has been relatively little attention paid to the spatial and temporal contexts of physical activity participation, that is, on the when, where and how long of physical activity participation (Sener and Bhat 2012). Leisure-walking participation studies can provide important insight into the design of customized physically active lifestyle promotion strategies in different built-environment and time-of-day contexts.

5.3.2 *Forecast-Oriented Behavior Theory*

Travel modal choices form the skeleton of activity patterns. Choices in walking, transit or driving are discrete by nature. Discrete choice models are thus very helpful for this kind of analysis, which is based on the assumption that choice alternatives can be represented as bundles of attributes. Individuals are assumed to derive some utility from these attribute values and combine them into an overall measure of utility (Ben-Akiva and Lerman 1985). Because of measurement errors and taste variations, these utilities are assumed to comprise a systemic measurable component and a random term. Depending on the assumptions made in these error terms, choice probabilities can be derived.

However, the forecast-oriented travel behavior theory tends to limit the variables included in models. Rather than the larger set of variables, only factors that can be forecasted and that researchers believe might affect travel behavior are employed. McFadden recognizes the importance of the perceptions and attitudes of individuals, but argues that such factors cannot be forecasted and hence should be excluded from forecasting models (McFadden 1974). As a result, even today, forecasting models incorporate relatively few attributes (Handy 2005). In addition, we can only detect the revealed effect of a built environment on walking due to the calibration nature of discrete choice models. Calibration is difficult because revealed behavior – what residents actually do – is not necessarily the same as preferred behavior – what residents would choose to do given a desired set of alternatives (Handy and Niemeier 1997). Revealed behavior is shaped by the specific alternatives available. For example, if a community does not have good pedestrian access, residents will make few pedestrian trips; however, this does not imply they would not make such trips if pedestrian access were better. Hence, an understanding-oriented theory that defines specific factors influencing travel behavior could provide a constructive process that explores the real concerns of individuals.

5.3.3 Understanding-Oriented Behavior Theory

Social psychology theories have been used widely in physical activity studies, including empirical studies that apply the TPB to walking behavior. The theory of reasoned action (TRA) and its extension, the TPB, together provide a framework to decipher individuals' actions by identifying, measuring and combining their beliefs, allowing us to understand their own reasons that motivate the behavior of interest (Montano and Kasprzyk 2008). The TRA and TPB assume that the best predictor of a behavior is behavioral intention (Ajzen 1991). In the TPB, the first determinant of intention, attitude, is an individual's positive or negative evaluation of the behavior. The second determinant, subjective norm, is the person's perception of the social pressure to perform or not perform the behavior under consideration. The third determinant, perceived behavioral control, is the sense of self-efficacy accompanying the ability to perform the behavior of interest. It has been argued that spatial cognition of the environment is equal to the salient beliefs of the TPB and the preference is the same as intention (Van Acker et al. 2010). If more theoretical foundations and empirical evidence on this point could be proven, the TPB may be used to normalize subjective measurements of that environment.

Relevant behavioral outcomes, referents and control beliefs are likely to be different for different populations and behaviors (Ajzen and Driver 1992). Belief elicitation study is a critical step when using the TPB to establish the cognitive foundation of a population's salient behavioral, normative and control beliefs (Ajzen and Fishbein 1980). Only after the belief elicitation procedure can the general TPB theory be applied to understand a specific behavior for a specific population. Although the elicitation phase of TPB studies is important, it is often neglected by researchers (Hagger et al. 2002; Downs and Hausenblas 2005). In the TPB, these background factors are insignificant when constructing the determinants of behavior (Ajzen 2005). In contrast, in transportation studies, it is doubtful that an individual's socio-demographic and socio-economic background factors have real effects on travel behavior, and land use indicators may be surrogates of those factors. Ajzen argues that although a given background factor may influence behavioral, normative or control beliefs, there is no necessary connection between background factors and beliefs. Background factors, including demographic and economic characteristics, are assumed to operate through model constructs but do not independently contribute to explaining the likelihood of a behavior being performed. Whether or not a given belief is affected by a particular background factor is an empirical question. Hence, the TPB leaves us spaces to complement the determinants of walking behavior if the necessity arises.

5.3.4 Utility Maximization and "Rationality"

Clarifying basic assumptions will help us understand the theories. Activity-based approaches and discrete choice models assume that human decision making is based on the utility maximization principle (Koenig 1980). Discrete choice models and the TPB hypothesize that human behavior is rational. It should be noted that the common use of the term "rational behavior" is based on the beliefs of an observer, not the individual, about what the outcome of a decision should be. The beliefs and objective functions of different observers may vary.

It has been recognized that individuals do not always act rationally when maximizing their utility. This is not to say that individuals are irrational; rather, it means that rationality is not always simple (Golledge and Stimson 1997; Montano and Kasprzyk 2008). A fundamental assumption of the TPB is that individuals are "rational actors" who process information and have underlying reasons that determine motivations to perform a behavior. These reasons constitute one's behavioral, normative, and control beliefs, and determine his or her attitudes, subjective norms and perceived control, regardless of whether those beliefs are rational, logical or correct by some objective standard (Fishbein 2007; Montano and Kasprzyk 2008). Some researchers argue that behavior is "irrational" when the behavior changes to a habit (Verplanken et al. 1998). However, measurements of habit strength are neither reliable nor provable. Studies have found that a prior behavior (habit) doesn't predict a later behavior effectively when the built environment has changed (Bamberg et al. 2003).

5.4 Conceptual Model and Study Design

Many studies signify the need for better conceptual models to guide future studies (Handy 2005; Handy et al. 2002; Ewing et al. 2003; Owen et al. 2004). Other than the cross-sectional study design, researchers must undertake longitudinal design to achieve a "deeper examination of direct and indirect relationships, interactions, and hypothesized paths of causality" (Saelens and Handy 2008). To fully explore the built environment's influence on walking behavior, both forecast-oriented and understanding-oriented theories should be applied.

5.4.1 Cross-Sectional Designs Examining the "Association" Relationship

Cross-sectional designs provide continuous empirical studies to quantify the relation between the built environment and walking behavior by comparing different neighborhood types. The examination of the relationship usually runs in the following kinds of regression analysis. (1) Objective measures regress on walking behavior: most studies in transportation are based on this method, especially with the aid of GIS. (2) Subjective measures regress on walking behavior: most studies in public health adopt this method. (3) "Objective and subjective" measures regress on walking behavior: First, both objective and subjective variables are employed to run the regression, respectively. The two measures are then combined to check if the predictability can be improved. This method identifies whether the objective measure, subjective measure or their combination explains the walking behavior more effectively. (4) "Spatial distortions" and walking behavior: Analyses explore the "mismatch" between the objective measures and their subjective counterparts and the effect of this mismatch on walking behavior outcomes (Gebel et al. 2009). The regression results usually show that the built environment has a significant effect on individuals' walking behavior. However, in reality, it is only statistically significant rather than remarkable and the correlation coefficient is relatively small. In addition, although the cross-sectional design is effective in controlling potentially confounding effects (e.g. socio-economic or socio-demographic), it poses challenges in identifying the causal mechanisms involved.

5.4.2 Longitudinal Designs Examining the "Causality" Relationship

Several longitudinal design studies have also been conducted. Krizek (2003) uses longitudinal household travel data in Seattle to examine the relationship between changes in neighborhood forms and household travel behavior. The results show that in controlling for changes in lifestyle, relocating households to neighborhoods with more accessibility could effectively reduce vehicle miles traveled, but it has no significant effect on trip generation or mode splitting (Krizek 2003). Cao et al. (2007) examine the relationship between the residential environment and non-work travel frequencies by automobile, transit and walk/bicycle modes in Northern California. Their study uses quasi-longitudinal data from 547 movers and assumes the movers' residential preferences and travel attitudes remain constant (Cao et al. 2007). Through a structural equation model, they detect more promising neighborhood characteristic effects than those found in previous studies after controlling for "self-selection." However, the reliability of the movers' memories leaves room for argument.

Longitudinal designs exploring causal relationships include two kinds of perspectives or efforts. (1) Examination of the changes in behavior before and after an effort to change human behavior beliefs: "Intervention" studies of physical activity typically take this line of methodology, but it is difficult to say how long this behavior change could last (Sallis et al. 2006). (2) Examination of the changes in behavior before and after a change in the built environment: Urban planning and transport policy studies focus on this method. It is feasible only if there is variation in the environmental conditions detailed by respondents. A key step involves conducting a longitudinal before-and-after survey using the same samples that experienced the changes while working on the assumption that individual attitudes towards travel remain stable. However, physical changes in the built environment are difficult, especially in Western developed cities. In contrast, Chinese cities are experiencing rapid urban growth with built environment changes holding large potential. This will be illustrated in Sect. 5.5.

5.4.3 A Combination of the Discrete Choice Model and the Theory of Planned Behavior

This study proposes a conceptual model (Fig. 5.1). A discrete choice model is calibrated from travel survey data with objective and subjective environmental measures employed jointly. The TPB is used to guide and normalize the process of constructing the subjective measures. A TPB questionnaire is designed to identify key behavioral, normative and control beliefs affecting behaviors equal to perception, cognition and preference. The discrete choice model forecasts the built environment's influence on walking behavior, with more attention paid to the reasons behind the behavior and the underlying preferences of residents. While the TPB is used to construct the perception or cognition process, it can also predict the behavior via the inherent structural equation models between its constructs.

A longitudinal survey of the same respondents that experienced a change in built environment is encouraged. Through the "before" measurements, the relationship

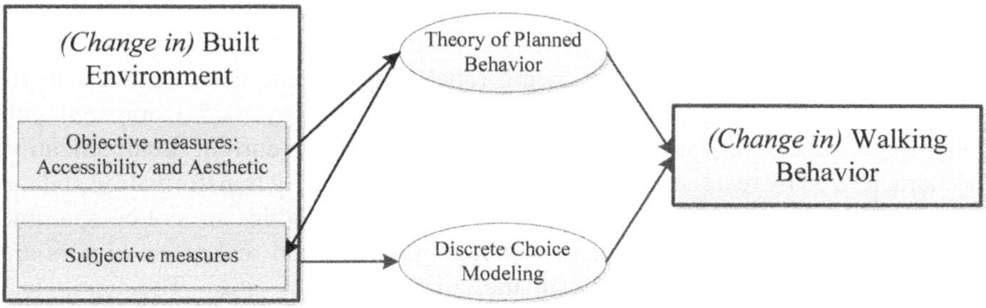

Fig. 5.1 A conceptual model combining discrete choice model and TPB

between the built environment and travel behavior is analyzed. Based on that, the situation after the change in built environment will be predictable. The "after" measurement can act as a comparison with the predictions. It is instructive to examine the degree to which the measures obtained in the first wave predict those obtained in the second wave and, more importantly, to compare the temporal stability of the measures in terms of different modes of transportation. To assess the overall impact of the change, studies can also compare attitudes, subjective norms, perceived behavioral control, intentions, past behavioral tendencies and reported behaviors before and after the change in built environment.

5.5 Study Progress for Walking Behavior in Chinese Cities

It is noteworthy that most of the current studies in this field have been carried out in developed Western countries and regions (e.g. Australia, the United States and Western Europe). The experiences of developed countries may not be applicable to other contexts, especially those undergoing fast-paced urbanization and motorization like China. There is a great potential for China to provide fresh evidence and significant insight.

5.5.1 Chinese Cities Versus Western Cities

5.5.1.1 Developed and Developing

Since the Second World War, Western city urban growth patterns have been dominated by low-density development and employment decentralization. Cities in developed countries are largely built up with an automobile-oriented structure, within which the society is dominated by an automobile culture. After the 1980s, North American and European countries experienced "new suburbanization." United States studies focus on whether and how the addition of a compact city development strategy in a low-density context can reduce automobile-based travel and improve air quality (Cervero and Duncan 2006). Empirical studies in Western cities tend to confirm the transport and health benefits of land use strategies such as densification, infill development, mixed land use and the job-housing balance.

China, however, is still experiencing rapid urbanization and motorization. Its urbanization rate will increase from 46 % in 2010 to 60 % by 2025. Compared with employment decentralization in Western countries, China's current suburbanization is characterized by residential decentralization due to old city transformation and the suburban housing market (Zhao et al. 2011). Job opportunities are not comparable with housing decentralizations. The finance, insurance and real-estate sectors are concentrated and growing rapidly in the central areas of cities. The featureless expansion of the central built-up area will most likely form a mono-centric rather than a multi-nuclei city (Jiawen Yang et al. 2011). Fortunately, most Chinese cities are still in the early stages of motorization, with an average of just 47 vehicles per 1,000 people in 2009 compared with 802 in the United States (World Bank 2011). However, China's automobile industry boom and highway construction will further encourage private motor vehicle ownership. It is critical to avoid the formation of car-dependent travel attitudes and lifestyles in this stage of city "restructuring" (Zacharias and Tang 2010).

5.5.1.2 Public Ownership and Institutional Intervention

Different from Western cities, urban land in China is under public ownership. Although the transformation of land use from a planning economy to a market economy has replaced numerous urban administrative procedures, the municipal government is still fundamentally in charge of making and implementing detailed land use plans and infrastructure investment strategies, taxing users of urban land and licensing various kinds of urban activities (Shen 1997). Yang et al. (2011) point out that the continuous expansion of Beijing's central built-up area and the absorption of the suburbs are not purely market-based processes but processes conditioned by government-formulated spatial planning and infrastructure investment strategies. The highest quality infrastructure and social services are supplied in the central part, while suburban communities are relatively neglected in terms of receiving municipal investment (Jiawen Yang et al. 2011). If polices are centralized, it is hard to form decentralized self-sustainable centers with working, living, education, shopping and entertainment units, all of which are believed beneficial to encouraging short trips and non-motorized movement.

In the transformation process of Chinese cities, old institutions and new institutional factors coexist and have a mixed effect on individual travel patterns. *Danwei,* with its successors *xiaoqu* and *shequ*, atomizes the structure of trip sets and urban life at the neighborhood scale (Yang and Gakenheimer 2007; Wang and Chai 2009; Bray 2005). Zhao and Lu (2010) argue that in the case of China, the housing provision system and labor mobility management have played important roles in influencing job accessibility and hence commuter behavior (Zhao and Lu 2010). They conclude that in addition to controlled household income, individuals' occupations and the transport mode, the interaction of the housing provision system (welfare-oriented housing versus market-oriented housing (Li 2000)), the market system (labor market institution), the *Hukou* system (urban *Hukou* versus rural *Hukou*) and the urban life unit (*xiaoqu* or *shequ*) have had a significant effect on individual commuting behavior. More attention should be paid to these institutional factors when researching the built environment's influence on walking behavior in China.

5.5.1.3 High Density and Overconcentration

Contemporary urban China is conducting the same strategies as Western cities, such as new urbanism, smart growth and TOD, but in high-density urban areas.

Many Chinese metropolitan areas average 200–250 persons per hectare, with local urban densities reaching much higher figures. Megacities like Shanghai and Beijing are among the densest urban areas in the world (Yang and Gakenheimer 2007). In Hong Kong's urbanized areas, dwelling density reaches over 1,250 units per hectare; this is in stark contrast to European cities, where inner-city areas rarely exceed 125 residential units per hectare (Cerin et al. 2011). In such a high density, however, less than 10 % of the urban land is used for road infrastructure in most pre-reform Chinese cities. Hence, overconcentration is a potential problem affecting individual travel and urban transport. The distance-saving benefits of high densities will probably be offset by the congestion stemming from overconcentration (Jiawen Yang et al. 2011). Yang et al. (2011) have confirmed that neighborhoods with relatively high densities do not necessarily have lower drive commute percentages. It seems difficult to strike a balance between high density and overconcentration in China.

5.5.2 *Pilot Studies in Chinese Cities*

5.5.2.1 Pilot Studies and Stage Results

Typical empirical studies in China are summarized in Table 5.1, and several findings can be observed. First, pilot studies are mainly conducted in megacities such as Beijing and Shanghai. Fewer cases are found in medium-sized cities. Second, the built environment is usually measured objectively. Regression models are adopted to reveal and forecast the behavior. Internal human psychology factors are rarely considered. Third, job-housing balance and commuting behaviors are the major focuses rather than non-work behavior. The mismatch of jobs and housing spaces in these studies is believed to be a main reason for traffic congestion during rush hour (Zhou and Liu 2010; Zhao et al. 2011). Finally, most of these studies are carried out by Western-trained and foreign scholars.

5.5.2.2 Learning from China's History

China has a longer urban development history than most developed countries. Rather than borrowing ideas from the West, China may be better off learning from its own history. Recent urban development has departed significantly from the traditional compact, mixed-use settings with most activities within walking distance that used to characterize Chinese cities. Neighborhoods in China have continuously evolved and undergone several major societal and institutional transformations, including the declined work-unit compounds (*danwei*), the courtyard in Beijing, the *Linong* in Shanghai, and the newly raised *xiaoqu* and *shequ*. In China, walking and bicycling have a long tradition. Jiang found that old neighborhood forms

Table 5.1 Typical empirical studies of walking behavior in China

Reference	Sample	Study design	Environmental attributes	Walking scenario	Scale	Method	Results
Pan et al. (2009) ***Shanghai***	1,709 individuals from four selected neighborhoods (conducted by on-street interviews)	CS, O	(1) *Distance to city center*: Distance to CBD and the nearest city sub-center; (2) *Diversity of land use*: Commercial, industrial, etc.; (3) *Density*: Population density, average block size, road length, road density; (4) *Distance to transit*: Number of bus stops and rail transit stations within 1 km radius of the neighborhood centroid	Travel modal choice: Non-motorized modes (walking and cycling), transit and driving	N	Multiple logistic regressions	In traditional neighborhood with relatively high densities and small street blocks, trip distance is shorter and people are more likely to choose walking or cycling
Cervero and Day (2008) ***Shanghai***	A before-and-after survey of 900 recent movers (containing 2,840 inhabitants) in three suburban neighborhoods	LD, O (relocation)	(1) *Distance to transit*: Metro rail services, levels of mixed-use activities; (2) *After relocation*: In superblock, less walking-friendly, mixed-use environments than residents' previous neighborhoods	Mode change: Non-motorized to motorized transport; bus to other motorized transport	R & N	Descriptive statistics and binary logistic models	Relocating to a suburban area near a metro-rail station encourages commuters to switch from non-motorized and bus transit to rail. Proximity to regional rail networks has stronger influence on commuting behavior than neighborhood street designs and land-use patterns

(continued)

Table 5.1 (continued)

Reference	Sample	Study design	Environmental attributes	Walking scenario	Scale	Method	Results
Wang and Chai (2009) ***Beijing***	736 heads of households who are employed	CS, O	(1) House provided by *danwei* (self-contained work unit compound); (2) House provided by private market	Commuting behavior; walking is included in non-motorized transport	R & N	Structural equation model	*Danwei* (work-unit) housing commuters have shorter commuting trips and higher usage of non-motorized transport modes than those in market housing
Yang (2010) ***Beijing***	1,499 workers from 48 neighborhoods, with 25 households surveyed in each neighborhood	CS, O	Neighborhood location: Distance to Beijing's geometric center (*Tian an men* Square), distance to the closest job centers and distance to the closest subway station	Mode choice (percentages of commutes by various transport modes)	R & N	Simple regression model	Commuting time and mode split vary according to the age of the neighborhood. The newer the neighborhood, the higher the percentage of non-motorized modes (including walking and cycling) and the shorter the commuting time
Zhou and Deng (2010), Zhou and Liu (2010) ***Guangzhou***	1,006 individuals from 800 household in 11 "street blocks" that are enclosed by roads with 1 km^2 area	CS, O	(1) *Housing types:* House provided by *danwei*, low-priced welfare housing, rental housing, commercial housing, urban village (*chengzhongcun*), affordable housing (*jingjishiyongfang*), etc.; (2). *Housing locations:* Traditional city centers, transaction zones, fringes of the city	Daily activities with mode choice, walking may be included but not specific	R & N	Distance buffer analysis	The spatio-temporal activity patterns vary with different classes (income). The activity space of the lower class is smaller, concentrated mainly in the inner city and the area around their residential communities, opposite to the higher income class, who have free choice of housing

Zhou and Yan (2006) ***Guangzhou***	1,428 individuals from 827 households in 8 "street blocks"	CS, O & P	House types and locations (same as aforementioned), population density and service facilities	Walking is included in commuting behavior	R & N	Correlation statistics, GIS	Residents in the city center area have shorter commute times, and more walking and bicycling may be adopted. Perception is collected roughly but not employed in travel behavior analysis
Yang Jiang et al. (2012) ***Jinan***	1,233 user surveys at 19 bus rapid transit (BRT) stations along three BRT corridors	CS, O & P	(1) *BRT station context*: Terminal, transfer or typical; (2) *Corridor types*: Arterial edge, integrated boulevard, below expressway	Distance people will walk to access the BRT	N	Ordinary least squares regression	People would walk farther to BRT stations when the walking environment has certain features (median transit-way station location, shaded corridors, busy and interesting)

CS cross-sectional design, *LD* longitudinal design, *O* objective, *P* perceived, *N* neighborhood, *R* regional

are associated with 65–80 % less travel energy use than the "superblock" (Jiang 2010). The energy reduction is probably due to the mixed land uses, implicit traffic calming measures and parking restrictions. Today, mid-rise and high-rise residential towers have gradually replaced neighborhoods with unique local characteristics. To improve automobile accessibility using the transportation infrastructure investment, only the motor-vehicle ways have been widened and extended, while sidewalks and cycle tracks have been narrowed in most cases (Yang and Gakenheimer 2007). Nevertheless, historical design principles could be revisited to inspire China's policymakers and urban planners to establish guidelines for new neighborhood forms in future urban development.

5.5.2.3 Potential for Longitudinal Design Studies

Economic growth is the major driving force of development. In addition to the urban growth and plan reform movements, land use pattern changes, transportation and design occur almost every day in China. Relocation is common in Chinese cities, from inner cities to the outskirts, freely or as a result of enforcement, with goals of housing ownership, good living conditions and physical proximity to workplace (Cervero and Day 2008). Examining longitudinal design in China to track movers and conducting a before-and-after survey of their walking behavior will provide not only fresh experiences, but also rigorous evidence for the built environment's causal influence on walking behavior.

5.5.2.4 Transit-Oriented Development May Be the Future

Given China's large population and limited land size, the ideal choices for urban China are compact and transit-friendly cities rather than American-style automobile-oriented urban landscaping (J. Yang and Gakenheimer 2007). Cervero and Day (2008) suggest that TOD holds considerable promise for placing rapidly suburbanizing Chinese cities on more sustainable pathways (Cervero and Day 2008). Opportunities for creating sustainable city forms by bundling land use investments and TOD in large Chinese cities are quite substantial and largely untapped, and to date, China has not formed this interconnection (Pan et al. 2007). Today, increasing numbers of large, rail-served Chinese cities consider TOD an alternative form of urbanism that reduces over-reliance on private automobiles. Many new communities developed along TOD corridors could become veritable bedroom communities and deserve more attention. Even if the TOD concept is favorable, the kind of TOD we should pursue remains an urgent question (Zhang 2007). This cannot be answered without empirical investigation in the local context.

5.6 Conclusions

This study reviews and analyzes the experiences and results of the influence of built environment on walking behavior conducted in Western countries. It argues that while accessibility as an integrated objective measure could improve comparability across studies, more subjective measures are needed. It suggests that the TPB can help us understand the built environment's influence on walking behavior and normalize the procedure to subjectively measure the built environment. In addition, introducing greater specificity to the measures would improve the explanatory ability of the developed model. A theoretical basis for the examination of the influence is offered from two perspectives: a forecast-oriented perspective by a discrete choice model and an understanding-oriented perspective by the TPB. Discrete choice models mostly adopt objective measures, while the subjective measures constructed by the TPB can be systematically employed into a discrete choice model as complements. A conceptual model combining a discrete choice model and the TPB is proposed to explore the built environment's influence on walking behavior.

Different from Western cities, Chinese cities are experiencing rapid motorization and urbanization. Driven by economic growth and transportation infrastructure investment, contemporary urban China is departing from a history marked with high-density, mixed-land-use and walking traditions. The built environment changes every day. Residence relocation is ongoing either freely or under enforcement, making longitudinal design studies very possible. Although the effect of the market economy on the urban form is becoming more significant, institutional factors still play an important role in the modern city transformation process. Governments are still pivotal in making urban planning polices in China. Opportunities for creating sustainable city forms through bundling land use investment and TOD in large Chinese cities will be quite substantial in the future. Most Chinese studies focus on the job-housing balance and less on the social differentiation and stratification caused by the job-housing separation. More empirical studies that move beyond Western experiences and policies are required.

Acknowledgement This research is sponsored by Direct Grant from CUHK (CUHK 2021094) and partially supported by National Natural Science Foundation of China (Grant NO. 41171146 and 41101370).

References

Ajzen, I. (1991). The theory of planned behavior. *Organizational Behavior and Human Decision Processes, 50*(2), 179–211.

Ajzen, I. (2005). *Attitudes, personality and behavior*. Maidenhead: Open University Press.

Ajzen, I., & Driver, B. (1992). Application of the theory of planned behavior in leisure choice. *Journal of Leisure Research, 24*, 207–224.

Ajzen, I., & Fishbein, M. (1980). *Understanding attitudes and predicting social behavior* (278). Englewood Cliffs: Prentice-Hall.

Bamberg, S., Ajzen, I., & Schmidt, P. (2003). Choice of travel mode in the theory of planned behavior: The roles of past behavior, habit, and reasoned action. *Basic and Applied Social Psychology, 25*(3), 175–187.

Ben-Akiva, M. E., & Lerman, S. R. (1985). *Discrete choice analysis theory and application to travel demand* (9). Cambridge, MA: The MIT Press.

Besser, L. M., & Dannenberg, A. L. (2005). Walking to public transit: Steps to help meet physical activity recommendations. *American Journal of Preventive Medicine, 29*(4), 273–280.

Boarnet, M., & Crane, R. (2001). The influence of land use on travel behavior: Specification and estimation strategies. *Transportation Research Part A: Policy and Practice, 35*(9), 823–845.

Bray, D. (2005). *Social space and governance in urban China: The danwei system from origins to reform*. Stanford: Stanford University Press.

Cao, X., Mokhtarian, P. L., & Handy, S. L. (2007). Do changes in neighborhood characteristics lead to changes in travel behavior? A structural equations modeling approach. *Transportation, 34*(5), 535–556.

Cerin, E., Chan, K., Macfarlane, D. J., Lee, K., & Lai, P. (2011). Objective assessment of walking environments in ultra-dense cities: Development and reliability of the Environment in Asia Scan Tool – Hong Kong version (EAST-HK). *Health & Place, 17*(4), 937–945. doi:10.1016/j.healthplace.2011.04.005.

Cervero, R., & Day, J. (2008). Suburbanization and transit-oriented development in China. *Transport Policy, 15*(5), 315–323.

Cervero, R., & Duncan, M. (2006). Which reduces vehicle travel more: Jobs-housing balance or retail-housing mixing? *Journal of the American Planning Association, 72*(4), 475–490.

Cervero, R., & Kockelman, K. (1997). Travel demand and the 3Ds: Density, diversity, and design. *Transportation Research Part D: Transport and Environment, 2*(3), 199–219.

Chapin, F. S. (1974). *Human activity patterns in the city: Things people do in time and in space*. New York: Wiley.

Chin, G. K. W., Van Niel, K. P., Giles-Corti, B., & Knuiman, M. (2008). Accessibility and connectivity in physical activity studies: The impact of missing pedestrian data. *Preventive Medicine, 46*(1), 41–45.

Crane, R. (1996). On form versus function: Will the new urbanism reduce traffic, or increase it? *Journal of Planning Education and Research, 15*(2), 117–126.

Cunningham, G. O., & Michael, Y. L. (2004). Concepts guiding the study of the impact of the built environment on physical activity for older adults: A review of the literature. *American Journal of Health Promotion: AJHP, 18*(6), 435–443.

Downs, D. S., & Hausenblas, H. A. (2005). Elicitation studies and the theory of planned behavior: A systematic review of exercise beliefs. *Psychology of Sport and Exercise, 6*(1), 1–31.

Duncan, M. J., Badland, H. M., & Mummery, W. K. (2009). Applying GPS to enhance understanding of transport-related physical activity. *Journal of Science and Medicine in Sport, 12*(5), 549–556.

Duncan, M. J., Winkler, E., Sugiyama, T., Cerin, E., Leslie, E., & Owen, N. (2010). Relationships of land use mix with walking for transport: Do land uses and geographical scale matter? *Journal of Urban Health, 87*(5), 782–795.

Ettema, D., & Timmermans, H. J. P. (1997). Theories and models of activity patterns. In D. F. Ettema & H. J. P. Timmermans (Eds.), *Activity-based approaches to travel analysis* (pp. 1–36). Oxford: Pergamon Press.

Ewing, R., Schmid, T., Killingsworth, R., Zlot, A., & Raudenbush, S. (2003). Relationship between urban sprawl and physical activity, obesity, and morbidity. *American Journal of Health Promotion, 18*(1), 47–57.

Fishbein, M. (2007). A reasoned action approach: Some issues, questions, and clarifications. In I. Ajzen, D. Albarracin, & R. Hornik (Eds.), *Prediction and change of health behavior: Applying the reasoned action approach* (pp. 281–296). Hillsdale: Lawrence Erlbaum & Associates.

Gärling, T., Gillholm, R., & Gärling, A. (1998). Reintroducing attitude theory in travel behavior research: The validity of an interactive interview procedure to predict car use. *Transportation, 25*(2), 129–146.

Gebel, K., Bauman, A., & Owen, N. (2009). Correlates of non-concordance between perceived and objective measures of walkability. *Annals of Behavioral Medicine, 37*(2), 228–238.

Giles-Corti, B., Timperio, A., Bull, F., & Pikora, T. (2005). Understanding physical activity environmental correlates: Increased specificity for ecological models. *Exercise and Sport Sciences Reviews, 33*(4), 175–181.

Golledge, R. G., & Stimson, R. J. (1997). *Spatial behavior: A geographic perspective*. New York: The Guilford Press.

Hägerstrand, T. (1970). What about people in regional science? *Papers in Regional Science, 24*(1), 6–21.

Hagger, M. S., Chatzisarantis, N. L. D., & Biddle, S. J. H. (2002). A meta-analytic review of the theories of reasoned action and planned behavior in physical activity: Predictive validity and the contribution of additional variables. *Journal of Sport & Exercise Psychology, 24*(1), 3–32.

Handy, S. (2005). *Critical assessment of the literature on the relationships among transportation, land use, and physical activity*. Washington, DC: Transportation Research Board and the Institute of Medicine Committee on Physical Activity, Health, Transportation, and Land Use.

Handy, S. L., & Clifton, K. J. (2001). Local shopping as a strategy for reducing automobile travel. *Transportation, 28*(4), 317–346.

Handy, S. L., & Niemeier, D. A. (1997). Measuring accessibility: An exploration of issues and alternatives. *Environment and Planning A, 29*, 1175–1194.

Handy, S. L., Boarnet, M. G., Ewing, R., & Killingsworth, R. E. (2002). How the built environment affects physical activity: Views from urban planning. *American Journal of Preventive Medicine, 23*(2), 64–73.

Hansen, W. G. (1959). How accessibility shapes land use. *Journal of the American Institute of Planners, 25*(2), 73–76.

Heath, G. W., Brownson, R. C., Kruger, J., Miles, R., Powell, K. E., & Ramsey, L. T. (2006). The effectiveness of urban design and land use and transport policies and practices to increase physical activity: A systematic review. *Journal of Physical Activity & Health, 3*, 55–76.

Hoehner, C. M., Brennan Ramirez, L. K., Elliott, M. B., Handy, S. L., & Brownson, R. C. (2005). Perceived and objective environmental measures and physical activity among urban adults. *American Journal of Preventive Medicine, 28*(2), 105–116.

Humpel, N., Owen, N., Leslie, E., Marshall, A. L., Bauman, A. E., & Sallis, J. F. (2004). Associations of location and perceived environmental attributes with walking in neighborhoods. *American Journal of Health Promotion, 18*(3), 239–242.

Jiang, Y. (2010). *Does energy follow urban form? An examination of neighborhoods and transport energy use in Jinan, China*. Thesis completed towards the Degree of Master of City Planning and Master of Science in Transportation. Cambridge, MA: MIT.

Jiang, Y., Zegras, P. C., & Mehndiratta, S. (2012). Walk the line: Station context, corridor type and bus rapid transit walk access in Jinan, China. *Journal of Transport Geography, 20*(1), 1–14. doi:10.1016/j.jtrangeo.2011.09.007.

Johnson-Kozlow, M., & Matt, G. E. (2004). What respondents recall about walking and what self-report items elicit about walking. *Preventive Medicine, 38*(2), 227–236.

Kockelman, K. M. (1997). Travel behavior as function of accessibility, land use mixing, and land use balance: Evidence from San Francisco Bay Area. *Transportation Research Record: Journal of the Transportation Research Board, 1607*, 116–125.

Koenig, J. G. (1980). Indicators of urban accessibility: Theory and application. *Transportation, 9*(2), 145–172.

Kremers, S., de Bruijn, G.-J., Visscher, T., van Mechelen, W., de Vries, N., & Brug, J. (2006). Environmental influences on energy balance-related behaviors: A dual-process view. *International Journal of Behavioral Nutrition and Physical Activity, 3*(1), 1–10. doi:10.1186/1479-5868-3-9.

Krizek, K. J. (2003). Residential relocation and changes in urban travel: Does neighborhood-scale urban form matter? *Journal of the American Planning Association, 69*(3), 265–281.

Lee, C., & Moudon, A. V. (2004). Physical activity and environment research in the health field: Implications for urban and transportation planning practice and research. *Journal of Planning Literature, 19*(2), 147–181.

Li, S. (2000). Housing consumption in urban China: A comparative study of Beijing and Guangzhou. *Environment and Planning A, 32*(6), 1115–1134.

Lin, L., & Moudon, A. V. (2010). Objective versus subjective measures of the built environment, which are most effective in capturing associations with walking? *Health & Place, 16*(2), 339–348.

Maat, K., Van Wee, B., & Stead, D. (2005). Land use and travel behaviour: Expected effects from the perspective of utility theory and activity-based theories. *Environment and Planning B: Planning and Design, 32*(1), 33–46.

McCormack, G. R., Cerin, E., Leslie, E., Du Toit, L., & Owen, N. (2008). Objective versus perceived walking distances to destinations. *Environment and Behavior, 40*(3), 401–425.

McFadden, D. (1974). The measurement of urban travel demand. *Journal of Public Economics, 3*(4), 303–328.

McGinn, A. P., Evenson, K. R., Herring, A. H., Huston, S. L., & Rodriguez, D. A. (2007). Exploring associations between physical activity and perceived and objective measures of the built environment. *Journal of Urban Health, 84*(2), 162–184.

Montano, D. E., & Kasprzyk, D. (2008). Theory of reasoned action, theory of planned behavior, and the integrated behavioral model. In K. Glanz, B. K. Rimer, & K. Viswanath (Eds.), *Health behavior and health education: Theory, research, and practice* (pp. 67–96). San Francisco: Jossey-Bass.

Moudon, A. V., Lee, C., Cheadle, A. D., Garvin, C., Johnson, D., Schmid, T. L., et al. (2006). Operational definitions of walkable neighborhood: Theoretical and empirical insights. *Journal of Physical Activity & Health, 3*, 99–117.

Owen, N., Humpel, N., Leslie, E., Bauman, A., & Sallis, J. F. (2004). Understanding environmental influences on walking: Review and research agenda. *American Journal of Preventive Medicine, 27*(1), 67–76.

Pan, H., Ren, C., & Yang, T. (2007). 上海道交通站点地区土地使用影响的研究 [A study on the impact on the land use of station areas brought by urban rail transport in Shanghai]. 城市划学刊 [Urban Planning Forum], 170(4), 92–97.

Pan, H., Shen, Q., & Zhang, M. (2009). Influence of urban form on travel behaviour in four neighbourhoods of Shanghai. *Urban Studies, 46*(2), 275–294.

Pereira, M. A., FitzerGerald, S., Gregg, E., Joswiak, M., Ryan, W., Suminski, R., et al. (1997). A collection of physical activity questionnaires for health-related research. *Medicine & Science in Sports & Exercise, 29*(6 Suppl), 1–205.

Saelens, B. E., & Handy, S. L. (2008). Built environment correlates of walking: A review. *Medicine and Science in Sports and Exercise, 40*(Suppl 7), 550–566.

Sallis, J., & Saelens, B. (2000). Assessment of physical activity by self-report: Status, limitations, and future directions. *Research Quarterly for Exercise and Sport, 71*(Suppl 2), 1–14.

Sallis, J. F., Cervero, R. B., Ascher, W., Henderson, K. A., Kraft, M. K., & Kerr, J. (2006). An ecological approach to creating active living communities. *Annual Review Public Health, 27*, 297–322.

Sener, I. N., & Bhat, C. R. (2012). Modeling the spatial and temporal dimensions of recreational activity participation with a focus on physical activities. *Transportation, 39*(3), 627–656. doi:10.1007/s11116-011-9356-7.

Shen, Q. (1997). Urban transportation in Shanghai, China: Problems and planning implications. *International Journal of Urban and Regional Research, 21*(4), 589–606.

Smith, G., Gidlow, C., Davey, R., & Foster, C. (2010). What is my walking neighbourhood? A pilot study of English adults' definitions of their local walking neighbourhoods. *The International Journal of Behavioral Nutrition and Physical Activity, 7*, 34.

Tudor-Locke, C., Williams, J. E., Reis, J. P., & Pluto, D. (2004). Utility of pedometers for assessing physical activity: Construct validity. *Sports Medicine, 34*(5), 281–291.

Van Acker, V., Van Wee, B., & Witlox, F. (2010). When transport geography meets social psychology: Toward a conceptual model of travel behaviour. *Transport Reviews, 30*(2), 219–240.

van Wee, B. (2011). Evaluating the impact of land use on travel behaviour: The environment versus accessibility. *Journal of Transport Geography, 19*(6), 1530–1533. doi:10.1016/j.jtrangeo.2011.05.011.

Verplanken, B., Aarts, H., Knippenberg, A., & Moonen, A. (1998). Habit versus planned behaviour: A field experiment. *British Journal of Social Psychology, 37*(1), 111–128.

Wang, D., & Chai, Y. (2009). The jobs–housing relationship and commuting in Beijing, China: The legacy of Danwei. *Journal of Transport Geography, 17*(1), 30–38. doi:10.1016/j.jtrangeo.2008.04.005.

World Bank. (2011). *World development indicators.*http://data.worldbank.org/indicator/IS.VEH.NVEH.P3. Washington, DC: The World Bank.

Yang, J. (2010). Spatial and social characteristics of urban transportation in Beijing. *Transportation Research Record: Journal of the Transportation Research Board, 2193*, 59–67.

Yang, J., & Gakenheimer, R. (2007). Assessing the transportation consequences of land use transformation in urban China. *Habitat International, 31*(3–4), 345–353.

Yang, J., Shen, Q., Shen, J., & He, C. (2011). Transport impacts of clustered development in Beijing: Compact development versus overconcentration. *Urban Studies, 49*(6), 1315–1331. doi:10.1177/0042098011410336.

Zacharias, J., & Tang, Y. (2010). Restructuring and repositioning Shenzhen, China's new mega city. *Progress in Planning, 73*(4), 209–249.

Zhang, M. (2007). Chinese edition of transit-oriented development. *Transportation Research Record: Journal of the Transportation Research Board, 2038*, 120–127.

Zhao, P., & Lu, B. (2010). Exploring job accessibility in the transformation context: An institutionalist approach and its application in Beijing. *Journal of Transport Geography, 18*(3), 393–401.

Zhao, P., Lu, B., & de Roo, G. (2011). The impact of urban growth on commuting patterns in a restructuring city: Evidence from Beijing. *Papers in Regional Science, 90*(4), 735–754.

Zhou, S., & Deng, L. (2010). 基于 T-GIS 的广州市居民日常活动时空关系 [Spatio-temporal pattern of residents' daily activities based on T-GIS: A case study in Guangzhou, China]. 地理学报 [Acta Geographica Sinica], *65*(12), 1454–1463.

Zhou, S., & Liu, Y. (2010). 转型期广州城市居民居住与就业地区位选择的空间关系及其变迁 [The situation and transition of jobs-housing relocation in Guangzhou, China]. 地理学报 [Acta Geographica Sinica], *65*(2), 191–201.

Zhou, S., & Yan, X. (2006). 基于居民通勤行为分析的城市空间解读–以广州市典型街区为案例 [The impact of commuters' travel pattern on urban structure: A case study in some typical communities in Guangzhou]. 地理学报 [Acta Geographica Sinica], *61*(2), 179–189.

Chapter 6
Spatio-temporal Networks: Modeling, Storing, and Querying Temporally-Detailed Roadmaps

Michael R. Evans, KwangSoo Yang, Viswanath Gunturi, Betsy George, and Shashi Shekhar

6.1 Introduction

Given a spatial network and its variations over time (e.g., time-varying travel times on road networks) this chapter discusses how to model, query and store spatio-temporal networks. This problem has application in several domains such as transportation networks, emergency planning, knowledge discovery from sensor data, and crime analysis. Adequately representing the temporal nature of spatial networks would potentially allow us to raise interesting questions (e.g., eco-routing, non-FIFO behavior) and find efficient solutions. In transportation networks, travelers are often interested in finding the best time to start so that they spend the least time on the road. Crime data analysts may be interested in finding temporal patterns of crimes at certain locations or the routes in the network that show significantly high crime rates. Modeling the time dependence of sensor network data would certainly improve the process of discovering patterns such as hot spots. In these application domains, it is often necessary to develop a model that captures both the time dependence of the data and the underlying connectivity of the locations. There are significant challenges in developing a model for spatio-temporal networks. The model needs to balance storage efficiency and expressive power and provide adequate support for the algorithms that process the data.

M.R. Evans (✉) • K. Yang • V. Gunturi • B. George • S. Shekhar
Department of Computer Science and Engineering, University of Minnesota, 4-192 Keller Hall 200 Union Street SE, Minneapolis, MN 55455, USA
e-mail: mevans@cs.umn.edu; ksyang@cs.umn.edu; gunturi@cs.umn.edu; bgeorge@cs.umn.edu; shekhar@cs.umn.edu

M.-P. Kwan et al. (eds.), *Space-Time Integration in Geography and GIScience*, DOI 10.1007/978-94-017-9205-9_6

6.2 Illustrative Application Domains

In the present discussion, modeling is not the same as predicting in this context. Incidents that happen without any warning and against all forecasts are part of the *unpredictable future*. Random road accidents and consequent changes in traffic patterns fall under this category. In the absence of such events, changes in the network largely follow known patterns. In such *predictable future* scenarios, we assume that the changes in network parameters and topology are known in advance. This allows for the development of clear and concise representations for modeling.

6.2.1 Transportation Network Routing

Transportation networks are the kernel framework of many advanced transportation systems such as the Advanced Traveler Information System and Intelligent Transportation Systems. These transportation networks are spatio-temporal in nature and require significant database support to handle the storage of their large amounts of multi-dimensional data. Many important applications based on transportation networks, including travelers' trip planning, consumer business logistics, and evacuation planning are built upon spatio-temporal networks. A transportation network displays *non-stationary* characteristics when route preferences change with time, possibly due to non-uniform congestion levels at different parts of the network or a time-dependent availability of certain road segments or faster services in a public transport system.

With the increasing use of sensor networks to monitor traffic data on spatial networks and the subsequent availability of time-varying traffic data, it becomes important to incorporate this data into the models and algorithms related to transportation networks. As an example, Fig. 6.1 shows a layout of traffic sensors in the Twin Cities of Minneapolis-St Paul and the congestion measurements at two different times of a day.

6.2.2 Emergency Traffic Management

A key step in emergency management is the evacuation of a population from areas affected by disasters to safe locations. One significant challenge in this step comes from the time-dependence of the transportation network. Travel times on the road segments and the available capacities of the roads are both time-dependent. The dynamic nature of the networks raises some interesting questions (as given in Table 6.1) and the model for the transportation networks should provide support for such queries.

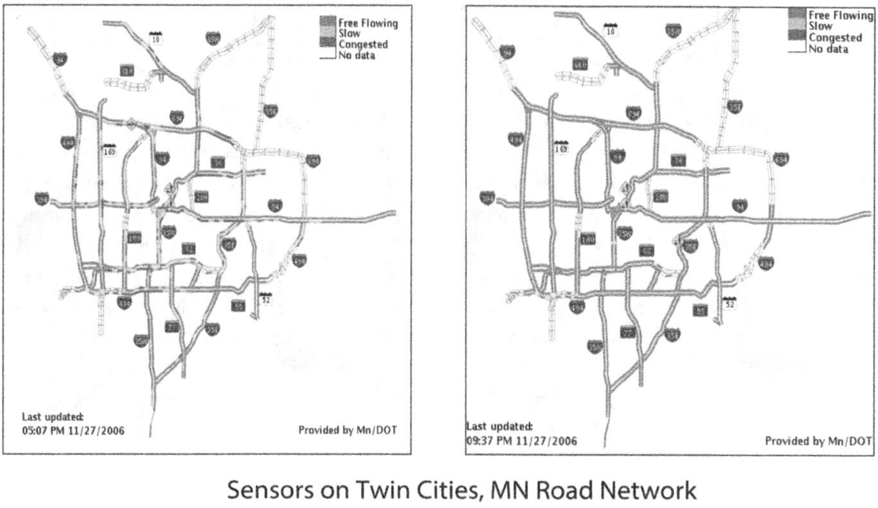

Fig. 6.1 Sensors periodically report time-variant traffic volumes on Minneapolis-St Paul highways (Best viewed in color, Source: Mn/DOT) (Color figure online)

Table 6.1 Example queries in time-varying networks

Category	Example questions
Snapshot	Where are the evacuee groups located at present?
Node/edge centric	Which evacuee groups are at a node
Time-aggregated	At different time instants?
Trajectory based	Find the route of a person on a given day

6.2.3 Knowledge Discovery in Spatio-temporal Sensor Networks

Many scientific domains collect sensor data in outdoor environments with underlying physical interactions. A collection of sensors may be represented as a sensor graph where the nodes represent the sensors and the edges represent selected relationships.

For example, sensors upstream and downstream in a river may have physical interactions via water flow and related phenomena such as plume propagation. Relationships can also be geographical in nature, such as proximity between the sensor units. A sensor graph is spatio-temporal in nature since the relative locations of the sensor nodes and the time-dependence of their characteristics are significant. Developing a model that facilitates the representation and knowledge discovery on sensor data is extremely critical in mining useful information from data.

6.3 History, Related Work and New Directions

The history of spatio-temporal network modeling arguably begin in 1988 with the landmark paper 'Framework for Temporal Geographic Information' (Langran and Chrisman 1988). This paper illustrated the challenges associated with visualizing temporal information and introduced the idea of world time and database time. More importantly, it introduced a system to visualize spatio-temporal changes of an image (e.g., a map). Each change would be layered on top of an old change, so that the layers could be tracked through to detect where changes occurred (Shekhar and Xiong 2007). This work established the field of temporal geography and laid the foundation for new research in temporal and spatial representations.

Though much work has been done in conceptualization and modeling (Christakos et al. 2001; Frank 2003), there is a fundamental struggle between spatial and temporal databases, the former representing things with graphical snapshots and the latter doing so with time-stamps. It is difficult to merge the two in an efficient way and current research is moving slowly toward a unified theory for these representations (Langran 1992; Egenhofer and Golledge 1998; O'Sullivan 2005; Peuquet and Duan 1995; Goodchild et al. 2007; Frank et al. 1999).

Conceptually, processes and events can be used to help create unified spatio-temporal representations (Galton and Worboys 2005). This may eventually lead to logical models useful for implementation in spatio-temporal databases. In the meantime, these ideas led to an important new area of study in temporal geography known as geographic dynamics. Geographic dynamics can be described as a series of processes and events that allow for descriptive capabilities to illustrate spatio-temporal changes. This idea of geographic dynamics is not new. It can be thought of in the same light as describing the dynamics of weather; only here we are talking about geographic changes such as land use change over time. These dynamics have been used and represented in various ways, starting with animated maps and moving towards more interactive models (DiBiase et al. 1992; Monmonier 1990; Harrower 2004). Current techniques combine visualization with the modeling itself. For example, cellular automata models are made up of a large number of cells that must change based upon a simple set of rules (Batty 2005). These rules allow for easy creation of models which can then provide interesting insights into the real world, and demonstrates the changes in action as the cells move and change. Visualization has proven historically to be a powerful tool for understanding complex spatio-temporal data (Kwan 2004; Torrens 2006; McIntosh and Yuan 2005).

Extensive work has also been done analyzing the human use of time in relation to geography, the primary components being space-time paths and space-time prisms (Burns 1979; Hägerstrand 1970; Janelle 2004; Lenntorp 1977; Miller 2005; Pred 1977; Timmermans et al. 2002). The focus here is on individual spatio-temporal paths, allowing for interesting decisions and analyses to be made. Both approaches represent a person's position over time by plotting geographic space on the X and Z axis, and time on the Y axis. The path based graph shows where a person traveled over time, while a prism shows their potential travels based upon maximum travel

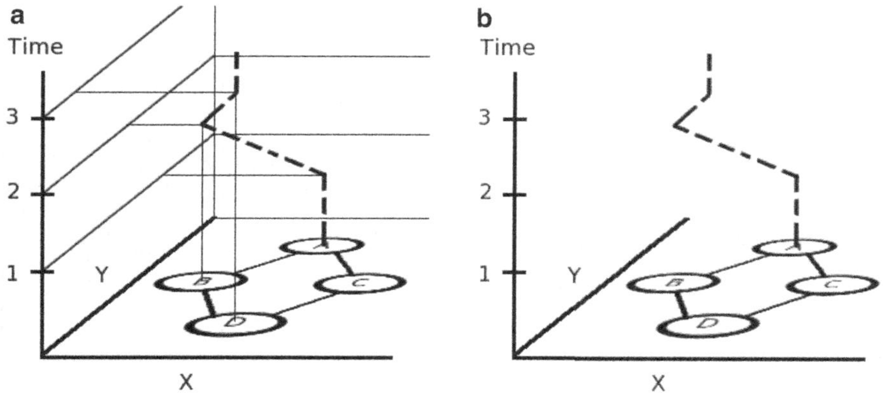

Fig. 6.2 Space-time trajectories. (**a**) with axis information. (**b**) without axis information

velocity at their current location constrained by some amount of time. These ideas were translated to networks with network time prisms that show how far one could get from a starting point on a network on any given path (Miller 1999).

The movement of individuals has also been represented using space-time trajectories, another broad area of study in GIS (Miller 2005). We will be using an example network to represent many different types of visual representations throughout this paper, starting with space-time trajectory in Fig. 6.2a. The dotted line represents trajectory of a person. The X/Y plane shows a spatial graph with four nodes, A, B, C and D, possibly representing a road network. The Z-axis represents time. At $t = 1$, the person is at node A, at $t = 2$, he is at node B, and at $t = 3$, he is at node D. Figure 6.2a, b show the same information, but Fig. 6.2a has a grid to help with depth perception.

6.4 Change in Reference Frame

Navigation and routing services can benefit from a traveler's frame of reference. In other words, candidate routes should be evaluated from the perspective of a person moving through a transportation network. In other fields, such a frame of reference is defined as a Lagrangian frame of reference (Herrera and Bayen 2009).

A Lagrangian representation of a roadmap can be described through a model called a time-expanded graph (TEG) (Ford and Fulkerson 1962). To illustrate, Fig. 6.3a provides a snapshot graph representation of a simple roadmap with four consecutive time-instants. In the figure, the travel-time for edge A to C is 2 for time = 1 but decreases to 1 for time = 2, 3, and 4. An equivalent TEG is shown in Fig. 6.3b. Note that all nodes are replicated across time points. For example, node A at time 1, 2, 3, and 4 in Fig. 6.3a is represented as a set of nodes (A1, A2, A3,

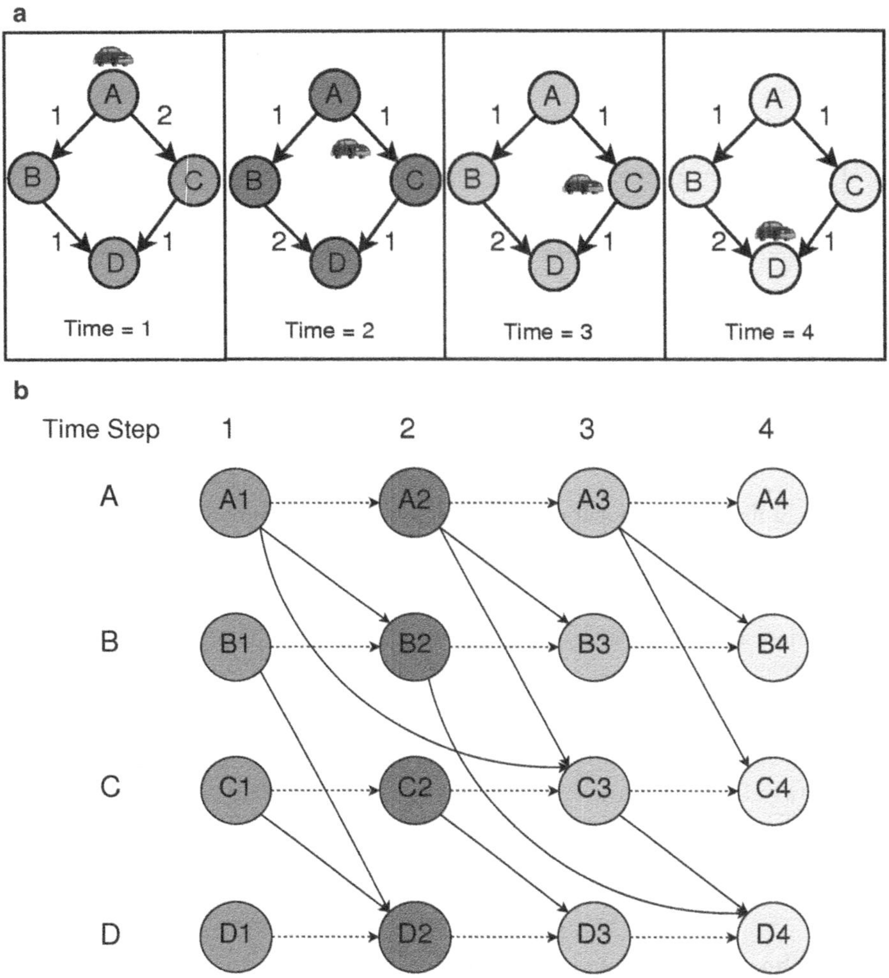

Fig. 6.3 Two representations of spatio-temporal graphs. (Colors encode time steps). (**a**) Snapshot model of a spatio-temporal graph. (**b**) Time-expanded graph (Color figure online)

and A4, respectively) in Fig. 6.3b. Edge AC in Fig. 6.3a is represented as a set of edges (i.e., A1–C3, A2–C3, A3–C4, ...) in Fig. 6.3b, encoding the time-series of travel-times. A1–C3 indicates that the travel time for edge AC is 2 at start-time 1. Similarly, A2–C3 indicates a travel-time of 1 for start-time 2.

Unlike snapshot graphs, time-expanded graphs model the actual travel time experience of a moving object. For example, when traveling from node A to node D via node B starting at time = 1, the object will reach node B at time = 2, since edge A–B has a travel-time of 1 at start-time = 1. The object will reach node D at time = 4, since travel-time for edge B–D is 2 at time = 2. Thus, the total travel

Table 6.2 Snapshot and Lagrangian travel-time

	Lagrangian travel-time				Snapshot travel-time			
2*Route	Start t = 1	Start t = 2	Start t = 3	Start t = 4	t = 1	t = 2	t = 3	t = 4
A–B–D	3	3	3	–	2	3	3	3
A–C–D	3	2	2	–	3	2	2	2

time is 3, which is different than the total travel-time of 2 based on the snapshot graph for time = 1. It also models a waiting time, i.e., lack of motion. For example, edge A1–A2, A2–A3, etc., indicate staying at the same location over time. Table 6.2 displays the travel time for various routes via both Lagrangian and Snapshot models for Fig. 6.3.

6.5 Conceptual Model of Spatio-temporal Networks

Spatio-temporal networks have wide applications in domains such as crime analysis, sensor networks, and transportation science (George and Shekhar 2007a, b; George et al. 2007a, b). Models of these networks need to capture the possible changes in topology and values of network parameters with time and provide the basis for the formulation of computationally efficient and correct algorithms. In this section we discuss the basics of conceptual models used to represent time dependent spatial networks. A spatio-temporal network (STN) can be represented as a spatial graph with temporal attributes. Spatial elements of the graph are a finite set of nodes and a finite set of edges connecting the nodes. A graph $G = (N,E)$ consists of a finite set of nodes N and edges E between the nodes in N. If the pair of nodes that determine the edge is ordered, the graph is directed; if it is not, the graph is undirected. In most cases, additional information is attached to the nodes and the edges. Here we explain how the time dependence of these edge/node parameters is handled in various conceptual models.

Given spatio-temporal networks (e.g., roadmaps with traffic speed reported as a time-series in 5 min increments over a typical day for each road-segment) and operators (e.g., network snapshot, shortest path or path evaluation), a Spatio-Temporal Network Model (STNM) provides a computer representation to facilitate reasoning, analysis and algorithm design for important societal applications. For example, next generation routing services are estimated to save consumers hundreds of billions of dollars in terms of time and fuel saved by 2020. Developing a model for spatio-temporal networks is challenging due to potentially conflicting requirements of expressiveness and model simplicity. Related work in Time Geography models spatio-temporal movement and relationships via dimension-based representations such as space-time prisms and space-time trajectories. These representations are not adequate for many STN use-cases, such as spatio-temporal routing queries. To address these limitations, we discuss a novel model called time-aggregated graphs (TAG) that allows the properties of the network to be modeled as a time series. This model retains spatial network information while reducing the temporal replication needed in other models, thus resulting in a much more efficient model for several computational techniques for routing problems. In this chapter, we discuss spatio-temporal networks as represented by time-aggregated graphs at a conceptual, logical, and physical level. This chapter also focuses on shortest path algorithms for spatio-temporal networks. We develop the topics via case studies using TAGs in context of Lagrangian shortest-path queries and evacuation route planning.

6.6 Snapshot Model

A network that changes over time can be modeled as a series of networks, each associated to a time stamp. Figure 6.4 shows a snapshot representation of the information presented before as a space-time trajectory in Fig. 6.2a. As we can see in Fig. 6.4, each time instant is represented as a full model of the network. This representation type is beneficial when trying to compare differences between different states of the complete network. Instantaneous locations of person P are shown by bold nodes (and a P symbol) in each snapshot. However, in this approach, a copy of the network is required for every time step, making it prohibitively expensive to represent long time series and to use it for computational modeling.

6.7 Time-Expanded Graphs

To highlight some of these problems with traditional spatio-temporal network representations we will use a quick example. Operations Research uses a model called the time expanded network to represent temporal changes in networks

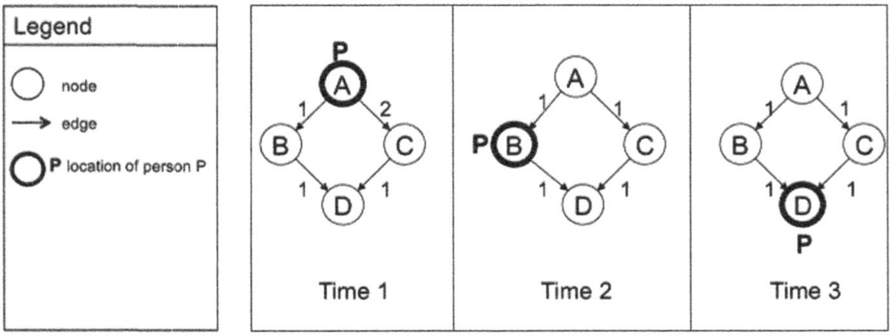

Fig. 6.4 Snapshot model illustrating three time steps

(Dreyfus 1969; Ford and Fulkerson 1958, 1962; Kaufman and Smith 1993; Kohler et al. 2002; Orda and Rom 1991; Pallottino and Scuttella 1998). This model duplicates the original network for each discrete time unit $t = 0,1,\ldots,T$ where T represents the extent of the time horizon. The expanded network has edges connecting a node and its copy at the next instant in addition to the edges in the original network, replicated for every time instant. This significantly increases the network size and is very expensive with respect to memory. Because of the increased problem size due to replication of the network, the computations become expensive. In addition, time expanded graphs (TEG) have representational issues when modeling non-flow networks. This model incorporates the time dependent edge attributes into the graph in the process of graph expansion, making it more application-dependent, and thus making physical data independence harder to achieve.

As noted previously, in order for TEG to show the changes in edge connections over time, they must duplicate the network for every time step. This repetition is excessive in models with many time-steps. A TEG has one copy of the set of nodes for each discrete time instant. Figure 6.4 shows a TEG representation of the example path presented before as a space-time trajectory (Fig. 6.2a). The location of person P at different time-instants is shown via the bolded nodes and edges. This representation may be quite natural for the last question in Table 6.1, however it will require more exerted effort to answer the other questions.

Corresponding to each edge with transit time t in the original network, there is a copy of an edge (called the cross edge) between each pair of copies of nodes separated by the transit time t (Ford and Fulkerson 1958; Hamacher and Tjandra 2002; Kohler et al. 2002). Thus, a time-dependent flow in a dynamic network can be interpreted as a static flow in a time expanded network. This allows application of static algorithms on such networks to solve dynamic flow problems. Apart from the "enormous increase in the size of the underlying network" (Kohler et al. 2002), this model may be suitable for some application domains of linear programming (Luenberger 1973).

A time expanded graph assumes that the edge weight represents a flow parameter, and it represents the time taken by the flow to travel from the source node to the end node. This is represented by the cross edges between the copies of the graph. Since the cross edges in a time expanded graph represent a flow across the nodes, the representation of non-flow networks using this model is not obvious.

In most spatio-temporal networks, the length of the time period (indicated by T in this paper) might not be known in advance since data arrives as a sequence at discrete time instants. For example, sensors in transportation networks collect data at a rate of about once every 30 s. Crimes are reported whenever an incident occurs. It is not clear how they would handle time series of infinite length. In addition, the need for prior knowledge of T may lead to problems in the algorithms based on time expanded networks since an underestimation of T can result in failure to find a solution, and lead to high storage and run time costs that would adversely affect the scalability of the algorithms.

Finally, since time expanded graphs model the time-dependence of edge parameters through the cross edges that connect the copies of the nodes, they lack the means to separate data (for example, an edge attribute series) from its physical representation, which, again, can adversely affect physical data independence.

6.8 Time-Aggregated Graphs

Many of the limitations of TEG can be eliminated by a time aggregated graph (TAG) approach. A Time Aggregated Graph (TAG) is defined as

$$\begin{aligned} TAG = \Big(&N, E, TF, \\ &f_1 \ldots f_k, g_1 \ldots g_l, \\ &(nw_1, ne_1) \ldots (nw_k, ne_p), \\ &(ew_1, ee_1) \ldots (ew_p, ee_p), \,| \\ &f_i : N \rightarrow R^{TF};\ g_i : E \rightarrow R^{TF}; \\ &nw_i : N \rightarrow R^{TF},\ ne_i : N \rightarrow PD, \\ &ew_i : E \rightarrow R^{TF},\ ee_i : E \rightarrow PD\Big), \end{aligned}$$

where N is the set of nodes, E is the set of edges, TF is the length of the entire time interval, $f_1 \ldots f_k$ are the mappings from nodes to the time-series associated with the nodes (for example, the time instants at which the node is present), $g_1 \ldots g_l$ are mappings from edges to the time series associated with the edges, and $(ew_1, ee_1) \ldots (ew_p, ee_p)$ indicate the time dependent attribute on the edges. PD indicates a probabilistic error. These attributes are the quantitative descriptors of the physical relationship between the nodes. To represent the stochastic nature of the measured values of physical phenomena, each attribute is a pair that represent the measured value and the associated error.

Time aggregated graphs (TAG), at their simplest, show a network with a time series of attributes on nodes, edges or both. These time series represent the state of an object at each time step. If we were to compare it to a more familiar temporal geography idea, such as space-time paths, we can easily represent the travel of one person over time in a TAG. Figure 6.5 contains a TAG representation of the sample network shown earlier as a space-time trajectory (Fig. 6.4), snapshots (Fig. 6.4) and a TEG (Fig. 6.6). Each node, A, B, C and D has a time-series to represent sets of people associated at different time instants. For example, node A has person P at t = 1, and no other people at t = 2 or t = 3. Similarly, node B has person P at t = 2, etc.

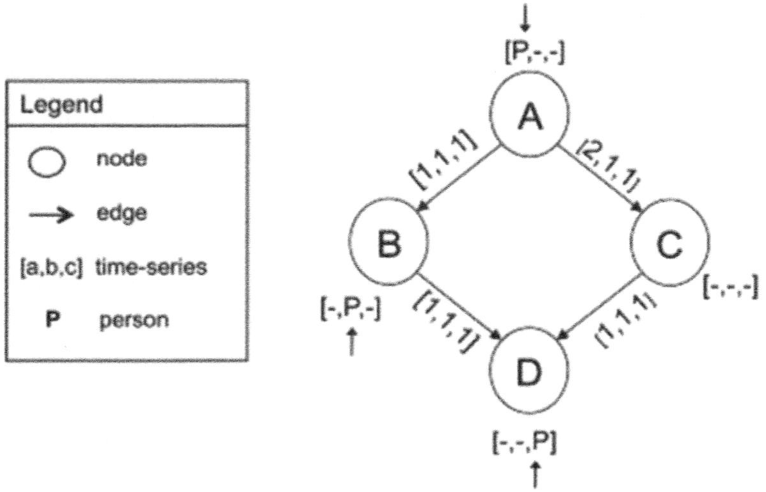

Fig. 6.5 Time-aggregated graph

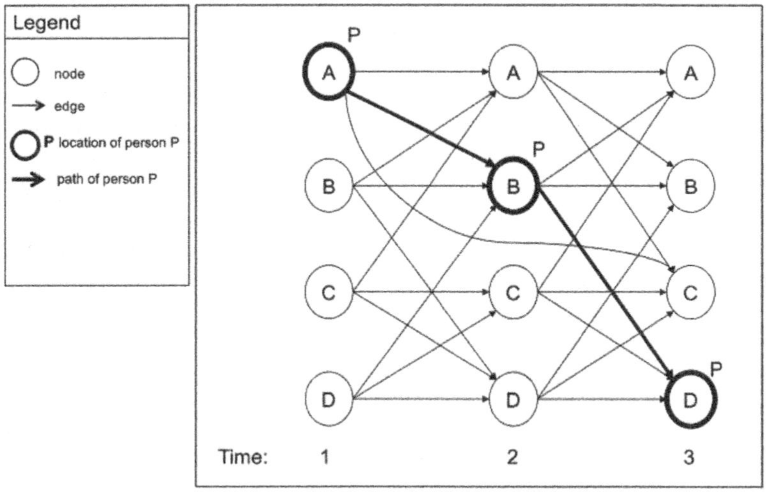

Fig. 6.6 Time expanded graph

6.9 Logical Model of Spatio-temporal Networks

The logical model is based on spatial static graph model, which is further extended to incorporate the time dependence of the network. The framework of the model consists of two dimensions (1) graph elements, namely nodes, edges, routes, graph,

Table 6.3 Examples of operators in the accessor category

	at_time	at_all_time	at_earliest
Node	get(node,time)	get_node(node)	get_node_earliest(node,time)
Edge	getEdge(node1,node2,time)	get_edge(node1,node2)	get_edge_earliest(node1, node2,time)
Route	getRoute(node1,node2,time)	getRoute(node1,node2)	get_route_earliest(node1, node2,time)
Route	getSP_Route(node1,node2, time)	getSP_Route(node1, node2)	
Flow	get_max_Flow(node1,node2, time)	get_max_Flow(node1, node2)	
Graph	get_Graph(time)	get_Graph()	–

etc. and (2) the operator categories which consist of accessors, modifiers and predicates. A representative set of operators for each operator category is provided in Tables 6.3, 6.4 and 6.5.

6.10 STN Operations

Table 6.3 lists a representative set of 'access' operators. For example, the operator *getEdge(node1,node2,time)* returns the edge properties of the edge from node 1 to node 2, such as the edge identifier (if any) and associated parameters at the specified time instant. For example operator *getEdge(N1,N2,1)* on the time-aggregated graph shown in Fig. 6.5 would return the attribute value of the edge N1−N2 at $t = 1$, that is (1, 0.5). Similarly, *get_edge(node1,node2)* returns the edge properties for the entire time interval. In Fig. 6.5, the operator *get_edge(N1,N2)* would result in ((1,0.5),(1,0.5),∞). *get_edge_earliest(N3,N4,2)* returns the earliest time instant at which the edge N3−N4 is present after $t = 2$ (that is $t = 3$).

Table 6.4 shows a set of modifier operators that can be applied to the time aggregated graphs. We can also define two predicates on the time-aggregated graph.

exists_at_time_t: This predicate checks whether the entity exists at the start time instant *t*.

exists_after_time_t: This predicate checks whether the entity exists at a time instant after *t*.

Table 6.5 illustrates these operators. For example, node *v* is adjacent to node *u* at any time *t* if and only if the edge (*u*,*v*) exists at time *t* as shown in the table. *exists (N1,N2,1)* on the time aggregated graph in Fig. 6.7 returns a "true" since the edge N1-N2 exists at $t = 1$.

Table 6.4 Examples of operators in the modifier category

	insert		delete		modify	
	at_time	at_all_time	at_time	at_all_time	at_time	at_all_t
Node	insert(node,time,value)	insert(node,valueseries)	delete(node,time)	delete(node)delete (node)	update(node,time, value)	update(node,series)
Edge	insert(node1,node2,time, value)	insert(node1,node2, valueseries)	delete(node1, node2,time)	delete(node1,node2), node2)	update(node1,node2, time,value)	update(edge,series)
Route	insert(node1,node2,time)	insert(node1,node2)	delete(node1,node2,time)	delete(node1,node2)		
Graph	insert(graph,time)	insert(graph)	delete(graph,time)	delete(graph)	update(graph,time)	update(graph)

Table 6.5 Predicate operators in time-aggregated graphs

	exists_at_time_t	exists_after_time_t
Node	exists(node u,at_time_t)	exists(node u,after_time_t)
Edge	exists(node u,node v,at_time_t)	exists(node u,node v,after_time_t)
Route	exists(node u,node v,a_route r,at_time_t)	exists(node u,node v,a_route r,after_time_t)
Flow	exists(node u,node v,a_flow r,at_time_t)	exists(node u,node v,a_flow r,after_time_t)

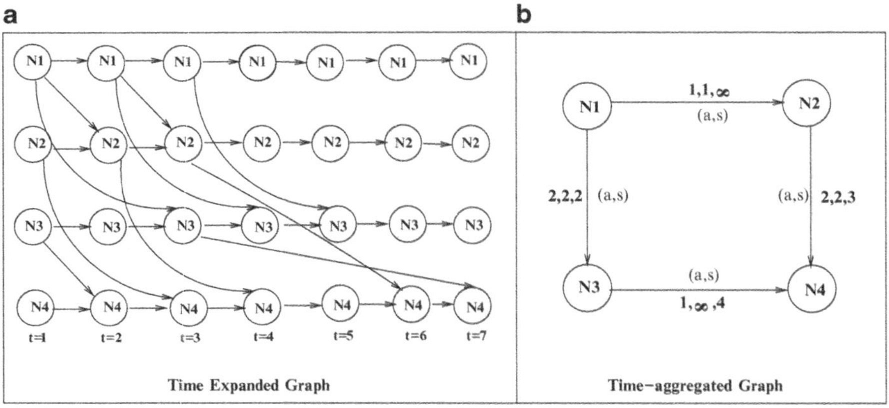

Fig. 6.7 Time-aggregated graph vs. time expanded graph

6.11 Case-Study: STN Queries

SBD magnifies the already partial and ambiguous nature of a traditional routing query. This is because a typical routing query specified by a start location and a destination may result in multiple answers. The apparent ambiguity in the shortest path query is clearly visible in the multiple candidate routes returned for a single start and destination pair in common web-based routing services like [Google Maps, http://maps.google.com, Mapquest, http://www.mapquest.com/, Maps, http://www.bing.com/maps/]. For example, consider Fig. 6.8a, which shows two candidate shortest travel time routes from the University of Minnesota to MSP International Airport. This ambiguity increases tremendously with the availability of SDB datasets, resulting in increased computational costs due to re-computation that may be necessary for thousands of possible start times. Let us consider this issue in the context of TD roadmaps via the all start-time Lagrangian shortest paths (ALSP) problem.

Given a TD roadmap, a source, a destination, and a start-time interval, an ALSP problem determines a route collection which includes the shortest path for every start time in the interval. The ALSP output includes both the shortest paths and the corresponding set of time instants when the paths are optimal. For example, for the problem of determining the shortest travel time path between the University of Minnesota and the MSP International Airport during the interval from 7:00 AM to 10:30 AM. Figure 6.8a shows two different routes between the University and the Airport. The 35 W route is preferred outside rush-hour, whereas the route via Hiawatha Avenue is preferred during rush-hour (i.e., 7:00 AM–9:30 AM) (see Fig. 6.8b). Thus, the ALSP route collection may be a set of two routes (one over 35 W and one over Hiawatha Avenue) and their corresponding time intervals.

Computing ALSP is computationally challenging for three reasons. First, many links in the network may violate the property of first-in-first-out (FIFO) behavior. This is illustrated in Table 6.6, which shows the flight schedule for Delta airways

a

Fig. 6.8 Sample query for spatial big data. (Best in color). (a) Different routes between University and Airport (Google Maps). (b) Preferred route varies over time. (c) Total travel time of candidate paths between University and Airport (Color figure online)

b

Time	Preferred Routes
7:30am	Via Hiawatha
8:30am	Via Hiawatha
9:30am	via 35W
10:30am	via 35W
......	....
......	....
....	....
....	....

c

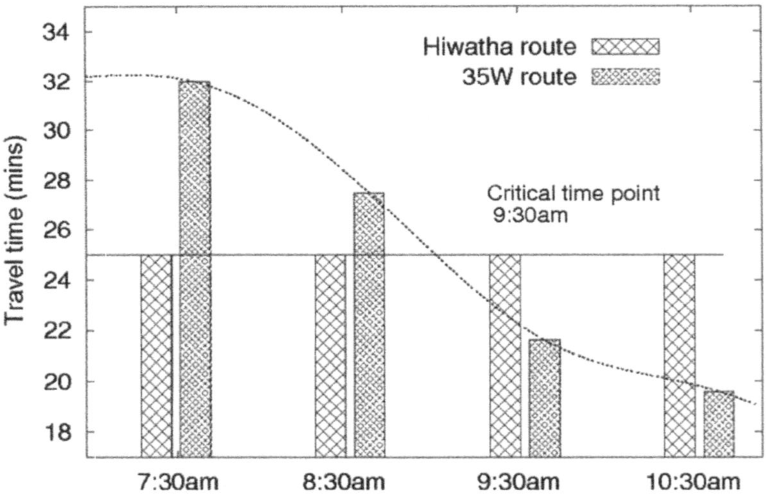

Fig. 6.8 (continued)

Table 6.6 Minneapolis–Austin, TX flight schedule

Start time	Route	Travel time	Arrival time	Earliest arrival time
8:30	Via Detroit	6 h 31 min	15:01	12:01
9:10	Direct flight	2 h 51 min	12:01	12:01
11:00	Via Memphis	4 h 30 min	15:38	15:38
11:30	Via Atlanta	6 h 28 min	17:58	17:21
14:30	Direct flight	2 h 51 min	17:21	17:21

[Airlines, http://www.delta.com] between Minneapolis and Austin, TX. Here, the travel time at 8:30 is 6 h 31 min, whereas waiting 40 min for the 9:10 flight would yield a quicker route 2 h and 51 mins. This violation of first-in-first-out (FIFO) is called non-FIFO behavior. Surface transportation networks such as road networks also exhibit such behavior. For example, UPS (Joel Lovell 2007; Deutsch 2007) minimizes the number of left turns in their delivery routes during heavy traffic conditions. This leads to faster delivery and fuel savings. Second, the ranking of alternate paths between any particular source and destination pair in the network is not stationary. In other words, the optimal path between a source and destination for one start time may not be optimal for other start times. In our previous example of shortest route between the university and airport, different routes were optimal at different times. The principle of stationarity states that, if two reward sequences $R_1, R_2, R_3, \ldots$ and $S_1, S_2, S_3, \ldots$. begin with the same reward then the sequences should be preference ordered the same way as the sequences $R_2, R_3, \ldots$. and $S_2, S_3, \ldots$. (Bertsekas 1987; Russel and Norwig 1995]. This means that, in shortest route computation, if two journeys $e1, e2, e3, \ldots$ and $f1, f2, f3, \ldots$ start at the same node, then the preference order of the journeys should not change for another start time. This lack of stationarity in TD roadmaps eliminates the possibility of using dynamic programming based techniques. Lastly, due to the potentially large and ever growing sizes of SBD such as TD roadmaps, efficient computational methods need to developed which can scale to large datasets.

Relatively little work has been done in the area of route collections with minimization of travel time as the preference function (Chabini and Lan 2002; Kanoulas et al. 2006; Dehne et al. 2009) due to the FIFO assumptions. Other related work includes skyline based techniques (Kriegel et al. 2010) which assume independence among dimensions and do not incorporate a Lagrangian frame of reference. The challenges of ALSP can be addressed using the idea of earliest arrival-time transformation and critical time points.

6.11.1 Earliest Arrival-Time Transformation (EAT)

The first challenge for solving ALSP is the non-FIFO behavior of the network. Our work attempts to address this by converting travel information associated with an edge into earliest arrival time (at destination) information (George et al. 2008; Orda and Rom 1990). This is a two step process. First, the travel time information is converted into arrival time information. Second, the arrival time information is converted into earliest arrival time information. The second step captures the possibility of arriving at an end node earlier by waiting (non-FIFO behavior). For example, consider again Table 6.6 which shows the flight schedule between Minneapolis and Austin, TX. Here, the result of the first step, shown in fourth column, was obtained by adding the start time with the travel time (flight time). Now, we would scan the arrival time information (in the fourth column) to capture any benefits associated with waiting. For example, we can observe that a quicker path for start time 8:30 am can be obtained by waiting for 40 mins. The last column of the table shows the result of the second step. To date, we have applied this approach only to the travel time preference function.

6.11.2 Critical Time Points

A naive approach for solving the non-FIFO ALSP problem would involve determining the shortest path for each start time in the interval using techniques (George et al. 2007b; Ding et al. 2008; Demiryurek et al. 2011; Orda and Rom 1990; Chabini 1998; Delling 2008; Delling and Wagner 2007; Demiryurek et al. 2010] developed for single-start time shortest paths. This leads to redundant re-computation of the shortest path across consecutive start times sharing a common solution. Some efficiencies can be gained using a time series generalization of a label-correcting algorithm (George et al. 2007b). However, this approach still entails large number of redundant computations (Gunturi et al. 2011). Here, we propose the use of critical time points to reduce this redundancy. For instance, consider again the problem of determining the shortest path between University of Minnesota and MSP international airport over a time interval of 7:30 am through 10:30 am. Figure 6.8b shows the preferred paths at some time instants during this time interval, and Fig. 6.8c shows the travel-times for the candidate paths for all start-times during this interval. The dotted lines across bars are drawn for ease of understanding only.

As can be seen, the Hiawatha route is faster for times in the interval [7:30 am 8:30 am], whereas the 35 W route is faster for times in the interval [9:30 am 10:30 am]. This shows that the shortest path changed at some instant inside the interval [8:31 9:30]. For sake of simplicity we assume that the shortest path changed at 9:30 am. We define this time instant as a *critical time point*. Critical time points can be determined by computing the earliest intersection points between the functions representing the total travel time of paths. For example, the earliest intersection point of the Hiawatha route was at 9:30 am (with 35 W route function). Therefore, it would be redundant to recompute shortest paths for all times in interval [7:31 am 9:29 am] and [9:31 am 10:30 am] since the optimal path for times within each interval did not change. This approach is particularly useful in cases when there are a fewer number of critical time points.

Thus, critical time points inspire a divide and conquer based approach to handle network non-stationarity. In this approach we divide the time interval over which the network exhibits non-stationarity into smaller intervals which are guaranteed to show stationary behavior. These intervals are determined by computing the critical time points, that is, the time points at which the cost functions representing the total cost of the path as a function of time intersect. Now, within these intervals, the shortest path can be easily computed using a single run of a dynamic programming (DP) based approach (George et al. 2007b). In our previous example of determining the shortest paths between the university and the airport over an interval of [7:30 am 10:30 am], the critical time point was 9:30 am. This created two discrete sub-intervals [7:30 9:29] and [9:30 10:30]. Now, we can compute the ALSP using two runs of a DP based algorithm (George et al. 2007b) (one on [7:30 9:29] and another on [9:30 10:30]).

The advantages of critical time point based approaches are twofold. First, they allow us to compress the result set of the query to a set of routes and their corresponding time intervals instead of a single route for every start time. Second, they are computationally more efficient since they avoid redundant re-computation across the start-times sharing a common solution.

A key challenge in designing algorithms based on critical time points is to minimize the amount of time needed to compute them while ensuring correctness and completeness. Our preliminary work explored critical time point based approaches for shortest path (Gunturi et al. 2011) and minimum spanning tree problems (Gunturi et al. 2010). Preliminary results showed that these approaches are promising, but require further research.

6.12 Physical Model of Spatio-temporal Networks

Physical models of spatio-temporal networks need to capture the possible changes in topology and values of network parameters with time and provide the basis for the formulation of computationally efficient and correct algorithms. In this section we review the strengths and weaknesses.

6.13 Snapshot Partitioning

Snapshot Partitioning for spatio-temporal network storage can be represented by a snapshot graph, such as Fig. 6.4. Snapshot storage techniques are partitioned into data pages using geometry (Guttman 1984; Samet 1995) or connectivity (Shekhar and Liu 1997) methods. If a snapshot requires more space than a data page, it may be decomposed into multiple data pages using techniques such as CCAM (Shekhar and Liu 1997), R-Tree (Guttman 1984), etc. Fig. 6.9 shows a snapshot storage approximation, with the data page partitioning visualized with dashed lines and page numbering, using a small graph where multiple snapshots fit inside a data page. However producing a time stamped static graph at each time step leads to great I/O cost when executing queries such as *evalRoute*(*route*,*time*) in Table 6.3 due to the need to frequently access data pages as the STN is traversed. Calling *evalRoute*(ACD,1) requires first accessing the traversal time attribute of edge AC at $t = 1$, stored on Data Page 1 in this example. Next, edge CD at $t = 3$ is needed to complete the route evaluation, stored on Data Page 2. Note how the route evaluation operation had to access two different data pages due to the snapshot-based orthogonal partitioning.

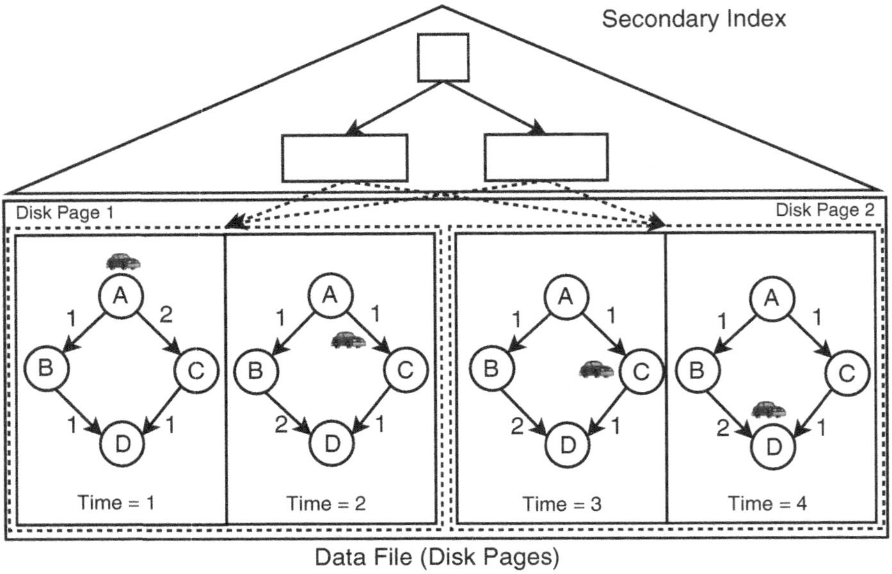

Fig. 6.9 Snapshot storage of a STN

As the name suggests, the snapshot family described in Table 6.1 is best represented with the snapshot model. Answering questions about current or past locations of objects are trivial. The model can easily show a network state at any point in time, but information such as trajectory of a person or aggregate information on a edge or node are less obvious.

6.14 Longitudinal Partitioning

Longitudinal Partitioning for a spatio-temporal network is based on the adjacency-list main memory storage structure used by (Ahuja et al. 1993; George et al. 2008). Each node is stored with its attribute information and all outgoing edges and their attribute information. This orthogonal storage solution, as shown in Fig. 6.10, suffers from the increasing disk I/O to evaluate routes in spatio-temporal networks using operators such as *evalRoute*(*route*,*time*) in Table 6.3. This example network has a short time series compared to its graph size, allowing multiple node records (with adjacency list) to fit inside a data page. If the time-series length was larger then the node record could become larger than a data page, that node record may be split into multiple pages. This is due to the long time series being stored with each node, resulting in a small number of nodes to be stored on each data page. Calling *evalRoute*(*ACD*,1) requires first accessing the traversal time attribute of edge *AC* at

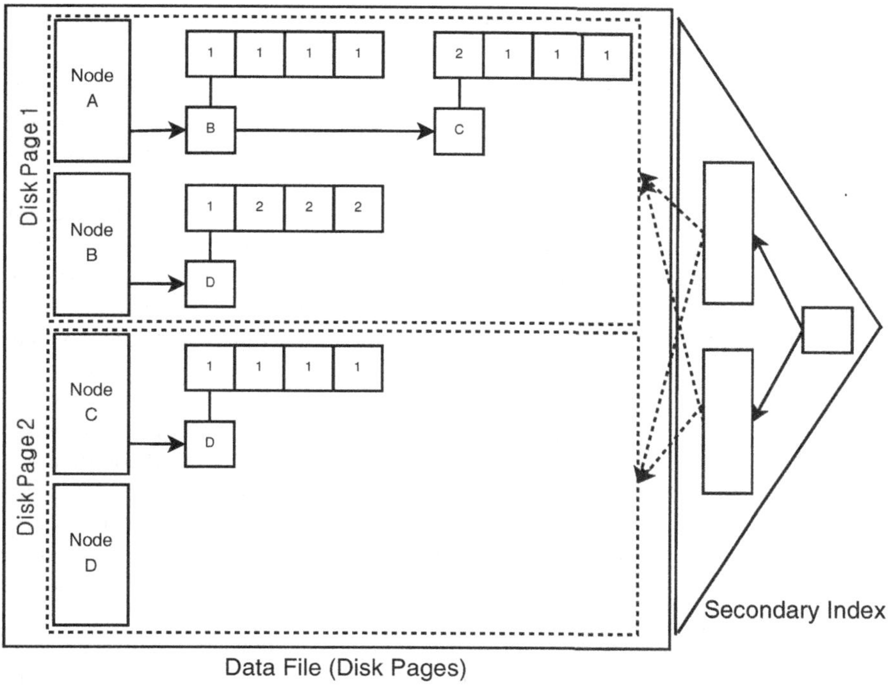

Fig. 6.10 Longitudinal storage of a STN

$t = 1$, stored on Data Page 1 in this example. Next, edge *CD* at $t = 3$ is needed to complete the route evaluation, stored on Data Page 2. Note how the route evaluation operation had to access two different data pages due to the node-based orthogonal partitioning. CCAM (Shekhar and Liu 1997) did not consider STNs, however, it uses node-centric storage techniques. Therefore, if CCAM was applied to a spatio-temporal network, it may end up with a storage method resembling Longitudinal Partitioning due to the long time series characteristic of STNs causing node-centric storage to fill entire data pages with a single node's information.

The orthogonal approaches such as Snapshot and Longitudinal do not capture the movement access patterns inherent in the route evaluation operation. This may be formalized via a Lagrangian frame of reference (Herrera and Bayen 2009) attached to a user moving through space over time. For example, evaluation of route *ACD* at different start times will retrieve the following subsets of edges: (*AC* at $t = 1$, *CD* at $t = 3$), (*AC* at $t = 2$, *CD* at $t = 3$), (*AC* at $t = 3$, *CD* at $t = 4$), etc. Note that these groups are not orthogonal. Our proposed use of a Lagrangian frame of reference in our approach intends to move toward capturing such non-orthogonal access patterns.

6.15 Lagrangian Partitioning

The Lagrangian-Connectivity Partitioning (LCP) method optimizes disk storage based upon traversal across spatio-temporal networks. Data is stored using the sub-node record design, allowing for non-orthogonal temporal information to be stored on a data page.

By representing a spatio-temporal network as a modified time-expanded graph, focusing on Lagrangian connections between nodes (movement edges), a min-cut graph partitioning (Abou-Rjeili and Karypis 2006) algorithm creates partitions clustering nodes by minimizing the cuts of these movement edges. This results in LCP collocating connected spatio-temporal nodes together on data pages, stored as sub-node records. We propose LCP will result in more efficient I/O when using STN operators from Table 6.3 or queries composed of them.

To illustrate, we call the route evaluation operation *evalRoute(ACD,1)* on each of the partitioning methods. With orthogonal partitioning, Snapshot or Longitudinal, whenever an edge is traversed, (e.g., $A1$ to $C3$ and $C3$ to $D4$), a disk I/O is needed to retrieve the data page containing the record for the next node. However, with Lagrangian-Connectivity Partitioning, traversing from node $A1$ to $C3$ and then $C3$ to $D4$ requires only one data page as all relevant sub-node records are collocated on the same data page.

Figure 6.11c illustrates the results of LCP applied to the same input STN as in the other orthogonal partitioning methods. A min-cut partitioning was run on the modified time-expanded graph (wait edges are removed to emphasize Lagrangian connectivity) and then stored as sub-nodes on four data pages. For efficiency, we used a bulk load operation, which sorts these statements along the block number and inserts them into data pages. Physically, sub-node data records are stored in data pages and a B + tree index is created to support retrieve operations.

6.16 Case Study: Evacuation Route Planning

Efficient tools are needed to identify routes and schedules to evacuate affected populations to safety in the event of natural disasters. Hurricane Rita and the recent tsunami revealed limitations of traditional approaches to provide emergency preparedness for evacuees and to predict the effects of evacuation route planning. Challenges arise during evacuations due to the spread of people over space and time and the multiple paths that can be taken to reach them; key assumptions such as stationary ranking of alternative routes and optimal substructure are violated in such situations. Algorithms for evacuation route planning were first developed by researchers in operations research and transportation science. However, these proved to have high computational complexity and did not scale well to large

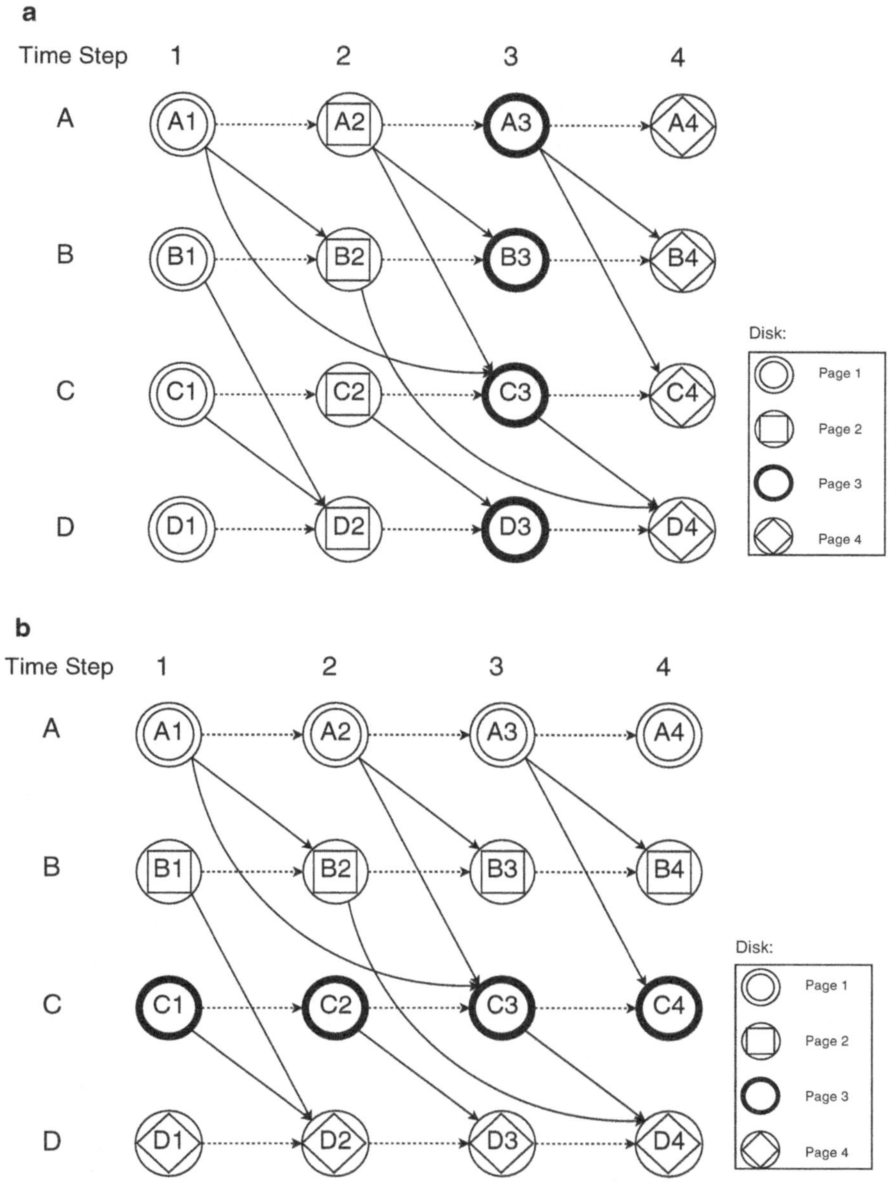

Fig. 6.11 Partitioning techniques for spatio-temporal networks. (**a**) Snapshot partitioning. (**b**) Longitudinal partitioning. (**c**) Lagrangian partitioning

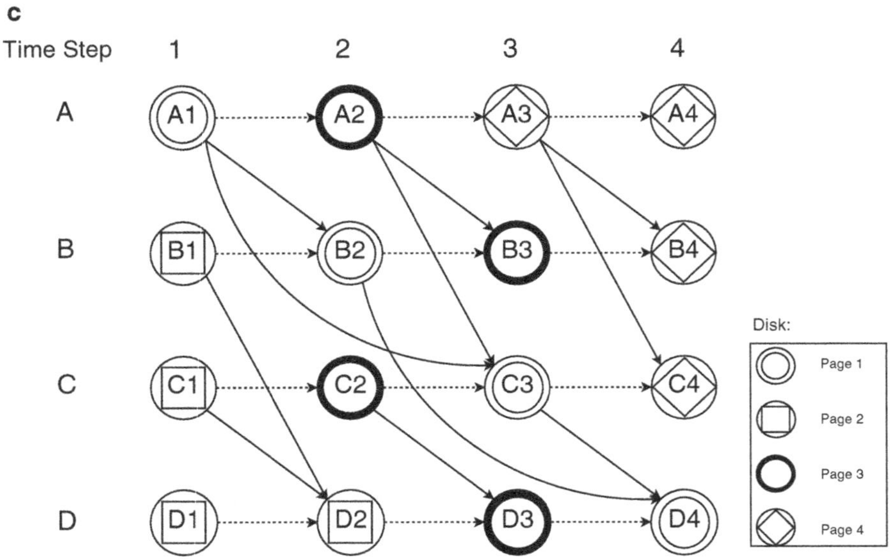

Fig. 6.11 (continued)

problems. Over the last decade, we developed a different approach, namely the Capacity Constrained Route Planner (CCRP), which generalizes shortest path algorithms by honoring capacity constraints and the spread of people over space and time. CCRP uses Time Aggregated Graphs to reduce storage overhead and increase computational efficiency. Experimental evaluation and field use in Twin-cities Homeland Security scenarios demonstrated that CCRP is faster, more scalable and easier to use than previous techniques.

There has been a considerable amount of research on route planning for evacuation scenarios. Recent work falls into three categories: (1) Network flow methods (Francis and Chalmet 1984; Kisko and Francis 1985; Kisko et al. 1998; Ahuja et al. 1993; Hamacher and Tjandra 2002) (2) Simulation methods (Ben-Akiva 2002; Mahmassani et al. 2004), and (3) Heuristic methods (Hoppe and Tardos 1994; Lu et al. 2003, 2005, 2007). In contrast to previous approaches, heuristic approaches use approximate methods to find near-optimal solutions with minimizing the computational cost. A well known approach that falls in this category is the Capacity Constrained Route Planner (CCRP) (Lu et al. 2003, 2005, 2007). These methods use time aggregated graphs (George and Shekhar 2006) and evaluate a shortest route with capacity constraints to find the evacuation route at each time step. It is useful for medium sized networks (e.g., 1-mile evacuation zone), but has a limitation for large scale networks (e.g., 50-mile evacuation zone). (Kim et al. 2007) present a new scalable heuristic by reusing shortest routes based on bottleneck saturation checking. It showed a 95 % reduction in computational time with small degradation of solution quality.

In this section we describe the Evacuation Route Planning (ERP) problem from a computational perspective. In a typical evacuation planning scenario, the spatial structure is represented and analyzed as a network model with non-negative integer capacity constraints on the nodes and edges. With additional information pertaining to initial locations of all evacuees and their final destinations, the ERP problem produces a set of origin and destination routes for evacuees. Consider a simple building ERP problem in Fig. 6.12a. Each room, corridor, staircase, and exit of the building is represented as a node and each pathway from one node to another node is represented as an edge. Every node has two attributes: maximum node capacity and initial node occupancy. For example, the maximum capacity of node $N1$ is 50, indicating that the node can hold at most 50 evacuees at any time step. The initial occupancy is shown to be 10, which means 10 evacuees prepare to move out of the node. Every edge has two attributes: maximum edge capacity and travel time. For example, the maximum edge capacity of $N4 - N6$ is 5, indicating that at most 5 evacuees can traverse the edge. The travel time of this edge is 4, which means it takes 4 time steps to traverse the edge. Suppose we initially have 10 evacuees at node $N1$, 5 at node $N2$, and 15 at node $N8$, the ERP produces evacuation routes as shown in Fig. 6.12b. The objective of the ERP problem is to minimize the computational cost of producing the evacuation plan while minimizing evacuation time.

6.17 Capacity Constrained Route Planner

CCRP is based on an iterative approach for creating a complete evacuation plan. In each iteration, the algorithm searches for a route R with the earliest arrival time to any destination node from any source node, taking previous reservations and possible wait times into consideration. Then, CCRP computes the actual number of evacuees that will travel through R. This number is affected by the available capacity of R and the remaining number of evacuees. The maximum number of evacuees to be sent on R is then determined as the minimum of the available capacities on the component edges in R; CCRP reserves the node and edge capacity on R for those evacuees. The algorithm terminates when all the evacuees have been given an evacuation route to any of the destinations.

A key step in CCRP is to determine the route R with the earliest arrival time. A naive way to obtain the route R could involve executing Dijkstra's algorithm (Dijkstra 1959; Cormen et al. 2001) (generalized to work with edge capacities and travel times) for every of source and destination nodes, followed by selecting the minimum. However, this becomes a major performance bottleneck and adversely effects the scalability of the algorithm (Lu et al. 2005, 2007). This bottleneck is handled in CCRP by adding a pseudo source node S_0 and edges with zero travel time and infinite capacity between S_0 and all other source nodes.

a

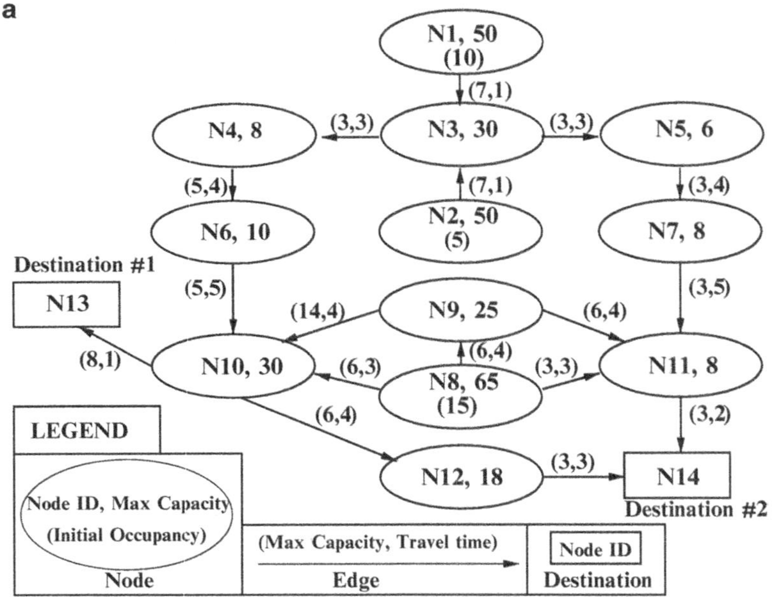

b

Group of Evacuees			Route with Schedule	Start Time	Dest. Time
ID	Source	Group Size			
A	N8	6	N8(T0)–N10(T3)–N13	0	4
B	N8	6	N8(T1)–N10(T4)–N13	1	5
C	N8	3	N8(T0)–N11(T3)–N14	0	5
D	N1	3	N1(T0)–N3(T1)–N4(T4)–N6(T8)–N10(T13)–N13	0	14
E	N1	3	N1(T0)–N3(T2)–N4(T5)–N6(T9)–N10(T14)–N13	0	15
F	N1	1	N1(T0)–N3(T1)–N5(T4)–N7(T8)–N11(T13)–N14	0	15
G	N2	2	N2(T0)–N3(T1)–N5(T4)–N7(T8)–N11(T13)–N14	0	15
H	N2	3	N2(T0)–N3(T3)–N4(T6)–N6(T10)–N10(T15)–N13	0	16
I	N1	3	N1(T1)–N3(T2)–N5(T5)–N7(N9)–N11(T14)–N14	1	16

Fig. 6.12 Building evacuation problem. (**a**) Network model for simple building evacuation scenario. (**b**) Example of evacuation route scheduling

6.18 Monticello Power Plant Case Study

In the wake of the Fukushima nuclear disaster in Japan, the US Nuclear Regulatory Commission is considering annual revisions of evacuation plans for nuclear power plants (USA-Today 2011). The commission's interest in regularly updated plans due to changes in nearby populations or transportation networks is likely to accelerate adoption of CCRP-like computer-based selection of evacuation routes over manual methods (Vasconez and Kehrli 2010). As shown in Fig. 6.13, the Monticello nuclear

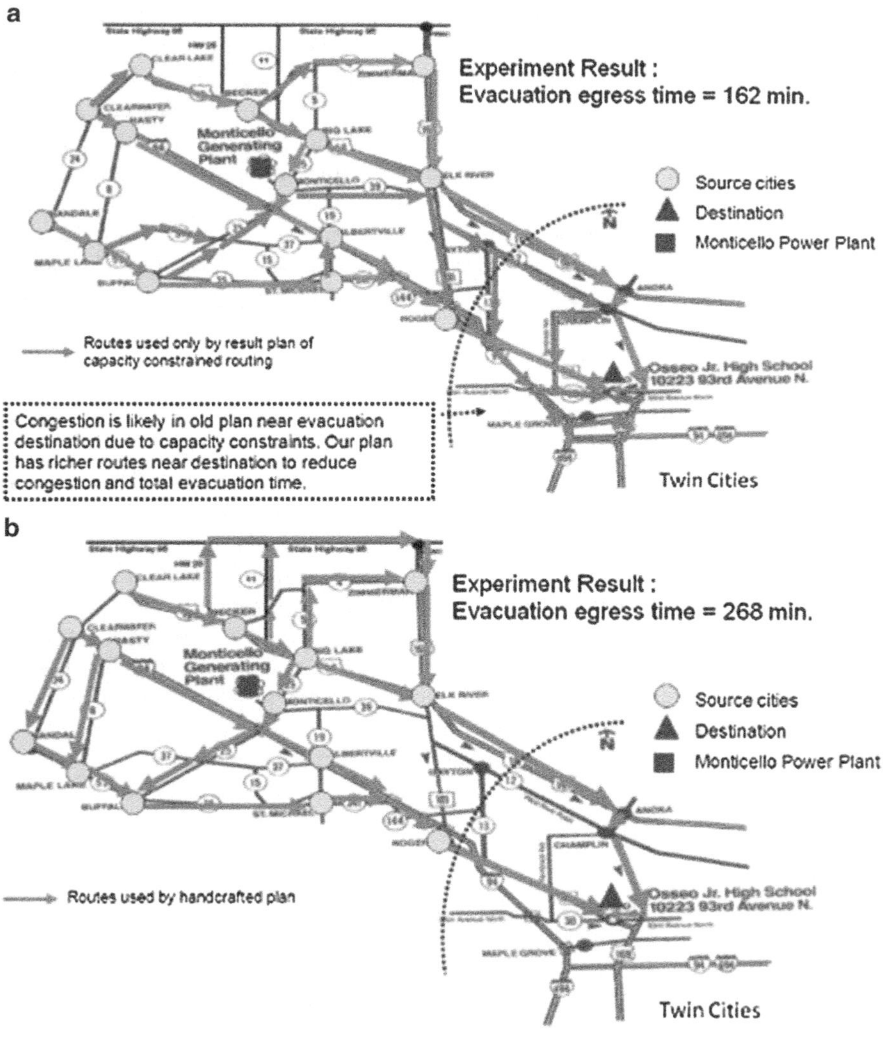

Fig. 6.13 Overlay of (**a**) CCRP generated and (**b**) hand-crafted evacuation routes for the Monticello power plant

power plant is located about 40 miles northwest of the Twin Cities of Minneapolis-St. Paul. Evacuation plans need to be in place in case of accidents or terrorist attacks. The evacuation zone is a 10-mile radius around the nuclear power plant as defined by Minnesota Homeland Security and Emergency Management (Lu et al. 2007). An initial hand-crafted evacuation route plan called for the affected population to travel to a nearby high school. The plan did not consider the capacity of the road network in the area, nor the high traffic load that would be put on two nearby highways.

We experimentally tested the CCRP algorithm using the road network around the evacuation zone provided by the Minnesota Department of Transportation (Lu et al. 2007) and the Census 2000 population data for each affected city (circle in Fig. 6.13). The total number of evacuees was about 42,000. As can be seen in Fig. 6.13, our algorithm produced a much better evacuation route plan (a) by selecting shorter paths to reduce evacuation time and (b) by utilizing richer routes i.e., routes near the evacuation destination to reduce congestion. As a result, evacuation egress time was reduced from 268 min under the old plan to only 162 min with CCRP. This experiment demonstrated the effectiveness of our algorithm in real evacuation planning scenarios to reduce evacuation time and improve existing plans.

6.19 Conclusion

Spatio-temporal networks are a key component of critical applications such as transportation networks, sensor data analysis, and crime analysis. We discuss existing approaches in modeling these networks and proceeds to present an approach that is based on time aggregation of graphs. Though spatio-temporal trajectories have been proposed to model temporal variations in spatial data, there is limited work on spatio-temporal networks. Another approach based on replication of networks leads to networks of enormous sizes and consequently computationally expensive algorithms. Moreover, these models do not provide adequate support for node/edge centric operations which might be essential in applications such as evacuation planning, routing, and temporally enhanced ER diagrams.

References

Abou-Rjeili, A., & Karypis, G. (2006). Multilevel algorithms for partitioning power-law graphs. In *20th international parallel and distributed processing symposium, 2006. IPDPS 2006* (p. 10). IEEE.

Ahuja, R. K., Magnanti, T. L., & Orlin, J. B. (1993). *Network flows*. Englewood Cliffs: Prentice Hall.

Airlines. http://www.delta.com/

Batty, M. (2005). Agents, cells, and cities: New representational models for simulating multiscale urban dynamics. *Environment and Planning A, 37*, 1373–1394.

Ben-Akiva, M. (2002). *Development of a deployable real-time dynamic traffic assignment system: Dynamit and dynamit-p user's guide*. Technical Report. Cambridge: Massachusetts Institute of Technology.

Bertsekas, D. P. (1987). *Dynamic programming: Deterministic and stochastic models*. Englewood Cliffs: Prentice-Hall.

Burns, L. (1979). *Transportation, temporal, and spatial components of accessibility*. Lexington: Lexington Books.

Chabini, I. (1998). Discrete dynamic shortest path problems in transportation applications: Complexity and algorithms with optimal run time. *Transportation Research Record: Journal of the Transportation Research Board, 1645*(-1), 170–175.

Chabini, I., & Lan, S. (2002). Adaptations of the A* algorithm for the computation of fastest paths in deterministic discrete-time dynamic networks. *IEEE Transactions on Intelligent Transportation Systems, 3*(1), 60–74.

Christakos, G., Bogaert, P., & Serre, M. (2001). *Temporal GIS: Advanced functions for field-based applications*. New York: Springer.

Cormen, T. H., Leiserson, C. E., Rivest, R. L., & Stein, C. (2001). *Introduction to algorithms*. Cambridge, MA: MIT Press.

Dehne, F., Omran, M. T., & Sack, J.-R. (2009). Shortest paths in time-dependent FIFO networks using edge load forecasts. In *Proceedings of the second international workshop on computational transportation science, IWCTS'09* (pp. 1–6). New York, NY: ACM.

Delling, D. (2008). Time-dependent SHARC-routing. In *Proceedings of the 16th annual European symposium on algorithms, ESA'08* (pp. 332–343). Berlin/Heidelberg: Springer.

Delling, D., & Wagner, D. (2007). Landmark-based routing in dynamic graphs. In *Experimental algorithms* (pp. 52–65). Berlin/Heidelberg: Springer.

Demiryurek, U., Banaei-Kashani, F., & Shahabi, C. (2010). A case for time-dependent shortest path computation in spatial networks. In *Proc. of the ACM SIGSPATIAL Intl. Conf. on Advances in GIS, GIS'10* (pp. 474–477).

Demiryurek, U., Banaei-Kashani, F., Shahabi, C., & Ranganathan, A. (2011). Online computation of fastest path in time-dependent spatial networks. In *Advances in spatial and temporal databases*. LNCS 6849 (pp. 92–111). Berlin/Heidelberg: Springer.

Deutsch, C. (2007). *UPS embraces high-tech delivery methods*. www.nytimes.com/2007/07/12/business/12ups.html

DiBiase, D., MacEachren, A., Krygier, J., & Reeves, C. (1992). Animation and the role of map design in scientific visualization. *Cartography and Geographic Information Science, 19*(4), 201–214.

Dijkstra, E. (1959). A note on two problems in connexion with graphs. *Numerische Mathematik, 1*(1), 269–271.

Ding, B., Yu, J., & Qin, L. (2008). Finding time-dependent shortest paths over large graphs. In *Proceedings of the 11th international conference on Extending database technology: Advances in database technology* (pp. 205–216). New York: ACM.

Dreyfus, S. (1969). An appraisal of some shortest path algorithms. *Operations Research, 17*, 395–412.

Egenhofer, M., & Golledge, R. (1998). *Spatial and temporal reasoning in geographic information systems*. New York: Oxford University Press.

Ford, L., & Fulkerson, D. (1958). Constructing maximal dynamic flows from static flows. *Operations Research, 6*(3), 419–433.

Ford, L., & Fulkerson, D. (1962). *Flows in networks*. Princeton: Princeton University Press.

Francis, R. L., & Chalmet, L. G. (1984). *A negative exponential solution to an evacuation problem* (Technical Report Research Report No. 84–86) Washington, DC: National Bureau Of Standards, Center for Fire Research.

Frank, A. (2003). 2: Ontology for spatio-temporal databases. *Lecture Notes in Computer Science, 2520*, 9.

Frank, A., Grumbach, S., Güting, R., Jensen, C., Koubarakis, M., Lorentzos, N., Manolopoulos, Y., Nardelli, E., Pernici, B., Schek, H., et al. (1999). Chorochronos: A research network for spatiotemporal database systems. *ACM SIGMOD Record, 28*(3), 12–21.

Galton, A., & Worboys, M. (2005). Processes and events in dynamic geo-networks. *Lecture Notes in Computer Science, 3799*, 45.

George, B. & Shekhar, S. (2006). Time-aggregated graphs for modeling spatio-temporal networks. In *Advances in conceptual modeling – Theory and practice* (pp. 85–99). Berlin/Heidelberg: Springer.

George, B., & Shekhar, S. (2007a, November). Modeling spatio-temporal network computations – A summary of results. In Proceedings of second international conference on GeoSpatial Semantics (GeoS 2007), Mexico City, Mexico.

George, B., & Shekhar, S. (2007b). Time-aggregated graphs for modeling spatio-temporal networks – An extended abstract. *Journal on Data Semantics*, XI. 191–212. Berlin/Heidelberg: Springer.

George, B., Kang, J., & Shekhar, S. (2007a, August). Spatio-temporal sensor graphs: A data model for discovery of patterns in sensor data. In Proceedings of first international workshop on knowledge discovery in Sensor Data (SensorKDD) in connection with KDD'07.

George, B., Kim, S., & Shekhar, S. (2007b, July). Spatio-temporal network databases and routing algorithms: A summary of results. In Proceedings of international symposium on Spatial and Temporal Databases (SSTD'07), Boston, MA.

George, B., Shekhar, S., & Kim, S. (2008). *Spatio-temporal network databases and routing algorithms* (Technical Report 08–039). Minneapolis: University of Minnesota – Computer Science and Engineering.

Goodchild, M., Yuan, M., & Cova, T. (2007). Towards a general theory of geographic representation in GIS. *International Journal of Geographical Information Science, 21*(3), 239–260.

Google Maps. http://maps.google.com

Gunturi, V., Shekhar, S., & Bhattacharya, A. (2010). Minimum spanning tree on spatio-temporal networks. In *Proceedings of the 21st international conference on database and expert systems applications: Part II*, DEXA'10 (pp. 149–158). Berlin/Heidelberg: Springer.

Gunturi, V. M. V., Nunes, E., Yang, K., & Shekhar, S. (2011). A critical-time-point approach to all-start-time Lagrangian shortest paths: A summary of results. *Advances in spatial and temporal databases*. LNCS 6849 (pp. 74–91). Berlin/Heidelberg: Springer.

Guttman, A. (1984). R-trees: A dynamic index structure for spatial searching. In *Proceedings of the 1984 ACM SIGMOD international conference on management of data* (pp. 47–57). New York: ACM.

Hägerstrand, T. (1970). What about people in regional science? *Papers in Regional Science, 24*(1), 6–21.

Hamacher, H., & Tjandra, S. (2002). Mathematical modeling of evacuation problems: A state of the art. In *Pedestrian and evacuation dynamics* (pp. 227–266). Berlin/Heidelberg: Springer.

Harrower, M. (2004). A look at the history and future of animated maps. *Cartographica: The International Journal for Geographic Information and Geovisualization, 39*(3), 33–42.

Herrera, J., & Bayen, A. (2009). Incorporation of Lagrangian measurements in freeway traffic state estimation. *Transportation Research Part B: Methodological, 44*, 460–481.

Hoppe, B., & Tardos, E. (1994). Polynomial time algorithms for some evacuation problems. In *Proceedings of the fifth annual ACM-SIAM symposium on discrete algorithms, SODA'94* (pp. 433–441). Society for Industrial and Applied Mathematics. Philadelphia, PA.

Janelle, D. (2004). Impact of information technologies. In *The geography of urban transportation* (p. 86). New York, NY: Addison-Wesley.

Joel Lovell (December 9, 2007). Left-hand-turn elimination. *New York Times*. http://goo.gl/3bkPb

Kanoulas, E., Du, Y., Xia, T., & Zhang, D. (2006). Finding fastest paths on a road network with speed patterns. In *Proceedings of the 22nd International Conference on Data Engineering (ICDE)* (p. 10). IEEE.

Kaufman, D., & Smith, R. (1993). Fastest paths in time-dependent networks for intelligent vehicle highway systems applications. *IVHS Journal, 1*(1), 1–11.

Kim, S., George, B., & Shekhar, S. (2007). Evacuation route planning: Scalable heuristics. In *Proceedings of the 15th annual ACM international symposium on Advances in geographic information systems (ACM-GIS)* (pp. 1–8). New York, NY: ACM.

Kisko, T. M., & Francis, R. L. (1985). Evacnet+: A computer program to determine optimal building evacuation plans. *Fire Safety Journal, 9*(2), 211–220.

Kisko, T., Francis, R., & Nobel, C. (1998). *EVACNET4 user's guide*. University of Florida.

Kohler, E., Langtau, K., & Skutella, M. (2002). Time-expanded graphs for flow-dependent transit times. In *Proceedings 10th annual European symposium on algorithms* (pp. 599–611). Berlin/Heidelberg: Springer.

Kriegel, H.-P., Renz, M., & Schubert, M. (2010). Route skyline queries: A multi-preference path planning approach. In *26th international conference on data engineering (ICDE), 2010 IEEE* (pp. 261–272). IEEE. Long Beach, CA.

Kwan, M. (2004). GIS methods in time-geographic research: Geocomputation and geovisualization of human activity patterns. *Geografiska Annaler: Series B, Human Geography, 86*(4), 267–280.

Langran, G. (1992). *Time in geographic information systems*. London: Taylor & Francis.

Langran, G., & Chrisman, N. (1988). A framework for temporal geographic information. *Cartographica: The International Journal for Geographic Information and Geovisualization, 25*(3), 1–14.

Lenntorp, B. (1977). Paths in space-time environments: A time-geographic study of movement possibilities of individuals. *Environment and Planning A, 9*, 961–972.

Lu, Q., Huang, Y., & Shekhar, S. (2003). Evacuation planning: A capacity constrained routing approach. In *First NSF/NIJ symposium on intelligence and security informatics* (pp. 111–125). Berlin/Heidelberg: Springer.

Lu, Q., George, B., & Shekhar, S. (2005). Capacity constrained routing algorithms for evacuation planning: A summary of results. In *Proceedings of 9th International Symposium on Spatial and Temporal Databases (SSTD™05)* (pp. 22–24). Berlin/Heidelberg: Springer.

Lu, Q., George, B., & Shekhar, S. (2007). Evacuation route planning: A case study in semantic computing. *International Journal of Semantic Computing, 1*(2), 249. World Scientific Publishing.

Luenberger, D. (1973). *Introduction to linear and nonlinear programming*. Reading: Addison-Wesley.

Mahmassani, H. S., Sbayti, H., & Zhou, X. (2004). *Dynasmart-p version 1.0 user's guide*. (Technical Report). Maryland Transportation Initiative, University of Maryland.

Mapquest. http://www.mapquest.com/

Maps. http://www.bing.com/maps/

McIntosh, J., & Yuan, M. (2005). Assessing similarity of geographic processes and events. *Transactions in GIS, 9*(2), 223–245.

Miller, H. (1999). Measuring space-time accessibility benefits within transportation networks: Basic theory and computational procedures. *Geographical Analysis, 31*(2), 187–212.

Miller, H. (2005). A measurement theory for time geography. *Geographical Analysis, 37*(1), 17–45.

Monmonier, M. (1990). Strategies for the visualization of geographic time-series data. *Cartographica: The International Journal for Geographic Information and Geovisualization, 27*(1), 30–45.

O'Sullivan, D. (2005). Geographical information science: Time changes everything. *Progress in Human Geography, 29*(6), 749.

Orda, A., & Rom, R. (1990). Shortest-path and minimum-delay algorithms in networks with time-dependent edge-length. *Journal of the ACM (JACM), 37*(3), 607–625.

Orda, A., & Rom, R. (1991). Minimum weight paths in time-dependent networks. *Networks, 21*, 295–319.

Pallottino, S., & Scuttella, M. G. (1998). Shortest path algorithms in transportation models: Classical and innovative aspects. In *Equilibrium and advanced transportation modelling* (pp. 245–281). Kluwer. Springer US.

Peuquet, D., & Duan, N. (1995). An event-based spatiotemporal data model (ESTDM) for temporal analysis of geographical data. *International Journal of Geographical Information Science, 9*(1), 7–24.

Pred, A. (1977). The choreography of existence: Comments on Hagerstrand's time-geography and its usefulness. *Economic Geography, 53*, 207–221.

Russel, S., & Norwig, P. (1995). *Artificial intelligence: A modern approach*. Upper Saddle River: Prentice-Hall.

Samet, H. (1995). Spatial data structures. In *Modern database systems, the object model, interoperability and beyond* (pp. 361–385). New York, NY: Addison-Wesley.

Shekhar, S., & Liu, D.-R. (1997). CCAM: A connectivity-clustered access method for networks and network computations. *IEEE Transactions on Knowledge and Data Engineering, 9*(1), 102–119.

Shekhar, S., & Xiong, H. (2007). *Encyclopedia of GIS*. Berlin: Springer.

Timmermans, H., Arentze, T., & Joh, C. (2002). Analysing space-time behaviour: New approaches to old problems. *Progress in Human Geography, 26*(2), 175.

Torrens, P. (2006). Simulating sprawl. *Annals of the Association of American Geographers, 96*(2), 248–275.

USA-Today. (2011, June). Evacuation plans age as population near nuclear plants soars.

Vasconez, K., & Kehrli, M. (2010). *Highway evacuations in selected metropolitan areas: Assessment of impediments* (Technical Report FHWA-HOP-10-059). Washington, DC: Federal Highway Administration, Office of Transportation Operations.

Chapter 7
Spatiotemporal Critical Opportunity and Link Identification for Joint Participation Scheduling

Zhixiang Fang, Shih-Lung Shaw, Wei Tu, and Qingquan Li

7.1 Background

Individuals commonly participate in work activities, social activities, and cultural events. Joint participation is an essential part of individuals' social activities. How to schedule spatiotemporal activities for social communication comfortably is a key issue for human beings. To address this question, modern information and telecommunication technologies provide diverse communication approaches to help decision-making in activity scheduling. For example, these technologies have been integrated to facilitate an activity-based location-based service (LBS) for travelers, such as advanced traveler information systems (Ben-Elia and Shiftan 2010) or infectious disease control (Angulo et al. 2012). Normally, to schedule comfortable and acceptable joint participation, an LBS needs to identify critical opportunities and links among travelers based on cost considerations like travel distance and time, congestion, and others.

Z. Fang (✉)
State Key Laboratory for Information Engineering in Surveying, Mapping, and Remote Sensing, Wuhan University, Wuhan 430079, People's Republic of China
e-mail: zxfang@whu.edu.cn

S.-L. Shaw
Department of Geography, University of Tennessee, 304 Burchfiel Geography Building, Knoxville, TN 37996-0925, USA
e-mail: sshaw@utk.edu

W. Tu • Q. Li
Shenzhen Key Laboratory of Spatial Smart Sensing and Services, College of Civil Engineering, Shenzhen University, 3688 Nanhai Avenue, Nanshan District, Shenzhen 518060, People's Republic of China
e-mail: tuwei@szu.edu.cn; liqq@szu.edu.cn

M.-P. Kwan et al. (eds.), *Space-Time Integration in Geography and GIScience*, DOI 10.1007/978-94-017-9205-9_7

On the basis of time geography theory (Hägerstrand 1970), several approaches (Yu and Shaw 2008; Fang et al. 2011; Chen and Kwan 2012; Scott and He 2012; Yoon et al. 2012; Shaw 2012) have been proposed to identify an opportunity set of activities which is able to support activity scheduling. These approaches focus on the space-time constrained opportunities within the potential path area of a (dynamic) transportation network. For example, Fang et al. (2011) defined a time-varying network-based prism to identify the joint participation space of multiple people within space and time dimensions. The opportunities within the joint participation space constituted their candidate activity opportunities. Scott and He (2012) used the potential path area to determine an individual's destination choice set for shopping. Chen and Kwan (2012) proposed four models to identify the choice set with multiple flexible activities under space-time constraints (Newell et al. 1994; Sips et al. 2007; Wentz et al. 2010; McQuoid and Dijst 2012; Sui 2012) confined by a space-time movement environment (Hägerstrand 1970; Lenntorp 1976; Islam 2010; Ellegård and Svedin 2012). These four models conceptualized trip-chaining behavior based on a fixed set of activities and a fixed number of origin and destination stations. Yoon et al. (2012) also confirmed that the space-time prism concept can account for variations in choice sets.

However, the choice sets for activities limited by a space-time prism should be investigated based on knowledge of the transportation network as well as of time-space constraints. To accumulate transportation knowledge for urban management, many cities have equipped their transportation networks with various sensors to collect plentiful video data and loop signals for future data mining. Beside this traditional knowledge, trajectory data collected by Global Positioning System (GPS)-equipped vehicles also provide much useful information for mining transportation network knowledge, such as traffic speeds in occupied links and the spatiotemporal frequency of often-used links. This kind of information can facilitate activity scheduling by identifying critical links. Therefore, this chapter proposes an approach which identifies critical links and opportunities for joint participation based on analysis of GPS-recorded taxi trajectories within the framework of time geography theory.

This chapter is organized as follows. In the next section, the dynamic network-based space-time prism is reviewed. Then, in Sect. 7.3, critical links and opportunities are identified given the dynamic network-based space-time prism and vehicle occupation and frequency data from the transportation network. A multi-objective approach to scheduling the joint participation of multiple individuals is proposed in Sect. 7.4. Section 7.5 describes a scenario study demonstrating the feasibility of identifying critical links and opportunities for scheduling multiple trips for multiple people. Brief conclusions are drawn in Sect. 7.6.

7.2 Dynamic Network-Based Space-Time Prism

A space-time prism representing a reachable space-time regime for individuals is commonly used to identify available activity opportunities between a given set of origins and destinations. To identify real-life available activity opportunities, several

Table 7.1 Congestion index used by the Beijing Transportation Research Center

Index	[0,2)	[2,4)	[4,6)	[6,8)	[8,10]
Transportation status	No congestion	Very little congestion	Slight congestion	Congestion	Heavy congestion
Factor for additional travel time	0	[0.3,0.5)	[0.5,0.8)	[0.8,1.0)	>1.0

Source: http://www.bjtrc.org.cn/PageLayout/IndexReleased/IndexReader.aspx?menuid=li4

researchers have calculated the space-time prism in a (dynamic) transportation network environment. For example, Miller (1991) was the first to develop a network-based algorithm for computing a space-time prism. Kwan and Hong (1998) also implemented this approach in a transportation network. Wu and Miller (2001) proposed a more realistic space-time prism by taking congestion into account. Based on the 3D polygon representation of the space-time prisms proposed by Kuijpers and Othman (2009) and Kuijpers et al. (2010), Fang et al. (2011) proposed a time-varying network-based prism to identify available activity opportunities for scheduling joint participation. This chapter formulates a dynamic network-based space-time prism.

A dynamic transportation network is denoted as $G(t) = \{V, E, S(t)\}$, where V is a set of nodes, $V = \{v_i = (x_i, y_i) | 0 \leq i \leq n\}$, n is the number of nodes in the network, E is a set of edges, $E = \{e_i = (v_i, v_j) | 0 \leq i \leq m, 0 \leq j \leq m, i \neq j\}$, m is the number of edges in the network, $S(t)$ is a collection of transport indices on all edges at time t, $S(t) = \{s(e_i, t) | 0 \leq i \leq m\}$, and $s(e_i, t)$ is a congestion index of edge e_i at time t. For example, Table 7.1 lists observed values of the congestion index used here and as defined by the Beijing Transportation Research Center (Xu et al. 2008). The congestion index is calculated according to the combination of travel speed derived from moving car data, road status, and traffic flow on the corresponding link. A detailed approach to calculating the congestion index can be found on the Web site (http://www.bjtrc.org.cn) of the Beijing Transportation Research Center. The ratio of additional travel time on edge e_i means that travelers need more time to pass through this link. For example, say that it normally takes 10 min for a traveler to pass through e_i. When the congestion index of e_i is 3, the evaluated time that a traveler will need to pass through e_i is $10 \times (1 + 0.4) = 14$ min.

Fang et al. (2011) defined a time-varying network-based prism (TVNBP) between origin O and destination D within a dynamic transportation network $G(t)$ as follows:

$$TVNBP\,(O, D, T) = \sum_{e_i \in \overline{E}} f\left(e_i^{[t'_s, t'_e]}, v\left((x, y), t\right)\right) \tag{7.1}$$

$$f\left(e_i^{[t'_s, t'_e]}, v\left((x, y), t\right)\right) = \int_{l'_s}^{l'_e} \int_{t'_{min}}^{t'_{max}} l^{t\prime}_{(x,y)} dldt, t'_s \leq t'_{min} \leq t'_{max} \leq t'_e, \left[l'_s, l'_e\right] \in e_i \tag{7.2}$$

subject to:

$$\int_{t'_s}^{t'_e} v\left((x,y),t\right) dt = Length\left(l'_s, l'_e\right) \tag{7.3}$$

$$l_{(x+\Delta x, y+\Delta y)}^{t'_i+\Delta t} = l_{(x,y)}^{t'_i} + v\left((x,y),t'_i\right)\Delta t \tag{7.4}$$

$$v(x,y) = \begin{cases} \frac{1}{n}\sum_{i=1}^{n} v_i, & \text{if a moving car has passed} \\ \overline{v}, & \text{otherwise} \end{cases} \tag{7.5}$$

Equations (7.1) and (7.2) represent a time-varying network-based prism and a time-varying accessible vertical area derived from edge e_i, where t'_s, t'_e, $t'_{\min}$, and $t'_{\max}$ are time- and location-dependent variables and $v((x, y), t)$ is an instantaneous travel speed at location (x, y) at time t. Equation (7.3) defines the condition that the total travel distance is covered within a time interval $[t'_s, t'_e]$. $l^t_{(x,y)}$ is a space-time point in 3D space. $Length(l'_s, l'_e)$ denotes the accessible length of edge e_i in TVNBP. Equation (7.4) illustrates the dynamic motion process from point $l^{t'_i}_{(x,y)}$ to point $l^{t'_i+\Delta}_{(x+\Delta x,y+\Delta y)}$ along edge e_i, where Δt is a variable indicating a small time slice. Equation (7.5) defines the travel speed used to evaluation the congestion index. This speed is the average of the speed values in all trajectories in which there are some moving cars passing over edge e_i. If no moving cars are present, the speed on edge e_i is the average value of the speeds on all transportation links with the same road function and hierarchical position as edge e_i. Figure 7.1 illustrates a time-varying network-based prism as represented by lines (Fig. 7.1a) and three-dimensional polygons (Fig. 7.1b).

The TVNBP can be extended by considering the influence of the congestion index as follows:

$$TVNBP'(O,D,T) = \sum_{e_i \in \overline{E}} f\left(e_i^{[t'_s,t'_e]}, v\left((x,y),t,s\left(e_i,t\right)\right)\right) \tag{7.6}$$

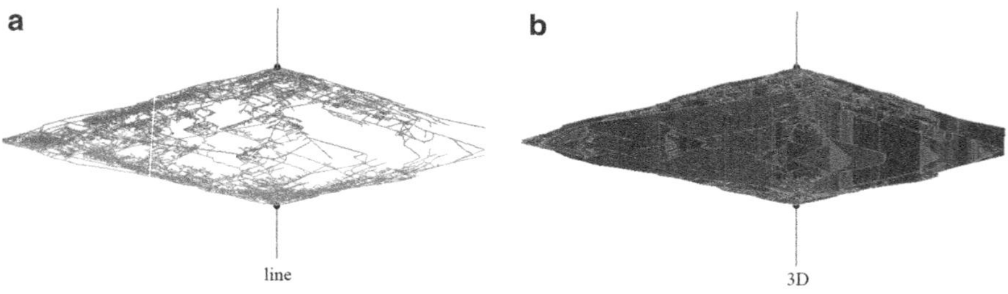

Fig. 7.1 Time-varying network-based prism

$$f\left(e_i^{[t_s',t_e']}, v\left((x,y), t, s\left(e_i, t\right)\right)\right) = \int_{l_s'}^{l_e'} \int_{t_{min}'}^{t_{max}'} l_{((x,y),s(e_i,t))}^{t'} dldt \tag{7.7}$$

$$l_{((x+\Delta x, y+\Delta y),s(e_i,t))}^{t_i'+\Delta t} = l_{((x,y),s(e_i,t))}^{t_i'} + v\left((x,y), t_i'\right)\left(1 + s\left(e_i, t\right)\right)\Delta t \tag{7.8}$$

Equations 7.6, 7.7, and 7.8 describe a dynamic transportation network-based prism incorporating the congestion index. Equation 7.8 represents the set of space-time points constrained by the congestion index, which reflects the influence of the congestion index on travel time.

7.3 Critical Links and Activity Opportunities

Unlike the network structure analysis approach used for critical links in transportation, this chapter uses the frequency of taxi trajectories on links to identify critical links for activity scheduling because taxi drivers usually choose high-utility (low-cost and fast) routes in an urban environment. In other word, the critical links with higher frequency are easier and more convenient for travelers to follow when they carry out their activities. This chapter introduces several definitions related to critical links and opportunities.

Let EUN denote the number of taxi trajectories using an edge within a time period $[t_s, t_e]$, which is defined as follows:

$$\mathrm{EUN}\left(\mathrm{e}, [t_s, t_e]\right) = \sum_i \sum_t \gamma\left(i, e, t\right), \quad \mathrm{t_s} < \mathrm{t} < \mathrm{t_e} \tag{7.9}$$

subject to:

$$\gamma\left(i, e, t\right) = \begin{cases} 0 & \text{if taxi } i \text{ did not pass link } e \text{ at time } t \\ 1 & otherwise \end{cases} \tag{7.10}$$

Equation 7.10 represents the individual frequency value derived from the situation that taxi i either passed over link e at time t or not. Figure 7.2 illustrates the number of taxi trajectories using an edge in a dynamic transportation network. The thick black lines highlight the critical links of taxi trajectories in this network during a specific time period.

The critical index of a link e in a space-time environment is denoted as CI, which is defined as:

$$CI\left(e, [t_s, t_e]\right) = \frac{EUN\left(e, [t_s, t_e]\right)}{max\ UN\left(e_i, [t_s, t_e]\right)}, e_i \in E \tag{7.11}$$

Fig. 7.2 Critical links in a dynamic transportation network

The critical index is used to identify the most popular links for taxi drivers in a dynamic transportation network. This index is one of the objectives used in the optimization approach for scheduling joint participation described in the following section.

The common approach to identify available activity opportunities is illustrated in Fig. 7.3. For example, Fig. 7.3a represents the two dynamic network-based space-time prisms of two persons (A and B). Figure 7.3b illustrates their joint participation space within the prisms. The activity opportunities within this joint participation space in Fig. 7.3c represent available opportunities for the two persons to participate. Once an available activity opportunity has been chosen as a joint activity, their corresponding travel paths can be found within this prism (Fig. 7.3d). However, this common approach does not reflect the influence of critical links on the selection of activity opportunities.

Based on the selection of a common approach, several available opportunities can be found. This chapter presents a utility index which can evaluate the use of critical links while choosing an available opportunity in the prism. This utility index for an opportunity is defined as:

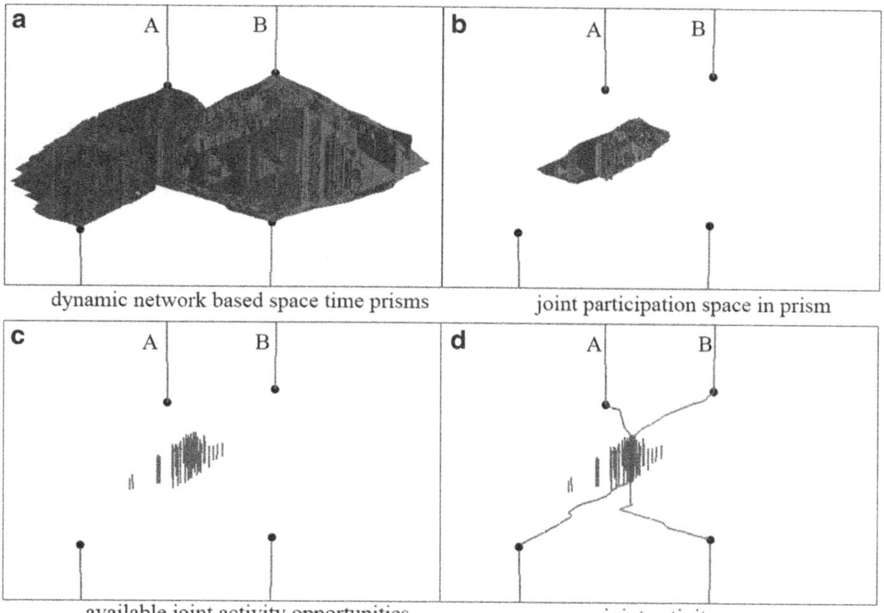

Fig. 7.3 Joint activity opportunities

$$U\left(opp_k, G(t)\right) = max\left\{\left(\sum_{i \in STP_j} CI\left(e_i, [0, t]\right)\right) / linknum\left(STP_j\right) \middle| STP_j \in ST\overline{P}\left(k, kst\right)\right\} \tag{7.12}$$

where $linknum(STP_j)$ represents the link number of the *j*th shortest space-time path for opportunity *k*, STP_j is a space-time path in the shortest space-time path set $ST\overline{P}\,(k, kst)$, and *kst* is a user-defined constant which indicates the number of the *kst* shortest space-time paths for opportunity *k*.

An opportunity with a utility index value within the first *kst* maximum utility index values $U(opp_k, G(t))$ of its corresponding path set $ST\overline{P}\,(k, kst)$ is referred to here as a critical activity opportunity.

Therefore, two steps are required to identify critical activity opportunities for scheduling joint participation. The first step is to identify the first *kst* shortest space-time paths for each opportunity in the TVNBP. The second step is to find the critical activity opportunity by finding the activity with the highest ratio $U(opp_k, G(t))/criterion(STP_j)$ among all available activity opportunities, where $criterion(STP_j)$ is the value of the specific evaluation criterion (e.g., travel distance, travel time) for opportunity *k*'s critical activity.

7.4 Modified Multi-objective Optimization Approach for Scheduling Joint Participation

Several researchers have focused on activities for joint activity scheduling. For example, Ronald et al. (2012) modeled social interactions between individuals for joint activity scheduling. Nijland et al. (2012) extended the need-based model by incorporating planned activities and events into a dynamic multi-day activity agenda generator. Zhou and Golledge (2007) proposed a unified data collection framework for real-time tracking of activity scheduling and schedule execution. Auld and Mohammadian (2012) represented the activity planning process in an activity-based microsimulation model called ADAPTS (Agent-based Dynamic Activity Planning and Travel Scheduling). This model can dynamically simulate activity and travel planning and scheduling. Most work on activity scheduling procedures has been done from a transportation planning perspective, for example, SCHEDULER (Gärling et al. 1989), ALBATROSS (Arentze and Timmermans 2003), FRBS (fuzzy rule based system) (Olaru and Smith 2002), TASHA (travel activity scheduler for household agents) (Miller and Roorda 2003; Roorda et al. 2009), and AURORA (an agent for utility-driven rescheduling of routinized activities) (Joh et al. 2001; Joh 2004). Very few research studies have attempted to solve the activity scheduling problem under constraints from time geography theory.

This chapter modifies the multi-objective approach of Fang et al. (2011) to schedule joint participation using the extended TVNBP and with the help of the critical link and activity opportunity concepts.

The modified multi-objective optimization problem for scheduling joint activities for multiple persons can then be defined as follows:

$$min\left(\mathrm{D} = \sum_{p_i \in P, k_j \in K} \sum \mathrm{path\_}dist\left(opp_{k_j}, p_i\right)\right) \tag{7.13}$$

$$min\left(\mathrm{T} = \sum_{p_i \in P, k_j \in K} \sum travel\_time\left(opp_{k_j}, p_i\right)\right) \tag{7.14}$$

$$max\left(\mathrm{R} = \sum_{p_i \in P, k_j \in K} \sum \mathrm{desire\_DT}\left(opp_{k_j}, p_i\right)\right) \tag{7.15}$$

$$min\left(\mathrm{AT} = \sum_{p_i \in P, k_j \in K} \sum \mathrm{added\_T}\left(opp_{k_j}, p_i\right)\right) \tag{7.16}$$

$$max\left(\mathrm{U} = \sum_{p_i \in P, k_j \in K} \sum \mathrm{U}\left(opp_{k_j}\right)\right) \tag{7.17}$$

subject to

$$desire\_DT\left(opp_{k_j}, p_i\right) = \gamma\,(time\_of\_day)\exp\left(-\lambda\left|\frac{DT\left(opp_{k_j} - DT\_min\right)}{DT\_min}\right|\right) \tag{7.18}$$

$$added\_T\left(opp_{k_j}, p_i\right) = \text{alternative\_}time\left(opp_{k_j}, p_i\right) - travel\_time\left(opp_{k_j}, p_i\right) \tag{7.19}$$

$$opp_{k_j} \in TVNBP\,(O\,(p_i)\,, D\,(p_i)\,, T\,(p_i)) \tag{7.20}$$

where p_i is person i, P is a set of people who would like to participate jointly, opp_{k_j} is the critical activity opportunity k_j available to p_i, K is a critical activity opportunity set constrained by TVNBP of p_i, and $path\_dist\left(opp_{k_j}, p_i\right)$ represents the travel distance over the space-time path for critical activity opportunity opp_{k_j}. Similarly, $travel\_time\left(opp_{k_j}, p_i\right)$ indicates the travel time for this corresponding critical activity opportunity opp_{k_j}, and $desire\_DT\left(opp_{k_j}, p_i\right)$ represents participation desirability by time of day. Equation 7.18 defines the desirability for person p_i to participate in critical activity opportunity opp_{k_j} and $\gamma(time\_of\_day)$ is a parameter indicating the degree of satisfaction of time-of-day requirements. DT_min is the minimum expected participation activity time of this person, $\text{alternative\_}time\left(opp_{k_j}, p_i\right)$ represents the added travel time for an alternative space-time path over the shortest space-time path for critical activity opportunity opp_{k_j} in this prism, and $\text{added\_T}\left(opp_{k_j}, p_i\right)$ is the added travel time for alternative space-time paths over the shortest space-time path. Equations 7.13, 7.14, 7.15, 7.16, and 7.17 constitute the objective function of the optimization problem for scheduling joint activities for multiple persons. They involve: (i) minimizing travel distance; (ii) minimizing travel time; (iii) maximizing expected participation activity time; (iv) minimizing the added travel time relative to alternative space-time paths; and (v) maximizing the utility index of the selected critical activity opportunity. Unlike the approach proposed by Fang et al. (2011), this modified problem is solved by limiting the search space using critical activity opportunities.

The proposed approach uses the (NSGA-II) nondominated sorting genetic algorithm (Deb et al. 2002) to solve this multi-objective optimization problem. Detailed descriptions of NSGA-II can be found in Deb et al. (2002) and Murugan et al. (2009). The pseudocode for the algorithm as implemented is given in Fig. 7.4. The selection operation is based on tournament selection (Murugan et al. 2009). The crossover operation is based on a single-point crossover scheme (Hájek et al. 2010). The mutation operator is the same as that of the standard genetic algorithm. The operation space of selection, recombination, and mutation in this algorithm is limited within the critical activity opportunities of the dynamic transportation

```
-Initialize parameters
   1. Build up dynamic transportation network G
   2. Assign time-varying average travel speeds to each link
   3. Generate the activity demand of persons
-Initialize population
   1. Select critical activity opportunities randomly
   2. Create space-time prisms for the opportunities
   3. Generate population from activity opportunities within space-time prisms
   4. Evaluate objective values in Equations (13–17)
   5. Assign rank based on Pareto dominance
   6. Generate child population
       6.1 Binary tournament selection
       6.2 Recombination and mutation within critical activity opportuni-ties
-Do
   1. With parent and child population
       1.1 Assign rank based on Pareto dominance
       1.2 Generate sets of nondominated fronts
       1.3 Find M populations beginning at the "first" front and insert them into
           the next generation
   2. Select critical activity opportunities on the lower front (with lower rank)
      which are outside a crowding distance
   3. Create space-time prisms of these opportunities
   4. Select critical activity opportunities within space-time prisms as candidates
   5. Create next generation
       5.1 Binary tournament selection
       5.2 Recombination and mutation within candidate activity opportunities
       5.3 Evaluate objective values in Equations (13–17)
   6. Increment generation index
While (current generation index is not the maximum number)
```

Fig. 7.4 Pseudocode of implemented NSGA-II algorithm

network-based prism. The solution of the problem includes several critical activity opportunities and their corresponding space-time paths. In this way, an activity participation plan for each person can be scheduled.

7.5 Experiments

This section describes an activity scheduling scenario which states that four people in Wuhan, China, need to meet each other within a week. They are a professor, a postgraduate student, a government official, and a retailer. Table 7.2 gives their detailed attributes and Table 7.3 their requirements for joint activity participation. There are three activities in which they want to participate:

1. Persons P1 and P2 need to carry out a business survey in five business centers among nine candidate business centers ("Qunguang Square", "Xudong shopping mall", "Hanzheng Street", "Han Street", "Luxiang Square", "Wangjiawan" in Hanyang, "Central South shopping mall", "Wuhan Tiandi", and "Wuhan shopping mall"), which are distributed in Wuhan, China.
2. Persons P1, P2, P3, and P4 need to have a meeting to discuss the business survey. This activity should be scheduled after the first activity.
3. Persons P2, P3, and P4 want to enjoy a weekend entertainment together.

Table 7.2 Attributes of the four people

Person	Attributes
P1	A professor teaching at HuaZhong University of Science and Technology
P2	Home is in GuanShan
	A post-graduate student at HuaZhong University of Science and Technology
P3	Home is in Hankou
	A government official working in Hanyang. Home is in Hanyang
P4	A retailer. Home is in Xudong

Table 7.3 Requirements for joint activity participation

ID	Activity	Participants	Day	Time
A1	Business survey	P1, P2	Workday	In the morning
A2	Meeting	P1, P2, P3, P4	Saturday	In the morning
A3	Entertainment	P2, P3, P4	Sunday	In the morning

Table 7.4 Preferences of people for starting times and durations of activities

People	Activity	Preference for starting time $\gamma(time\_of\_day)$	DT_min (/min)
P1	A1	0.9	150
P2	A2	0.8	60
	A1	0.95	160
P3	A2	0.95	80
	A3	0.85	55
	A2	0.7	70
P4	A3	0.9	50
	A2	0.8	50
	A3	0.8	60

Table 7.4 lists preferences of people for starting times and durations of activities. Based on the above activity assumptions, an activity scheduling problem for four people over one week has been clearly described.

Before scheduling their joint activity participations, it is necessary to describe two types of important background information about the activity scheduling:

1. The first type of information concerns the routing mobility of the four people. Figure 7.5 illustrates their routing mobility and their available activity time for joint participation. Each space-time path represents the routing movements of one person. The time window between the disconnected lines represents each person's available activity time window for joint participation with others.
2. The second important type of information is the dynamic transportation network. This research uses the transportation network shown in Fig. 7.2, which contains 26,429 directed links. The travel speed over each transportation link for every 5-min period was derived from the trajectory data of 12,325 monitored taxis

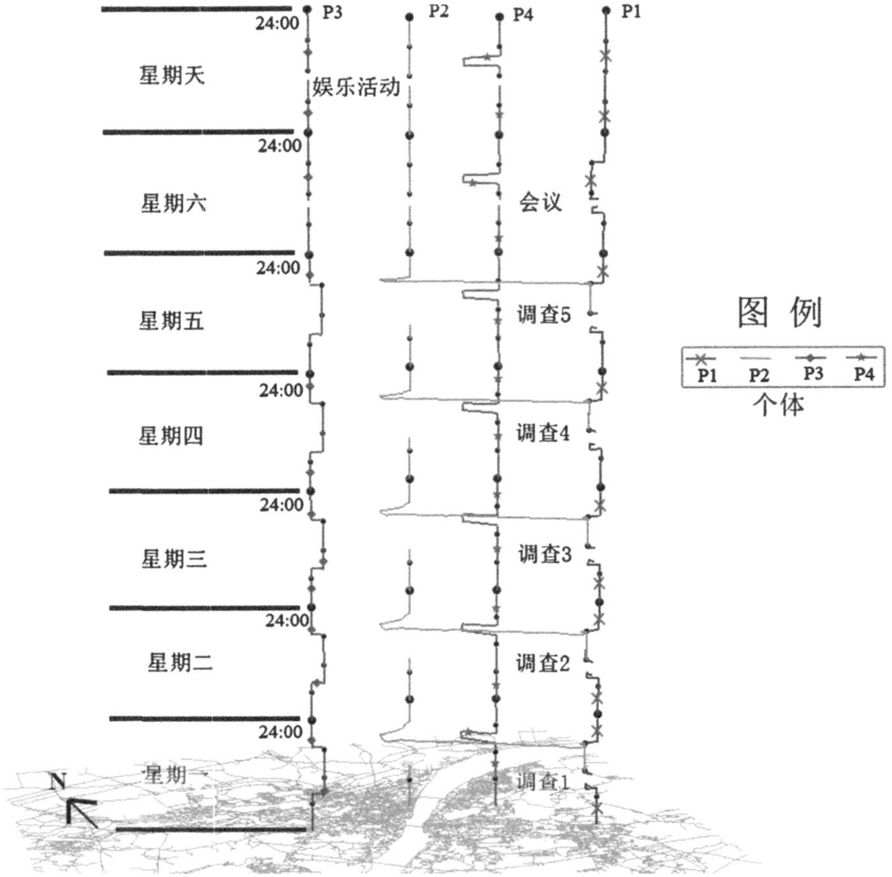

Fig. 7.5 Routing mobility and available activity times for joint participation

during one week between August 31, 2009, and September 6, 2009, if there were taxis in this link and within this time window. Otherwise, the travel speed was assigned as an acceptable speed related to the kind of road because not every road was occupied by taxis at all times. The detailed method for deriving travel speed from taxi trajectories can be found in Fang et al. (2012).The extended TVNBP used in this modified algorithm was implemented in this dynamic transportation network.

In this research, the modified NSGA-II algorithm has been used to solve the multi-objective optimization problem of scheduling joint activity participation in a VC++6.0 development environment. All required activities of four individuals

Table 7.5 Details of scheduled joint participation

	Activity			Objective				
Date	Joint activity	Location	Time	D (/km)	T (/min)	R	AT	U
Monday	Business survey	Qunguang Square	9:27–11:52	36.946	42.7	1.84	4.37	1.79
Tuesday	Business survey	Han Street	9:33–11:44	47.038	60.8	1.82	6.51	1.81
Wednesday	Business survey	Luxiang Square	9:26–11:58	25.604	31.8	1.84	3.20	1.91
Thursday	Business survey	Zhong County in Hanyang	9:42–11:39	58.660	71.7	1.80	7.64	1.74
Friday	Business survey	Central South Shopping mall	9:31–11:46	44.027	54.2	1.82	5.63	1.72
Saturday	Business meeting	New Kaixinyi Hotel	9:54–10:30	87.293	111.8	3.22	22.23	3.80
Sunday	Entertainment	Marshland in Hankou	10:29–11:20	48.826	62.0	2.66	16.46	2.86
Sum				348.394	435.0	15.00	46.04	15.63

were optimized by using this algorithm. Table 7.5 lists the details of scheduled joint participations in a week. Different starting times and activity durations of business survey in five days were affected by not only the preferences of people, but also the travel speed and the congestion index of transportation network. Figure 7.6 shows the four space-time paths of individuals, including the optimized joint activity opportunities generated by the modified algorithm. The red dots on the network are the business survey locations for persons P1 and P2, The blue dot on the network is the location of the meeting for persons P1, P2, P3, and P4, and the pink dot on the network is the location of the entertainment activity for P2, P3, and P4. The highlighted cylinders represent the activity time windows of the corresponding opportunities for persons to participate. Moreover, the cylinder's connected space-time paths are unbroken, which means that one person can finish his tasks one by one in the order of activities in each complete space-time path. For example, persons P1 and P2 can do their business survey in five locations one by one from Monday to Friday. The solutions in the Pareto fronts provide feasible combinations of joint activity participation to users for making comfortable decisions among them. The space-time path projected on the transportation network suggests the travel path for the corresponding person. Figure 7.7 illustrates the joint activity participation for the meeting of the four persons P1, P2, P3, and P4 on Saturday. The hotel highlighted in Fig. 7.7 represents their joint activity opportunity. Figures 7.6 and 7.7 demonstrate that the proposed approach is capable of scheduling joint activities for multiple persons in a real situation. This capability can support spatiotemporal travel decision-making services in commercial traveler information systems.

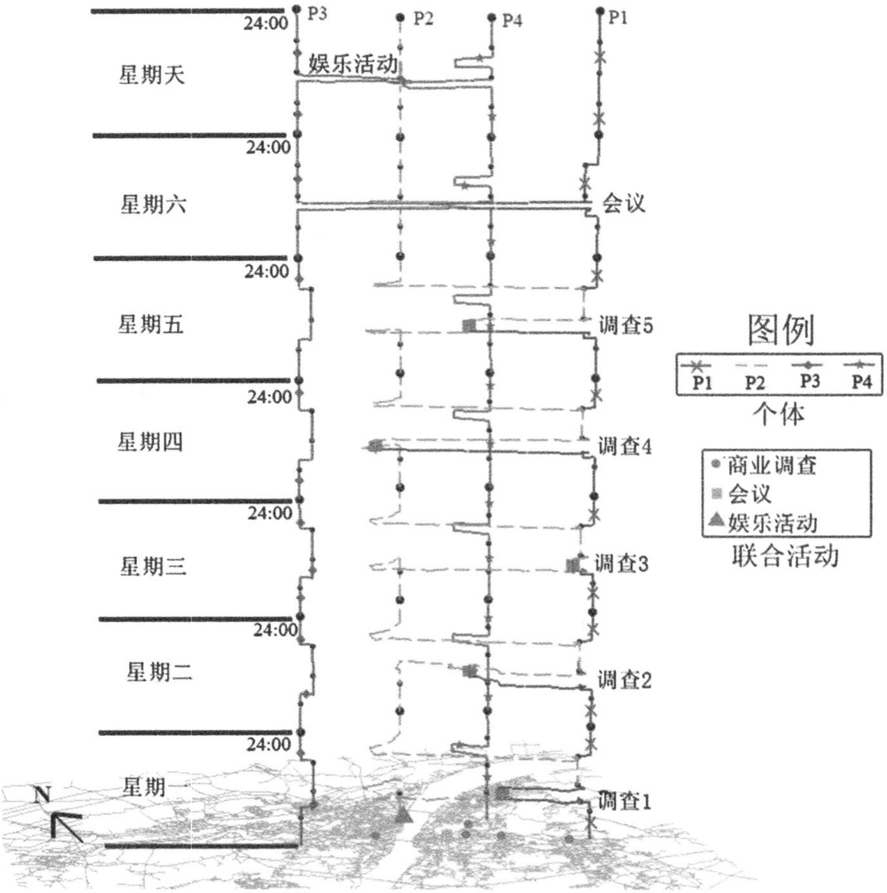

Fig. 7.6 Space-time paths including optimized joint activity participation

7.6 Conclusions

In this chapter, the computational analysis of a space-time prism in the context of time geography theory has been discussed. The concept of a time-varying network-based prism has been extended by considering the influence of the congestion index on estimated travel speed over links, which is the extended TVNBP. This type of prism can be viewed as an estimated space-time prism of origins and destinations rather than the strict space-time prism of these in time geography.

From the perspective of activity scheduling, concepts of critical link and activity opportunity have been proposed for scheduling of joint activity participation. Spatiotemporal analysis of alternative space-time paths between origin–destination pairs helps to identify the time-dependent critical links of the joint participants.

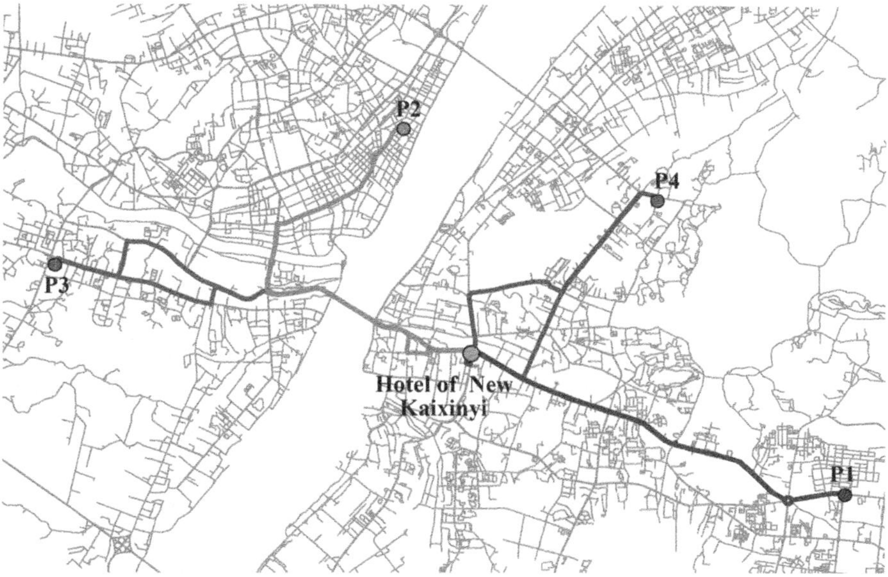

Fig. 7.7 Joint activity participation on Saturday (meeting: P1, P2, P3, and P4)

The critical index concept of a link has been proposed to evaluate the use priority of links in real taxi services. This definition helps to choose a feasible and comfortable space-time path for scheduling joint activity participation. In addition, the concept of the utility index of an opportunity has been presented to evaluate the use of critical links while choosing an available opportunity in the prism. A high index value means that the opportunity has a higher probability of using the popular paths of taxi drivers. These two concepts can facilitate activity scheduling services using a feasible mechanism of incorporating taxi drivers' choices of travel paths.

With the help of the extended TVNBP, the critical links and opportunities were identified in two steps. The first step is to find all available activity opportunities between origins and destinations within the extended TVNBP. The second step is to filter these available activity opportunities by analyzing a number of alternative space-time paths to find the shortest space-time path for the opportunity connecting an origin and a destination with optimal values of the link critical index and the utility index of opportunities.

The joint activity participation scheduling problem was modeled as a multi-objective optimization problem. Five objectives were used in this problem: (i) minimizing travel distance; (ii) minimizing travel time; (iii) maximizing expected participation activity time; (iv) minimizing the added travel time over alternative space-time paths; and (v) maximizing the utility index of the selected critical activity opportunity. The problem was solved by modifying the NSGA-II algorithm to limit the search space to focus only on critical links and available activity opportunities in the extended TVNBP. In this way, the joint activity participation scheduling problem can be solved for a dynamic transportation network using knowledge of the most-used links in taxi trajectories in this network. A scenario of joint participation among four people was designed and implemented to demonstrate the feasibility of this approach.

A large number of geography researchers are very interesting in computational analysis based on time geography theory, for example, Harvey J. Miller at the University of Utah, Mei-Po Kwan at the University of Illinois, Urbana-Champaign, Shih-Lung Shaw at the University of Tennessee, Harry Timmermans at the Eindhoven University of Technology, and Otto Huisman at the University of Twente. These researchers have carried out much important research to accelerate the development of time geography theory. In addition to theoretical research studies on time geography, further research should also focus on potentially popular applications such as traveler assistance services which could expand the research domain of time geography. This study has tried to promote the application of research based on time geography theory. The activity scheduling problem is one such promising application which makes it easy for people to acknowledge the importance of time geography and to improve the efficiency of their activities. This deserving research avenue could make much contribution to improve quality of life for human beings.

Acknowledgements This research was supported in part by the National Science Foundation of China (grants #40971233, #41231171, #60872132, #91120002), the project from State Key Laboratory of Resources and Environmental Information Systems, CAS of China (#2010KF0001SA), LIESMARS Special Research Funding, and the Funding for Excellent Talents in Wuhan University.

References

Angulo, J. M., Yu, H.-L., Langousis, A., Madrid, A. E., & Christakos, G. (2012). Modeling of space–time infectious disease spread under conditions of uncertainty. *International Journal of Geographical Information Science, 26*(10), 1751–1772.

Arentze, T. A., & Timmermans, H. J. P. (2003). A learning-based transportation oriented simulation system. *Transportation Research Part A: Policy and Practice, 38*(7), 613–633.

Auld, J., & Mohammadian, A. (2012). Activity planning processes in the Agent-based Dynamic Activity Planning and Travel Scheduling (ADAPTS) model. *Transportation Research Part A: Policy and Practice, 46*(8), 1386–1403.

Ben-Elia, E., & Shiftan, Y. (2010). Which road do I take? A learning-based model of route-choice behavior with real-time information. *Transportation Research Part A: Policy and Practice, 44*(4), 249–264.

Chen, X., & Kwan, M.-P. (2012). Choice set formation with multiple flexible activities under space–time constraints. *International Journal of Geographical Information Science, 26*(5), 941–961.

Deb, K., Pratap, A., Agarwal, S., & Meyarivan, T. (2002). A fast and elitist multi-objective genetic algorithm: NSGA-II. *IEEE Transactions on Evolutionary Computation, 6*(2), 182–197.

Ellegård, K., & Svedin, U. (2012). Torsten Hägerstrand's time-geography as the cradle of the activity approach in transport geography. *Journal of Transport Geography, 23*, 17–25.

Fang, Z. X., Tu, W., Li, Q. Q., & Li, Q. P. (2011). A multi-objective approach to scheduling joint participation with variable space and time preferences and opportunities. *Journal of Transport Geography, 19*(4), 623–634.

Fang, Z. X., Shaw, S.-L., Tu, W., Li, Q. Q., & Li, Y. G. (2012). Spatiotemporal analysis of critical transportation links based on time geographic concepts: A case study of critical bridges in Wuhan, China. *Journal of Transport Geography, 23*, 44–59.

Gärling, T., Brännäs, K., Garvill, J., Golledge, R. G., Gopal, S., & Holm, E. (1989). Household activity scheduling. In World Conference on Transport Research (Ed.), *Transport policy management and technology towards 2001* (Vol. IV, pp. 235–248). Ventura: Western Periodicals.

Hägerstrand, T. (1970). What about people in regional science? *Papers of the Regional Science Association, 24*(1), 7–24.

Hájek, J., Szöllös, A., & Šístek, J. (2010). A new mechanism for maintaining diversity of Pareto archive in multi-objective optimization. *Advances in Engineering Software, 41*(7–8), 1031–1057.

Islam, M. S. (2010). Measuring people's space–time accessibility to urban opportunities – An activity-based spatial search algorithm in a GIS. *International Journal of Urban Sustainable Development, 2*(1–2), 107–120.

Joh, C. H. (2004). *Measuring and predicting adaptation in multidimensional activity travel patterns*. Den Haag: CIP-Data Koninklijke Bibliotheek.

Joh, C. H., Arentze, T. A., & Timmermans, H. J. P. (2001). Understanding activity scheduling and rescheduling behaviour: Theory and numerical illustration. *GeoJournal, 53*(4), 359–371.

Kuijpers, B., & Othman, W. (2009). Modeling uncertainty of moving objects on road networks via space–time prisms. *International Journal of Geographical Information Science, 23*(9), 1095–1117.

Kuijpers, B., Miller, H. J., Neutens, T., & Othman, W. (2010). Anchor uncertainty and space-time prisms on road networks. *International Journal of Geographical Information Science, 24*(8), 1223–1248.

Kwan, M. P., & Hong, X. D. (1998). Network-based constraints-oriented choice set formation using GIS. *Geographical Systems, 5*(1), 139–162.

Lenntorp, B. (1976). *Path in space-time environments: A time-geographic study of the movement possibilities of individuals* (Lund studies in geography B). Lund: CWK Gleerup.

McQuoid, J., & Dijst, M. (2012). Bringing emotions to time geography: The case of mobilities of poverty. *Journal of Transport Geography, 23*, 26–34.

Miller, H. J. (1991). Modelling accessibility using space-time prism concepts within geographical information systems. *International Journal of Geographical Information Systems, 5*(3), 287–301.

Miller, E. J., & Roorda, M. J. (2003). Prototype model of household activity/travel scheduling. *Transportation Research Record: Journal of the Transportation Research Board, 1831*, 114–121.

Murugan, P., Kannan, S., & Baskar, S. (2009). NSGA-II algorithm for multi-objective generation expansion planning problem. *Electric Power Systems Research, 79*(4), 622–628.

Newell, K. M., Carlton, L. G., & Kim, S. (1994). Time and space-time movement accuracy. *Human Performance, 7*(1), 1–21.

Nijland, L., Arentze, T., & Timmermans, H. (2012). Incorporating planned activities and events in a dynamic multi-day activity agenda generator. *Transportation, 39*(4), 791–806.

Olaru, D., & Smith, B. (2002). Modelling daily activity schedules with fuzzy logic. In: *The 24th conference of Australian Institutes of Transport Research (CAITR2002), Sydney, Australia* (pp. 1–9).

Ronald, N., Arentze, T., & Timmermans, H. (2012). Modeling social interactions between individuals for joint activity scheduling. *Transportation Research Part B: Methodological, 46*(2), 276–290.

Roorda, M. J., Carrasco, J. A., & Miller, E. J. (2009). An integrated model of vehicle transactions, activity scheduling and mode choice. *Transportation Research Part B, 43*(2), 217–229.

Scott, D. M., & He, S. Y. (2012). Modeling constrained destination choice for shopping: A GIS-based, time-geographic approach. *Journal of Transport Geography, 23*, 60–71.

Shaw, S.-L. (2012). Guest editorial introduction: Time geography – Its past, present and future. *Journal of Transport Geography, 23*, 1–4.

Sips, M., Schneidewind, J., & Keim, D. A. (2007). Highlighting space–time patterns: Effective visual encodings for interactive decision-making. *International Journal of Geographical Information Science, 21*(8), 879–893.

Sui, D. (2012). Looking through Hägerstrand's dual vistas: Towards a unifying framework for time geography. *Journal of Transport Geography, 23*, 5–16.

Wentz, E. A., Peuquet, D. J., & Anderson, S. (2010). An ensemble approach to space–time interpolation. *International Journal of Geographical Information Science, 24*(9), 1309–1325.

Wu, Y. H., & Miller, H. J. (2001). Computational tools for measuring space–time accessibility within transportation networks with dynamic flow. *Journal of Transportation and Statistics, 4*(2/3), 1–14.

Xu, L., Yang, X., AI, S., & JIN, B. (2008). Study the service level of the city roads in Beijing. *Traffic and Transportation, 5*(1), 43–45 (in Chinese).

Yoon, S. Y., Deutsch, K., Chen, Y., & Goulias, K. G. (2012). Feasibility of using time–space prism to represent available opportunities and choice sets for destination choice models in the context of dynamic urban environments. *Transportation, 39*(4), 807–823.

Yu, H., & Shaw, S.-L. (2008). Exploring potential human activities in physical and virtual spaces: A spatio-temporal GIS approach. *International Journal of Geographical Information Science, 22*(4), 409–430.

Zhou, J., & Golledge, R. (2007). Real-time tracking of activity scheduling/schedule execution within a unified data collection framework. *Transportation Research Part A: Policy and Practice, 41*(5), 444–463.

Chapter 8
Robustness Analysis of City Road Network at Different Granularities

Yingying Duan and Feng Lu

8.1 Introduction

Urban road system undertakes the major transport tasks of human being and goods which makes it the fundamental infrastructure of a city. This means most of the socioeconomic activities in cities should use road networks, thus the performance of road networks is essential to the operation of the entire city (Xie and Levinson 2009). Given the importance of the urban road networks, understanding its robustness is becoming an important issue of urban planners and transportation managers at all levels. Robustness analysis of road networks can help identify the most important streets or locations that will greatly affect the efficiency of the entire road network and require more attention or planned protection strategies (Yin et al. 2009). Furthermore, a better understanding of road network structural robustness can help improve the robustness of existing road networks by considering network robustness in road planning and traffic guidance (Zhang and Levinson 2008; Santos et al. 2010).

Unfortunately, the robustness of urban road networks is very difficult to evaluate directly. Firstly, the robustness of urban road networks has multiple types of performances including the connections and transport efficiency among various locations in cities. Secondly, the traditional urban road robustness analysis requires accurate socioeconomic data such as traffic demands, land use, population in different locations (Jenelius et al. 2006). These socioeconomic data are inconvenient to get for researchers in developing counties, and the quality of the data is not always assured.

Y. Duan • F. Lu (✉)
State Key Laboratory of Resources and Environmental Information System, Institute of Geographical Sciences and Natural Resources Research, Chinese Academy of Sciences, 11A Datun Road, Chaoyang District, Beijing 100101, People's Republic of China
e-mail: duanyy@lreis.ac.cn; luf@lreis.ac.cn

M.-P. Kwan et al. (eds.), *Space-Time Integration in Geography and GIScience*, DOI 10.1007/978-94-017-9205-9_8

Previous researches have proved that the efficiency and robustness of urban road networks highly depend on the network structure (Demšar et al. 2008). To provide a practical robustness analysis procedure for urban road networks, we aim at evaluating the robustness of urban road networks by analyzing urban road network structure.

Originating from studies of complex networks, the concept of network structural robustness denotes the decrease of network performance due to the selected removal of nodes or edges (Holme et al. 2002). Many robustness analyses are performed in the real world artificial networks using complex network methods such as North American power grid (Albert et al. 2004), water distribution networks (Yazdani and Jeffrey 2011). Although the urban road network is already a network intuitionally, it is different from the other networks because the granularity of the road networks varies according to the demands of different applications. In traditional GIS applications and navigation systems, the urban road network contains road segment and junctions. From the point of view of urban planning, the urban network is consisted of roads and their intersections where several road segments are grouped as one element. Structural diversity occurs at different granularities of road network. Whether this diversity will cause bias judgment for robustness analysis remains an important research question.

Thus the fundamental issue about using complex network theory in robustness analysis in urban road networks is exploring the structural differences at multiple granularities. Our goal is inquiring the robustness of urban road networks at three different granularities based on complex network theory to figure out whether and how different granularities affect the robustness evaluation through an empirical study in Beijing. Three successive hypothetical attack strategies are carried out to observe the performances of road network under pressure. In each simulated attack procedure, nodes are successively removed from the network and the performances of the network as well as the spatial form of urban road network changes resulted from the removal of nodes are recorded. These series of spatial forms provide a clear conception of a spatial change procedure of urban road networks under attacks. The results show that the performances under attacks vary in different granularity. Segment-based road network lacks the sensitivity under intentional attacks, and stroke-based road network cannot represent the split procedure of the urban road network under attacks. Neither of them is appropriate to evaluate the robustness of city road networks. Community-based road network can capture the traffic interactions between adjacent road segments and use local impact area to evaluate network robustness. We argue that community-based can reflect the network change situation under attack at a mesoscopic point of view and is suitable for robustness evaluation of city road networks.

This chapter is organized as follows: In the second section, we will introduce three models of road network representation in different granularities. In Sect. 8.3, the experimental procedures to evaluate the robustness of urban road network at multiple granularities are described. The fourth section is devoted to the main results and we summarized our conclusions and the implications for the further research in the last section.

8.2 Representation of Urban Road Network in Multiple Granularities

To provide an explicit representation of the relationships between roads, we use the graph model to represent the urban road network in this paper. A graph is defined as $G = (N, E)$ in graph theory, where N and E are two disjoint finite sets. N is the set of all nodes in G, and E is the set of all edges connecting two elements in N. Based on this conception, we use three types of graph to represent urban road networks in different granularity.

8.2.1 *Segment Based Dual Graph*

In traditional GIS network analysis and navigation system, the road networks are always organized as primal graph: the nodes represent the junctions while the road segment between two junctions is tread as the edge (Porta et al. 2006a). This data structure is from the geographic view and easy to understand. For example, the way finding algorithms in navigation systems are based on the segment-based primal graph. So guidance to users are always like "Go 0.4 miles and turn right to ** Avenue" which is clear and easy to follow for users.

However, further relationship between road segments such as "Does certain road segments have more neighbors than others?" or "Do neighbors of one road segments tend to connect with each other?" are difficult to provide directly from the primal graph. To overcome this shortage, dual graph, where road segments are nodes and the junction between road segments map to edges between nodes is constructed to represent the relationship between road segments (Porta et al. 2006b). In Fig. 8.1 we present a simple example of a piece of urban road network mapped to such dual graph.

Figure 8.1a shows a sketch map of initial urban road network where nodes are intersections and edges are road segments between intersections. It can provide both location and geometric shape for road segments. In Fig. 8.1b, road segments are shown as nodes and the direct connections between segments are shown as edges. It can provide more information about the connections than Fig. 8.1a and these information is valuable for robustness analysis. In the following part of this article, "road-segment based road network" refers to road-segment based dual graph.

Road segments in Fig. 8.1 show variation in the network structure. This variation can be used in robustness analysis. For example, if segment 1 is attacked, the rest of the network can still be functional. But if segment 5 is attacked, vehicles from segment 1–4 to segment 7–10 can only take segment 6, which can put great pressure on segment 6. If Segment 6 is malfunctioned too, the network is split into two components and becomes malfunction. This suggests that attack on different segment can cause different damage.

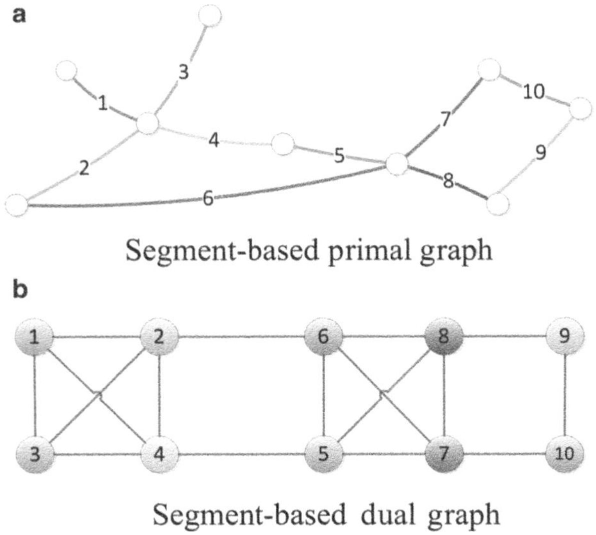

Fig. 8.1 Segment-based road network

8.2.2 Stroke Based Dual Graph

A stroke is defined as a linear geographic entity that stretches in two dimensional spaces. Named roads are more difficult to implement due to missing or wrong road segment names. Previous research proved that stroke can overcome this shortage and provide a natural way to represent self-organized natural road (Jiang 2007). Different from segment based network, the stroke is not fractured by junctions. In this article, we generate the stroke from road segments using an enhanced stroke generating algorithm originated from every-best fit algorithm proposed by (Jiang et al. 2008). In our algorithm, not only the angle of the two segments is considered, the grade of the segments is used as well. That means even the angle between two segments is acceptable, segments with different grade cannot be connected. After generating the strokes, we use the stroke based dual graph as the second representation of urban stroke networks where the strokes map to nodes and the intersection between strokes map to edges between nodes. A caricature urban road network shown in Fig. 8.2 presents an example of generating procedure of a stroke based dual graph.

In Fig. 8.2a, road segments are generated into strokes based on their angle and grade. The dual graph of stroke-based urban road network is shown in Fig. 8.2b where nodes are strokes and edges are the corresponded connections. Stroke-based graph in Fig. 8.2 show a different structure with segment-based graph in Fig. 8.1. For example, none of these strokes can cause the network split into pieces. But if we delete stroke A or F, the route choices for vehicles are reduced. The damage can also affect urban road network robustness. This indicates that robustness analysis using stroke-based graph can be different from that using segment-based graph.

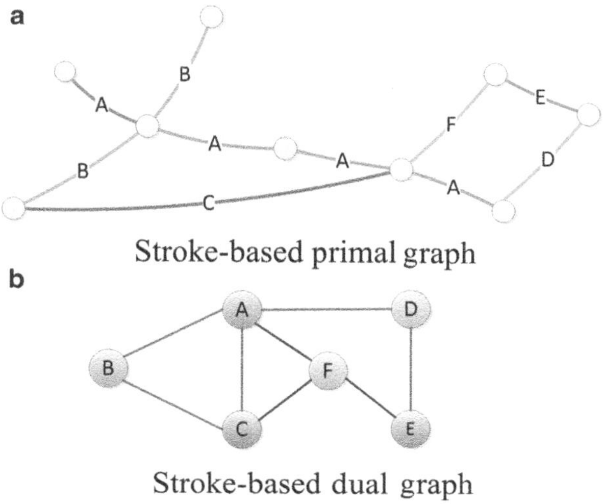

Fig. 8.2 Stroke-based urban road network

8.2.3 Community Based Dual Graph

Recently, some researchers found that the group of highly related nodes can provide more structural information beyond single nodes (Fortunato 2010). Although the stroke-based road network combines the topological related road segments into a group, the combining rules make the group less correlated for the end nodes if the street contains a long series of segments. Here we proposed a community based dual graph to represent the groups of highly connected road segments.

Communities, also called clusters or modules, in complex network theory refers to a densely connected subset of nodes that is only sparsely linked to the remaining network (Gulbahce and Lehmann 2008; Fortunato 2010). The procedure to group these highly related nodes in the network is called community detection. By network clustering, nodes with many connections are grouped from original networks. In this way, one attains a coarse-grained description of the original network from a modular view (Fortunato 2010). Communities in social networks such as families, friendship circles, and scientific collaborations have been studied for a long time (Moody and White 2003).

However, little attention has been paid in community structure in urban road networks yet. Actually, city road network also contains modules like traffic zones or business districts. Road segments in these areas are densely connected with each other. Communities provide a novel way to gather topologically densely connected road segments. In this paper, we use a label propagation based community detection algorithm proposed by Raghavan et al. (2007) to identify the communities in city road networks and get more insights into the structural properties of road networks. The generating procedure of community-based dual graph is shown in Fig. 8.3.

Segments clustered in the same community are rendered with the same color in Fig. 8.3a. Figure 8.3b is the corresponding dual graph. Communities in Fig. 8.3

Fig. 8.3 Community-based urban road network

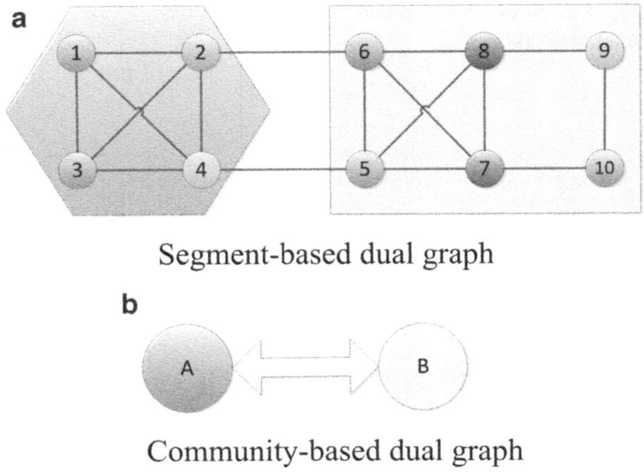

imply that community structure in urban road networks can provide novel perspective for network robustness analysis. Different from segment-based and stroke-based urban road networks, if a blue segment fails, we consider all blue segments in community C_1 is affected. Thus we can see that destroy of green community can be more harmful than blue community to the entire network since the malfunction of green community will isolate the blue and the red community. Furthermore, the community perspective can help improving road network robustness. As we can see in Fig. 8.3, if another community is built between blue and red community and provides more connections, the damage of green community can be less harmful to the entire network. From Fig. 8.3, we can see that community can provide a novel and important perspective on urban road network's structure and help enhance its robustness analysis.

To evaluate the feasibility of city road network robustness analysis at multiple granularities, we provide a series experiments with real world urban road network.

8.3 Data and Methodology

8.3.1 Experimental Data

In this article, we use the Beijing road network to test our method. This network consists of 16,491 road segments which cover the downtown of Beijing. Figure 8.4 shows the initial road segments.

Based on the initial road segments, we generate three types of Beijing road network including segment-based, stroke-based and community-based dual graph using the methods described in Sect. 8.2. To display the differences in three dual graphs, we use open source software GUESS (Adar and Kim 2007) to visualize the dual graphs in topological spaces in Fig. 8.5.

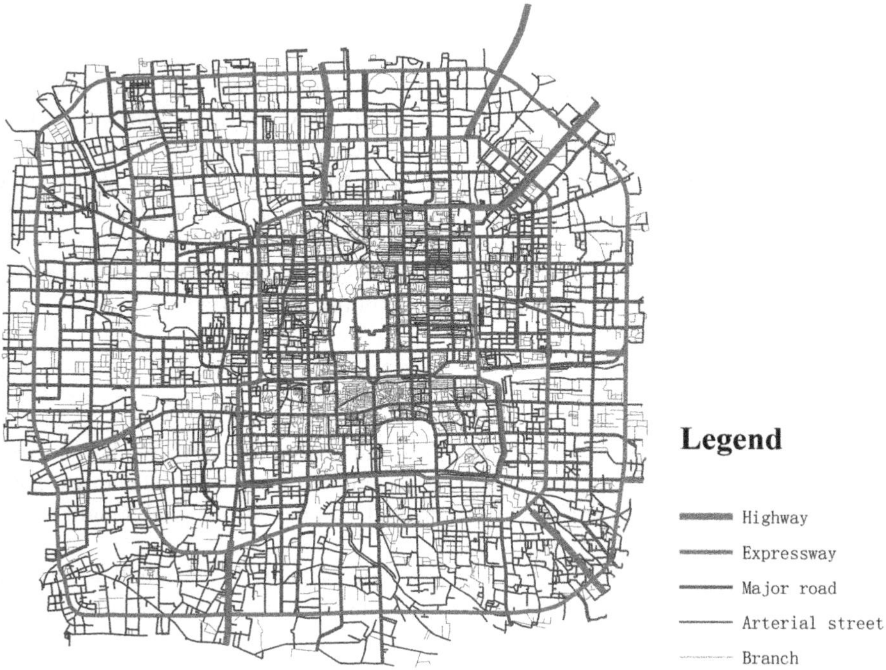

Fig. 8.4 Beijing road segments

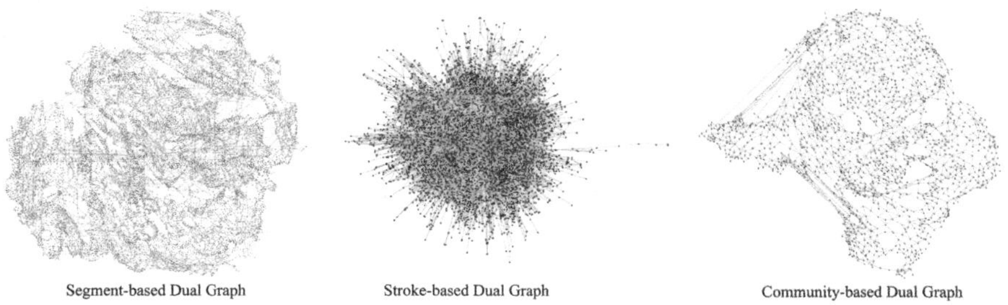

Fig. 8.5 Three dual graphs of Beijing road network

In Fig. 8.5, the position of each segment/stroke/community is not based on its location in the real world. In this figure, we use GEM (Generalization Expectation-Maximization) (Dempster et al. 1977) algorithm provided by GUESS to layout the nodes and edges. The layout principle of this algorithm is to minimize the overlapping of edges. The significant variations of the three graphs imply that dual graph can provide more insight on the urban road structure. To compare three dual graphs quantitatively, we characterize the network structure with several metrics of complex network theory.

8.3.2 Beijing Road Network Characterization

The analysis, discrimination, and synthesis of complex networks therefore rely on the use of measurements capable of expressing the most relevant topological features (Newman 2010). The efficiency of a network is the basic function of a transportation network. It is usually described by the average length of the shortest paths L which is regarded as a sign of global efficiency and clustering coefficient C which is considered as the local efficiency (Watts and Strogatz 1998). Here L measures the typical separation between two nodes in a graph while C measures the cliquishness of a typical neighborhood. The definition is as follows:

$$L = \frac{1}{n(n-1)} \sum_{v \in V} \sum_{w \neq v \in V} d(v, w) \tag{8.1}$$

Where n is the number of nodes and d(v, w) denotes the length of the shortest path between node v and w which means the number of edges in the shortest path connecting the two. If the graph is not connected, L is the average path length of the largest connected component of the graph.

L measures the global transport efficiency in the network. A small L implies a short path length between nodes and the dynamic flow of the network can be transmitted faster. For instance, the social networks are known to have very short average path length, with the "Six degree of separation", where $L = 6$. Comparing with the population of the world, the length is quite small, thus a network with a small L is called small world network (Watts and Strogatz 1998).

In contrast, cluster coefficient which measures the local transport efficiency is defined as:

$$C = \frac{1}{n} \sum_{v \in V} C_v \tag{8.2}$$

Where n is the number of nodes and C_v denotes the fraction of the actually existed edges between the neighbors of C_v compares to the all allowable edges. A higher C means a more clustered neighborhood and the flow can be transmitted more efficiently in the local area.

Table 8.1 shows these two basic characteristics of the Beijing road network represented in three dual graphs.

Where N is the node number in the network and E is the edge number. L and C are calculated using the formulas introduced before. Table 8.1 confirms that the different representation can provide different structural information on the same initial road

Table 8.1 Characterization of Beijing road network

	N	*E*	*L*	*C*
Segment-based	16,491	35,575	45.14	0.43
Stroke-based	377	833	4.86	0.18
Community-based	1,594	4,124	13.65	0.46

network. Specifically, stroke-based network has a much higher L and lower C than the other two networks. But the drop of L is less than the size difference. This implies there are more hidden structural variations between the three types of networks rather than just size differences. This suggests that all three networks should be considered to provide a comprehensive robustness analysis of city road networks.

8.3.3 Experimental Design

The robustness of the network relies on how the network reacts to attacks. To evaluate the road networks quantitatively, we designed a series of experiments to simulate the reaction of real road networks to attacks.

Urban road malfunctions can be caused by many reasons such as natural disasters, traffic accidents, congestion or regulations on certain roads. The non-recurrent failures, caused by natural disasters such as floods, hurricanes and traffic accidents, are more likely to occur randomly on roads. Other incidents such as terrorist attacks or big events may specifically happen on important roads. Thus, we chose two types of strategies to test our networks: random attack and intentional attack. The random attack strategy is easy to perform on randomly chosen nodes; however, the intentional attacks on important nodes are difficult to determine because the structural importance of a node cannot be simply evaluated by single measurements. The most widely used node structural importance measurements based on graph theory are degree (Diestel 2005) and betweenness centrality (Barthelemy 2004).

The degree of node v, denoted as d_v in this paper, is the number of edges linked to node v. It is one of the most fundamentally important measurements in graph theory. With a higher degree, the node connects to more nodes and contributes to increased accessibility in the local area.

Betweenness centrality is a structural importance measurement based on the global information of the shortest paths connecting all pairs of nodes. It refers to the proportion of the number of shortest paths between nodes that contain a particular node to the total number of shortest paths in the network. The definition is as follows:

$$\mathrm{Bc}\,(\mathrm{v}) = \sum_{\mathrm{u} \neq \mathrm{w} \in \mathrm{V}} \frac{\sigma_{\mathrm{uw}}\,(\mathrm{v})}{\sigma_{\mathrm{uw}}} \tag{8.3}$$

Where σ_{uw} represents the number of shortest paths between nodes u and w and $\sigma_{\mathrm{uw}}(\mathrm{v})$ represents the number of shortest paths between nodes u and w that include node v. A high betweenness of a node represents more of the shortest paths via this node, which suggests a higher selective probability of this node when effectively transmitting material in the network.

Both degree and betweenness, to some extent, show the importance of a node but measure different importances. Thus we should consider both measurements to evaluate the importance of a node. Based on these analyses, we choose one random attack strategy and two intentional attack strategies: a degree-based attack strategy and a betweenness-based attack strategy.

After determining the attack strategies, the performances of network under attack should be measured. Since the performances under attack are difficult to judge with single measurement, we argue that the concept of robustness is based on two types of performance of networks to attacks: efficiency behavior and fragmental fraction. To provide a comprehensive perspective on the network efficiency behavior, we evaluate the efficiency from both global and local views. The global efficiency of urban road networks can be reflected from average path length L introduced in formula (8.1) while clustering coefficient introduced in formula (8.2) is used to evaluate the local efficiency. Network fragmentation is also an important property to show the performance of city road network under attacks. We use the relative size of the largest connected component S which is shown as a fraction of the total network size to investigate the impact of attacks on the network structure.

Single attack targeting on certain node can only provide the network performance when the observed node is removed. However, the robustness of the entire network does not depend on only one node. To quantitatively measure the overall road network robustness under pressure, we used a successive attack strategy to destroy the network until the capacity reaches its maximum and record the network performance during each attack.

For the random attack strategy, we randomly choose one node at each step and delete the node and all edges linked to it from the network. The loop is performed until the whole network collapses when $S \approx 0$ where the network is isolated into single nodes only. The performance of a network under attack, measured as efficiency and fragmentation, is quantified at each loop to observe the reaction of the revised network.

For the degree-based attack, we remove the node that has the highest degree at each step, calculate the parameters L, C, S and recalculate the degree of each node. The removal procedure is repeated until the network collapses. The betweenness-based attack is similar, except we remove the node that has the highest betweenness and recalculate the betweenness for each node at every loop.

8.4 Experiment Results

8.4.1 Network Performances Under Successive Attacks

In this section, we study the performances of the Beijing road network under successive random attacks in three dual graphs. Figure 8.6 shows the changes of L, C and S during the attacks.

In Fig. 8.6, the x axis shows the fraction of the removal nodes while the y axis shows the change of L, C or S. Figure 8.6a–c show the changes of the Beijing road

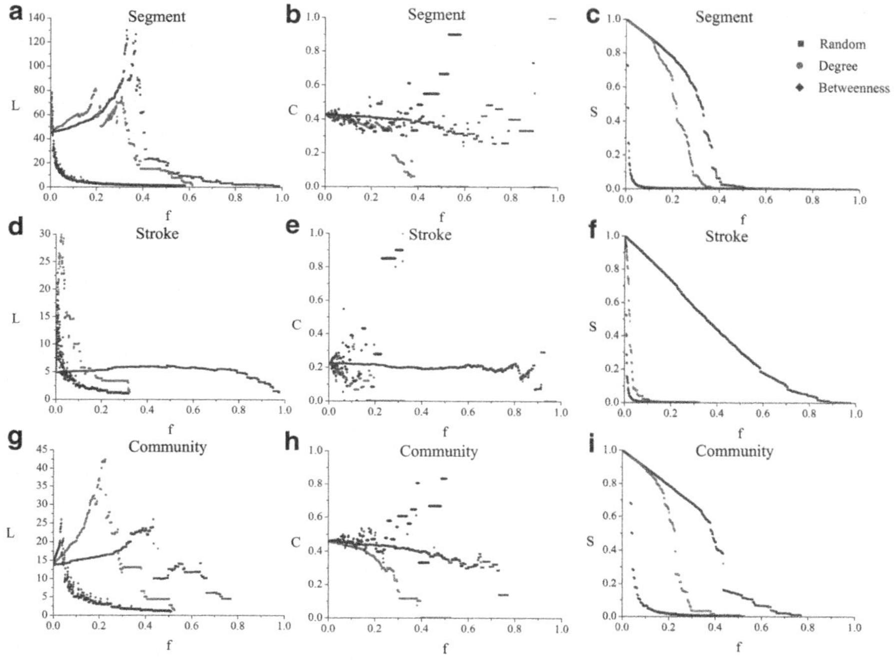

Fig. 8.6 Network changes under attacks

network represented as a segment-based network. Figure 8.6d–f show changes of the network represented as stroke-based network. And Fig. 8.6g–i show changes of the network represented as a community-based network. In each sub graph, three types of points stand for the changes under different attack strategies. Red cubes show the changes of network under random attacks while the green circles show the changes of network under degree-based attacks and blue diamonds show the changes of network under betweenness-based attacks.

Although represented as three dual graphs with multiple granularities, the reaction of the network under attacks are similar. The changes of network performances under successive attacks can all be divided into three stages. In the first stage, the removal of target single nodes and some small isolated components caused by node removal slowly decreased the size of the largest component. At this stage, *L* increases, whereas *C* slightly decreases. This finding indicates that at the beginning of the attack procedure, network efficiency, including both global and local efficiency, slowly decreases. In the second stage, major changes in network structure occur. The abrupt decrease in *S* shows that the network split into several components with similar sizes. At this stage, *L* begins to drop sharply. However, the decrease of *L* does not indicate the increase of global efficiency because the size of the giant component dramatically decreases. Thus after the first stage, the information provided by L is useless to the entire network. Different from L and S, the change of C shows significant differences between different attack strategies. The changes of C under random attacks display a minor decrease while changes of C under degree based attacks display a remarkable decrease and changes of C under betweenness based attacks display a noticeable increase. This proves that different attack strategies targets to road with different importance and cause various damages to the network. The third stage contains the final changes of the network. Although the network is not totally destroyed, it has malfunctioned and become valueless to this research.

The inflections to divide the stages occur at different removal rate of nodes. The shorter the first stage is, the faster the network becomes malfunction. Compare with the three attack strategies, we can find out that all dual graphs are robustness to random attacks but vulnerable to intentional attacks especially betweenness based attacks. According to the red points in Fig. 8.6, the network only malfunctions once more than 50 % of the nodes are removed. However, degree-based attacks are more dangerous, especially to local efficiency. According to the green points, although removal of nearly 20 % of the nodes will cause fragmentation, local traffic in the attacked area will soon become paralyzed. As blue points show, attacks on nodes with high betweenness cause the network to split after only several nodes are removed. This is because the attack strategy targets on nodes with a high betweenness, such as bridges or cloverleaf junctions in the city. These nodes hold the whole network together, and when they are removed the city will separate into several parts. Local efficiency under betweenness attack increases with the removal of high betweenness nodes.

Comparing the results of the three types of dual graphs under attacks, we can see some structure differences and how the differences affect the network vulnerability under attack. In Fig. 8.6a–c, the green points referring to network performances under degree based attacks are similar with red points referring to network performances under random attacks. This means the network represented by segment-based dual graph lacks of sensitivity under degree based attack. Stroke-based network performances under attacks are shown in Fig. 8.6d–f. Unlike other two dual graphs, the first stage in successive intentional attacks is so short that is difficult to observe. This finding indicates that the network represented by stroke-based dual graph splits at the very beginning of the attack procedure. Thus robustness of stroke-based dual graph is tolerant under random attacks yet fragile under target attacks including degree based attacks as well as betweenness based attack. Community based dual graph under attacks in Fig. 8.6g–i shows a clear three stage procedure and sensitive to both intentional attacks. It implies that community-based network can capture the structural diversity from the modular perspective of the road network. The structural diversity makes the community-based city road network sensitive to intentional attacks than the other two models of road network and provides a more reliable robustness evaluation of network structure.

8.4.2 Network Spatial Form Changes Under Attacks

Since the second stage provides the most valuable information on how the network splits into pieces, we provide a series of figures on spatial form changes of the initial network when split. These series of spatial forms provide a clear conception of spatial changes under attacks.

To cause a similar level of damage, the removal rate of nodes varies in different network models. Figure 8.7 shows the spatial forms of Beijing road network models

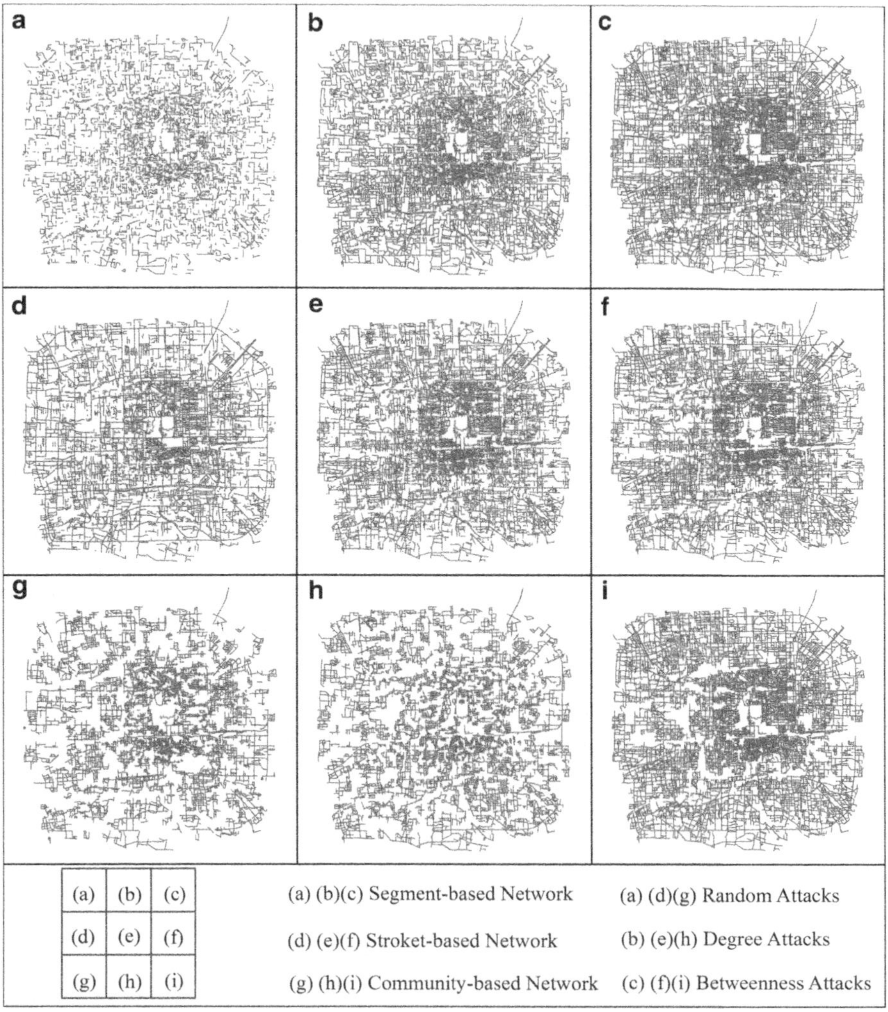

Fig. 8.7 Spatial form of Beijing road network under attacks

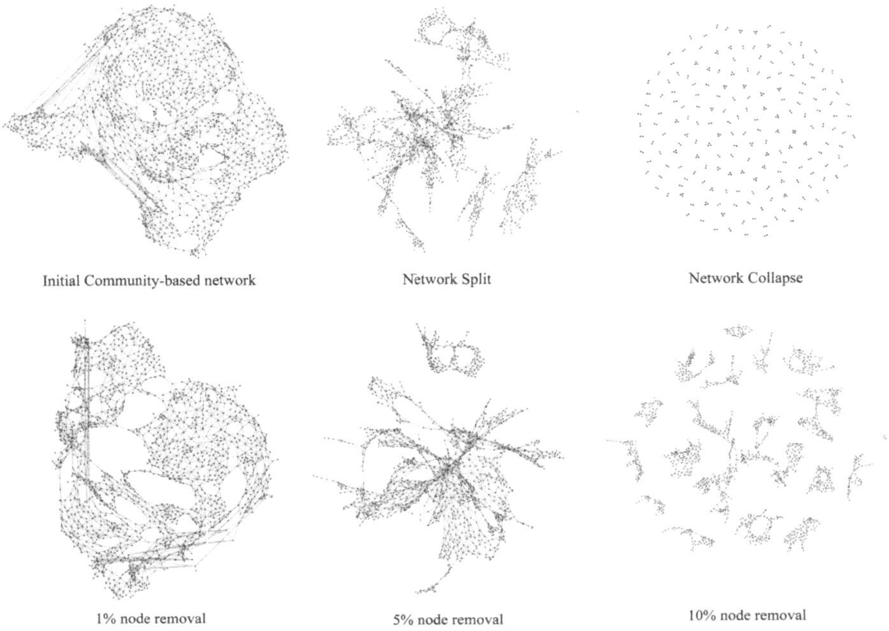

Fig. 8.8 Topological form of community-based road network under attacks

under attacks till split into disconnected components. The variation in road network sparseness between different columns indicate that road network can still operate even the network is sparse under random attacks. But under intentional attacks especially betweenness based attacks, even the network seems density, the split will cause the network malfunction.

To provide the topological changes of network under attacks, we visualize the topological form of community-based urban road at different level of the betweenness based attack process in Fig. 8.8.

The above sub figures in Fig. 8.8 presents the layouts of community-based dual graph under attacks according to the three stages of the network changes. We can see that in the second stage, the network split into three disconnect components. The network becomes malfunction by collapsing into small pieces in the third stage. The below sub figures in Fig. 8.8 show the layouts of community-based dual graph of different node removal rate. When 5 % nodes are removed from the initial network, the network haven't split although the topological form of the network changes. But when 10 % nodes are removed, the network is split into several pieces.

Figures 8.7 and 8.8 provide a clear conception of spatial change of urban road networks under attacks. From which we can figure out how the structure of network changes under different strategic of attacks.

8.5 Discussions and Conclusions

Evaluating the robustness of city road networks have been the focus of various disciplines for a long time. Most of existing researches focus on the role of single road segment or street and try to evaluate how the failure of single road may affect the urban road network. Our results prove that the robustness evaluation varies in different granularity network models due to the structural differences. This implies that choose the appropriate granularity for network is essential for the urban road network structure analysis.

In this article, we proposed a comprehensive robustness analysis procedure using three types of urban road network model in different granularities in this paper. Moreover, three successive attack strategies are carried out to evaluate urban road network performances under pressure. This procedure can provide the concept of structural robustness of the entire urban road network rather than single road perspective. In addition, a series figures are used to describe the structural forms of the urban road network under attacks. These figures can provide visuals on the structural changes in the urban road network.

The comparisons between different models show that the three dual graphs provide different network structural information and properties. According to the comparison, the homogeneous degree distribution makes it difficult to identify the important nodes in the segment based dual graph. The lack of sensitivity of segment based dual graph under degree based attack suggests this granularity of view cannot provide the network performance correctly under degree based attacks. The stroke based dual graphs have a diverse degree and betweenness distribution, and sensitive to the target attack. But stroke based dual graphs show no remarkable separation occurring under intentional attacks that cannot truly reflect the reaction of urban road networks. In addition, in the real world, a stroke especially the main roads are always very long and cross over several areas. Under normal circumstances, the incidents only cause several road segments and junctions blocked. Even the extreme situation rarely leads to the malfunction of the whole stroke. Thus the stroke-based network is not appropriate to evaluate network robustness.

The community based dual graph groups the topological correlated road segments into subsets, which takes the spread of traffic jam into consideration. The attack to a node implies the possible response when the real urban street network is attacked to a road segment or junction and affects the segments near it. The experiments show sensitivity under target attacks, suggest a way to identify the crucial community in urban road and imply a framework to compare the robustness between cities. So we recommend the community based dual graph for urban road networks to evaluate the network structure robustness.

Further work beyond this topic can be considered in different point of views: Firstly, although we recommend the community based dual graph, the community detection is an NP hard problem, how to choose the appropriate algorithms requires more work. Secondly, different application requires different network model, how to find the right model for a certain situation is still an open question. Thirdly, since we can identify the critical locations in the city through robustness analysis, can we guide the traveler to choose their way wisely to protect the critical locations?

Acknowledgement This research is supported by the Chinese NSFC Program under grant No. 41271408, the Chinese Hi-tech Research and Development Program under grant No. 2012AA12A211, and State Key Laboratory of Resources and Environmental Information System open foundation No. 088RA500KA.

References

Adar, E., & Kim, M. (2007). SoftGUESS: Visualization and exploration of code clones in context. *International Conference on Software Engineering: Proceedings of the 29th International Conference on Software Engineering, 20*(26), 762–766. doi: 10.1109/ICSE.2007.76.

Albert, R., Albert, I., & Nakarado, G. L. (2004). Structural vulnerability of the North American power grid. *Physical Review E, 69*(2), 025103. doi:10.1103/PhysRevE.69.025103.

Barthelemy, M. (2004). Betweenness centrality in large complex networks. *The European Physical Journal B-Condensed Matter and Complex Systems, 38*(2), 163–168. doi:10.1140/epjb/e2004-00111-4.

Dempster, A. P., Laird, N. M., & Rubin, D. B. (1977). Maximum likelihood from incomplete data via the EM algorithm. *Journal of the Royal Statistical Society: Series B: Methodological, 39*, 1–38.

Demšar, U., Špatenková, O., & Virrantaus, K. (2008). Identifying critical locations in a spatial network with graph theory. *Transactions in GIS, 12*(1), 61–82. doi:10.1111/j.1467-9671.2008.01086.x.

Diestel, R. (2005). *Graph theory*. Heidelberg: Springer.

Fortunato, S. (2010). Community detection in graphs. *Physics Reports, 486*(3–5), 75–174. doi:10.1016/j.physrep.2009.11.002.

Gulbahce, N., & Lehmann, S. (2008). The art of community detection. *Bioessays, 30*(10), 934–938. doi:10.1002/bies.20820.

Holme, P., Kim, B. J., Yoon, C. N., & Han, S. K. (2002). Attack vulnerability of complex networks. *Physical Review E, 65*(5), 056109. doi:10.1103/PhysRevE.65.056109.

Jenelius, E., Petersen, T., & Mattsson, L. G. (2006). Importance and exposure in road network vulnerability analysis. *Transportation Research Part A: Policy and Practice, 40*(7), 537–560. doi:10.1016/j.tra.2005.11.003.

Jiang, B. (2007). A topological pattern of urban street networks: Universality and peculiarity. *Physica A: Statistical Mechanics and Its Applications, 384*(2), 647–655.

Jiang, B., Zhao, S., & Yin, J. (2008). Self-organized natural roads for predicting traffic flow: A sensitivity study. *Journal of Statistical Mechanics: Theory and Experiment*, P07008. doi:10.1088/1742-5468/2008/07/P07008.

Moody, J., & White, D. R. (2003). Structural cohesion and embeddedness: A hierarchical concept of social groups. *American Sociological Review, 68*, 103–127. doi:10.2307/3088904.

Newman, M. E. J. (2010). *Networks: An introduction*. New York: Oxford University Press.

Porta, S., Crucitti, P., & Latora, V. (2006a). The network analysis of urban streets: A primal approach. *Environment and Planning B: Planning and Design, 33*(5), 705–725. doi:10.1016/j.physa.2005.12.063.

Porta, S., Crucitti, P., & Latora, V. (2006b). The network analysis of urban streets: A dual approach. *Physica A: Statistical Mechanics and Its Applications, 369*(2), 853–866. doi:10.1016/j.physa.2005.12.063.

Raghavan, U. N., Albert, R., & Kumara, S. (2007). Near linear time algorithm to detect community structures in large-scale networks. *Physical Review E, 76*(3), 036106. doi:10.1103/PhysRevE.76.036106.

Santos, B. F., Antunes, A. P., & Miller, E. J. (2010). Interurban road network planning model with accessibility and robustness objectives. *Transportation Planning and Technology, 33*(3), 297–313. doi:10.1080/03081061003732375.

Watts, D. J., & Strogatz, S. H. (1998). Collective dynamics of "small-world" networks. *Nature, 393*(6684), 440–442. doi:10.1038/30918.

Xie, F., & Levinson, D. (2009). Topological evolution of surface transportation networks. *Computers, Environment and Urban Systems, 33*(3), 211–223. doi:10.1016/j.compenvurbsys.2008.09.009.

Yazdani, A., & Jeffrey, P. (2011). Complex network analysis of water distribution systems. *Chaos, 21*(1), 016111. doi:10.1063/1.3540339.

Yin, Y., Madanat, S. M., & Lu, X. Y. (2009). Robust improvement schemes for road networks under demand uncertainty. *European Journal of Operational Research, 198*(2), 470–479. doi:10.1016/j.ejor.2008.09.008.

Zhang, L., & Levinson, D. (2008). Investing for reliability and security in transportation networks. *Journal of the Transportation Research Board, 2041*(1), 1–10.

Chapter 9
Spatial Analysis of Census Mail Response Rates: 1990–2010

Daniel A. Griffith and Yongwan Chun

9.1 Introduction

As early as 1990, the United States (US) Census Bureau recognized the importance of analyzing the geographic variation in census mail response rates (MRRs) at the county level across the country. Keller (1990) notes the importance of: understanding where MRRs decline; gaining insights in preparation for the next decennial census; and, the usefulness of developing a predictive equation that allows a more accurate forecast of the MRRs for the subsequent census.

Although the MRRs for a given census can serve only as a covariate for subsequent census (i.e., these rates are not a causative indicator, but rather a covariate), they are a very valuable piece of predictive information to include in an analysis. In a multiple linear regression analysis using county level data, Keller (1990) found that the MRR covariate ranked second only to the percentage of occupied housing units. His equation contains 36 additional variables (e.g., household size, percent post office boxes), with the total set of 38 variables accounting for roughly 69 % of the variation in the MRR across US counties.

The geospatial data for an analysis of MRRs consist of a county map shapefile for the coterminous US and county-level MRRs for the decennial census years 1990, 2000, and 2010.[1] Restricting attention to the coterminous US for analysis purposes here, 25 of the 3,067 counties are classified as not being parts of states, and

[1]A US county map shapefile is available at http://www.census.gov/geo/www/, and county level data for the three decennial census MRRs are available at http://2010.census.gov/2010census/take10map/ and http://www.census.gov/dmd/www/mailresp.html

D.A. Griffith (✉) • Y. Chun
School of Economic, Political and Policy Sciences, University of Texas at Dallas, Richardson, TX, USA
e-mail: dagriffith@utdallas.edu; ywchun@utdallas.edu

M.-P. Kwan et al. (eds.), *Space-Time Integration in Geography and GIScience*, DOI 10.1007/978-94-017-9205-9_9

Table 9.1 Summary of US county-level MRRs

Year	n	Mean	Standard deviation	Minimum	Maximum
1990	2,431	66.7	8.23	18	87
2000	3,015	70.0	10.28	13	90
2010	3,042	70.4	11.12	20	95

essentially lack MRR data. The 1990 data were complied for only those counties with substantial populations, resulting in and additional 611 missing values, and hence a total of 636 missing values. The 2000 data contain an additional 25 missing values, for a total of 50 missing values for that year. The 2010 data are complete for 3,042 counties. The data are reported as percentages, resulting in standardization to a denominator of 100 vis-a-vis the geographic variation in population across the US. Table 9.1 tabulates basic summary statistics for these percentages. Figure 9.1 portrays the three geographic distributions of MRRs.

9.2 A Conceptual Framework for the Data Analysis

Because MRRs are percentages, they constitute a binomial random variable. Their proper analysis involves generalized linear model or logistic regression techniques that account for both spatial and temporal dependencies, as well as space-time interactions. This paper extends eigenvector spatial filtering principles and conceptualizations to eigenvector space-time filtering of binomial random variables through the employment of a random effects term, a novel methodological development at the research frontiers of space-time geography and GIScience.

9.2.1 Eigenvector Spatial Filtering

Eigenvector spatial filtering (see Griffith and Chun 2012) uses a set of synthetic proxy variables, which are extracted as eigenvectors from an n-by-n spatial weight matrix **C** that ties geographic objects together in space, and then adds these vectors as control variables to a model specification in order to account for spatial autocorrelation by filtering it out of the model residuals. These control variables identify and isolate the stochastic spatial dependencies among the georeferenced observations, thus allowing model building to proceed as if the observations are independent.

One popular implementation of eigenvector spatial filtering utilizes an eigenfunction decomposition of the matrix version of the numerator of the Moran Coefficient index of spatial autocorrelation:

$$\left(\mathbf{I} - \mathbf{1}\mathbf{1}^{\mathrm{T}}/\mathrm{n}\right)\mathbf{C}\left(\mathbf{I} - \mathbf{1}\mathbf{1}^{\mathrm{T}}/\mathrm{n}\right), \tag{9.1}$$

a

b

c

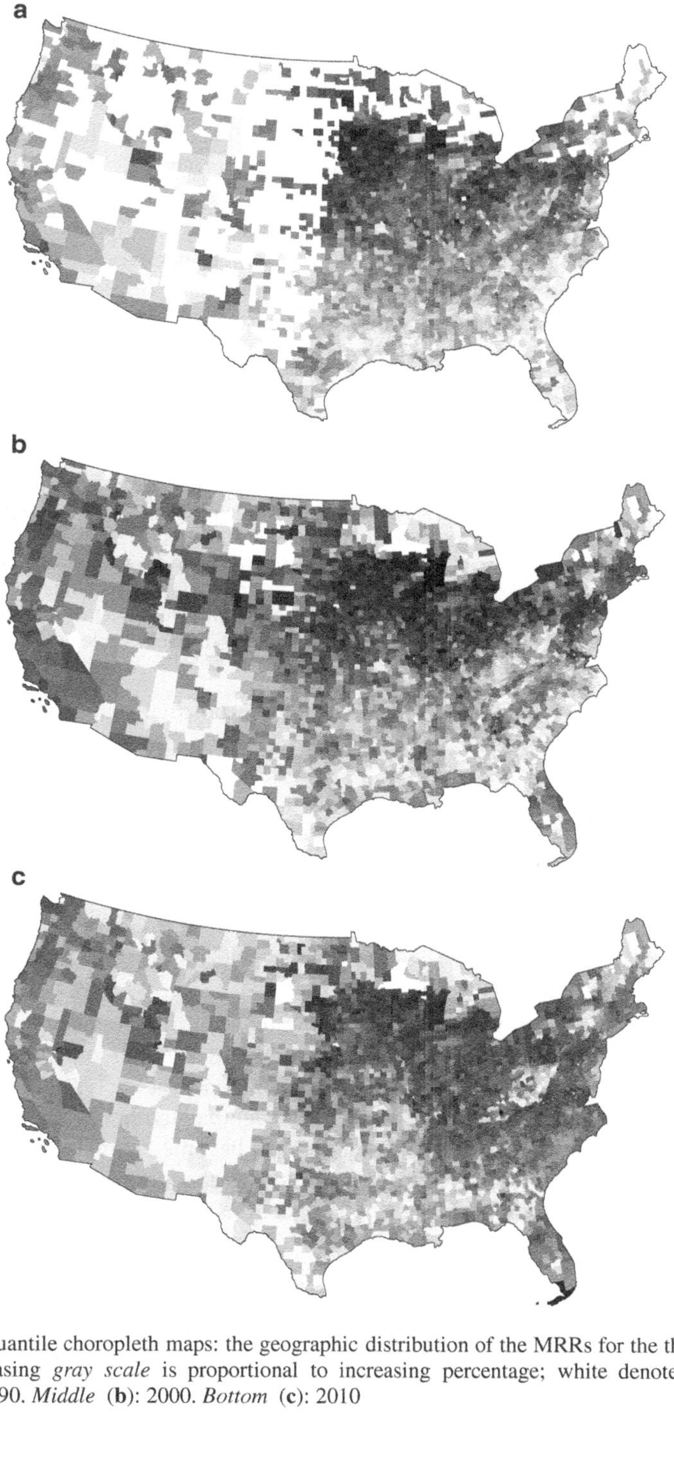

Fig. 9.1 Quantile choropleth maps: the geographic distribution of the MRRs for the three census year. Increasing *gray scale* is proportional to increasing percentage; white denotes no data. *Top* (**a**): 1990. *Middle* (**b**): 2000. *Bottom* (**c**): 2010

where **1** is an n-by-1 vector of ones, and T denotes the matrix transpose operator. The decomposition generates n eigenvectors and their associated n eigenvalues. In descending order, the n eigenvalues, usually are denoted by $\boldsymbol{\lambda} = (\lambda_1, \lambda_2, \lambda_3, \ldots, \lambda_n)$, where the subscript represents descending order of value. These eigenvalues quantify the value of the Moran Coefficient for their corresponding n-by-1 eigenvectors, which usually are denoted by $\mathbf{E} = (\mathbf{E}_1, \mathbf{E}_2, \mathbf{E}_3, \ldots, \mathbf{E}_n)$ – i.e., $MC_j = \frac{n}{\mathbf{1}^T\mathbf{C}\mathbf{1}} \cdot \lambda_j$, for $\mathbf{E}_j$.

The eigenfunctions have some important properties: the eigenvectors are mutually orthogonal and uncorrelated (Griffith 2000); the eigenvectors portray distinct map patterns exhibiting a specified level of spatial autocorrelation when mapped onto the n areal units associated with the corresponding spatial weight matrix **C** (Tiefelsdorf and Boots 1995); and, given a spatial weight matrix **C**, the feasible range of the Moran Coefficient values is determined by the largest and smallest eigenvalues, λ_1 and λ_n (de Jong et al. 1984). Based upon these properties, the eigenvectors can be interpreted as (Griffith 2003) follows:

> The first eigenvector, $\mathbf{E}_1$, is the set of real numbers that has the largest MC value achievable by any set of real numbers for the spatial arrangement defined by the spatial weight matrix **C**; the second eigenvector, $\mathbf{E}_2$, is the set of real numbers that has the largest achievable MC value by any set that is uncorrelated with $\mathbf{E}_1$; the third eigenvector, $\mathbf{E}_3$, is the set of real numbers that has the largest achievable MC value by any set that is uncorrelated with both $\mathbf{E}_1$ and $\mathbf{E}_2$; the fourth eigenvector is the fourth such set of values; and so on through $\mathbf{E}_n$, the set of real numbers that has the largest negative MC value achievable by any set that is uncorrelated with the preceding (n−1) eigenvectors.

Frequently the eigenfunctions of interest relate to positive spatial autocorrelation, with a threshold value of $MC_j/MC_{max} > 0.25$ indicating noticeable positive spatial autocorrelation.

Griffith (2004) formulates an eigenfunction spatial filter binomial regression model.

9.2.2 Random Effects

A random effects term accounts for within-areal-unit heterogeneity, and is time invariant for space-time data. Spatially structured random effects models are increasing in popularity (see, for example, Griffith and Paelinck 2011; Hughes and Haran 2012), partially because they can be conceptualized in terms of eigenvector spatial filtering, and partially because they have become estimable. One common specification is for the intercept term to be written as a random effects term, resulting in it representing variability about the conventional single-value, constant mean. One role of a random effects term when analyzing space-time data is to account for correlation between repeated measures in time for each areal unit. With n very short time series, each constituting a non-random sample, a random effects term represents a common constant across time that introduces correlation among values. In other words, it captures temporal correlation effects.

A random effects term captures both spatially structure and spatially unstructured variation. Once estimated, it can be decomposed into these two components, dramatically simplifying an eigenvector spatial filtering analysis that holds spatial structure constant across time. Otherwise, the space-time structure can be posited as either lagged or contemporaneous and estimated simultaneously with a spatially unstructured random effects term.

9.3 Pure Spatial Effects: Initial Eigenvector Spatial Filtering Results

Estimation of a generalized linear mixed model (using SAS NLMIXED) for the 3,067-by-3 dataset results in a random effects term that accounts for considerable geographic variation in MRRs across counties. This term has a mean of 0.00281, a variance of 0.15047, and a statistically significant Kolmogorov-Smirnov test statistics for the null hypothesis that it conforms to a bell-shaped curve. In other words, this term deviates from the assumptions of a zero mean and normality.

The number of positive spatial autocorrelation eigenvectors is 770, and the number of negative spatial autocorrelation eigenvectors is 1,057. The random effects term yields no evidence of negative spatial autocorrelation. In contrast, the spatially structured component of the random effects term comprises 48 positive spatial autocorrelation eigenvectors that account for roughly 47.3 % of its variance.[2] The resulting eigenvector spatial filter (Fig. 9.2), whose Moran Coefficient (MC_{SF}) is 1.00774 (indicating marked positive spatial autocorrelation; $MC_{SF}/MC_{max} = 1.00774/1.08718 = 0.92693$), reduces excess binomial variance, and accounts for about one-third of the variability in the MRRs (Table 9.2). Its conspicuous map pattern (Fig. 9.2a) reveals a stark differentiation between the south and Rocky Mountains regions, on the one hand, and the mid-west and mid-Atlantic regions, on the other hand. Heterogeneity accounts for at least twice as much variance, and essentially further reduces the overdispersion nearer to its expected level of 1 (although the 1990 and 2000 MRRs may contain some underdispersion now). The SURE component exhibits a random map pattern (Fig. 9.2b).

9.4 Spatial Effects in the Presence of a Time Lag

The random effects term in the presence of time lagged MRRs has a mean of 0.00278, a variance of 0.04531, and exhibits significant deviation from a normal distribution. Again the random effects term yields no evidence of

[2]A stepwise linear regression was employed that used the Bonferroni adjusted selection significance level of 0.10/770.

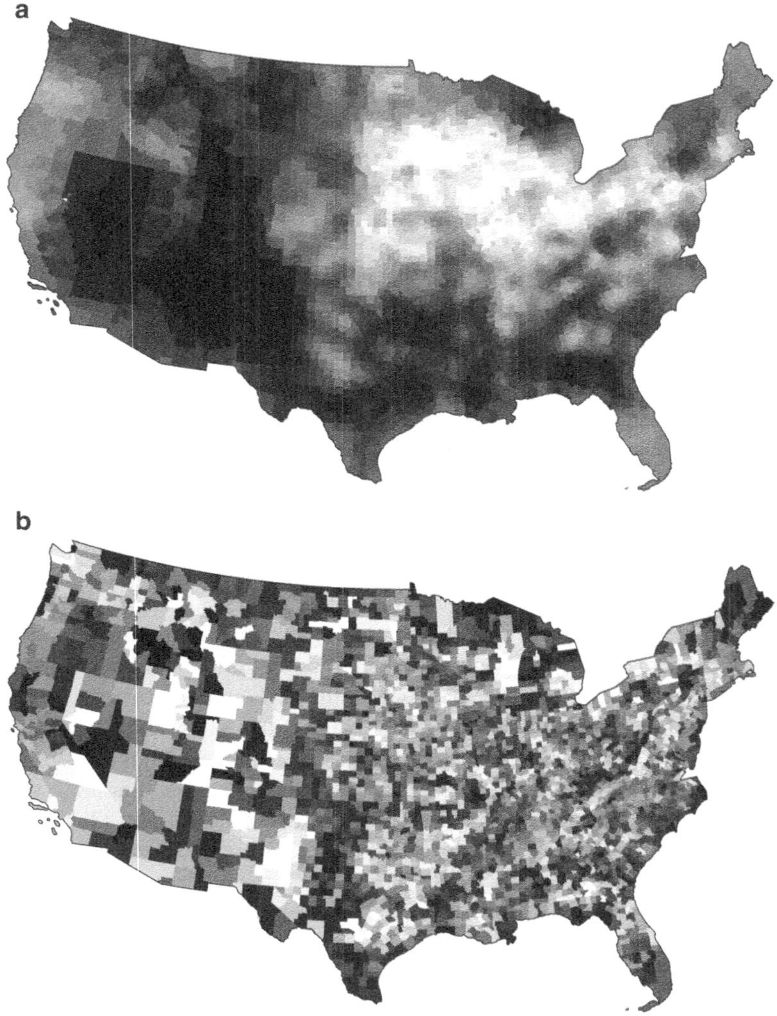

Fig. 9.2 Quantile choropleth maps: components of the random effects term. Increasing *gray scale* is proportional to increasing values. *Top* (**a**): spatially structured random effects. *Bottom* (**b**): spatially unstructured random effects

negative spatial autocorrelation. Time lagged MRRs account for more than half of the variability. Here the spatially structured component of the random effects term comprises 38 positive spatial autocorrelation eigenvectors that account for roughly 36.3 % of its variance. The resulting eigenvector spatial filter (Fig. 9.3), whose MC_{SF} is 1.01125 (indicating marked positive spatial autocorrelation; $MC_{SF}/MC_{max} = 1.01125/1.08718 = 0.93016$), modestly reduces excess binomial

Table 9.2 Results from including a random effects term in the binomial model specification

	MRRs	Spatially structured (SSRE)		SSRE + spatially unstructured (SURE)	
Year	Deviance	Deviance	Pseudo-R^2	Deviance[a]	Pseudo-R^2
1990	3.0755	1.8451	0.3951	0.8359	0.7212
2000	5.0397	3.0754	0.3771	0.6830	0.8649
2010	5.7721	3.7449	0.3411	1.0892	0.8161

[a]This value assumes that the random effects terms has only 2 degrees of freedom

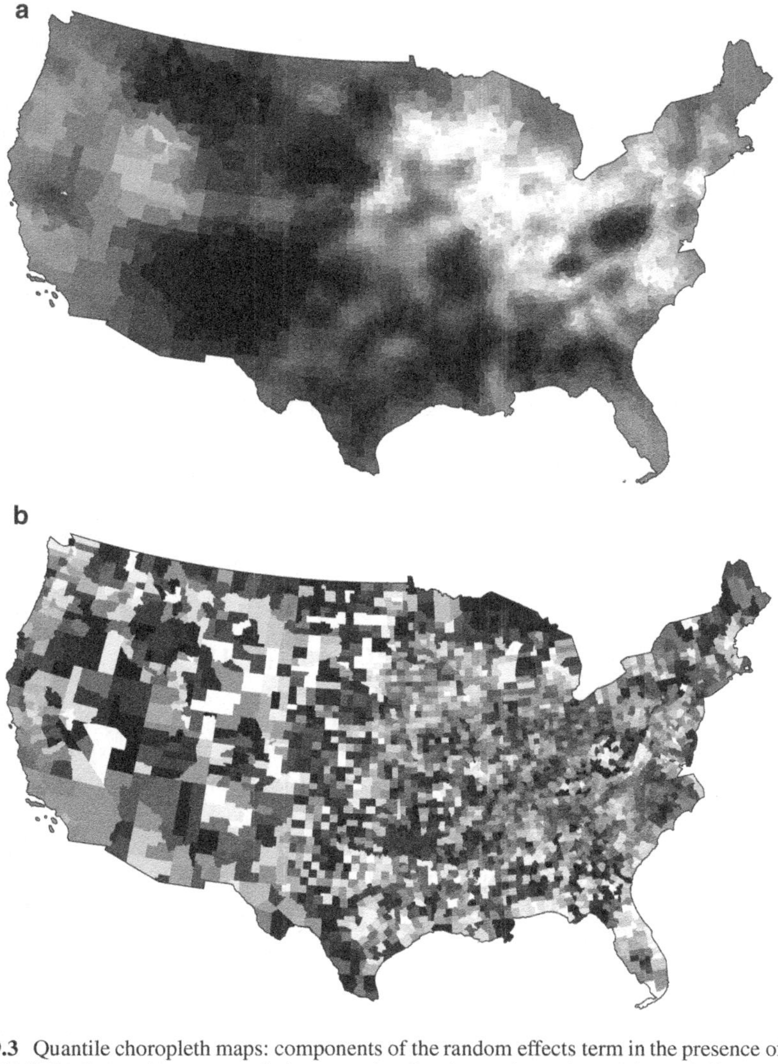

Fig. 9.3 Quantile choropleth maps: components of the random effects term in the presence of time lagged MRRs. Increasing *gray scale* is proportional to increasing values. (**a**): spatially structured random effects. (**b**): spatially unstructured random effects

Table 9.3 Results from including a random effects term in the binomial model specification containing time lagged MRRs

	MRRs		Lag + SSRE		Lag + SSRE + SURE	
Year	Deviance	Pseudo-R^2	Deviance	Pseudo-R^2	Deviance	Pseudo-R^2
2000	2.1512	0.5095	1.9023	0.5647	0.9296[a]	0.7886
2010	2.4977	0.5584	1.9617	0.6564	0.5317	0.9128

[a]This value assumes that the random effects terms has only 2 degrees of freedom

variance, and accounts for only a modest amount (i.e., 5–10 %) of the variability in the MRRs (Table 9.3; i.e., the immediately preceding decennial map captures much of the spatial structure effect). Its conspicuous map pattern (Fig. 9.3a) is very similar to that portrayed in Fig. 9.2a. Once more, heterogeneity accounts for far more of the variance, being second to the time lagged MRRs, and essentially further reduces the overdispersion; this case more strongly suggests the presence of underdispersion. The SURE component exhibits a random map pattern (Fig. 9.3b).

9.5 Eigenvector Space-Time Filter Results

Griffith (2012) extends the eigenvector spatial filtering methodology to space-time contexts, exploiting the space-time structure latent in the space-time Moran Coefficient (Cliff and Ord 1981; Griffith 1981). One of his extensions is for contemporaneous spatial dependence (i.e., a value at a given location for a particular point in time is a function of the value at that location for the previous point in time and the values of nearby locations for the same point in time). The space-time structure for this situation is given by

$$\left(\mathbf{I} - \mathbf{1}\mathbf{1}^{\mathrm{T}}/\mathrm{n}\right)\left(\mathbf{I}_{\mathrm{T}} \otimes \mathbf{C}_{\mathrm{s}} + \mathbf{C}_{\mathrm{T}} \otimes \mathbf{I}_{\mathrm{s}}\right)\left(\mathbf{I} - \mathbf{1}\mathbf{1}^{\mathrm{T}}/\mathrm{n}\right), \tag{9.2}$$

where: $\otimes$ denotes Kronecker product, $(\mathbf{I} - \mathbf{1}\mathbf{1}^{\mathrm{T}}/\mathrm{n})$ now is an nT-by-nT projection matrix, $\mathbf{C}_{\mathrm{s}}$ is the n-by-n binary spatial weights matrix appearing in expression (9.1), and $\mathbf{C}_{\mathrm{T}}$ is a binary T-by-T time series weights matrix, with 1s in its upper and lower off-diagonals and 0 elsewhere.

With three points in time (i.e., 1990, 2000, and 2010) and 3,067 counties, the space-time matrix given by expression (9.2) is 9,201-by-9,201. This matrix has 2,264 eigenfunctions with relative Moran Coefficients of at least 0.25. Of these, the resulting eigenvector space-time filter (ESTF) comprises 723 vectors yielding individual year Moran Coefficients of 0.86589 (1990), 0.83668 (2000), and 0.89125 (2010). This synthetic variate offers an alternative covariate to those based upon only spatial structure (see Table 9.2) for predicting MRRs. When compared with the time-lagged model specification, it offers the advantage of being able to more fully incorporate the first time period (i.e., 1990), which is very appealing in short time series, especially ones containing only three points in time.

The constructed ESTF yields a pseudo-R^2 value of 0.6868, and produces a deviance statistic value of 1.6468. These results are superior to those for the two model specifications presented in the preceding section of this paper. The individual time components of this ESTF produce the following results:

Year	Pseudo-R^2
1990	0.6685
2000	0.7075
2010	0.6622

Estimating a random effects term with this ESTF specification renders a SURE geographic variate, which should account solely for data heterogeneity, that is constant through time; none of the eigenvectors from expression (9.1) should, or do, covary with it. The estimated SURE has a mean of −0.00122, but deviates from a normal frequency distribution. The addition of this variate increases the pseudo-R^2 value to 0.8785, which is a substantial increase over the ESF and time-lagged model specifications. The individual time components of this ESTF-plus-SURE specification produce the following results:

Year	Pseudo-R^2
1990	0.7899
2000	0.9109
2010	0.8932

The affiliated deviance statistic is 0.64810, again raising the question of whether or not the SURE term has only 2 degrees of freedom associated with it.

Figure 9.4 portrays the ESTF and the SURE. The cloud of highest values shifts eastward through time, whereas the cloud of lowest values basically covers Arizona, New Mexico and Texas. Keeping in mind that eigenvectors are unique to a multiplicative factor of -1, these patterns resemble a mirror image of those portrayed in Figs. 9.2 and 9.3. The heart of Appalachia is visible in all of these maps. The maps in Fig. 9.4 more closely resemble their MRR counterparts in Fig. 9.1 than do these maps in Figs. 9.2 and 9.3. The SURE looks like the random pattern that it is. It looks more chaotic than the SURE patterns appearing in Figs. 9.2 and 9.3.

9.6 Implications and Concluding Comments

This paper summarizes an analysis of census MRRs from the three US decennial censuses conducted for 1990–2010 whose most successful descriptions employ a random effects model framework. Especially, eigenvector spatial filtering combined with a random effects model furnishes a flexible way to account for spatial as well as non-spatial random effects simultaneously in a model specification. Further, eigenvector spatial filtering has been extended to eigenvector space-time filtering

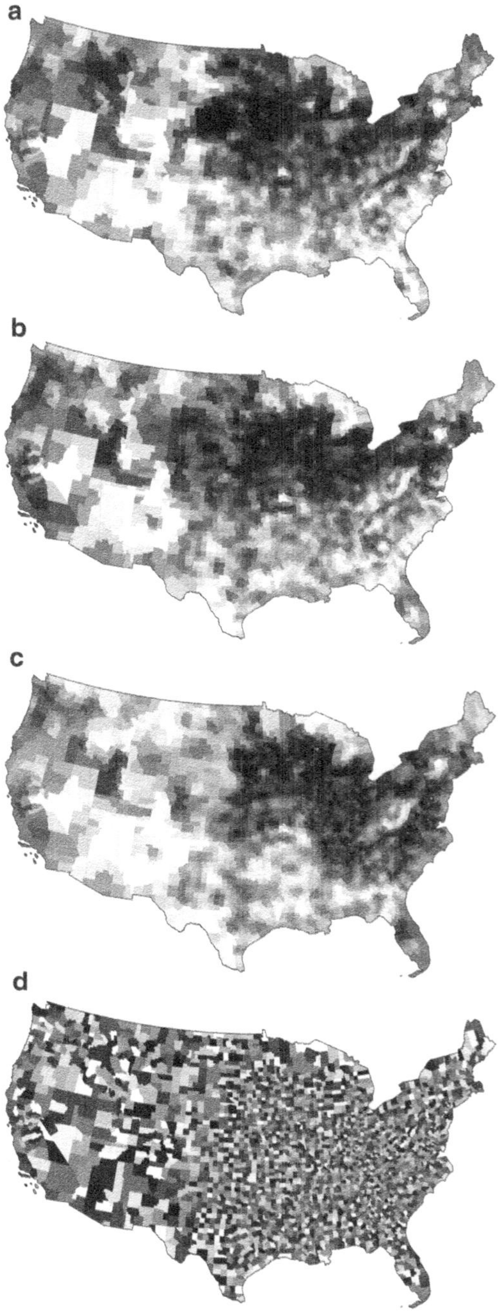

Fig. 9.4 Quantile choropleth maps: the ESTF for the three points in time, and the SURE term. Increasing *gray scale* is proportional to increasing values. *Top* (**a**): ESTF for 1990. *Top middle* (**b**): ESTF for 2000. *Bottom middle* (**c**): ESTF for 2010. *Bottom* (**d**): SURE

in order to model space-time structure; this extension has been utilized for this paper to accommodate a space-time dependency structure latent in the MRRs via an augmented space-time weights matrix.

Estimated spatial random effects from the different model specifications show some common spatial structure in the MRRs, part of which relates to time lagged MRRs, with a large difference in the estimated random effects between the south and Rocky Mountain regions, and the Midwest and mid-Atlantic regions. In conjunction with response rates in Fig. 9.1, this outcome indicates that the Midwest and mid-Atlantic regions have high MRRs overall, while the south and Rocky Mountain regions have relatively low MRRs. Another noticeable pattern is that the heart of the Appalachian region has low MRRs surrounded by high rates in the Midwest and mid-Atlantic regions. This island pattern centering on the core of the Appalachian mountains is more clearly shown by the spatial effects in the model specification with temporal components (i.e., spatial effects in the model specifications with a time lagged or an eigenvector space-time filter). Specifically, eigenvector space-time filters uncover this space-time autocorrelation structure for each of the three points in time. In addition, the time lagged MRRs account for a considerable amount of the variability in the MRRs over space and time. Hence, either the MRRs or the eigenvector space-time filter (whose performance is superior to that of the time-lagged MRRs) can be utilized to establish overall county-level expectations of MRRs for subsequent census, although neither is a causative covariate for any subsequent census. However, this result is somewhat limited because the estimated random effects term reported in this paper is based upon a time sequence of only 2–3 periods (especially the 1990 MRRs constitute an incomplete spatial series having numerous missing values). This result may well be too specific to the time periods under study. Nevertheless, because the eigenvector space-time filter accounts for such a sizeable amount of the variability in the MRRs (i.e., roughly 69 %), with its accompanying random effects term accounting for roughly an additional 19 %, replicating this study in the future should reveal very similar results, even if the variance percentage for the estimated random effects term decreases somewhat. The eigenvector space-time structure accounts for the same amount of variability as Keller's (1990) 38 substantive variables, but without having to collect that information [which now would have to come from the American Community Survey (ACS)].

Consequently, exploratory results reported here indicate that the potential exists for developing a useful descriptive model of the census MRRs, one that exploits the space-time structure of the data rather than relying on time-lagged MRRs or other covariates (i.e., a rural–urban classification, percentage of occupied housing units). Not only could one be developed to help with the 2020 census planning and administration, but, if so desired, one also could be developed to help address response rate concerns associated with administering the annual US ACS.

Acknowledgements We wish to thank Parmanand Sinha for helping check the data retrieved from selected internet sources. D. Griffith is an Ashbel Smith Professor.

References

Cliff, A. D., & Ord, J. K. (1981). Spatial and temporal analysis: Autocorrelation in space and time. In N. Wrigley & R. Bennett (Eds.), *Quantitative geography: A British view* (pp. 104–110). London: Routledge & Kegan Paul.

de Jong, P., Sprenger, C., & van Veen, F. (1984). On extreme values of Moran's I and Geary's c. *Geographical Analysis, 16*(1), 17–24.

Griffith, D. A. (1981). Interdependence in space and time: Numerical and interpretative considerations. In D. A. Griffith & R. MacKinnon (Eds.), *Dynamic spatial models* (pp. 258–287). Alphen aan den Rijn: Sijthoff and Noordhoff.

Griffith, D. A. (2000). Eigenfunction properties and approximations of selected incidence matrices employed in spatial analyses. *Linear Algebra & Its Applications, 321*, 95–112.

Griffith, D. A. (2003). *Spatial autocorrelation and spatial filtering: Gaining understanding through theory and scientific visualization*. Berlin: Springer.

Griffith, D. A. (2004). A spatial filtering specification for the auto-logistic model. *Environment & Planning A, 36*, 1791–1811.

Griffith, D. A. (2012). Space, time, and space-time eigenvector filter specifications that account for autocorrelation. *Estadística Española, 54*(177), 7–34.

Griffith, D. A., & Chun, Y. (2012). Spatial autocorrelation and eigenvector spatial filtering. In M. Fischer, & P. Nijkamp (Eds.), *Handbook of regional science*. Berlin: Springer.

Griffith, D. A., & Paelinck, J. (2011). *Non-standard spatial statistics and spatial econometrics*. Berlin: Springer.

Hughes, J., & Haran, M. (2012). Dimension reduction and alleviation of confounding for spatial generalized linear mixed models. *Journal of the Royal Statistical Society, Series B: Statistical Methodology, 75*(1), 139–159.

Keller, J. K. (1990). *1990 census mail response and related variables: An examination at the county level using regression analysis*. Unpublished paper, U.S. Bureau of the Census, Suitland, MD, November.

Tiefelsdorf, M., & Boots, B. (1995). The exact distribution of Moran's I. *Environment and Planning A, 27*, 985–999.

Chapter 10
Creating Spatiotemporal Semantic Maps from Web Text Documents

Wei Wang and Kathleen Stewart

10.1 Introduction

With the continued development of the Internet, text information is widely available in the form of Web articles, news reports, blogs, Twitter feeds, and other formats. Because of the massive amount of information provided by unstructured texts, the need to acquire relevant information automatically from text sources is necessary for web users. Geographic information is commonly referenced in these Web documents. In addition, many documents describe dynamic occurrences (e.g., the track of a storm) of natural phenomena such as severe storms as well as human activities (e.g., the movement of relief supplies in the wake of a natural disaster). This chapter discusses methods for combining natural language processing techniques and ontologies to improve the extraction of semantics of hazard-related events sourced from web texts. Developing methods to systematically represent the movement of dynamic entities such as severe storms and their associated spatiotemporal behaviors is important for next generation information system design where details extracted automatically from text can be presented to users for their own domain applications.

Information retrieval (IR) provides valuable opportunities for users to obtain information automatically from Internet search engines or digital library applications. For example, when a user types "2012 London Olympics" into a search engine (e.g., Google), the query terms are used for accessing relevant web documents that contain this text string using IR algorithms, and a set of documents related to London Olympics will be retrieved. While the field of traditional IR has contributed solutions for helping users to automatically find information based

W. Wang (✉) • K. Stewart
Department of Geography, The University of Iowa, Iowa City, IA 52240, USA
e-mail: wei-wang-1@uiowa.edu; kathleen-stewart@uiowa.edu

M.-P. Kwan et al. (eds.), *Space-Time Integration in Geography and GIScience*, DOI 10.1007/978-94-017-9205-9_10

on their interests, the possibility of automatically tracking spatiotemporal and semantic change relating to events is a more recent challenge for IR, one that will improve understanding of event-based information (e.g., hazard events) described in collections of unstructured web documents from spatiotemporal and semantic perspectives. As the techniques of IR have improved over the past decade, IR has begun to accommodate and exploit a broader range of information, for example, geographic information. Geographic information retrieval (GIR) stems from the discipline of IR, and involves not only many IR methods, such as indexing, searching, browsing and querying documents, but also considers the geographic scope of documents (Kemp et al. 2007; Jones and Purves 2008; Keller et al. 2008; Teitler et al. 2008). With GIR, natural language processing (NLP) techniques are used to extract geo-references such as zipcodes, addresses, well-known landmarks, toponyms, telephone and area codes from web pages (Purves and Jones 2011).

In geographic information science (GIScience), modeling geographic dynamics based on spatiotemporal information extracted from the web is a growing field of research. Geographic dynamics refers to the change or movement of an event with spatial and temporal characteristics, and involves understanding "the fundamental functions of relevant forces and their interrelationships in space and time" (Yuan and Stewart Hornsby 2008, p. 7). The topic of modeling dynamics of geographic domains includes: calculating the characteristic patterns of movement of individuals or groups (Laube et al. 2007; Dodge et al. 2008; Yuan and Stewart Hornsby 2008; Wood and Galton 2010; Dodge et al. 2012); time geography analyses that investigate patterns of people's activities from spatial and temporal perspectives (Miller 1991; Kwan 2000; Raubal et al. 2004; Yu 2006; Shaw et al. 2008; Chen and Kwan 2012); and modeling the trajectories of moving objects such as people, vehicles or natural phenomena (Dodge et al. 2008; Stewart Hornsby and Li 2009; Demsar and Virrantaus 2010). Automatically capturing spatiotemporal and semantic information from the web and representing the underlying relationships through geographic information systems (GIS) not only transfers text content to a visual representation that preserves informational characteristics from the documents, but also offers people a better understanding of spatial and temporal characteristics that are described in the immense resource of web documents and provides a clearer overview of dynamic change patterns and trends over space-time.

In this work we discuss a hybrid method of combining the result of natural language processing techniques and ontologies to improve the extraction of semantics of hazard-related events sourced from web texts. This work combines principles from the field of NLP and GIR with ongoing work in the field of temporal GIS where there is an interest in modeling geographic dynamics. For this research, an open source software tool, General Architecture for Text Engineering (GATE) (http://gate.ac.uk/), is used as the primary tool for spatiotemporal and semantic information extraction. The motivation for this research is to automatically represent the spatiotemporal information and track moving entities over space-time to show the spatial extent and trends of the dynamics of hazard-related events. Also, semantics are automatically processed to make the extracted spatiotemporal information more meaningful. Methods for automatically processing spatiotemporal and semantic information from text documents are discussed in this chapter.

The rest of this chapter is organized as follows: The next section reviews related work in the field of GIR and GIScience. In the section that follows, the main features of an approach that automatically extracts spatiotemporal information from text documents are introduced, and algorithms for combining spatiotemporal information to model the geographic dynamics are demonstrated. A hazard ontology is presented that can be used for modeling the semantics associated with storm events, followed by a section that demonstrates the work for a tornado outbreak in central US in April 2012 as described in web news reports. Evaluation is conducted based on comparing results obtained from manual and automated processed performances. Conclusions are discussed in the final section of the chapter.

10.2 Background

In this research, a hybrid method is developed to combine gazetteers and ontologies for extracting semantics of hazard-related events sourced from web texts, such that not only spatiotemporal information is automatically represented and tracked over space-time, but also text-based semantics are automatically processed, adding meaning to spatiotemporal information. It should be noted that the term 'gazetteer' is applied more broadly in NLP than the traditional usage of the term in Geography, i.e., a gazetteer refers to a dictionary that contains lists of geographic references. Instead, gazetteer refers to a list of specific terms that are used to match the corresponding information from text documents. In this work, a spatial gazetteer refers to a list of spatial terms commonly found in text, and a temporal gazetteer contains a list of temporal phrases.

A key objective of GIR is to detect and capture location-based information from natural language text. Most GIR systems are based on detecting spatial references in text, for example, latitude, longitude, state, county, and city names. To extract geographic information from text documents, a spatial gazetteer is a key element for data processing and affects the accuracy of extraction results. Specifically, the references in the documents are compared with the terms in the gazetteers and, if a match is found, those words or phrases from text documents are annotated by the NLP system. Numerous systems have been developed based on GIR techniques. For example, *GIPSY*, a geo-referenced information processing system, supports automatic geographic indexing of text documents (Woodruff and Plaunt 1994) by matching terms in the document to terms in a gazetteer. Since *GYPSY*, the idea of designing and applying gazetteers has become a standard component in many GIR systems. *NewsStand*, another GIR system, detects geographic-related information from RSS feeds using a custom-built geotagger (Teitler et al. 2008). The extracted locations are displayed via a map viewer that dynamically displays the primary location associated with each news article based on the frequency of the term in the article (Lieberman et al. 2007; Teitler et al. 2008). More recently *TwitterStand* automatically imports breaking news from tweets posted by Twitter users and provides a map interface for reading this news (Sankaranarayanan et al. 2009; Jackoway et al. 2011).

In the field of GIScience, researchers are growing increasingly interested in incorporating GIR techniques into their studies. Methods are being investigated for extracting route directions automatically from text (Klippel et al. 2008; Zhang et al. 2010), and research involving oral histories and biographies combine narrative analysis with a GIS-based time-geographic framework for interactive interpretation, analysis, and visualization (Kwan and Ding 2008). In other recent work, humanitarian terms are extracted from articles describing the long-term humanitarian crisis situation in the Sudan and visualized using an earth-based visualization (Tomaszewski 2008; Tomaszewski and MacEachren 2010).

Ontologies play an important role in the development of GIR techniques with regard to providing a knowledge base to improve semantic understanding of text with respect to searching and extracting appropriate information (e.g., geographic information). Ontologies deal with the actual nature of the phenomena themselves, focusing on "the nature and the organization of reality" (Ping and Yong 2009, p. 306). In information science, ontologies have been applied as a tool for knowledge management and are defined as a concept-based framework using formal methods that represent entities, attributes, relationships, and values for a specific domain (Wiegand and Garcia 2007; Ping and Yong 2009). Well-developed ontologies can serve as a standard for conceptualizing and understanding domains of interest, and ontologies support semantic interoperability. For example, associations between different semantics can be achieved by applying ontologies with classification schemes and hierarchies (Kemp et al. 2007). Ontologies can be applied for disambiguation of geographic names and improving gazetteer interaction (Volz et al. 2007; Janowicz and Kebler 2008; Machado et al. 2011). Applying ontology in GIR applications can help to capture natural facts and related information from a human perspective, and represent extracted text information in a hierarchical form.

In this chapter, a hybrid method of combining gazetteers and ontologies is presented to model the dynamics of hazard-related events (in this work, tornadoes) sourced from web texts (Fig. 10.1) through spatiotemporal semantic information. The process can be divided into three main parts: gazetteers and ontology creation, text processing, and geovisualization. GIR and NLP techniques are used mainly in the part of text processing. The model aims at providing spatiotemporal semantic information representation and hidden patterns obtained from text documents.

10.3 Extracting Spatiotemporal Information from Web Documents

In this section, we employ a framework for modeling movement based on combined spatiotemporal information extended from web documents and investigate the role of GIS for representing spatiotemporal details of movement expressed in the

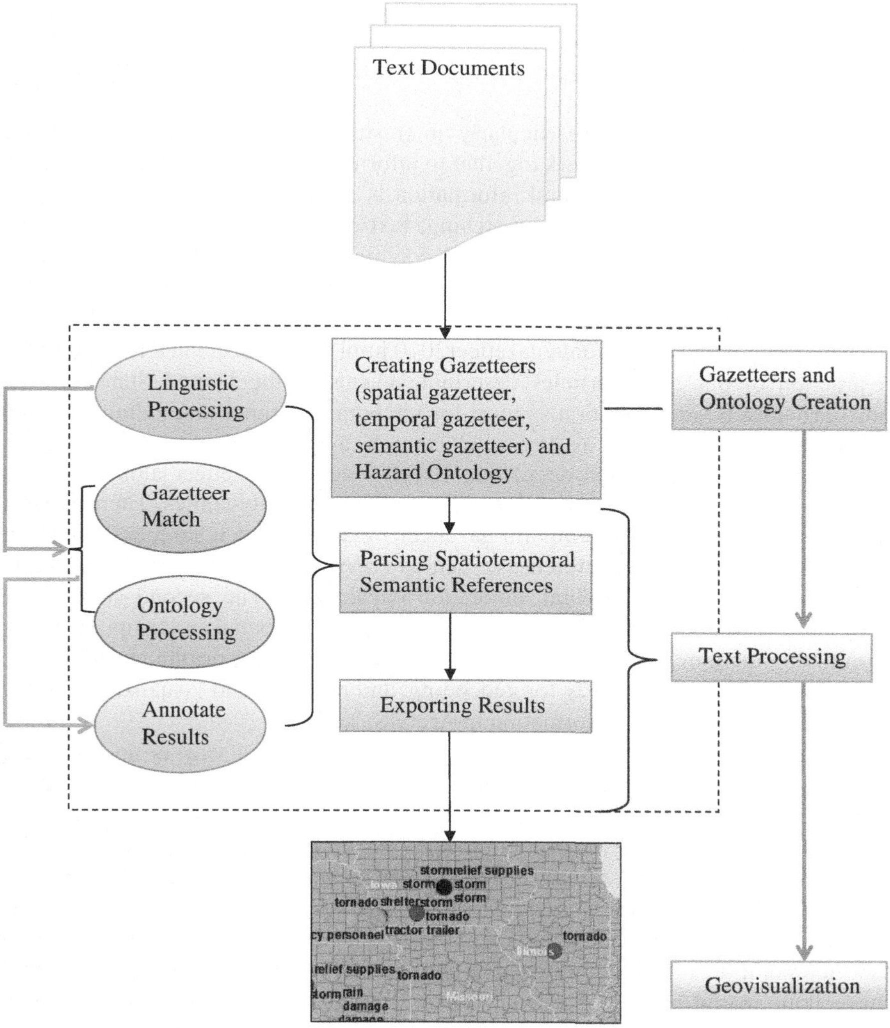

Fig. 10.1 A framework for spatiotemporal semantic information extraction

documents. As a result of information extraction, a list of location names and time references is obtained. The *x*,*y* coordinates are assigned to each extracted location using Google's geocoder (http://www.batchgeo.com/). The mapping of extracted spatiotemporal information is implemented using ESRI's ArcMap software.

Each document is ordered by its published date during spatiotemporal parsing. The date of the document provides a time frame for the phenomena being analyzed. Generally, each web-based news article includes a header specifying its published date. This date is used to distinguish documents and sort them in the order that they are published. In this way, document date can be used to give a basic temporal order to the extracted spatial locations. However, as presented in the following discussions, this information can be improved further when additional processing is applied to text information.

For this study, we are particularly interested in how spatial and temporal references available in text work together to inform us about the dynamics of events. The extraction of spatiotemporal information is accomplished by creating spatial and temporal gazetteers for text matching, text processing, and geovisualization. The default spatial gazetteer for GATE contains general world references plus mostly UK-based places, and so an additional gazetteer using data from StreetMap Premium for ArcGIS and the US Gazetteer Files for the 2010 Census (https://www.census.gov/geo/maps-data/data/gazetteer2010.html) has been created to support the extraction of places for articles describing events in the United States. GATE also provides basic annotation support for temporal references including common temporal references, such as day, week, month, and year. Documents also often include additional temporal detail that GATE's default gazetteer alone does not match, and that enriches our understanding of the event's dynamics. Therefore, we have extended the default temporal gazetteer by adding a new temporal gazetteer with additional temporal references, such as *early Sunday morning* or *late night* to provide additional temporal annotation capabilities. The second step of the extraction process involves the parsing and annotation of spatial and temporal terms, combining them in a meaningful way. The final step is to export the sorted results. In this way, the data are ready for geocoding, and mapping and visualization of the extracted spatiotemporal information in ArcGIS.

Combining spatiotemporal information in a meaningful way is an important objective for this research. Spatiotemporal data are extracted to capture the salient details of spatiotemporal dynamics in articles. For many documents, geographical references are associated with temporal expressions. For example, *More than 1,000 volunteers rushed to fill sandbags Wednesday as many in Fargo tried to protect themselves from a historic flood that is expected to swamp the area.* In this example, *Fargo* is associated with the temporal expression *Wednesday,* affording both spatial and temporal details about an event of interest. However, not every sentence will have both spatial and temporal information. Five possible cases can arise with respect to spatiotemporal information in a sentence (Stewart Hornsby and Wang 2010) (Fig. 10.2):

- When only spatial information is present in a sentence
- Sentences with one spatial term and one temporal reference
- Sentences with one spatial term and multiple temporal references
- Sentences with multiple spatial terms and a single temporal reference
- Sentences with multiple spatial and multiple temporal references

These cases provide a basis for automatically combining spatiotemporal information retrieved from text documents in order to use that information to model dynamic entities in a GIS. It is also possible for text documents to present only spatial information and not contain *any* temporal information. In these cases, the document

Algorithm

1. **Input:** Document D, Sentence E, Spatialterm S_E, Temporalterm T, Publisheddate Tp
2. **Output**: Combine (S_E, T)
3. **begin**
4. Parse D
5. **for** *Each sentence E in D* do
6. **If** only S_E exist in D, **then**
7. add Tp to S_E combine(S_E, Tp)
8. **If** one S_E and one T in E, **then**
9. assign T to S_E combine(S_E,T)
10. **If** one S_E and multiple T in E, **then**
11. assign each T to S_E combine(S_E,$T1$), combine(S_E, $T2$)…
12. **If** multiple S_E and one T in E, **then**
13. assign T to each S_E combine($S1$, T), combine($S2$,T)….
14. **If** multiple S_E and multiple T in E, **then**
15. Check left and right context of $S1$,
16. **if** a comma is found, **then** searching is stopped
17. **If** T exists, **then** assign T to $S1$
18. **Else** jump to $S2$, keep checking left and right context of $S2$,
19. …..
20. **Else**
21. Jump to the next sentence
22. Return combine(S_E,T)
23. **End**

Fig. 10.2 Algorithm for combining extracted spatiotemporal information from text documents

date in the headline of each news article is assigned to spatial locations mentioned in the article, providing some temporal information. All temporal information needs to be converted to standard time format (i.e., 08/06/2012) for further processing. When combining spatiotemporal information, additional processing is undertaken to manage temporal information. For example, the published date of news reports serves as a criterion to compare other temporal expressions in the document so that these expressions may be correctly ordered.

All results are exported in a database that includes a set of tuples $\{S_{ID}, S_E, S_G, X, Y, T\}$, where (1) S_{ID} is the id number of S_E (a spatial term) – GATE marks each annotated term with a specific *ID* number after the results are sorted. The spatial terms can be mapped based on their *ID* number that determines the order of positions according to where spatial terms occur in text documents; (2) S_E corresponds to the spatial term; (3) S_G describes spatial granularity of the spatial term (e.g., city, state, or county name); (4) X and Y represent x,y coordinates for the location of the term S_E; and (5) T refers to the associated temporal information (e.g., 08/09/2012). After all records are stored in ArcGIS, the locations are mapped with an x,y display function.

10.4 Integrating Ontologies with Gazetteers

In the previous section, we described how spatiotemporal extraction can reveal the information contained in web documents, e.g., the evolution of storm events. Ontologies can be used to provide additional semantic information about events. To capture semantic information from the text documents, general domain-related concepts (e.g., *tornado, flood, landslide,* and *earthquake*) from upper level ontologies, such as *WordNet* (http://wordnet.princeton.edu/), *DOLCE* (http://www.loa.istc.cnr.it/DOLCE.html), and *SUMO* (http://www.ontologyportal.org/), can be adopted. For this work, we use open source software *NeOn,* drawing from domain concepts in *OpenCyc* (http://sw.opencyc.org/), *DOLCE, WordNet, GeoNames* (http://www.geonames.org/ontology/documentation.html), and *Proton Ontology* (http://proton.semanticweb.org/) to develop a hazard ontology. The ontology is organized as a hierarchy specifying relations between classes (e.g., *is_a* relation). Each class has one superclass and a set of subclasses that form the class hierarchy.

Once complete, the hazard ontology is imported into GATE as a language resource, OWLIM ontology. A plugin tool named OntoGazetteer is used to link terms from the gazetteers with classes in the ontology. Two steps are necessary to create a connection between the gazetteers and ontology classes. First, domain-related references in a set of *training text documents* are manually annotated and stored as a linear list in a new gazetteer created for this task, the *semantic gazetteer.* The training set is used for collecting domain-related references described in news reports. Based on the results of processing the training set, an automatic mapping between the gazetteers and the ontology is constructed. As a second step, rules are made to link the newly created lists in the semantic gazetteer and the ontology. The lists are treated as classes in the ontology. This step assigns references in the gazetteer with terms in the ontology by reasoning over the hierarchical relations. For example, assume the term *supercell* is identified in one of the training documents, and that this term also exists in the ontology as a subclass of *thunderstorm,* and *thunderstorm, ice storm, dust storm, and blizzard* are subclasses of *storm. Supercell* therefore *is_a storm.* The spatial gazetteer includes a set of lists, such as lists of cities, counties, states, and countries. Each list in the gazetteer contains a set of spatial references, and these places are treated as instances of classes in the ontology. With rules implemented in GATE, the semantics are processed, integrating the gazetteers and the domain ontology. In this way, not only are spatial and temporal terms extracted from text documents, but semantics of the spatiotemporal events based on their relations are also captured. *Supercell* is associated with a geographic location, and the semantics (e.g., *thunderstorm* or more generally, *storm*) can be automatically linked to this geographic location as well.

In the next section, articles describing tornadoes that impacted the central United States in April 2012 serve as a case study to illustrate how spatiotemporal and semantic information can be extracted automatically from web documents and visualized for users.

10.5 Case Study: Tornadoes in Oklahoma, US, April 14–16, 2012

The dataset for this case study includes Web news reports on tornadoes that ravaged a large portion of the US Midwest, especially Oklahoma, during April 2012. Articles about this event published from April 14 to April 16 have been obtained from http://www.cnn.com/ and processed for their spatial, temporal and semantic content.

10.5.1 Combining Spatiotemporal Information

Each document in the dataset includes a header giving the published date of the article and temporal information is extracted using GATE, allowing locations retrieved from the text to be ordered according to different document dates (i.e., April 14–16). A map shows a portion of the central United States and the Great Lakes region that were impacted by the tornadoes with the highest number of events occurring in Kansas (Fig. 10.3). Currently, for generalized geographic references, such as states, counties, and regional geographic places, geocoding is handled by selecting a central point of reference. In this case study, these types of locations

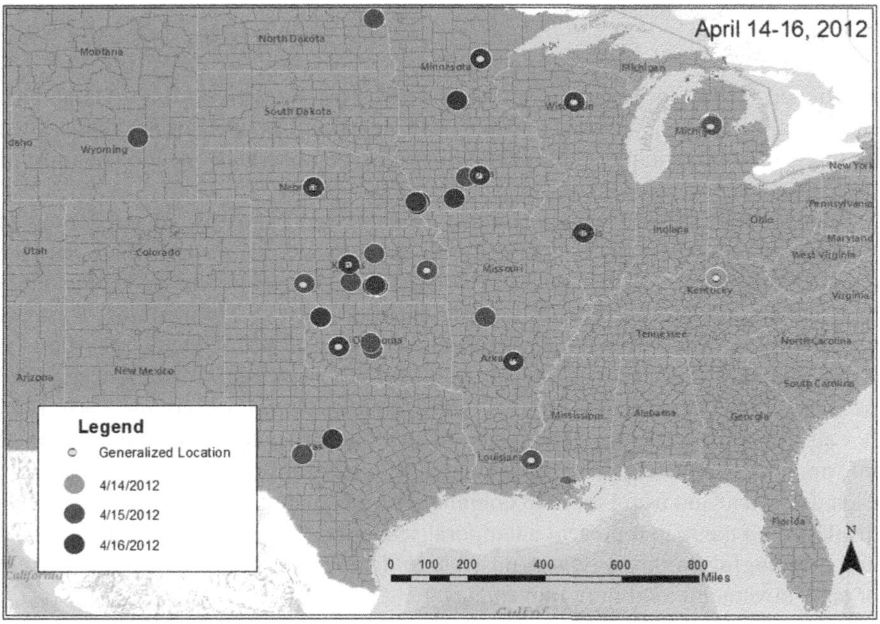

Fig. 10.3 Locations automatically extracted from news documents about tornadoes that struck Oklahoma, US in April 2012

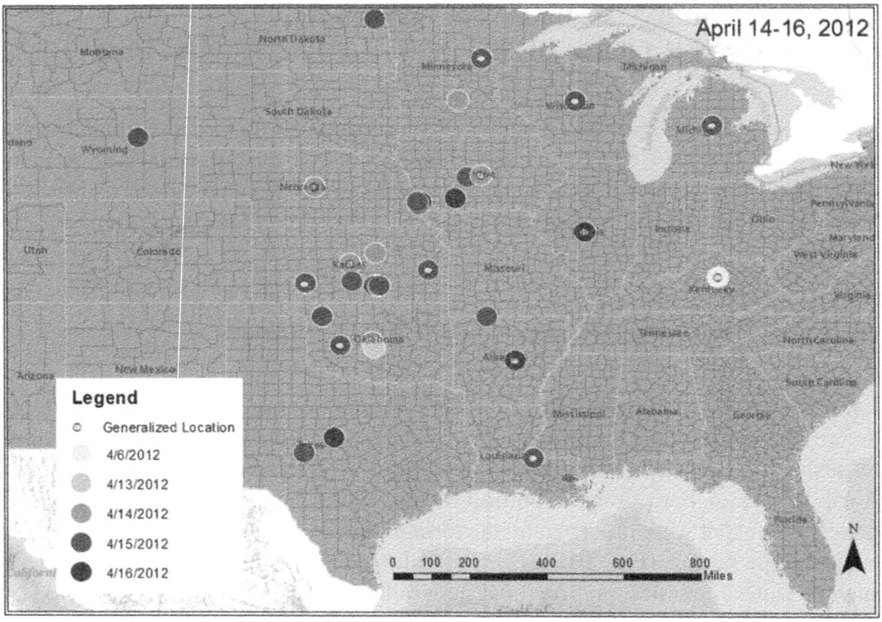

Fig. 10.4 Locations extracted from news reports about tornadoes in April 2012. Temporal information refines the spatial pattern of locations affected by the tornadoes associating events with more days as compared to the map in Fig. 10.3

mentioned in the documents (e.g., Kansas, western Kansas, northern Oklahoma, Nebraska, Michigan, Wisconsin, Minnesota, Iowa and Louisiana) are represented on the map with a unique symbol ○ to highlight the uncertainty of the locations of these events.

When temporal information about the tornado events is extracted and combined with spatial information, some of the locations identified in Fig. 10.3 where only the document date is used are actually associated with dates different to the document date and a greater range of dates is represented (Fig. 10.4). For example, Norman, Oklahoma, extracted from the news report on April 14, is associated with events on April 13th. Some state names, such as Kansas, Iowa, Nebraska, Oklahoma, reported on April 15th, are revealed to be actually associated with events on April 14th. It's also possible to detect some locations that are associated with future temporal information. After extracting and processing temporal references extracted from each document and using these in conjunction with spatial references, the locations displayed on the map represent a temporal sequence of events that is more realistic and meaningful. In this way, it is possible to distinguish information about current or past movements vs. future movements.

10.5.2 Incorporating Semantics

In this study, semantics about hazards relate mostly to two perspectives: natural phenomena and related human activities. Incorporating semantics of hazard-related events sourced from web texts describes weather-related topics as well as information about the human side of hazards. *A hazard ontology* is developed for this case study (Fig. 10.5). Two main upper-level classes are constructed for the ontology, *Object* and *Happening. Object* consists of three subclasses, *Place,*

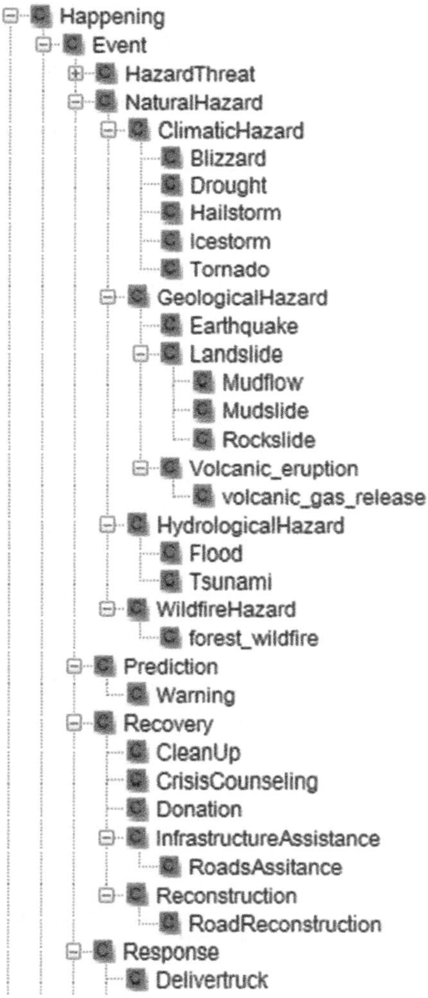

Fig. 10.5 A partial view of the classes in the hazard ontology

Agent, and *Vehicle*. *Happening* refers to the dynamic processes that occur during hazards. *Event*, is a subclass of *Happening*, and is associated with six subclasses *NaturalHazard*, *MeteorologicalPhenomena*, *Prediction*, *HazardThreat*, *Response*, and *Recovery*. Four subclasses are subsumed by class *NaturalHazard* including *Climatic Hazard*, *GeologicalHazard*, *HydrologicalHazard*, and *WildfireHazard*. *MeteorologicalPhenomena* subsumes weather classes including *Lightning*, *Thunder*, *Rain*, and *Wind*.

Prediction, *HazardThreat*, *Response*, and *Recovery* are related to classes of activities associated with hazards from a human perspective. *HazardThreat* subsumes subclasses *Injury*, *Fatality*, *HouseDamage*, *SchoolClosed*, *AirportClosed*, and *RoadClosed*. *Rescue*, *Evacuation*, *Search*, and *FoodDelivery* are subclasses of *Response*.

The hazard ontology is imported into GATE. By integrating gazetteers and ontologies, dynamics *and* semantic information about hazard events occurring at different locations is automatically extracted from news reports. Combining semantic information with the spatiotemporal mapped representation reveals the different kinds of events that unfold over space and time. The underlying relations (e.g., hierarchical relations) between events can be represented on maps, and with the relations, users can better comprehend details about the events described in documents.

Figure 10.6 shows the text terms associated with different classes in the ontology that are extracted from a news report about tornadoes in the Midwest

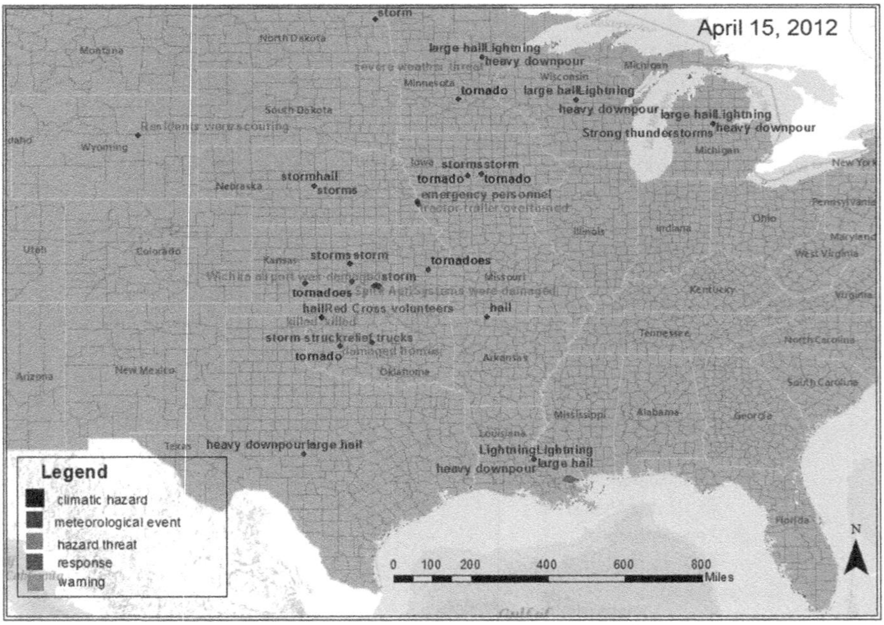

Fig. 10.6 Text terms from the news report about tornadoes that struck Oklahoma, US on April 15, 2012

on April 15th, 2012. The spatiotemporal pattern of terms is used to understand more about the dynamic tornado- related events across space. The map reveals that *large hail, lightning, heavy downpour,* and *strong thunderstorm* were associated with parts of Texas, Louisiana, Wisconsin, Michigan and Minnesota. Facilities were impacted by tornadoes, for example, *Wichita airport was damaged* in Kansas. In addition, relief activities were begun in Nebraska, Kansas, and Oklahoma. For example, *trucks were being sent out* for rescue in Oklahoma. Making sense of spatiotemporal information through extracted semantic information reveals details about the dynamic aspects of hazards for users.

The results can also be presented at a higher level of abstraction (Fig. 10.7). In this case, terms are presented at the granularity of super classes: *Warning, MeteorologicalPhenomena, ClimaticHazard, HazardThreat, Response* and *Recovery,* based on the hazard ontology. For example, *lightning, downpour, hail,* and *storm* are aggregated to the class: *MeteorologicalPhenomena* while *twister* and *tornado* are aggregated to the class of *ClimaticHazard.* It is possible to follow the sequence of these higher level semantics of a hazard (e.g., hurricane, response, recovery) automatically based on analysis of text. The map shows how the five classes vary spatially as per the report on April 15th, 2012. Based on this representation and for this document date, it is possible to see which areas are still experiencing storms (i.e., meteorological phenomena) and where response is beginning to occur.

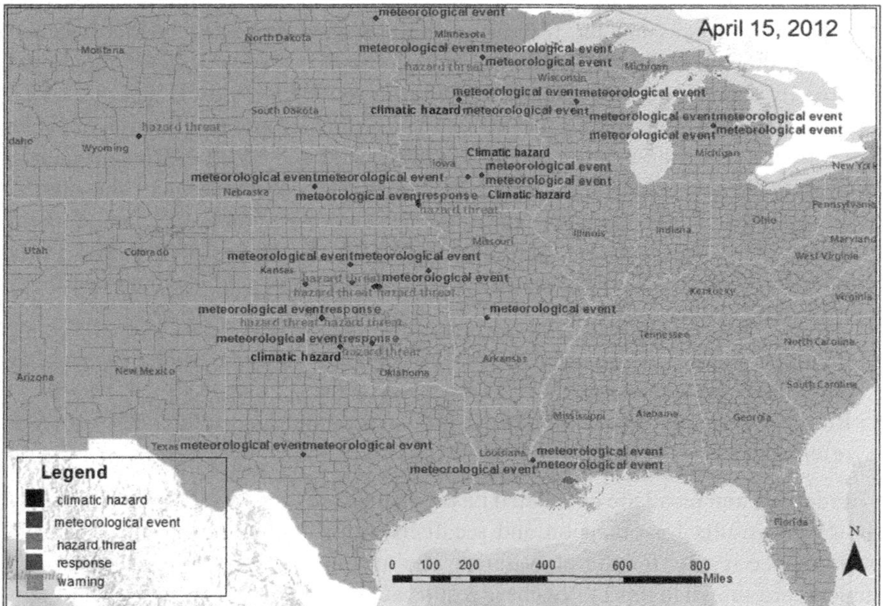

Fig. 10.7 Semantics extracted from a news report about tornadoes that struck Oklahoma, US on April 15, 2012

10.6 Evaluation of Combining Spatiotemporal Terms

The methods demonstrated in this research show how extracted spatiotemporal information can be integrated to represent the dynamics of events described in Web news reports. A preliminary evaluation of these results has been undertaken based on 20 news articles, selected as a sample data set for evaluating the methods. Kappa statistics are applied to provide a standard for which combining spatiotemporal information associated with events can be compared (Viera and Garrett 2005). In this evaluation, results automatically processed by the system are compared with results manually processed by humans. For human evaluators, five volunteers were recruited for evaluating the performance of system. None of the volunteers have GIR or NLP experience. The volunteers were trained to process spatiotemporal information from text documents by providing two sample news reports with instructions. Each volunteer manually annotated spatial and temporal terms from 20 news reports, and combined the annotated spatial and temporal results based on the context in the text documents. The spatiotemporal results from each volunteer are compared. An acceptable standard for assessing the results obtained from the manual test is where either the results by all volunteers agree, or four out of five of the results agree. All results that correspond to three out of five or two out of five are required to be rechecked by all volunteers, and results with 0 % agreement are excluded.

Manually-derived spatiotemporal combinations, and the results obtained from automatically processing the text using the approach presented in this research, are each compared with the standard. The number of correctly detected references, incorrectly detected references, and missed references for the users and the system are determined. A criterion for the evaluation of spatiotemporal extraction of events is developed based on adapting traditional precision and recall evaluation metrics as used in IR (Manning et al. 2008). Precision refers to the ratio of the number of correctly resolved spatiotemporal references and the number of spatiotemporal references that the system/or users attempts to resolve. Recall is the ratio of the number of correctly resolved spatiotemporal references and the number of all references.

- Precision = correctly detected references/ (correctly detected references + incorrectly detected references)
- Recall = correctly detected references/ (correctly detected references + missed references)

Precision and recall are calculated based on the metrics defined above. Table 10.1 shows the results for precision and recall based on human performance and the performance results for the algorithms. The average results annotated by human subjects are shown in this table. These results are based on 126 processed spatiotemporal references in the text documents. For this preliminary evaluation, there were an average of 116 correct references, 10 incorrect references, and 8 missed references. This results in precision and recall values of 0.92 and 0.94 respectively. For the

Table 10.1 Manually annotated and automatically processed precision and recall results based on numbers of solved spatiotemporal references

	Correct references		Incorrect references		Missed references		Precision		Recall	
Spatio-temporal references	Manual	Automated	Manual	Automated	Manual	Automated	Manual	Automated	Manual	Automated
20	19	17	1	3	1	2	0.95	0.85	0.95	0.89
50	47	41	3	9	4	6	0.94	0.82	0.92	0.87
80	74	65	6	15	6	10	0.93	0.81	0.93	0.87
126	116	108	10	18	8	15	0.92	0.81	0.94	0.89

same 126 processed spatiotemporal references, the system performed with 108 correct references, 18 incorrect references, and 15 missed spatiotemporal references. Based on this performance, precision and recall are calculated as 0.81 and 0.89 respectively. For this evaluation, for both the human evaluators and the system, it appears that as the total number of processed spatiotemporal references increases, the values of precision and recall decrease. Further testing is necessary to confirm the patterns observed here, but so far it appears the methods can lead to appropriate spatiotemporal assignments.

10.7 Conclusions

In this research, GIR methods and natural language processing techniques are combined such that spatiotemporal information is extracted automatically from sets of web documents to associate the semantics of events with spatial and temporal information. This offers a method to track the evolution of events over space-time described in web text and present this information in a geovisualization. A framework is presented for spatiotemporal semantic information extraction that is based on gazetteer and ontology creation, text processing, and geovisualization. Rules have been developed for combining extracted spatial and temporal terms in order to represent hazard events or other moving phenomena through space-time as they are described in text. Spatial and temporal gazetteers have been developed to aid the information retrieval process. These techniques are further extended through the use of ontologies that allow the semantics of events to be associated with spatial and temporal information. A hazard ontology and semantic gazetteer is invoked to support semantic representation of extracted events sourced from web texts. Including semantics contributes to a better understanding of spatiotemporal patterns of events from both natural and human perspectives. Using ontologies represents extracted text information in a hierarchical structure, and affords the opportunity to represent semantics at multiple granularities, giving further insights about events from a spatiotemporal perspective.

References

Chen, X., & Kwan, M.-P. (2012). Choice set formation with multiple flexible activities under space-time constrains. *International Journal of Geographic Information Science, 26*(5), 941–961.

Demsar, U., & Virrantaus, K. (2010). Space-time density of trajectories: Exploring spatio-temporal patterns in movement data. *International Journal of Geographic Information Science, 24*(10), 1527–1542.

Dodge, S., Weibel, R., & Lautenschutz, A.-K. (2008). Taking a systematic look at movement: Developing a taxonomy of movement patterns. *Information Visualization, 7*(3), 240–252. doi:10.1057/palgrave.ivs.9500182.

Dodge, S., Laube, P., & Weibel, R. (2012). Movement similarity assessment using symbolic representation of trajectories. *International Journal of Geographic Information Science, 26*(9), 1563–1588.

Jackoway, A., Samet, H., & Sankaranarayanan, J. (2011). Identification of live news events using Twitter. In *LBSN'11 proceedings of the 3rd ACM SIGSPATIAL international workshop on location-based social networks, Chicago, IL, USA* (pp. 25–32).

Janowicz, K., & Kebler, C. (2008). The role of ontology in improving gazetteer interaction. *International Journal of Geographical Information Science, 22*(10), 1129–1157.

Jones, C. B., & Purves, R. (2008). Geographical information retrieval. *International Journal of Geographical Information Science, 22*(3), 219–228.

Keller, M., Freifeld, C. C., Brownstein, J. S. (2008). *Expanding a gazetteer-based approach for geo-parsing text from media reports on global disease outbreaks*. In Proceedings of the 25th international conference on machine learning, Helsinki, Finland.

Kemp, Z., Tan, L., & Whalley, J. (2007). Interoperability for geospatial analysis: A semantics and ontology-based approach. In *ADC'07 proceedings of the eighteenth conference on Australasian database – Volume 63, Ballarat, VIC, Australia* (pp. 83–92).

Klippel, A., MacEachren, A., Mitra, P., Turton, I., Zhang, X., Jaiswal, A., Soon, K., Lyler, J., & Li, R. (2008). Geographic analysis of linguistically encoded movement patterns – A contextualized perspective. In *Extended abstracts for the 5th international conference GI Science 2008, Park City, Utah, USA* (pp. 113–117).

Kwan, M.-P. (2000). Interactive geovisualization of activity-travel patterns using three-dimensional geographical information systems: A methodological exploration with a large data set. *Transportation Research Part C: Emerging Technologies, 8*(1), 185–203.

Kwan, M.-P., & Ding, G. X. (2008). Geo-narrative geographic information systems for narrative analysis in qualitative and mixed-methods research. *The Professional Geographer, 60*(4), 443–465.

Laube, P., Dennis, T., Forer, P., & Walker, M. (2007). Movement beyond the snapshot–dynamic analysis of geospatial lifelines computers. *Journal of Environment and Urban Systems, 31*(5), 481–501.

Lieberman, M. D., Samet, H., Sankaranarayanan, J., & Sperling, J. (2007). STEWARD: architecture of a spatio-textual search engine. In *Proceedings of the 15th Annual ACM International Symposium on Advances in Geographic Information System, WA, USA* (pp. 186–193).

Machado, I. M. R., Alencar, R. O. D., Campos, R. D. O., & Clodoveu, A. D. (2011). An ontological gazetteer and its application for place name disambiguation in text. *Journal of the Brazilian Computer Science, 17*(4), 267–279.

Manning, C. D., Raghavan, P., & Schütze, H. (2008). *Introduction to information retrieval*. Cambridge: Cambridge University Press.

Miller, H. J. (1991). Modeling accessibility using space-time prism concepts within geographical information systems. *International Journal of Geographical Information Systems, 5*(3), 281–301.

Ping, D., & Yong, L. (2009). Building place name ontology to assist in geographic information retrieval. *International Forum on Computer Science-Technology and Applications, 1*, 306–309.

Purves, R., & Jones, C. (2011). Geographic information retrieval. *SIGSPATIAL Special, 3*(2), 2–4.

Raubal, M., Miller, H. J., & Bridwel, S. (2004). User-centred time geography for location-based services. *Human Geography, 86*(4), 245–266.

Sankaranarayanan, J., Samet, H., Teitler, B. E., Lieberman, M. D., & Sperling, J. (2009). TwitterStand: News in tweets. In *GIS'09 proceedings of the 17th ACM SIGSPATIAL international conference on advances in geographic information systems, Seattle, WA, USA* (pp. 42–51).

Shaw, S. L., Yu, H. B., & Bombom, L. S. (2008). A space-time GIS approach to exploring large individual-based spatiotemporal datasets. *Transaction in GIS, 12*(4), 425–411.

Stewart Hornsby, K., & Li, N. C. (2009). Conceptual framework for modeling dynamic paths from natural language expressions. *Transactions in GIS, 13*(1), 27–45.

Stewart Hornsby, K., & Wang, W. (2010). Representing dynamic phenomena based on spatiotemporal information extracted from web documents. *Extended abstracts, GIScience Conference 2010*. http://www.giscience2010.org/pdfs/paper_131.pdf. Accessed 31 May 2012.

Teitler, B. E., Lieberman, M. D., Panozzo, D., Sankaranarayanan, J., Samet, H., & Sperling, J. (2008). NewsStand: A new view on news. In *Proceedings of the 16th ACM SIGSPATIAL International Conference on Advances in Geographic Information Systems, Irvine, California, USA, 18*, 1–10.

Tomaszewski, B. (2008). Producing geo-historical context from implicit sources: A geovisual analytics approach. *The Cartographic Journal, 45*(3), 165–181.

Tomaszewski, B., & MacEachren, A. (2010). Geo-historical context support for information foraging and sensemaking: Conceptual Model, implementation, and assessment. In *IEEE Conference on Visual Analytics Science and Technology (IEEE VAST), Salt Lake City, Utah, USA* (pp. 139–146).

Viera, A. J., & Garrett, J. M. (2005). Understanding interobserver agreement: The Kappa statistic. *Family Medicine, 37*(5), 360–363.

Volz, R., Kleb, J., & Mueller, W. (2007). Towards ontology-based disambiguation of geographical identifiers. In *Proceedings of the WWW2007 Workshop Identity, Identifiers, Identification, Entity-Centric Approaches to Information and Knowledge Management on the Web, Banff, Canada, 25*, 1–8.

Wiegand, N., & Garcia, C. (2007). A task-based ontology approach to automate geospatial data retrieval. *Transaction in GIS, 11*(3), 355–376.

Wood, Z., & Galton, A. (2010). Zooming in on collective motion. In *Proceedings of Workshop 21, 19th European Conference on Artificial Intelligence, Lisbon, Portugal* (pp. 25–30).

Woodruff, A. G., & Plaunt, C. (1994). GIPSY: Automated geographic indexing of text documents. *Journal of the American Society for Information Science, 45*(9), 645–655.

Yu, H. B. (2006). Spatio-temporal GIS design for exploring interactions of human activities. *Cartography and Geographic Information Science, 33*(1), 3–19.

Yuan, M., & Stewart Hornsby, K. (2008). *Computation and visualization for understanding dynamics in geographic domains*. Boca Raton: CRC.

Zhang, X., Mitra, P., Klippel, A., & MacEachren, A. (2010). Automatic extraction of destinations, origins and route parts from human generated route directions. *Geographic Information Science Lecture Notes in Computer Science, 6292*, 279–294.

Chapter 11
The HGIS Experience of Drawing Sub-county Unit Boundaries in the Jiangnan Region of Late Imperial China

Ka-chai Tam and Kee-long Billy So

11.1 Introduction

During the late imperial period (1100–1911), the Chinese documented local information on almost every part of the imperial realm into generations of local gazetteers (difangzhi 地方志 or fangzhi 方志). "Local gazetteers are chronicles of the history, present conditions, and noted people of local areas, arranged by topic (Brook 1988: 49–72)." In addition to the detailed accounts of taxation and population records, local customs, governmental policies, biographies of native notables and local historical events, these gazetteers also include primitive maps of towns and other local landmarks. However, there are two obvious limitations to consulting these highly specialized sources when performing geographical and historical research. First, the primitive maps and diagrams of these gazetteers provide insufficient coordination for modern cartography, and give few hints related to spatial analysis (Fig. 11.1). In other words, it would not be worthwhile to handle them directly as secondary sources to be converted into Geographical Information System (GIS) raster data by computer scanning. Second, although much of the information provided by the gazetteers is extremely precise in spatial terms, without a reliable unit platform for sub-county divisions and landscape structures, they can only be analyzed verbally at the county (xian 縣), prefectural (fu 府) and even provincial levels. It used to be almost impossible to conduct a meso-level analysis of the individual households and the county, i.e., the basic units of Chinese imperial control.

K.C. Tam (✉)
Department of History, Hong Kong Baptist University, Kowloon Tong, Kowloon, Hong Kong
e-mail: kctam@hkbu.edu.hk

K.L. Billy So
Division of Humanities, Hong Kong University of Science and Technology, Clear Water Bay, Kowloon, Hong Kong

M.-P. Kwan et al. (eds.), *Space-Time Integration in Geography and GIScience*, DOI 10.1007/978-94-017-9205-9_11

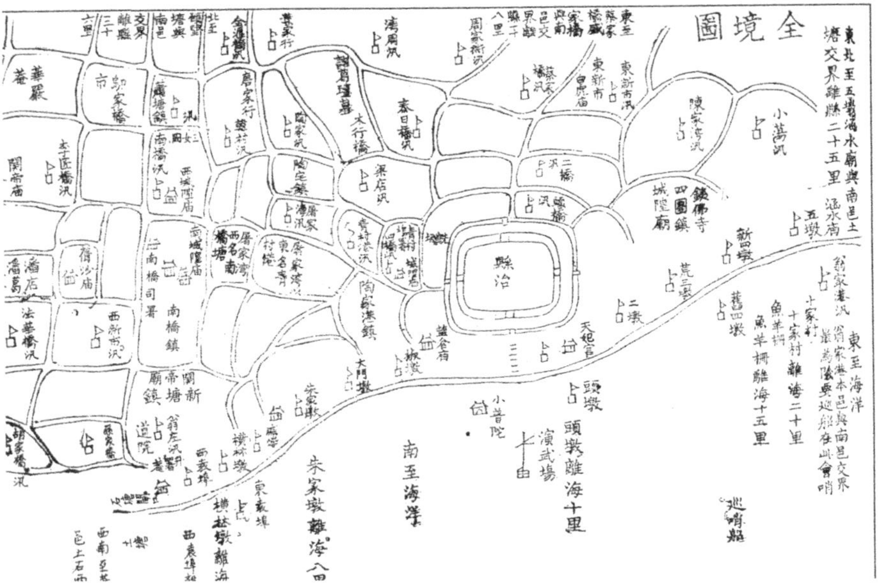

Fig. 11.1 A typical abstract map of traditional Chinese local gazetteers

In preparing the "Geographical Information System Database of Cotton Textile Industry of the Greater Songjiang Region from the Late Ming to the Mid-Qing (http://www.iseis.cuhk.edu.hk/songjiang/)," we used high scale digital maps of modern Shanghai as our base maps and GIS technologies to partially overcome the aforementioned shortcomings of the local gazetteers. We transformed the verbal information and primitive maps of these traditional sources into agreeable forms of modern cartography that were ready for spatial analysis.

In short, after plotting the exact locations of hundreds of towns, temples, rivers and other landmarks, we were then able to draw the boundaries of sub-county units or townships (*du* 都 or *bao* 保) beneath the county or prefectural levels. In many Jiangnan counties, these units nearly went unchanged during the 600 years of Ming-Qing rule, and thus they provide a perfect platform for conducting a meso-spatial analysis that would be far more precise than that restricted at the original county or prefectural levels.

In this study, the structure and characteristics of the Chinese local gazetteers are first introduced and practical hints are offered on how information related to the building of GISs is derived from the various categories of these specialized historical sources. This is followed by a step-by-step demonstration of our experiences in building a historical GIS for the Ming-Qing Songjiang region (modern Shanghai). The study concludes with a discussion on the use of sub-county polygons in historical and geographical research.

11.2 Traditional Chinese Local Gazetteers

No fewer than 8,000 sets of local gazetteers compiled prior to the People's Republic of China (1949) still exist today, and various levels of modern China's local governments continue the tradition by establishing local history offices to compile current gazetteers. They are by far the largest component of the history category (shibu 史部) of works in traditional Chinese bibliographies, and thus offer an indispensible collection of sources for spatial analyses of various parts of China, past and present, that are not found elsewhere.

Since the beginning of the Qing period (1644–1911), to avoid the restrictions imposed by the Manchu government on compiling general political histories of the Ming and earlier periods, some experienced historians shifted their concerns and participated in the compilation of these local histories. Among these historians was Zhang Xuecheng 章學誠 (1738–1801), who first suggested that a standard format of local gazetteers based on the many earlier works of its kind should be established for all future compilation projects to guarantee the quality of local history records all over the Qing realm (Fang 2011: 190–196). In other words, learning the ideas and characteristics of the standard format is the key to understanding the characteristics and therefore the use of the great literature of China's local gazetteers.

Zhang's structure includes up to 25 local gazetteer content categories, each of which can be further divided into some chapters (Wilkinson 2000: 154–158). All local gazetteers open with (1) the front matter, i.e., some prefaces and the general rules (fanli 凡例) of the compilers. The background and editorial concerns of the work are explained in this section. The front matter is followed by (2) the maps of the locality (yutu 輿圖), including the maps and diagrams of the county's boundaries, its water and irrigation systems, its towns and military garrison points, its tourist scenes, the city plan of the county capital and other important towns and so on. Various types of geographical information related to the county are then provided, sometimes in the form of map commentaries. These include (3) the county's transforming borders (jiangyu 疆域) and changing administration units (perhaps in table form) (yan'ge 沿革); (4) topographical features such as rivers and mountains (shanchuan 山川); (5) famous scenes (mingsheng 名勝), gardens (yuanlin 園林) and ruins (guji 古蹟); (6) official buildings (gongshu 公署), city walls and moats (chengchi 城池) and public works (jianzhi 建置); (7) passes, fords and bridges (guanjin qiaoliang 關津橋樑); and (8) water conservancy (hefang 河防), canals and rivers, hydraulics (hequ 河渠) and irrigation works (shuili 水利). The information provided by these sections is directly relevant to the respective study topics on the locality. Moreover, gazetteers usually supply the exact directions and distance counts in steps (bu 步, equivalent to a meter) of landmarks measured from the gates of the county capital or the front door of the magistrate hall, thus presenting some accurate coordination of the county's geographical features.

The information on county landmarks is followed by local history categories, including (9) a chronicle of natural and human disasters or omens (zaiyi 災異, xiangyi 祥異 or bing 兵) such as floods, droughts, hail, snow, locusts, uprisings, civil wars, high prices and other strange events. The educational and religious institutions of the locality are then presented, including (10) academics (private Confucian institutes or shuyuan 書院), (11) government schools (xuexiao 學校) and (12) Buddhist and Daoist temples (siguan 寺觀). In many cases, the exact coordinates of these schools and temples are included alongside other historical details.

Institutional information about the county is then offered, including information on its (13) officeholders or official posts (zhiguan 職官), (14) examinations and lists of successful candidates (xuangju 選舉) and (15) military institutions and men. These are followed by practical information categories such as (16) fiscal information related to household and head counts (hukou 戶口), land and other taxes (fushui 賦稅, tianfu 田賦); (17) granary reserves (cangchu 倉儲); (18) market towns, tolls and barriers (shizhen 市鎮) (in some gazetteers, the nature and specialties of these towns are mentioned along with their exact coordinates relative to the county capital); (19) products, produce and crops (wuchan 物產); and (20) customs and festivals (fengsu 風俗), including detailed accounts of the county's local cults and religious practices.

After these general institutional information categories, various types of notable native biographies are presented, including (21) biographies of upright officials, dignitaries and chaste women and (22) biographies of technicians such as doctors (fangji 方技) and Buddhist and Daoist monks (shilao 釋老).

The end of a local gazetteer sometimes includes up to three additional categories including (23) inscriptions and tombs (jinshi 金石), (24) bibliographies and selected excerpts of good essays from native writers (yimen 藝文) and (25) miscellaneous topics and records (zalu 雜錄), a section in which strange stories about ghosts and animals are occasionally recorded. These stories are very helpful for biographical research, as they may contain information about local figures not found elsewhere.

11.3 Historical GIS for the Songjiang Region During the Ming-Qing Periods

During the Ming and Qing periods (1368–1911), the region of Songjiang 松江 prefecture (which was established around 1277 during the Yuan period) and some adjacent counties, namely, Jiading 嘉定, Kunshan 崑山 and the Taicang 太倉 sub-prefecture of the Suzhou 蘇州 prefecture, as well as Jiashan 嘉善 and Pinghu 平湖 of the Jiaxing 嘉興 prefecture (i.e. the Greater Songjiang region of our HGIS), was the center of the Chinese cotton textile industry. In modern times, the Songjiang prefecture and Jiading county are incorporated into Shanghai Municipality (So and Tam 2011). From 1500–1800, it was one of the most densely populated areas in the world, with intervals appearing only during the catastrophic Ming-Qing transition (So and Tam 2011: 45–54). The local gazetteer of Songjiang prefecture published during Chongzhen's reign (1628–1644) at the end of the Ming Dynasty states that Su-Song-Jia-Hu (i.e., the four prefectures of Suzhou, Songjiang, Jiaxing and Huzhou 湖州) were "the most prominent prefectures in Southeast

China." In fact, the cotton-producing Songjiang, together with the nearby silk-textile center of Suzhou, constituted the core of the booming Jiangnan market economy in the late Ming and high Qing periods.

Chinese historical geography once applied only a qualitative method when reconstructing a historical horizontal cross-section to examine related sectional issues (So and Tam 2012). This method is inadequate in analyzing transitions via synchronic and diachronic comparisons (Li 2000). The increasing application of GISs in recent decades has had an impressive effect on the approaches of historical geographers, as it is now possible to analyze the spatial changes of geography over time using a computer (Zhang 2006). The combination of GISs and historical geography (i.e., historical GISs) has become an increasingly attractive research direction. Several historical GISs were prepared recently to provide a good deal of historical geography information (Lin et al. 2006). For example, Harvard University and Fudan University co-developed a Chinese historical GIS to establish a database of populated places and historical administrative units over the course of Chinese history from 222 B.C. to 1911. Moreover, the "Chinese civilization in time and space" set up by the Academia Sinica of Taiwan constructed a GIS application infrastructure for China across a timeframe of several thousand years (Zhang et al. 2007). These are important attempts to establish China's historical GISs at the national level. However, because they were created by scanning the secondary source of the Chinese historical maps prepared by Tan Qixiang 譚其驤 (originally done from base maps with 1:4,000,000 scales; Tan 1980) and provide only prefecture boundaries (Fig. 11.2), they can only create polygons at the prefectural level.

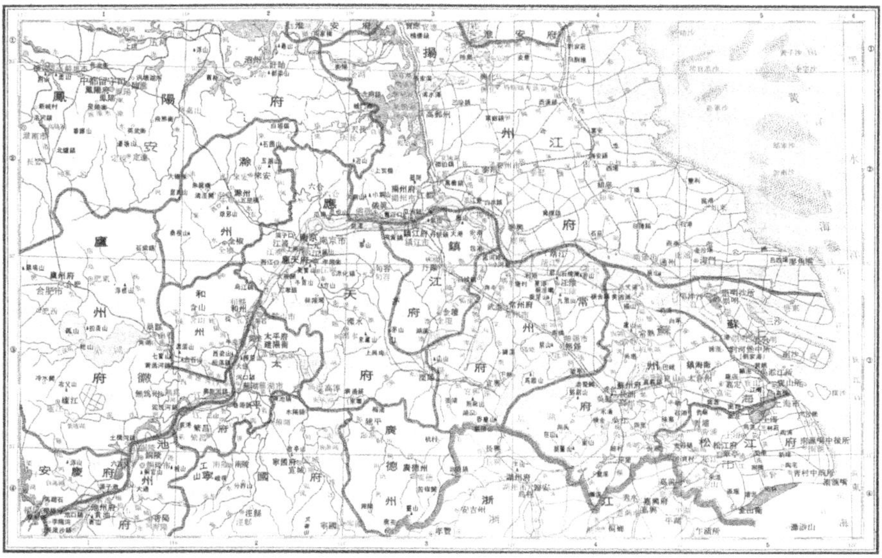

Fig. 11.2 A sample of the historical maps prepared by Tan Qixiang

Against this background, we designed our Hong Kong-based "Geographical Information System Database of Cotton Textile Industry of the Greater Songjiang Region from the Late Ming to the Mid Qing" as a GIS platform that uses quantitative and spatial methods to analyze the successful experiences of this cotton-textile production region over the last few centuries. To facilitate a precise intra-regional analysis within the Greater Songjiang region, this historical GIS identifies town and city locations and then draws the sub-county division boundaries (i.e., *bao* or *du*) in reference to the location data of temples and schools, trade routes, topology and so on. This is all executed according to information derived from the different categories of several generations of local gazetteers from the Greater Songjiang region.

To build a meaningful GIS that is appropriate for spatial analysis, it is critical to create spatial units at the highest resolutions possible so that the advantages of the computer system's calculation and display can be fully exploited. In a GIS, maps are normally converted into raster data by scanning (Gregory and Ell 2007: 43). We are well aware of the limitations in using Tan Qixiang's historical maps. Therefore, our first step was to lay our foundation on a modern Shanghai area map at a much higher resolution. We selected the digital version of the 1:150,000 map of Shanghai Municipality circa 1999 prepared by the Shanghai Institute of Surveying and Cartography (www.shsmi.cn) as our base map, with reference to historical maps collected from the *Shanghai lishi dituji* 上海歷史地圖集 (1999) and modern maps of nearby counties with scales of about 1:200,000.

Because the target of our research is a relatively small region the size of one or two prefectures (comprising seven to nine counties), drawing the administrative unit boundaries small enough to suit our meso-level research purpose required drawing the township boundaries, i.e. the *bao* of the Songjiang prefecture and the *du* of other counties in the Jiangnan region. The role of these micro divisions in traditional Chinese administrative systems can be considered as follows. Under a prefecture, there are counties; under the counties, there are *xiang* 鄉 units and under the *xiang* units are the *bao* or *du*. From the early Ming Dynasty to the Republican period, a varied number of taxation and administrative units of 110 households, i.e., tu 圖 or li 里, were assigned to each *bao* or *du*. In the traditional Chinese administrative systems, the townships were the principal units by which a county magistrate organized tax administration and were usually identified by a numbering system. The Ming Dynasty Founding Emperor Zhu Yuanzhang 朱元璋 (1368–1398) saw the townships in some regions as suitable units for setting up community granaries. He also ordered every unit to construct altars for conducting state-cult sacrifices. These units were also used to coordinate water control (Brook 2005). Although many new counties in the region were created over the course of the Ming-Qing periods by re-organizing the original county borders, the basic *bao* and *du* units remained basically unchanged until the end of the Qing Dynasty. As we used very similar measures to deal with the region's distinct parts, we now take the Songjiang prefecture as an example to explain our methods and procedures for drawing the sub-county unit boundaries.

Indeed, from the early Ming to the late Qing periods, generations of local gazetteers from the Songjiang prefecture, or the Songjiang fuji 松江府志 ([Chongzhen] Songjiang fuji (崇禎) 松江府志 1991; [Jiaqing] Songjiang fuji (嘉慶) 松江府志 1991; [Kangxi] Songjiang fuji (康熙) 松江府志 and [Zhengde] Songjiang fuji (正德) 松江府志 1990), all recorded information about market towns in category (18) (i.e., market towns) along with their locations in each *bao* unit, their coordinates relative to the county seat and their commerce and industry specialties. When grouping the towns of each *bao* together, it is clear that usually more than one town falls into the area of each division. In total, more than 160 towns with clear *bao* identities appear in different versions of the Songjiang prefecture's local gazetteers. In short, these market town points should be located within the boundaries of the relevant *bao* so that if the points of the towns in each division are plotted on our high resolution base map it would be possible to draft the rough borders of the *bao* that encircles those points.

The plotting of the Ming-Qing towns onto the base map began by discovering the modern residential points of towns with the same or very similar names. We also tried to compare the modern location of the towns against their locations on the Ming, Qing and high resolution Republican historical maps of the *Shanghai lishi dituji*. If any differences in location were found, we plotted both places, but tried to follow the earlier-period maps to mark the boundaries of the *bao*. In the due course of searching for the Ming-Qing town locations, we realized that some of them no longer exist in modern times, and we could only plot them with reference to the coordinates given by the local gazetteers and the aforementioned historical maps. The "Market Town" chapter of the *Songjiang fuji*, therefore, provides the coordinates of towns in the form of their relative directions and distances measured in Chinese li 里, equal to 460 m. After putting those towns on the base map, we then cross-checked their locations with the coordinate data by using the ArcGIS software measurement function. It is crucial to point out that our plotted points generally agreed with the coordination data of the local gazetteers.

The next procedure involved drawing the *bao* boundary lines around the market town points. In the GIS field, it is desirable to convert point data to polygons using the Thessian polygon (also known as the Voronoi polygon) method. These polygons are created from a point layer by drawing boundaries that separate each point from its neighbors. The boundaries are located such that they are equidistant to the two points. Each polygon theoretically contains only one point. This method makes it possible to recreate boundaries that have never been mapped according to modern standards, such as the township boundaries of the Songjiang region. Its limitation, however, is very obvious: the polygons are triangulated into a triangulated irregular network. In his attempt to recreate the 1971 and 1981 English and Welsh Enumeration District units, Walford (2005) found Thessian polygons unsatisfactory due to their unnatural appearance, and that they were not suitable for spatial analysis due to their inaccuracy. He used the Thessian polygons only when no other hybrid approaches were appropriate (Gregory and Ell 2007: 68–70).

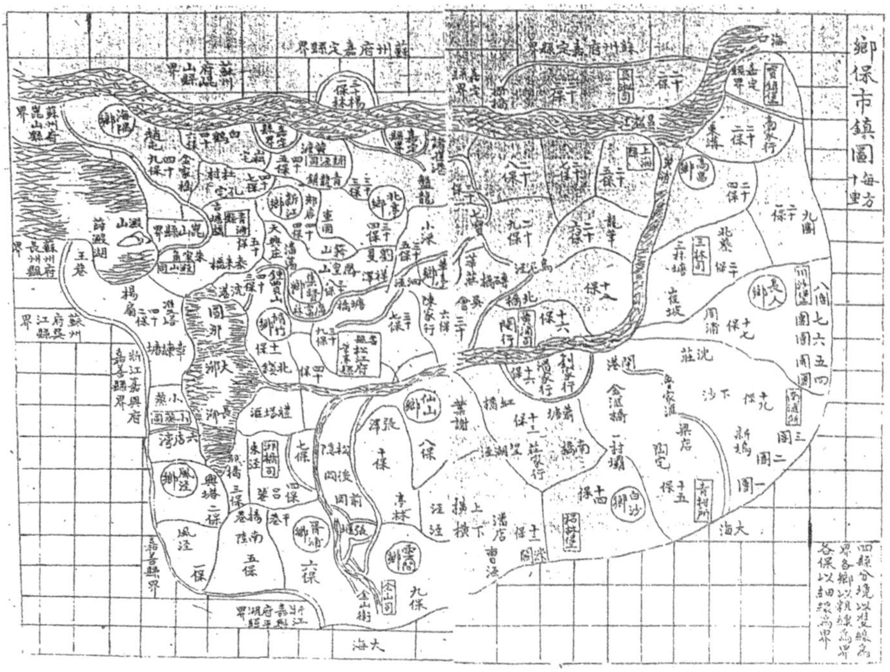

Fig. 11.3 The map of sub-county divisions and market towns of the *Songjiang fuji*

Nevertheless, there was more than one accurate town location alongside the other topological points and township boundary information in each of the *bao* units. As such, we abandoned the Thessian polygons and opted for recreating these boundaries according to our own method, which relied heavily on the information provided by the local gazetteers.

The abstract maps of the "Xiangbao shizhen tu 鄉保市鎮圖 (map of sub-county divisions and market towns)" found in category (2) of maps and diagrams of the *Songjiang fuji* published in the Jiajing reign (1792–1820) in the mid-Qing period) were our main references for determining the basic shape of the *bao* polygons (Fig. 11.3). However, because this map was not drawn according to modern cartography standards, further evidence was required to modify the drafted boundaries of the *bao*. Three more steps were therefore necessary to produce polygons that met our expectations.

First, we believe that when the *bao* was created, there were solid reasons for drawing their boundaries, and the best grounds for division were undoubtedly natural landmarks such as rivers and hills. Therefore, roughly following the shapes of the *bao* as shown in the local gazetteer maps, we moved their borders to nearby topological features shown on the base maps. Second, we modified the boundaries by consulting the waters displayed in the high resolution *Shanghai lishi dituji* maps and contemporary satellite photos of the Shanghai region.

Fig. 11.4 The *bao* divisions of the Songjiang prefecture (The points representing the towns of the Songjiang prefecture)

Last but not least, we used hundreds of temple points to boost the accuracy of the *bao* boundaries. In category (12) Buddhist and Daoist temples of different generations of the *Songjiang fuji*, the coordinates and *bao* identities of about 500 temples of different religions are given in detail. We applied the same principles behind encircling the town points to drawing the *bao* boundaries to address the temples. Thus, we created about 50 polygons for the Songjiang prefecture sub-county division on a base of over 650 points (Fig. 11.4).

One possible contribution of this sub-county division boundary drawing, which is the first and only drawing of its kind in the Chinese history field, to historical geography is that the infrastructure of these accurately drawn polygons are measurable in the ArcGIS system. It is because the sizes of the respective areas are measurable, we could now calculate and display the population density and taxation levels of each micro-division when the relevant population and fiscal figures were taken from category (16) (i.e., fiscal matters) of the local gazetteers.

The local gazetteers of the Songjiang region do not directly give the population figure of each *bao*. Instead, they provide the *li* or ding 丁 (taxable male) figures of all of the townships. These figures were originally assigned according to the actual population of the *bao* in the early Ming and early Qing periods, respectively. We may estimate that they represent the approximate percentages of each township against the total of the Songjiang prefecture. Amongst the versions of the *Songjiang fuji*, the most comprehensive figures of the *li* and *ding* are given in the fiscal matters category of the early Qing version, which was compiled during Kangxi's reign (1677–1722). Table 11.1 shows the number of *li* and *ding* of each *bao* and their population percentages (4 sig. fig.) against the total of the Songjiang prefecture.

Table 11.1 The number of *li* and *ding* of each *bao* and their population percentages against the total of the Songjiang prefecture

Bao	Number of *li*	Percentage (%)	Number of *ding*	Percentage (%)
1	26	1.935	4,041	1.942
2	12	0.893	1,167	0.561
3	29	2.158	3,987	1.916
4	32	2.381	2,838	1.364
5	25	1.860	2,423	1.164
6	22	1.637	3,092	1.486
7	41	3.051	5,191	2.494
8	22	1.637	2,785	1.338
9	14	1.042	2,087	1.003
10	15	1.116	2,557	1.229
11	15	1.116	2,553	1.227
12	30	2.232	5,268	2.531
13	38	2.827	5,961	2.864
14	41	3.051	6,656	3.198
15	63	4.688	10,747	5.164
16	30	2.232	4,633	2.226
17	34	2.530	10,196	4.899
18	43	3.199	5,323	2.558
19	89	6.622	19,325	9.285
20	27	2.009	6,360	3.056
21	25	1.860	4,874	2.342
22	43	3.199	7,952	3.821
23	15	1.116	2,714	1.304
24	41	3.051	6,506	3.126
25	12	0.893	1,137	0.546
26	14	1.042	2,853	1.371
27	11	0.819	1,907	0.916
28	17	1.265	3,428	1.647
29	6	0.446	2,026	0.973
30	11	0.819	2,423	1.164
31	8	0.595	647	0.311
33	7	0.521	503	0.242
34	12	0.893	1,057	0.508
35	32	2.381	4,697	2.257
36	53	3.944	6,453	3.100
37	31	2.307	2,867	1.378
38	47	3.497	5,837	2.804
39	20	1.488	2,184	1.049
40	25	1.860	3,000	1.441
41	68	5.060	10,872	5.224
42	47	3.497	7,516	3.611

(continued)

Table 11.1 (continued)

Bao	Number of *li*	Percentage (%)	Number of *ding*	Percentage (%)
43	50	3.720	9,059	4.353
44	4	0.298	374	0.180
45	23	1.711	1,198	0.576
46	14	1.042	1,840	0.884
47	11	0.819	737	0.354
49	19	1.414	2,429	1.167
50	30	2.232	3,854	1.852
Total	**1,344**	**100**	**208,134**	**100**

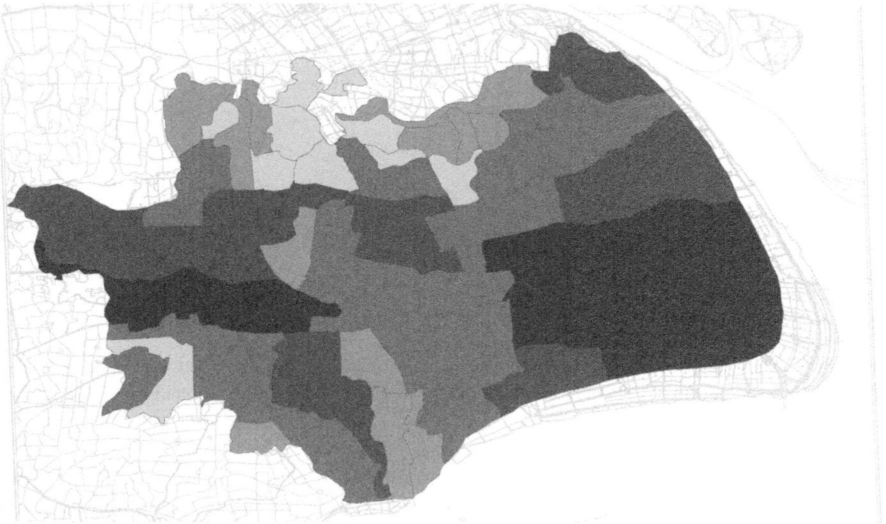

Fig. 11.5 A demonstration of the display of population density in our HGIS (The darker color representing higher population density of the sub-county units)

Applying this list of population percentages to the township polygon infrastructure allowed us to calculate the population figures of each *bao*, and therefore enquire into the population density and taxation issues of these micro-divisions (Fig. 11.5).

11.4 Concluding Remarks: Potential of Sub-county Units in Historical and Geographical Research

The aforementioned methods of drawing the *bao* boundaries of the Songjiang prefecture circa the Ming-Qing periods are only a few of the ways in which the verbal and qualitative information collected by traditional local gazetteers has been

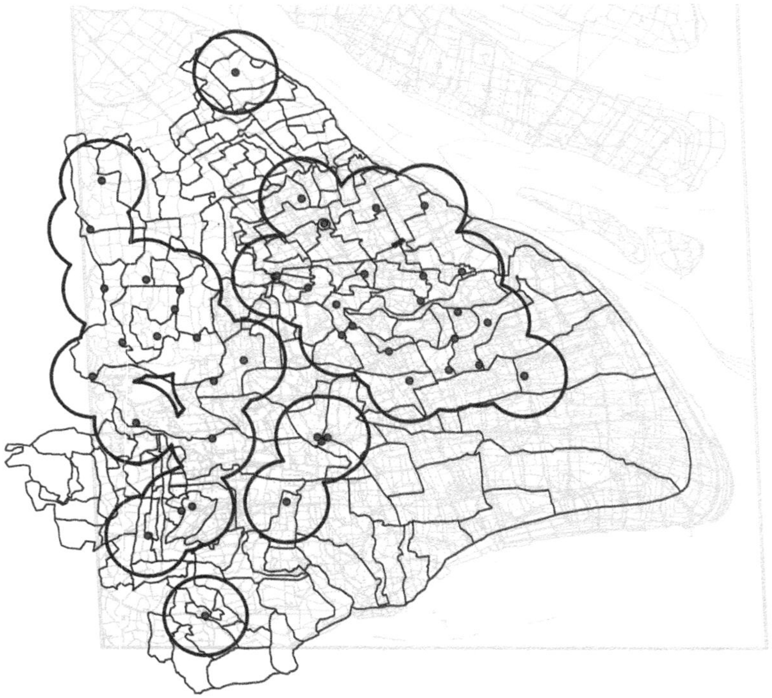

Fig. 11.6 The 8 km buffer zone of the cotton cloth textile towns of the Greater Songjiang region during the Ming period (The points of the circle centers representing the cotton cloth textile towns)

transformed into electronic displays and quantitative data in a modern HGIS (Tam 2012). In theory, because the methods used to recreate the Songjiang region's township boundaries can be applied to every part of China when relevant local gazetteers are available, it is possible to build a GIS infrastructure for China's historical geography at a much higher resolution than the prefectural level presently used by most of the country's HGISs. Moreover, the nature of each local gazetteer category should be carefully explored to generate ideas and methods for possible new GIS applications. It is also important to be aware that some GIS applications are more related to the data collected from certain local gazetteer categories.

For example, because the mode of cotton cloth production in the Songjiang region involved weaving performed by households residing in the villages surrounding the cotton towns, it was sensible to use the "buffer area" function of the GIS to create zones around the relevant towns so that the actual area of cotton cloth production could be identified (Fig. 11.6). To recognize the cotton towns, we consulted category (18) (i.e., market towns) of the local gazetteers in which the nature and specialties of these towns are mentioned. However, we then faced the question of how long the length of the buffer zone area's radius should be. As a general rule, the answers to similar economic and social questions are usually found in the last two categories of the gazetteers, namely, (19) products, produces and crops (wuchan 物產) and (20) customs and festivals (fengsu 風俗). In fact, the "fengsu" chapter of the late Qing version of the Jiading xianji 嘉定縣志 states that when carrying goods to trade in market towns, the villagers of the county shall not travel farther than 40 *li* so that they may return home within the same day (1991: 158b). In other words, a single trip should not exceed 20 *li* or 9.2 km, and thus it was safe to assume that the cotton-cloth-production buffer zone radius should be about 8–9 km. On the basis of the township unit boundaries, for instance, we could easily locate the *bao* in which cotton cloth production were concentrated in each stage of the Ming-Qing period, and cross-check whether their taxation responsibilities were changed accordingly.

Not all local gazetteers are compiled with all of the essential information prescribed in the perfect structure introduced by the Qing scholar Zhang Xuecheng. However, by incorporating the various kinds of available materials collected by these geographical sources into the historical GIS, we could create a much firmer ground on which to conduct multidisciplinary studies that combine socioeconomic history, local history, GISs, regional analysis and other social science methods to generate a new understanding of and new research strategies for the economic development of China's regions, past and present.

References

[Chongzhen] *Songjiang* fuji (崇禎) 松江府志. (1991), compiled by Fang Yuegong 方岳貢. Beijing: Shumu wenxian chubanshe.

[Guangxu] *Jiading* xianji (光緒) 嘉定縣志. (1991), compiled by Cheng Qiyu 程其. Shanghai: Shanghai shudian.

[Jiaqing] Songjiang fuji (嘉慶) 松江府志. (1991), compiled by Song Rulin 宋如林. Shanghai: Shanghai shudian.

[Kangxi] *Songjiang* fuji (康熙) 松江府志, part of the rare book collection of Shanghai Library, Shanghai, China.

[Zhengde] *Songjiang* fuji (正德) 松江府志. (1990), compiled by Gu Qing. Shanghai: Shanghai guji shudian.

Brook, T. (1988). *Geographical sources of Ming-Qing history*. Ann Arbor: Center for Chinese Studies, University of Michigan.

Brook, T. (2005). The spatial organization of subcounty administration. In T. Brook (Ed.), *The Chinese State in Ming Society*. Abingdon/New York: RoutledgeCurzon.

Fang, Z. (2011). *Zhongguo shixue gaiyao* 中國史學概要. Wuhan: Wuhan daxue chubanshe.

Gregory, I. N., & Ell, P. S. (2007). *Historical GIS: Technologies, methodologies and scholarship*. Cambridge: Cambridge University Press.

Li, B. (2000). *Jiangnan de zaoqi gongyehua* 江南的早期工業化. Beijing: Shehui kexue chubanshe.

Lin, H., Jie, Z., Ping, Y. & Jia, L. (2006). Kongjian zonghe renwenxue yu shehui kexue yanjiu jinzhan 空間綜合人文學與社會科學研究進展. In Diqiu xinxi kexue 地球信息科學 *8*(2), 31–32.

Shanghai lishi dituji 上海歷史地圖集. (1999). Shanghai: Shanghai renmin chuban she.

So, K.L. & Tam, K.C. (2011). Mingmo Qingchu Songjiangfu ji qi zhoubian diqu de renkou sunshi yu lishi renkou gusuan 明末清初松江府及其周邊地區的人口損失與歷史人口估算. In Qinghua daxue xuebao: zhuxue shehui kexue ban 清華大學學報:哲學社會科學版, *26*(6), 45–54.

So, K.L. & Tam, K.C. (2012). Ming Qing da Songjiang fu diqu mianye shizhen geju yu renkou fenbu: yige HGIS de guancha 明清大松江府地區棉業市鎮格局與人口分佈:一個HGIS的觀察. In Zou, C. & Huang, J. (eds.), Ming Qing yilei Jiangnan zhengshi fachan yu wenhua jiaoliu 明清以來江南城市發展與文化交流, Fudan shixue jikan 復旦史學集刊, Vol. 4 (pp. 207–223). Shanghai: Fudan daxue chubanshe.

Tam, K.C. (2012). Rixin yueyi: rang dianzi ditu shushuo defang shi 日新月異:讓電子地圖述説地方史. In Renwen jiangtang congshu bianji weiyuanhui 人文講堂叢書編輯委員會 (ed.), Tuxiang zhong de lishi shijie 圖像中的歷史世界 (pp. 124–133). Hong Kong: JPC.

Tan, Q. (1980). *The historical atlas of China*. Beijing: Zhongguo ditu chubanshe (eight volumes).

Walford, N.S. (2005). Connecting historical and contemporary small area geography in Britain: The creation of digital boundary data for 1971 and 1981 census units. *International Journal of Geographical Information Science*, *19*(7): 749–767.

Wilkinson, E. (2000). *Chinese history: A manual, revised and enlarged*. Cambridge, MA: Harvard University Press.

Zhang, X. (2006). GIS yu lishi dilixue GIS 與歷史地理學. In Diqiu xinxi kexue 地球信息科學, *8*(2), 38–39.

Zhang, P., Tam, K. C., Lin, H., & So, K. L. (2007). The study on spatio-temporal transition of population in Songjiangfu region of China between 1392 and 1816: A historical GIS approach. *The International Journal of the Humanities, 5*(4), 13–22.

Chapter 12
Beyond the Boundary: New Insights from Inside the Space-Time Prism

Ying Song and Harvey J. Miller

12.1 Introduction

Time geography is a perspective that focuses on the spatial and temporal constraints on an individual's ability to participate in activities in a given environment (Hägerstrand 1970). A core concept is the *space-time prism*: this captures spatial and temporal constraints on potential mobility and activity participation. The prism demarcates all locations that a mobile object can occupy in space and time given: (1) the locations of origin and destination anchors; (2) the earliest departure time at the origin and the latest arrival time at the destination; (3) the maximum velocity for moving in space; and (4) the minimum required time for conducting stationary activities (Lenntorp 1977). Advances in geospatial science and technologies have greatly improved the realism of the space-time prism. GPS and other *location-aware technologies* (LATs) such as cellular phones and *radiofrequency identification* (RFID) chips enable researchers and practitioners to collect high-resolution mobility data. A number of *database management systems* (DBMS) such as Oracle, IBM DB2, PostGIS and ESRI Spatial Database Engine have extensions that support geospatial and temporal applications (e.g., Abraham and Roddick 1999; Egenhofer 1993; Lin 2012; Shaw and Wang 2000). These technologies facilitate the wide application of the space-time prism in transportation science, urban planning, public health and other human sciences that are concerned with individuals' mobility and accessibility (Bodzina and Cirucci 2009; Dong et al. 2006; Ettema and Timmermans 2007; Kwan 1998, 1999; Miller 1999; Neutens et al. 2007; Nicholls and Shafer 2001).

Y. Song (✉) • H.J. Miller
Department of Geography, The Ohio State University, Columbus, OH, USA
e-mail: song.851@osu.edu; miller.81@osu.edu

M.-P. Kwan et al. (eds.), *Space-Time Integration in Geography and GIScience*, DOI 10.1007/978-94-017-9205-9_12

Most studies involving the space-time prism only employ the prism boundary to delimit individual's mobility and accessibility to the environment and the feasible activities within that environment. In contrast, little attention has been paid to the properties of prism interior and its implications for measuring accessibility and its benefits and costs. Although some researchers have developed methods for estimating the properties of activity locations that intersect with the prism (Geurs and Van 2004; Kwan 1998), intrinsic properties of the prism interior have not been well-explored. Intuitively, we should not expect properties such as the probabilities of visiting different locations and the distributions of feasible velocities to be uniform throughout the prism interior. For example, the locations near the space-time axis linking the two prism anchors are more likely to be visited than the locations near the prism boundary since there are more possible paths through the former locations (Winter and Yin 2010a, b). In addition, different paths with different velocity profiles may pass through a given location, implying a distribution of feasible velocities constrained by the prism's maximum velocity.

This chapter analyzes properties of the prism interior and the implications of these properties for measuring accessibility benefits and costs. We begin by reviewing the conceptual and analytical framework of time geography defined by Torsten Hägerstrand and his followers. Based on this, we explore the properties of prism interior, focusing on the visit probability and the velocity profile for each location within the prism. To illustrate the potential of estimating the properties of the prism interior, we provide two examples: (1) a modified utility-based accessibility benefit measure that considers the visit probabilities; and (2) a new measure for the potential environmental costs of accessibility based on the visit probabilities and velocity distributions within the prism.

12.2 Prisms in Classic Time Geography

Time geography (Hägerstrand 1970) represents individual activities considering both spatial and temporal constraints on mobility and activity participation. A *space-time path* represents historical (or simulated) movement of an individual in geographical space with respect to time. Figure 12.1a illustrates a space-time path for an individual's typical work-to-home journey. The path goes through three activity locations that are dispersed in space: a work place, home, and a grocery store. The former two locations are relatively fixed in an individual's daily routine, while the third location is flexible because the individual can possibly shop for groceries at other stores and at other times. In order to travel between activity locations, the individual needs to trade time for space. The velocities to travel between locations determine the shape of the space-time path. When an individual conducts stationary activities, s/he does not move in space but still moves with respect to time, which results in a path segment parallel to the vertical axis for time. When an individual travels between two locations, s/he moves along an inclined path

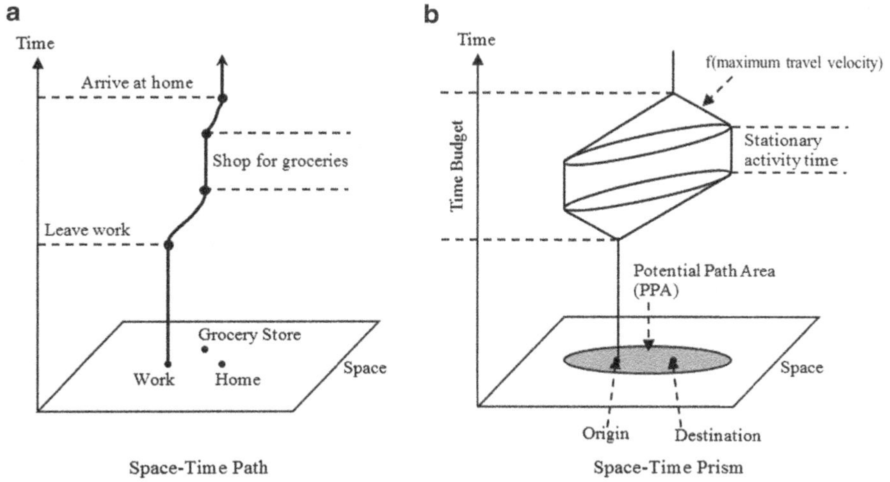

Fig. 12.1 Conceptual space-time path and prism

segment whose slope is a function of velocity (i.e., a lower velocity corresponds with a steeper angle and a higher velocity corresponds with a flatter incline).

Figure 12.1b illustrates a *space-time prism*: this is the envelope of all possible space-time paths between two fixed activity locations (e.g., home and work), a *time budget* for travel and activity participation (e.g., the time interval between earliest time a person can leave work and the latest time s/he must arrive at home), and the maximum velocity (speed) for travel.[1] If known, activity participation time can be netted from this prism (e.g., time spent for shopping groceries); this is the *stationary activity time* in Fig. 12.1b. The prism boundary delimits all locations in space-time that the mobile object can occupy; the spatial footprint or *potential path area* (PPA) delimits all accessible locations in geographical space.

Despite its simple geometry, it is difficult to describe a space-time prism analytically over the entire time interval of its existence. However, we can easily describe the spatial extent of a prism at an instantaneous time within the time budget (Fig. 12.2). Given an origin location (x_i, y_i), a destination location (x_j, y_j)., a maximum travel velocity V_m, and a minimum stationary activity time t_s, the spatial extent of the prism at any instant time t within the time budget can be defined analytically (Miller 2005a):

$$Z_{ij}(t) = \left\{(x,\ y) \,\middle|\, f_i(t) \cap p_j(t) \cap g_{ij}\right\} \tag{12.1}$$

[1] Strictly speaking from a physical perspective, the classic space-time prism is determined in part by a maximum "speed" (a scalar value) not a "velocity" (a vector with direction and magnitude). However, we will use the term "velocity" for consistency with the time geography literature.

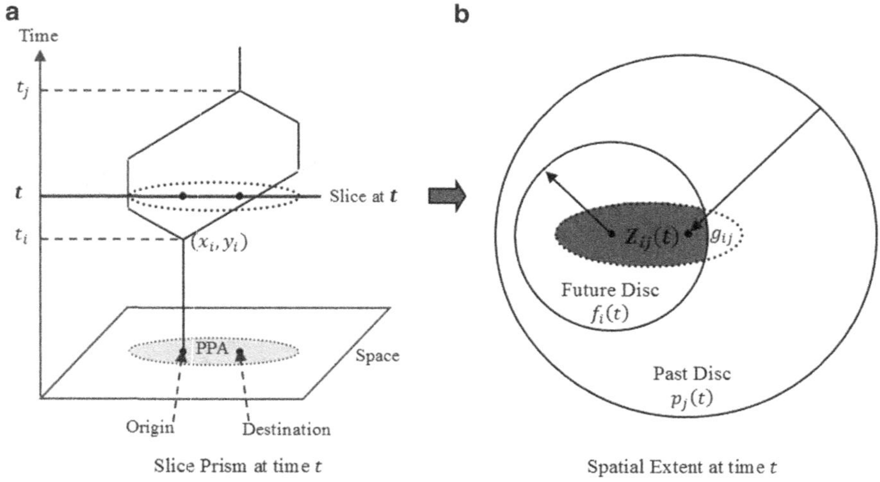

Fig. 12.2 Analytical space-time prism

where

$$f_i(t) = \left\{(x,\ y) \middle| \sqrt{(x - x_i)^2 + (y - y_i)^2} \le (t - t_i) \times V_m\right\} \tag{12.2}$$

$$p_j(t) = \left\{(x,\ y) \middle| \sqrt{(x - x_j)^2 + (y - y_j)^2} \le (t_j - t) \times V_m\right\} \tag{12.3}$$

$$g_{ij} = \left\{(x,\ y) \middle| \sqrt{(x - x_i)^2 + (y - y_i)^2} + \sqrt{(x - x_j)^2 + (y - y_j)^2} \le (t_j - t_i - t_s) \times V_m\right\} \tag{12.4}$$

This definition indicates that the spatial extend of the prism at time t is generally determined by two discs and the PPA (Fig. 12.2b): (1) the *future disc* (Eq. 12.2) delimiting all locations that can be reached from the first anchor by that moment; (2) the *past disc* (Eq. 12.3) delimiting all locations that can reach the second anchor given the remaining time; and (3) the PPA (Eq. 12.4) that constraints the prism locations to account for any stationary activity time.

We can also describe the prism from a locational perspective. Given any location (x, y) within the PPA (Eq. 12.4), the time interval $[t_{min}, t_{max}]$ available for activity participation at that location can be calculated (Fig. 12.3):

$$t_{min} = t_i + \frac{\sqrt{(x - x_i)^2 + (y - y_i)^2}}{V_m} \tag{12.5}$$

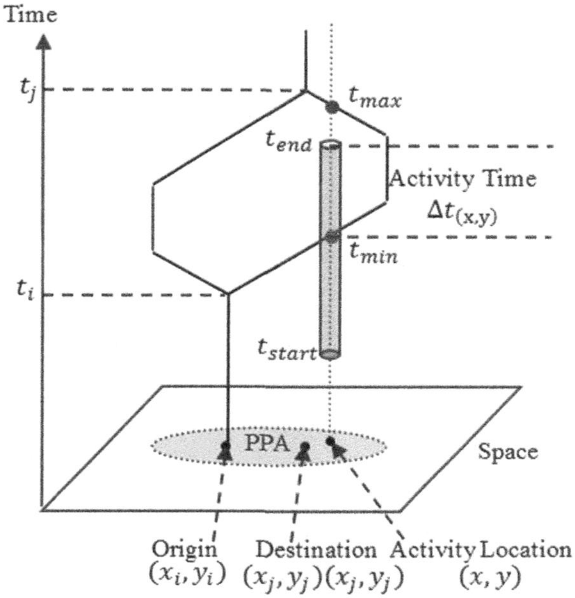

Fig. 12.3 Available time interval $[t_{min}, t_{max}]$ at an accessible location

$$t_{max} = t_j - \frac{\sqrt{(x - x_j)^2 + (y - y_j)^2}}{V_m} \tag{12.6}$$

where t_{min} is the earliest time an individual can reach that location from origin and t_{max} is the latest time s/he has to leave that location in order to reach destination on time. An opportunity is accessible to the individual only if it is located within the PPA and its own available time $[t_{start}, t_{end}]$ (e.g. opening hours of the grocery store) intersects the time available at that location $[t_{min}, t_{max}]$ to a sufficient degree, that is, no less than the minimum required time for activity participation t_s (Fig. 12.3):

$$\Delta t_{(x,\ y)} = \min(t_{end}, t_{max}) - \max(t_{start}, t_{min}) \geq t_s \tag{12.7}$$

In the real world, an individual's movements are often confined by the geometry and velocity limits allowed by transport networks (Miller 1991). Kuijpers and Othman (2009) extend the two definitions above to define the *network time prism* (NTP). Given a road network **RN**, the NTP at any instant time $t \in (t_i, t_j)$ can be defined mathematically as:

$$Z_{ij}(t) = \left\{(x,\ y) \middle| f_i(t) \cap p_j(t) \cap g_{ij} \cap \mathbf{RN}\right\} \tag{12.8}$$

$$f_i(t) = \left\{(x,\ y) \middle| t_{RN}((x_i,\ y_i),\ (x,\ y)) \leq t - t_i\right\} \tag{12.9}$$

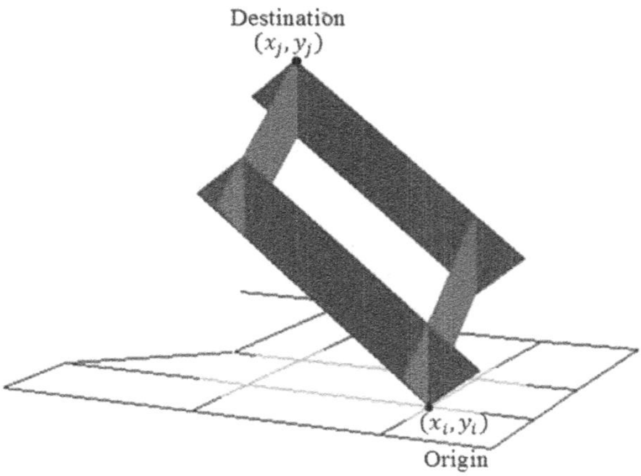

Fig. 12.4 A network time prism: the space-time prism in network space (Kuijpers and Othman 2009)

$$p_j(t) = \left\{(x, y) \middle| t_{RN}\left((x, y), (x_j, y_j)\right) \leq t_j - t\right\} \tag{12.10}$$

$$g_{ij} = \left\{(x, y) \middle| t_{RN}\left((x_i, y_i), (x, y)\right) + t_{RN}\left((x, y), (x_j, y_j)\right) \leq t_j - t_i - t_s\right\} \tag{12.11}$$

where $t_{RN}((x_i, y_i), (x, y))$ and $t_{RN}((x, y), (x_j, y_j))$ denote the minimum travel time from location (x, y) to the origin (x_i, y_i) and the destination (x_j, y_j), respectively, within the network. Figure 12.4 illustrates a network time prism: the green arcs comprise the prism's spatial footprint (Eq.12.11) and the red region is the full NTP whose boundary can be defined by the earliest arrival time t_{min} and the latest departure time t_{max} at each accessible location (x, y) along green arcs:

$$t_{min} = t_i + t_{RN}\left((x_i, y_i), (x, y)\right) \tag{12.12}$$

$$t_{max} = t_j - t_{RN}\left((x, y), (x_j, y_j)\right) \tag{12.13}$$

Based on the prism analytics in planar and network space discussed above, substantial progress have been made on techniques for calculating realistic prisms based on high-resolution mobility, transportation and other geographic data (e.g. Kuijpers and Othman 2009; Miller 1991; Miller and Bridwell 2009; Neutens et. al. 2007; O'Sullivan, Morrison and Shearer 2000; Wu and Miller 2001). In addition, path and prism concepts have been extended from the physical world to the virtual world implied by *information and communication technologies* (ICT) (Couclelis 2009; Miller 2005b; Yu and Shaw 2008). However, researchers have focused on the prism boundary as an indicator of the limits on an individual's mobility and accessibility, and ignored the properties of the prism's interior that relate to the individual's potential movements. Only recently has the interior structure of prism started to receive attention (Winter and Yin 2010a, b). In the following sections, we introduce analytical frameworks for two properties of locations within the prism interior, namely, visit probabilities and velocity distributions, respectively.

12.3 Properties of the Prism Interior

12.3.1 Visit Probabilities

At a moment in time $t \in [t_i, t_j]$, the spatial extent $Z_{ij}(t)$ delimits all locations that an individual can access in space at that moment (Eq. 12.1). For any two locations within this spatial extent, the number of possible paths going through them may be different, implying an unequal visit probability distribution $P((x, y)|t)$. Since an individual cannot reach a moment in time without passing an earlier moment, the probability to pass any time $t \in [t_i, t_j]$ equals to one; therefore, the total probability to visit all locations in $Z_{ij}(t)$ equals to one:

$$P\left[Z_{ij}(t)\right] = \iint\limits_{(x,y)\in Z_{ij}(t)} P\left((x,y)\middle|t\right) dxdy = 1 \ \ \forall\, t \in \left[t_i, t_j\right] \tag{12.14}$$

Winter and Yin (2010a) develop a mathematical foundation for modeling the distribution of visit probabilities based on random walk theory. They start with an undirected random walk from the origin anchor; this is a sequence of random steps that approximate a continuous stochastic diffusion process. Figure 12.5a shows the case for one-dimensional space: the number of paths through a location in the next step can be calculated as the sum of possible paths from itself and its neighbors in the current step.

Extending the calculation to two-dimensional discrete space, the object can move one step in one of the two dimensions, and the number of possible paths

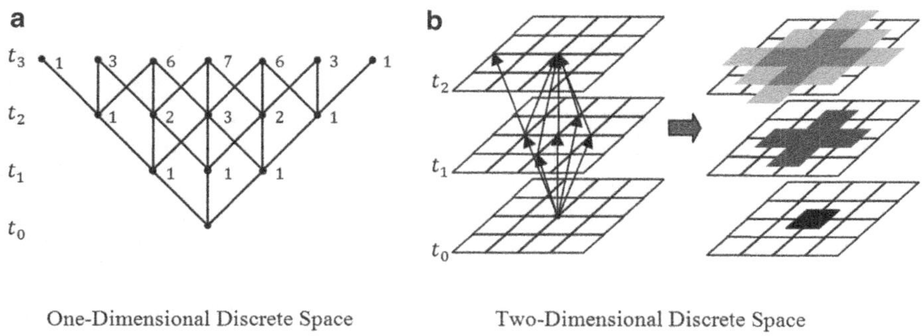

Fig. 12.5 Visit probability using random walk theory (Based on Winter and Yin 2010a)

through a location (x_k, y_k) at time t_k can be calculated by adding up the number of paths through its direct predecessor and four neighbors at time t_{k-1} (Fig. 12.5b) Therefore, the space-time cone between t_0 and t_n consists of $n+1$ layers can be computed as:

$$P_t\left((x, y)\,|t_k\right) = \begin{cases} 0 & \text{if } |x-x_0|+|y-y_0|>k \\ p\left((x,y)\ |t_{k-1}\right)\otimes \mathrm{h} & \text{if } |x-x_0|+|y-y_0|\le k \end{cases}, \quad k=1,\ 2,\ 3,\ \ldots..,\ n \tag{12.15}$$

where h is the normalized kernel for undirected/unbiased random walk:

$$h = \begin{bmatrix} 0.0 & 0.2 & 0.0 \\ 0.2 & 0.2 & 0.2 \\ 0.0 & 0.2 & 0.0 \end{bmatrix} \tag{12.16}$$

Based on this construction, the probability to visit accessible locations at each layer follows a bivariate multinomial distribution in discrete space and time.

Based on this framework, Winter and Yin (2010b) consider the case of directed movement from the first anchor to the second anchor in the prism. But, rather than modifying the fundamental random walk process to account for directionality, they assume that the uneven visit probabilities implied by the undirected case translates to the directed space-time axis connecting the two anchors. To describe visit probabilities in continuous space, they employ the continuous analogue of the multinomial distribution, that is, the multivariate normal distribution. In two-dimensional space, this is a bivariate normal distribution. However, these distributions are unbounded: this is incongruent with a bounded space-time prism. To resolve this, Winter and Yin (2010b) use the spatial extent of the prism at a given moment in time to modify the standard deviation at that moment and to clip the Bivariate normal distribution.

The conclusions reached by Winter and Yin (2010b) regarding visit probabilities within planar space prisms are intuitive, but their argument lacks a solid mathematical foundation. The transition from a discrete space random walk to continuous space is accomplished by analogy only: there is no explanation of why the expected location within the prism with respect to time (the mean of bivariate normal distribution) should be along the space-time axis linking the two anchors of the prism; this is simply assumed. Also, clipping the bivariate normal distribution using the spatial extent $Z_{ij}(t)$ is an artificial construction that is not based on fundamental principles. Bias may be introduced by cutting off part of the original distribution outside the extent, especially when the prism spatial extent is not symmetric with respect to the axis connecting the anchor points.

In the next subsections of this paper, we develop a more solid foundation for calculating visit probabilities within space-time prism in two steps. First, we maintain discrete space and time, but modify the undirected random walk to consider constraints imposed by the requirement to be at the second anchor by a specified time. This generates a directed random walk that obeys the space-time prism constraints. Second, we develop a foundation for directed movements in continuous space and time using *Brownian Bridges* (BBs) techniques.

12.3.1.1 Directed Random Walk

Keeping the basic mathematical framework based on random walk theory (Eq. 12.15), we modify the normalized kernel for undirected random walk (Eq. 12.16) to consider direction preference during the mobility process. Given the current location of an individual and the time left for travel, we maintain the basic assumption of the random walk theory: the probability to visit a location should increase with the number of paths/choices provided by that location. As the spatial distance to destination increases and/or the time left for travel decreases, the pressure to travel toward destination becomes higher. Therefore, the normalized kernel must be modified dynamically according to the relative location between current location (x, y) and the destination location (x_j, y_j), and the time left for travel $(t_j - t)$.

The kernel matrix still consists of five directions, which can be represented as:

$$H = \begin{bmatrix} 0.0 & P_{Up} & 0.0 \\ P_{Left} & P_{Stay} & P_{Right} \\ 0.0 & P_{Down} & 0.0 \end{bmatrix} \tag{12.17}$$

For each accessible location (x, y) at time step t, the weight of each direction is proportional to the total number of choices provided by that direction. For instance, if second anchor is below the current location (x, y), traveling upward means deviating from destination, and we get:

$$P_{Up} = \begin{cases} C_{T-1}^{D+1} \times k & T \geq D + 2 \\ 0 & T < D + 2 \end{cases} \tag{12.18}$$

where k is a scale factor for normalization purpose, and the kernel should satisfy:

$$P_{Stay} + P_{Up} + P_{Down} + P_{Left} + P_{Right} = 1 \tag{12.19}$$

After modifying the normalized kernel, we apply the basic transition formula (Eq. 12.15) for undirected random walk to calculate the visit probability at each time step. Figure 12.6 shows computed visit probability at selected time steps $\{t_k\}_{k \in [0, 500]}$ from an origin located at lower left to a destination located at upper right. These results support the intuition of highest visit probabilities near the axis connecting the prism anchors, as assumed by Winter and Yin (2010b).

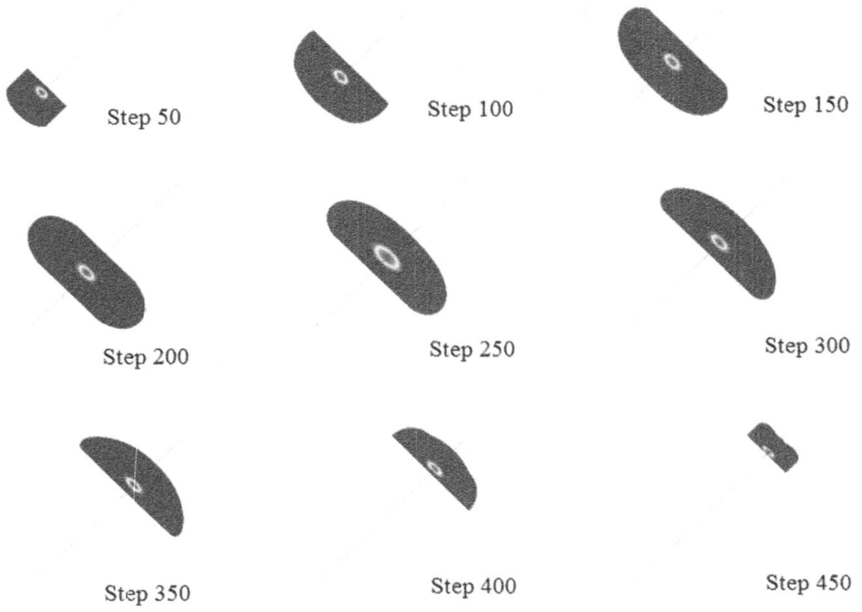

Fig. 12.6 RW simulation result with 500 time steps

12.3.1.2 Brownian Bridges Within Space-Time Prisms

We proceed to the continuous space and time and apply the technique of *Brownian Bridges* (BBs). A Brownian Bridge is a continuous-time stochastic movement between two known values, estimated at stationary increments. In economics and finance, BBs has been applied to analyze stock market prices (Chow and et al. 1995; Lo 1991; Rostek and Schöbel 2006). In ecological research, the *Brownian Bridge Movement Model* (BBMM) has been developed to estimate an animal's probability of occurrence at a given geographic location and time (Brillinger et al. 2002; Holmes et al. 1994; Horne et al. 2007).

A BB in one dimension from $a \in r$ to $b \in r$ on $[0, T]$ is:

$$B(t)^{a \to b} = a + \frac{b-a}{T} \times t - \left(W(t) - \frac{t}{T}\, W(T)\right), \; 0 \le t \le T \qquad (12.20)$$

where $\{W_t\}$ is a standard Wiener process (a.k.a. Brownian motion) that is indexed by nonnegative real numbers t, and should satisfy the following properties:

1. $W_0 = 0$.
2. With probability 1, the function $t \rightarrow W_t$ is continuous in t.
3. The process $\{W_t\}$ has stationary, independent increments.
4. The increment $W_{t+s} - W_s$ has the normal distribution $\mathcal{N}(0, t)$.

For the two dimension case, the probability at any time $t \in [t_i, t_j]$ follows bivariate normal distribution:

$$\begin{bmatrix} x(t) \\ y(t) \end{bmatrix} \sim \mathcal{N}\left(\begin{bmatrix} u_x(t) \\ u_y(t) \end{bmatrix}, \begin{bmatrix} \sigma_x^2(t) & 0 \\ 0 & \sigma_y^2(t) \end{bmatrix} \right) \tag{12.21}$$

where

$$u_x(t) = \frac{(t - t_i)\, x_i + (t_j - t)\, x_j}{t_j - t_i} \tag{12.22}$$

$$u_y(t) = \frac{(t - t_i)\, y_i + (t_j - t)\, y_j}{t_j - t_i} \tag{12.23}$$

$$\sigma_x^2(t) = \sigma_y^2(t) = \frac{(t - t_i)\,(t_j - t)}{t_j - t_i} \tag{12.24}$$

The Eqs. 12.21, 12.22, 12.23, and 12.24 are consistent with Winter and Yin's (2010b) assumption of the visit probability distribution within the prism: (i) the expected location at any time t is along the axis connecting the two anchors of the space-time prism, and (ii) the visit probabilities follow a bivariate normal distribution. However, our method has a firmer mathematical foundation, and the distribution is generated from first principles instead of introduced as an assumption.

Figure 12.7 shows 10,000 non-truncated BB trajectories, each with 50 or 100 interpolation time steps. Note trajectories fall outside of the generated PPA in both

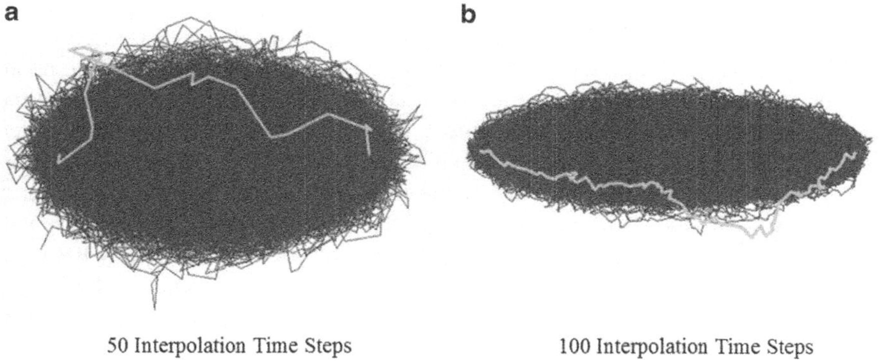

Fig. 12.7 Simulation results of non-truncated Brownian Bridges

cases (see the highlighted trajectory); this demonstrates the need for truncated functions to ensure that each interpolated location is always located within the accessible spatial extent at any time. An interesting finding is that with all else being equal, more interpolation time steps results in a smaller generated PPA. This is because with the same overall travel distance, when the number of interpolated time steps increases, the interval between two interpolation times will decrease; therefore, given the Property 4 of the Wiener process, the standard deviations for each interpolation will decrease and the generated PPA becomes smaller. Although this can eliminate many trajectories that are not feasible given the prism constraints, it may also exclude feasible trajectories. This suggests that the simulation of non-truncated Brownian bridges is sensitive to the number of interpolation times, and further investigation is required to find the relationship between the interpolation times and the accuracy of simulated trajectories.

To restrict BBs within the space-time prism, we apply the concept of a truncated distribution instead of artificially clipping the distribution by spatial extent of prism at time t (Eq. 12.1). A truncated distribution is a conditional distribution that results from restricting the domain of some other probability distribution. For any instant time $t \in [t_i, t_j]$, the x and y coordinate of the mobile individual follows the truncated normal distributions with density functions as:

$$Tr\left(x(t)\right) \sim \frac{\mathcal{N}\left(0, \sigma_x^2(t)\right)}{F\left(U_x\right) - F\left(L_x\right)} = \frac{2\sigma_x(t) \times \mathcal{N}\left(0, 1\right)}{\operatorname{erf}\left(\dfrac{U_x}{\sigma_x(t)\sqrt{2}}\right) - \operatorname{erf}\left(\dfrac{L_x}{\sigma_x(t)\sqrt{2}}\right)} \tag{12.25}$$

$$Tr\left(\left(y(t)\right) \middle| \tilde{x}(t)\right) \sim \frac{\mathcal{N}\left(0, \sigma_y^2(t)\right)}{F\left(U_y\left(\tilde{x}(t)\right)\right) - F\left(L_y\left(\tilde{x}(t)\right)\right)} = \frac{2\sigma_y(t) \times \mathcal{N}\left(0, 1\right)}{\operatorname{erf}\left(\dfrac{U_y\left(\tilde{x}(t)\right)}{\sigma_y(t)\sqrt{2}}\right) - \operatorname{erf}\left(\dfrac{L_y\left(\tilde{x}(t)\right)}{\sigma_y(t)\sqrt{2}}\right)} \tag{12.26}$$

where $erf(x)$ is the error function based on normal distribution, U_x and L_x is determined by the spatial extent of the prism at the time at time t, $Z_{ij}(t)$, and U_y and L_y are determined by $Z_{ij}(t)$ and a randomly simulated value for $x(t)$, $\tilde{x}(t)$.

Given these two truncated density functions, a BB within the space-time prism can be simulated by interpolating locations at a set of time steps $\{t_k\}$. Give a time series , the interpolated location at time t_s is calculated based on the interpolated locations at times t_u and t_t. However, since the interpolation of $(x(t_s), y(t_s))$ uses two random numbers that are independent from previous process (see Eqs. 12.25 and 12.26), the stochastic process has independent increments between any two locations, and the simulation process maintains the properties required by BBs.

The mathematical foundations demonstrated above applies and modifies BBs with consideration of general restrictions that define a space-time prism. Compared to Winter and Yin's (2010b) construction, it does not require assumptions about the expected location, and avoids the bias by artificially clipping an unbounded distribution by spatial extent of prism at time t. The truncated distribution preserves the general properties of a BB as a stochastic process with independent increments, and solves the issue of information loss outside the spatial extent caused by clipping process. Therefore, it provides more theoretically consistent and unbiased estimates of visit probabilities within the prism.

12.3.2 Velocity Distribution

A space-time path going through a given location in the prism has an instantaneous travel velocity that may be different from other paths through the same location. Hence, there is a velocity profile associated with each location in the prism. For a location (x, y) in the spatial extent $Z_{ij}(t)$., the probability to travel at velocity v among all feasible velocities can be represented as $P_t(v|(x, y))$, where $v \in [v_{min}, v_{max}]$. We start our analysis by assuming that an individual selected any velocity from feasible velocities equally possible. This implies the following uniform velocity distribution: (Note: "1" here indicates one unit of choice)

$$P_t\left(v\middle|(x, y)\right) = \begin{cases} \frac{1}{(v_{max}-v_{min})}, & \text{if } v_{max} > v > v_{min} \\ 1, & \text{if } v = v_{max} = v_{min} \\ 0, & \text{else} \end{cases} \quad (12.27)$$

In most cases, the individual can choose to stay still or move at maximum velocity; therefore, $v_{min} = 0$ and $v_{max} = V_m$ respectively. A special consideration is locations along the boundary of the prism, where individual has to travel at the maximum velocity V_m when not conducting stationary activities.

A uniform velocity assumption is straightforward but unsophisticated. A non-uniform velocity distribution takes into account time pressure imposed by the constraint of reaching the destination anchor by the required time t_j. With longer distance and less time to travel, the individual is more likely to travel with higher velocities. This suggests that the probability to select a velocity v is affected by three factors: (i) the total distance to travel from origin to destination passing location (x, y), $D_{(x,y)}$; (ii) the total time available for moving $T_m = (t_j - t_i - t_s)$; and, (iii) the maximum achievable travel velocity V_m:

$$P_t\left(v\middle|(x, y)\right) = f\left(D_{(x,y)},\ T_m,\ V_m\right),\ \text{where } (x, y) \in Z_{ij}(t) \quad (12.28)$$

There are several properties that this function should satisfy:

1. Since we have $v \in [0, V_m]$ for the entire space-time prism, given any location within the prism, the individual has to choose from this velocity range, that is:

$$\int_0^{V_m} P_t\left(v\middle|(x, y)\right) = 1,\ \text{for any } (x, y) \in Z_{ij}(t) \quad (12.29)$$

2. Given the same T_m and V_m, as the total moving distance $D_{(x,y)}$ increases, the individual intends to move faster at location (x, y);
3. Given the same $D_{(x,y)}$ and V_m, as the individual has more time to move (T_m increases), s/he is more likely to take his/her time and move with lower velocities.

Given these properties, a consistent probability function is the following normal distribution:

$$P_t\left(v \middle| (x, y) \in Z_{ij}(t)\right) \sim \mathcal{N}\left(\frac{D_{(x,y)}}{T_m}, \sigma_v^2(t)\right) \tag{12.30}$$

where $\sigma_v(t)$ is determined by the velocity range [0, V_m] and is uniform across all location within the prism except those along the prism boundary. This function is reasonable given that the individual intends to move smoothly between two locations.

Another approach to calculate the velocity distribution at a given location is to utilize the distribution of visit probabilities in the prism (see Sect. 12.3.1). Given a moment in time $t \in [t_i, t_j]$ and a later time $(t + \Delta t) \in [t_i, t_j]$, we can estimate all locations at time t that can reach a given location (x, y) at time $(t + \Delta t)$. In planar space, the spatial extent of all these location can be calculated as:

$$Z'_{ij}(t) = \left\{(x', y') \middle| Z_{ij}(t) \cap \left(\sqrt{(x' - x)^2 + (y' - y)^2} \leq \Delta t \times V_m\right)\right\} \tag{12.31}$$

If the time difference Δt is small enough, travelling from any location $(x', y') \in Z'_{ij}(t)$ at time t to location (x, y) at time $(t + \Delta t)$ generate an instantaneous velocity:

$$v_t\left((x, y) \middle| (x', y')\right) = \lim_{\Delta t \to 0} \frac{\sqrt{(x' - x)^2 + (y' - y)^2}}{\Delta t} \tag{12.32}$$

Along with the probability to visit location (x', y') at time t calculated by analytical framework discussed in Sect. 12.3.2, we can derive the velocity distribution for location (x, y) at time $(t + \Delta t)$. Although this method is sensitive to the method used to derive the visit probability, the estimated velocity distribution calculated has a more solid theoretical foundation than the former two methods and is consistent with the unequal visit probability case.

The visit probability and velocity distribution discussed in this section are the two example properties of locations within the prisms derived from the object's movement potential. At any instant time within the time budget, the spatial extent of all accessible locations can be calculated, and each location is expected to have a visit probability that follows binomial distribution for the discrete case and truncated bivariate normal distribution for the continuous case. The velocity distribution can be calculated based on the derived visit probability, and is expected to be nonuniform throughout the prism interior. The framework developed for the planar space can also be modified for the network space considering travel time within road network. This theoretical foundation can not only provide novel theoretical insights into the space-time prism, but also refine the prism as a measure of potential mobility. The next section provides two example applications to demonstrate the potentials of these novel insights derived from looking inside the space-time prism.

12.4 Implications of Insights from Inside the Prism

12.4.1 Accessibility Benefit Measures

In general, accessibility is a term that describes the degree to which facilities, services, or opportunities are available to people. In transportation planning and research, accessibility plays a significant role in evaluating current infrastructure and supporting political and planning decisions on future strategies. The space-time prism is an elegant and sensitive accessibility measure: it captures individual differences in scheduling constraints, the available locations and timing of activities, and the ability of the transportation system in meeting an individual's mobility and accessibility needs (Hägerstrand 1970).

Miller (1999) develops an analytical framework and computational procedures to derive accessibility benefit measures that are consistent with the rigorous Weibull (1976) axiomatic approach to the measurement of accessibility. This framework reconciles three complementary approaches to accessibility: (1) the constraints-oriented approach implemented by Hägerstrand's space-time prism; (2) the gravity-based approach that weights location's attractiveness against its required travel cost; and (3) a benefits measure based on the microeconomic theory of consumer surplus. In this section, we introduce and modify this space-time accessibility benefit measure by including visit probabilities in the first approach.

Considering the generic utility function (Burns 1979; Miller 1999)

$$u\,(a,\ T,\ t) = a^{\alpha} T^{\beta} \exp\,(-\lambda t) \tag{12.33}$$

where a indicates the attractiveness of the activity location, T is the time available for activity participation, t is required travel time from the first anchor to the activity location and subsequently to the second anchor, and α, β, and λ are parameters. Assume that an individual cannot leave a fixed activity location (x_i, y_i) until t_i and has to participate in another activity at location (x_j, y_j) by t_j, the utility of participating an activity at location (x_k, y_k) can be calculated by modifying Eq. 12.33 and the analytical prism with no stationary activity time (see Eqs. 12.1, 12.2, 12.3, and 12.4 and $t_s = 0$):

$$u\,(a_k,\ T_k,\ t_k) = a_k^{\alpha} T_k^{\beta} exp\,(-\lambda t_k) \tag{12.34}$$

where

$$t_k = \left[\sqrt{(x_k - x_i)^2 + (y_k - y_i)^2} + \sqrt{\left(x_k - x_j\right)^2 + \left(y_k - y_j\right)^2} \right] / V_m \quad (12.35)$$

$$T_k = \begin{cases} t_j - t_i - t_k > 0 \\ 0 \quad \quad \quad \quad \mathit{else} \end{cases} \quad (12.36)$$

Based on the consumer welfare theory that maximizing consumer surplus (Wilson 1976), the benefit from location (x_k, y_k) can be calculated as (Miller 1999):

$$B_k = \begin{cases} 0 & \mathit{if}\ a_k = 0\ \mathit{or}\ T_k \le 0 \\ \exp\left[\lambda\left(\frac{\alpha}{\lambda}\ln a_k + \frac{\beta}{\lambda}\ln T_k - t_k\right)\right] & \text{else} \end{cases} \quad (12.37)$$

and an additive accessibility provided by the choice set $k \in [1, n]$ can be measure as (Miller 1999):

$$AM = \sum_{k=1}^{n} B_k \quad (12.38)$$

However, this additive accessibility measure ignores the probability of choosing each option in the choice set, and may misestimate the actual benefits that can be provided to the individual. In order to solve this issue, we can incorporate the visit probability developed in Sect. 12.3.1, and define an accessibility measure for expected/potential benefits:

$$AM = \sum_{k=1}^{n} B_k P\left((x_k, y_k)\,\middle|\,t_k\right) \quad (12.39)$$

where $P((x_k, y_k)|t_k)$ is the probability to visit location (x_k, y_k) at time t_k. Whatever the actual method used to calculate the visit probability, it should always satisfy the unity property given in Eq. 12.14. As a result, the accessibility calculated using Eq. 12.39 can avoid overestimation by assigning probabilities to each locational benefit. These probabilities can be estimated using the stochastic simulation techniques discussed earlier in this chapter, modified to account for the varying attractiveness of the activity locations.

12.4.2 *Accessibility Cost*

Although the space-time prism is an effective measure of potential benefits from accessibility, there has been little attention paid to the cost of accessibility. When applying prism measures in accessibility analysis, there is an implicit assumption that, all else being equal, a larger prism is better since it affords greater potential mobility and therefore higher accessibility. However, there are costs required to achieve the accessibility; some of these burden broader society. Take the work-to-home journey using private owned vehicles as an example (see Fig. 12.1a): if using motorized conveyances such as the automobile, the individual consumes non-renewable resources (gasoline or diesel fuel) and generates noxious emissions to the environment. Therefore, a prism with larger size may also imply higher potential costs both to the individual and society at large.

In this section, we illustrate the general analytical framework to calculate expected environmental costs of accessibility. Specially, we focus on the engine-based costs, including consumption of non-renewable resources such as fossil fuels and emissions of harmful pollutants such as NO_X, PM-2.5, and *greenhouse gases* (GHG) such as CO_2.

The past two decades have witnessed increasing attention to the negative impacts of transportation and concerted efforts towards a sustainable transportation paradigm (Bannister 2008; Black et al. 2002; Ewing et al. 2008). The goal of transportation planning is no longer recognized as maximizing but rather managing mobility. However, widely-applied emissions models such as MOBILE or MOVES (United States Environmental Protection Agency (EPA) 2009) operate at the aggregate level; this is not appropriate for travel demand management policies that focus on individual-level activity and mobility patterns. Calculating the accessibility cost associated with a space-time prism considers the operation of an individual vehicle; this can evaluate microscale traffic scenarios and individual-level mobility and accessible patterns (Barth et al. 1996).

Barth et al. (1996) identify three essential parameters that determine the engine-based cost of vehicle movements: spatial reference, temporal reference, and velocity profile. The spatial reference and temporal reference cover the external factors that affect the engine performance such as the slope of road surface and the ambient temperature at each moment in time. Given these external factors, the engine-based cost of a vehicle is non-decreasing with respect to velocity. A general function $C(v|(x,y,t))$ to calculate the instantaneous vehicle cost should have the following properties: (i) $v < v' \Rightarrow C(v|(x,y,t)) \leq C(v'|(x,y,t))\ \forall\ (x,y,t)$ (non-decreasing with respect to velocity); and, (ii) $C(0|(x,y,t)) = 0$ (no cost while the vehicle is stationary). The cost function can be modified by characteristics of individual vehicles (e.g. engine type, weights, mileage), road conditions (e.g. slope and surface material), and environmental factors (e.g. humidity, temperature).

Given the cost function $C(v|(x,y,t))$, along with the velocity distribution and the visit probability at accessible locations within the prism, we can calculate engine-based cost of a prism as:

$$C_{ij} = \int_{t_i}^{t_j} \left\{ \iint_{(x,y)\in Z_{ij}(t)} P\left((x,y)\middle|t\right) \times \left[\int_{v_{min}}^{v_{max}} P_t\left(v\middle|(x,y)\right) C\left(v\middle|(x,y,t)\right) dv \right] dxdy \right\} dt \tag{12.40}$$

where $Z_{ij}(t)$ is the spatial extent of all accessible locations at time $t \in [t_i, t_j]$, $P((x,y)|t)$ is the probability for location (x,y) to be visited, and $P_t(v|(x,y))$ is the velocity profile at that location. The accessibility-related environmental cost calculated in this framework can be combined with the benefits accessibility measures, and used to develop a benefit-cost framework for evaluation within the sustainable transportation paradigm.

12.5 Conclusion

The potential of space-time prism in analyzing individual's potential mobility and accessibility has been widely acknowledged and applied in transportation science, urban planning, public health and other human sciences. A space-time prism captures all locations that can be reached in space and time given the origin and destination locations, the earliest departure time from the origin, the latest arrival time at destination, the maximum allowed travel velocity and the minimum time required for activities participation. The boundary of the prism is the envelope of all accessible locations, and has been recognized and proven as an effective measure of individual mobility and accessibility to the environment and opportunities in that environment.

However, little attention has been paid to the prism interior and its mobility-related properties. Visit probabilities and velocity distributions are two properties that should not be expected to be uniform distributed throughout the prism interior. This chapter analyzes these properties of the prism interior and the implications of these properties for measuring accessibility benefits and costs. Based on discrete space and time, we modify the classic undirected random walk theory to consider constraints imposed by the requirement to be at the second anchor by a specified time. We develop a framework for directed movements in continuous time and space using Brownian Bridges techniques. We also consider uniform and non-uniform velocity distributions, the latter based on temporal constraints and visit probability distributions. To illustrate the potentials of estimating the properties of the prism interior, we provide two examples: (1) a modified utility-based accessibility benefit measure that considers the visit probabilities; and (2) a new measure for the potential environmental costs of accessibility based on the visit probability and velocity distribution.

Novel insights into the internal structure of the space-time prism can improve not only the theoretical understanding of the space-time prism, but also the practical applications of it. By introducing the visit probability, existing prism-based accessibility measures can be modified to avoid the overestimate of individual accessibilities and location opportunities. In addition, the velocity distribution along with the visit probability at each accessible location enables us to estimate the potential environment costs to facilitate mobility at that location (e.g. energy consumption and waste emissions). The environment cost challenges the fundamental assumption that greater accessibility is an unmitigated benefit to individuals and society, and reveals the need and feasibility to incorporate the mobility-related costs along with the benefit to move towards a sustainable development paradigm.

The next steps in this research agenda are completing the theoretical construction of the prisms' interior, validating its effectiveness in practical applications using simulated and/or monitored data (e.g. estimating the theoretical carbon dioxide emission for a prism and compare it with the data collected from devices installed on vehicles), and then promoting its use to real-world projects by developing toolkits for planning and policy evaluation. Followed this, we can continue our research by adding more conditions to the theoretical framework of the space-time prism (e.g. limits on the linear and angular accelerations), investigate the changes on the boundary and internal structure of the prism, and explore and validate other applications accordingly.

Acknowledgements An award from the National Science Foundation (BCS – 1224102) supported this research.

References

Abraham, T., & Roddick, J. F. (1999). Incremental meta-mining from large temporal datasets. *Advances in Database Technologies, 1552*, 41–54. doi:10.1007/978-3-540-49121-7_4.

Banister, D. (2008). The sustainable mobility paradigm. *Transport Policy, 15*(2), 73–80.

Barth, M., An, F., Norbeck, J., & Ross, M. (1996). Model emission modeling: a physical approach. *Transportation Research Record: Journal of the Transportation Research Board, 1520*, 81–88. doi:10.3141/1520-10.

Black, J. A., Paez, A., Suthanaya, P.A., (2002). Sustainable urban transportation: performance indicators and some analytical approaches. *Journal of Urban Planning and Development, 128*(4), 184–209. doi: 10.1061/(ASCE)0733-9488(2002)128:4(184).

Bodzina, A. M., & Cirucci, L. (2009). Integrating geospatial technologies to examine urban land use change: A design partnership. *Journal of Geography, 108*(4–5), 186–197. doi:10.1080/00221340903344920.

Brillinger, D. R., Preisler, H. K., Ager, A. A., Kie, J. G., & Stewart, B. S. (2002). Employing stochastic differential equations to model wildlife motion. *Bulletin of the Brazilian Mathematical Society, 33*(3), 385–408. doi:10.1007/s005740200021.

Burns, L. D. (1979). *Transportation, temporal, and spatial components of accessibility*. Lexington: Lexington Books.

Chow, K. V., Denning, K. C., Ferris, S., & Noronha, G. (1995). Long-term and short-term price memory in the stock market. *Economics Letters, 49*(3), 287–293. doi:10.1016/0165-1765(95)00690-H.

Couclelis, H. (2009). Rethinking time geography in the information age. *Environment and Planning A, 41*, 1556–1575. doi:10.1068/a4151.

Dong, X. J., Ben-Akiva, M. E., Bowman, J. L., & Walker, J. L. (2006). Moving from trip-based to activity-based measures of accessibility. *Transportation Research Part A: Policy and Practice, 20*(2), 163–180. doi:10.1016/j.tra.2005.05.002.

Egenhofer, M. J. (1993). What's special about spatial?: Database requirements for vehicle navigation in geographic space. *ACM SIGMOD Record, 22*(2), 398–402. doi:10.1145/170036.170096.

Ettema, D., & Timmermans, H. (2007). Space–time accessibility under conditions of uncertain travel times: Theory and numerical simulations. *Geographical Analysis, 39*(2), 217–240. doi:10.1111/j.1538-4632.2007.00702.x.

Ewing, R., Bartholomew, K., Winkelman, S., Walters, J., & Chen, D. (2008). *Growing cooler: The evidence on urban development and climate change*. Washington, DC: Urban Land Institute.

Geurs, K. T., & Van Wee, B. (2004). Accessibility evaluation of land-use and transport strategies: Reviews and research directions. *Journal of Transport Geography, 12*(2), 127–140. doi:10.1016/j.jtrangeo.2003.10.005.

Hägerstrand, T. (1970). What about people in regional science? *Papers of the Regional Science Association, 24*(1), 6–21. doi:10.1007/BF01936872.

Holmes, E. E., Lewis, M. A., Banks, J. E., & Veit, R. R. (1994). Partial differential equations in ecology: Spatial interactions and population dynamics. *Ecology, 75*(1), 17–29. doi:10.2307/1939378.

Horne, J. S., Garton, E. O., Krone, S. M., & Lewis, J. S. (2007). Analyzing animal movements using Brownian bridges. *Ecology, 88*(9), 2354–2363. doi:10.1890/06-0957.1.

Kuijpers, B., & Othman, W. (2009). Modeling uncertainty of moving objects on road networks via space-time prisms. *International Journal of Geographical Information Science, 23*(9), 1095–1117. doi:10.1080/13658810802097485.

Kwan, M. P. (1998). Space-time and integral measures of individual accessibility: A comparative analysis using a point-based framework. *Geographical Analysis, 30*(3), 191–216. doi:10.1111/j.1538-4632.1998.tb00396.x.

Kwan, M. P. (1999). Gender and individual access to urban opportunities: A study using space-time measures. *The Professional Geographer, 51*(2), 211–227. doi:10.1111/0033-0124.00158.

Lenntorp, B. (1977). Paths in space-time environments: A time geographic study of movement possibilities of individuals. *Environment and Planning, 9*(8), 961–972. doi:10.1068/a090961.

Lin, H. Y. (2012). Using compressed index structures for processing moving objects in large spatio-temporal databases. *Journal of Systems and Software, 85*(1), 167–177. doi:10.1016/j.jss.2011.08.005.

Lo, A. W. (1991). Long-term memory in stock market prices. *Econometrica, 59*(5), 1279–1313.

Miller, H. J. (1991). Measuring accessibility using space-time prism concepts within geographic information systems. *International Journal of Geographical Information System, 5*(3), 287–301. doi:10.1080/02693799108927856.

Miller, H. J. (1999). Measuring space-time accessibility benefits within transportation networks: Basic theory and computational methods. *Geographical Analysis, 31*, 187–212. doi:10.1111/j.1538-4632.1999.tb00408.x.

Miller, H. J. (2005a). A measurement theory for time geography. *Geographical Analysis, 37*(1), 17–45. doi:10.1111/j.1538-4632.2005.00575.x.

Miller, H. J. (2005b). Necessary space-time conditions for human interaction. *Environment and Planning B: Planning and Design, 32*(3), 381–401.

Miller, H. J., & Bridwell, S. A. (2009). A field-based theory for time geography. *Annals of the Association of American Geographers, 99*, 49–75. doi:10.1080/00045600802471049.

Neutens, T., Witlox, F., Weghe, V. D., & Maeyer, P. H. D. (2007). Space-time opportunities for multiple agents: A constraint-based approach. *International Journal of Geographical Information Science, 21*(10), 1061–1076. doi:10.1080/13658810601169873.

Nicholls, S., & Shafer, C. S. (2001). Measuring accessibility and equity in a local park system: The utility of geospatial technologies to park and recreation professionals. *Journal of Park and Recreation Administration, 19*(4), 102–124.

O'Sullivan, D., Morrison, A., & Shearer, J. (2000). Using desktop GIS for the investigation of accessibility by public transport: An isochrones approach. *International Journal of Geographical Information Science, 14*(11), 85–104. doi:10.1080/136588100240976.

Rostek, S., & Schöbel, R. (2006). *Risk preference based option pricing in a fractional Brownian market, No 299*. Tübinger Diskussionsbeiträge, University of Tübingen, School of Business and Economics.

Shaw, S. L., & Wang, D. M. (2000). Handling disaggregate spatiotemporal travel data in GIS. *GeoInformatica, 4*(2), 1384–6175. doi:10.1023/A:1009824122914.
United States Environmental Protection Agency (EPA). (2010). Fast facts: U.S. Transportation sector greenhouse gas emissions. Resource document. EPA Office of Transportation and Air Quality. http://www.epa.gov/otaq/climate/basicinfo.htm. Accessed 3 July 2012.
Weibull, J. W. (1976). An axiomatic approach to the measurement of accessibility. *Regional Science and Urban Economics, 6*(4), 357–379. doi:10.1016/0166-0462(76)90031-4.
Wilson, A. G. (1976). Retailers' profits and consumers' welfare in a spatial interaction shopping model. In I. Masser (Ed.), *Theory and Practice in Regional Science (London Papers in Regional Science)* (pp. 42–57). London: Pion Ltd.
Winter, S., & Yin, Z. C. (2010a). The elements of probabilistic time geography. *Geoinformatica, 15*(3), 417–434. doi:10.1007/s10707-010-0108-1.
Winter, S., & Yin, Z. C. (2010b). Directed movements in probabilistic time geography. *International Journal of Geographical Information Science, 24*(9), 1349–1365. doi:10.1080/13658811003619150.
Wu, Y. H., & Miller, H. J. (2001). Computational tools for measuring space-time accessibility within transportation networks with dynamic flow. *Journal of Transportation and Statistics, 4*, 1–14.
Yu, H., & Shaw, S. L. (2008). Exploring potential human activities in physical and virtual spaces: A spatio-temporal GIS approach. *International Journal of Geographic Information Science, 22*, 409–430. doi:10.1080/13658810701427569.

Chapter 13
Detecting Activity Types and Trip Purposes from Passive GPS Data: A Data Mining Approach

Donggen Wang and Bingxia Sun

13.1 Introduction

The increasing demand for more accurate and more detailed data for activity-travel behavior studies has posted great challenges to the conventional data collection methods which suffer from under-reporting, inaccurate information on time and location and low response rate. The development of positioning technologies such as Global Positioning System (GPS) and the integration of GPS with ground-based wireless communication network have provided new opportunities for collecting activity-travel behavior data (Wolf 2000). Positioning technologies offer the possibility to trace an individual's spatiotemporal trajectory – .i.e., recording the locations of individuals second by second and the speed of movement between locations. As individuals' activity and time use patterns are imbedded in their daily spatiotemporal trajectories, since the 1990s, many have proposed to use GPS to collect activity and time use data (Timmermans et al. 2009; Chen et al. 2010). It is anticipated that the GPS-based method can not only collect behavior data with better accuracy and completeness, but also reduce the burden on respondents greatly (Draijer et al. 2000; Forrest and Pearson 2005; Stopher et al. 2006).

There are two major approaches of using GPS for collecting activity-travel behavior data: interactive GPS and passive GPS. The interactive approach combines GPS with the Internet or other methods and requests respondents to supplement or verify information collected by GPS. For examples, Stopher and Collins (2005) suggest using GPS together with Internet-based prompted recall survey to supplement each other in the collection of trip-making information. Bohte and Maat (2008)

D. Wang (✉) • B. Sun
Department of Geography, Hong Kong Baptist University, Kowloon Tong, Kowloon, Hong Kong, Hong Kong
e-mail: dgwang@hkbu.edu.hk; bingxiasun@hotmail.com

M.-P. Kwan et al. (eds.), *Space-Time Integration in Geography and GIScience*, DOI 10.1007/978-94-017-9205-9_13

adopted an interactive validation approach by providing a user friendly interface consisting of a map and a table that depict all trips that a respondent visited during a day and present to the respondent for verification. The interactive approach requires substantial involvement of respondents and thus may not be able to overcome the low response rate problem of traditional data collection methods. On the other hand, as the name implies, passive GPS does not require the active involvement of respondents. Respondents are passively involved in the survey by simply carrying the device along with them and turn it on/off in time without the need to input any travel-activity behavior data. As a result, burdens on respondents can be greatly reduced. However, passive GPS data includes information only on time, position as well as heading and speed of movement etc. at certain time intervals. Most of the activity-travel information, such as what (types of activities and trip purposes), where (place where activities are conducted as well as trip origins and destinations), when (starting time, duration as well as trip arrival and departure time) and what transportation modes are used, have to be detected. Among the most difficult travel behavior information to be derived from GPS data is that on activity types and trip purposes.

In the past years, many attempts have been made to derive activity-travel information from GPS data. For examples, Wolf et al. (2004) proposed and developed methods to derive from GPS data trip purpose information with the help of GIS data on public facilities, shopping centers, residential places, etc. A number of similar attempts in combining GPS tracking data with GIS land use data and analysis to detect transport mode and trip purposes have been reported in the literature (Chung and Shalaby 2005; Bohte and Maat 2008; Chen et al. 2010). While the GIS functionalities such as buffer and overlay and simple rules of thumb are used in these studies to derive from GPS data the needed information, in most cases no explicit methodologies are applied. The only exception is Chen et al. (2010), who employed a probability model to determine the probability that a trip had one of the predefined trip purposes. Recently, methods with self-learning capabilities have been proposed to mine GPS tracking data and detect information such as transport modes and trip purposes. Moiseeva et al. (2010) suggested combining Bayesian belief network with self-learning to detect trip mode and purpose. Activity-travel patterns that are confirmed or corrected by respondents are used to improve the conditional probabilities for classification. Deng and Ji (2010) proposed using a decision tree algorithm to derive trip purposes.

This study suggests applying a genetic algorithm, an artificial intelligent method, to identify trip purpose through mining GPS tracking, land use and socio-economic data. Specifically, this method establishes a set of classification models through self-learning to extract the information of trip purpose based on information concerning land use type of trip ends, trip duration and timing, socioeconomic characteristics of respondents, and other relevant information. The study seeks to contribute to recent attempts to pave the way towards the development of an entirely passive and automatic way of collecting data from individuals for activity-travel behavior studies. The next section will review the existing methods for trip purpose detection. Then the genetic algorithm and its application for trip purpose detection will be introduced. Explanations of the field study and data collection will follow, and the subsequent section reports the application of the algorithm and explains as well as discusses the results of trip purpose detection. The last section concludes the paper.

13.2 Methods for Detecting Trip Purposes from GPS Data

In order to derive activity types and trip purposes from GPS data, it is generally agreed that some additional information such as land use and point of interests (POIs, e.g., public facilities, shopping centers, and educational establishments) are needed (Bohte and Maat 2008; Wolf 2000). In addition, data like respondents' home address, work place and places that they frequently visit may also be required (Stopher et al. 2008; Wolf et al. 2004).

Wolf (2000) demonstrates that trip purposes might be detected by matching the land use types of trip ends with the possible trip purposes. A table of standardized land use types and corresponding trip purposes is developed. A primary trip purpose is identified for each land use description and two additional possible trip purposes are added to accommodate the cases that different types of trip purposes may be associated with the same type of land use. In a follow-up study, Wolf et al. (2004) propose to collect data on respondents' socio-demographics and the destinations of their habitual activities to supplement GPS and GIS data for detecting trip purposes. Some rules are developed. For example, trip ends within a radius of 200 m from home are assigned with the purpose of 'Home'. Other trip purposes were derived by matching the GPS data with available POI information and land use polygons in GIS. Stopher et al. (2008) suggest that information on workplace, home address and the location of the often visited grocery stores can help ascertain about 75 % of trip ends and over 60 % of trip purposes. As for the remaining trip purposes, some heuristic rules may be applied to deduce with the help of land use data and data on occupation. Stopher and Collins (2005) advocate using internet-based prompted recall survey by showing respondents the derived results and asking for additional information to improve trip identification and trip purpose detection. Similarly, Bohte and Maat (2008) make use of the Internet to help validate trips and trip purposes identified from GPS data. A user friendly web-based interface is developed to visualize the trajectories of movement and present trip information derived from GPS data to respondents. Respondents can adjust the information about the trips derived and add any missing trips or other information. Except for some heuristic rules, most of the studies reviewed so far did not employ any model or algorithm for processing the data to derive trip purposes.

Griffin and Huang (2005) propose to use a decision tree classification model to derive trip purpose. Trip characteristics such as spatial location, time of day, length of stay are used as attributes to build the decision tree, which leads to the classification of different trip purposes. Similarly, Deng and Ji (2010) suggest establishing an initial decision tree for purpose classification based on attributes like land-use and temporal information, duration of stay and socioeconomic characteristics. Instead of placing much emphasis on data preprocessing, they introduce an innovative approach called adaptive boosting to continuously train and improve the decision tree, in which, each classifier votes for its predicted class, and the one who gets most votes is considered to be the final class. 226 GPS trip records from 36 respondents are collected to test the method and results show that the application of decision tree with adaptive boosting can largely improve the accuracy of trip purpose detection. It is argued, however, that decision tree is more suitable for cases with relatively simple relationship with a small number of variables.

Chen et al. (2010) apply a probabilistic model (the multinomial logit model) to determine trip purpose. The model estimates the probabilities of a trip having one of several possible purposes from data on time of day, history dependence and land use characteristics. Moiseeva et al. (2010) conduct a pilot experiment of applying an artificial intelligent method with self-learning capability to detect transport modes and trip purposes. They employ the Bayesian network that establishes probabilistic relationships between input variables and outcomes. Transport modes and trip purposes are classified based on the conditional probabilities estimated from input variables on trip characteristics. The self-learning function is realized through updating the conditional probabilities. A web-based prompted recall is presented to respondents for confirming or correcting the derived activity-travel patterns, which are used to update the conditional probabilities for subsequent classifications. The results of an empirical application show that better accuracy of both mode and purpose detection can be achieved through the self-learning mechanism. A shortcoming of the Bayesian network is that it needs prior information from past experiences to derive the probability function. Nevertheless, this study demonstrates the potentials of artificial intelligent methods with self-learning for detecting trip purposes. Thus more research efforts to explore the potentials of the artificial intelligent methods for trip purpose detection are needed.

13.3 A Genetic Algorithm for Detecting Trip Purposes

To further explore the potentials of artificial intelligent methods for data mining passive GPS data, this study proposes to apply the genetic algorithm, which is characterized by strong capability to search for optimal solution through self-learning, to detect trip purposes.

Genetic algorithm (GA) is an evolutionary type of algorithms (EA), which provide effective ways to search for optimal solution to a problem through self-learning (Buckles and Petry 1992). GA has been widely used in a great variety of applications in the fields of science and engineering to solve classification-related problems such as feature selection and classification (Punch et al. 1993), remote sensing image classification (Yang et al. 2006), and classification and prediction of precipitation occurrence (Sen and Öztopal 2001). In the past decade, there have been a number of reported studies on GA applications in the field of transportation. For instance, Gen et al. (1999) present a genetic algorithm to optimize large-scale logistic distribution; Park et al. (1999) and Ceylan and Bell (2004) propose applying GA to the problem of traffic signal optimization; and Pattnaik et al. (1998) also use genetic algorithm to minimize the overall cost in urban bus route design.

13.3.1 Principles of Genetic Algorithm

The basic principle underlying genetic algorithm is the evolutionary ideas of natural selection and genetics as described by Charles Darwin's theory of "survival of the fittest". Every organism has a set of rules, describing how that organism is built up from cell, the smallest unit of life. These rules are encoded in the genes of cell nucleus, which in turn are connected together into long strings called chromosomes. Each gene has its unique position in the chromosome. The complete set of genetic material (all chromosomes) is known as genome. Particular set of genes in genome is called genotype, which can develop to a phenotype after birth, representing its physical and mental characteristics such as eye color, intelligence, etc. (Holland 1992).

In a genetic algorithm, a set of chromosomes called population (genome) encode candidate solutions (called individuals) to an optimization problem. Each chromosome in this population consists of several genes that carry the information representing particular functions. Traditionally, these chromosomes are coded in binary as strings of 0 and 1, other encodings like integer, floating type or character etc. are also widely used (Forrest 1993). The algorithm usually begins with a group of individuals which are randomly created as potential solutions to a problem. Performances of these individuals are measured by a predefined fitness function. The ones with higher fitness value are likely to be selected and modified randomly through crossover and/or mutation to generate a new population. This evolutionary iteration will continue until the stop condition such as when a satisfactory fitness level or a maximum number of generations or a number of generations without improvement has been reached (Holland 1992).

The success of GA depends on not only how the initial solutions (or chromosomes) are defined and formed, but also the evaluation mechanism (i.e. the fitness function) and the genetic operations including crossover and mutation (Nicolas and Hao 2001). The fitness function determines the chance of survival for each individual, whilst the genetic operations increase the population diversity while avoiding or at least reducing the chance of getting trapped from local maximum. Crossover operator exchanges the piece(s) of genes between chromosomes. Chromosomes selected as parents through crossover probability swap segments of their genes with each other to create new chromosomes as children and then their positions in the population would be replaced. During this process, new individuals can be generated to the population, which brings possibility to have fitter chromosome (Holland 1992). In the same way, mutation operator alters the chromosome selected by a predefined probability by modifying one or several of its gene(s) randomly within a certain range. The new population is then used in the next iteration of the algorithm until the algorithm terminates (Goldberg 1989). It can be seen in this evolution process that "survival of the fittest" is implemented by providing more chances to those better-fit individuals so that they can be carried over to the subsequent generation.

According to Buckles and Petry (1992), genetic algorithm has good self-learning and self-adaptation capabilities and needs neither prior knowledge nor artificial interference in the process of searching for the optimal solutions. Its selection, crossover and mutation operations increase the vitality of the search process and provide more chances to find the optimal solution. Nevertheless, like most optimization techniques, a major problem with GA is the risk of trapping into local optimal that we may not even know it. Over the past few decades, various methods has been proposed to design efficient and robust GA either from the aspect of algorithm itself like diversified encoding schemes, specific crossover or mutation operations etc. or by combining GA with other algorithms like neural network, ant colony algorithm and so on. For more details, readers are referred to Andre et al. (2001) and Roeva (2008).

13.3.2 A Genetic Algorithm for Trip Purpose Detection

We consider the detection of activity types or trip purposes from GPS tracking data as a classification problem because different trip purposes are associated with different conditions. In order to fulfill personal, family or social needs, individuals assume different roles at family, among friends, and at work through performing activities of different types associated with trips of different purposes in their daily life. There are different classifications of trip purposes or activity types (Bhat and Koppelman 1993). A general classification includes seven categories such as *work, study, shop, social visit, recreation, home and others* (Bohte and Maat 2008). Alternatively, one may use 'home' as the reference point and group different purposes of trip into 'home-based' or 'non-home based' such as *home-based work, home-based education, home-based shopping, non-home-based work,* etc. (Stopher et al. 2008).

Trip purposes or activity types differ on a number of aspects including where are the trip ends or activity locations, when and for how long the activities are conducted. Thus, the land use type of trip ends or where activities are conducted provides important information to identify trip purpose. Similarly, trip start/end time and the duration of activities are different for different trip purposes (Stopher et al. 2008; Wolf et al. 2004). On the other hand, individuals' socioeconomic characteristics such as age, gender and household composition determine the types of activities they perform (Lu and Pas 1999). For examples, Wang and Law (2007) find that employed persons made fewer trips and spent less time for recreation and maintenance activities than unemployed ones. Chen and Mokhtarian (2006) report that individuals with high income spent more time on maintenance activities than others did. According to the investigation of Curtis and Perkins (2006), male tend to be more active than females in recreation activities such as sports while women prefer to spend more time on maintenance activities such as shopping and child-care. Buliung (2005) reveal that the elder and married people with young children are less likely engaged in out-of-home recreational activities than others.

The above discussions show that trip purposes can be identified by the characteristics of the trip and the socio-economic characteristics of the person who makes the trip. Thus, the detection of trip purpose can be treated as a classification problem that match trip purposes with the corresponding trip characteristics (e.g. trip duration and timing) as well as individuals' socioeconomic characteristics. In this regard, the genetic algorithm is applied to establish the classification models. In the following, we shall explain how the GA may be used to establish the classification models. The classification models can be considered as a series of heuristic rules in the form of "IF, THEN". The "IF" part (i.e., the rule antecedent) contains a combination of conditions on attributes and values, whereas the rule consequent (the "THEN" part) contains goal attribute (Noda et al. 1999). In this case, land use type of trip ends, trip duration and timing as well as individuals' socioeconomic characteristics may be used as the attributes in the "IF" part, whereas categories of trip purpose are considered as the goal attribute in the "THEN" part.

13.3.2.1 Encoding Chromosome

The classification rules of GA are represented by chromosomes. There are two approaches, namely Michigan and Pittsburgh approaches that are commonly used to encode chromosomes. By the Michigan approach, each classification rule is represented by one chromosome, whereas by the Pittsburgh approach, a set of classification rules are encoded in a single chromosome (Freitas 2002). For the trip purpose detection problem, if the Michigan approach is applied, one chromosome is developed for each trip purpose separately. For example, the chromosome for the trip purpose 'work' may looks like:

IF (7:30 < start time < 9:00 + 17:00 < end time < 20:00 + 4 hours < duration < 10 hours . . . . . .) THEN "*work*".

On the other hand, if the Pittsburgh approach is used, a single chromosome will set the rules for detecting all the trip purposes. Such a chromosome may looks like:

IF (18:00 < start time < 23:00 + 7:00 < end time < 9:00 + 7 hours < duration < 12 hours . . . . . .) THEN "*home*"; IF (7:30 < start time < 9:00 + 17:00 < end time < 20:00 + 4 hours < duration < 10 hours . . . . . .) THEN "*work*"; IF (17:30 < start time < 20:00 + 19:00 < end time < 20:00 + 1 hour < duration < 5 hours . . . . . .) THEN "*recreation*" . . .).

While the Pittsburgh approach has the advantages of generating all classification rules at once and allowing interactions between classification rules and the evaluation of classification rules as a whole, its disadvantages are: the chromosomes are usually rather long and complex and adaptive genetic operators are needed (Janikow 1993). On the other hand, the Michigan approach generates simple and short chromosomes and has the advantages that the classification rules can be easily evaluated and the normal genetic operators can be applied. Nevertheless, since each rule is generated separately, it is necessary to run several times the algorithm to obtain all the classification rules. Furthermore, quality of the classification rules as a whole is hard to evaluate (Greene and Smith 1993). Based on the above discussions, the choice of which approach to use mainly depends on which types of classification rules one wants to obtain. Further, if the quality of the rules as a whole rather than that of single rules needs to be evaluated, the Pittsburgh approach seems more suitable, and vice versa (Freitas 2002).

13.3.2.2 Definition of Fitness Function

An important step of the GA is to evaluate the classification rules or chromosomes and based on which determine the chance of survival of the chromosomes into the next generation. Freitas (2002) suggests that the fitness of the chromosomes should be evaluated by their predictive accuracy, which can be measured by the product of the so-called confidence factor and completeness factor. The confidence factor is the percentage that the predictions of the chromosomes are correct, whereas the completeness factor refers to the percentage that the actual cases are correctly predicted by the chromosomes. Noda et al. (1999) proposed a fitness function measuring not only the predictive accuracy but also the degree of interestingness of the classification rules. Given that the user would tend to be more surprised and interested when he/she saw some seemingly irrelevant attributes were attributed into relevant ones, the computation of interestingness value is based on the following: the smaller the probability of the goal attribute value, the more interesting it is. Fitness function is essentially a weighted sum of two terms measuring predictive accuracy and degree of interestingness and the weight assigned to each term is set by user.

13.3.2.3 Genetic Operators

Chromosomes with high fitness values are selected from the population to be parents for crossover and mutation operations. According to Darwin's evolution theory, the best ones should survive and generate new offspring. Several methods have been proposed to select the best chromosomes for generating offspring including roulette wheel selection, rank selection, tournament selection (Razali and Geraghty 2011). The Roulette Wheel Selection, which is also known as fitness proportionate selection, allocates probabilities to be selected to the chromosomes in proportion to their fitness values. An advantage of this method is that all individuals in the population are given a chance to be selected so that diversity in the population is preserved. However, it may have problems when the fitness values differ very much. For example, if the best chromosome fitness takes 90 % of the roulette wheel, chances for the other chromosomes to be selected would be quite low. Rank selection first sort out chromosomes in the population according to their fitness values and then compute selection probabilities according to their ranks rather than fitness values. Hence, a uniform scaling across the population is introduced in the rank-based selection with no influence of super-individuals. Nonetheless, this method may lead to slower convergence because the best chromosomes do not differ much from other ones. In tournament selection, *n* individuals are selected randomly from the population and compete against each other. Winners with highest fitness value will be chosen. Although tournament selection has high efficiency because no fitness scaling or sorting is needed, diversity of the population may be lost.

After the chromosomes are selected, crossover and mutation operations are conducted to create new chromosomes that potentially have higher fitness values. Crossover operation can mainly be classified into three categories based on the above encoding methods, including one-point crossover, two-point crossover and uniform crossover. As the name implies, if one-point crossover is used, one gene (value of certain determinant of trip purpose) or one segment of genes (values of certain determinant) in a chromosome will be exchanged with the gene(s) of the same locus in another chromosome so as to generate two offspring. In two-point crossover, parents will exchange two genes (value of two determinants of trip purpose) or two segments (values of two determinants), so and so forth. Mutation operation can be conducted by inverting several continuous genes or change value of one/several gene(s) in the same chromosome in order to produce new chromosome (Holland 1992). Similar to the choice of selection approach, either of the above methods for crossover and mutation operation can be used, which largely depends on the representation of chromosome.

Through the above procedures, a new population is generated, evaluated, and modified. After hundreds or thousands of iterations, optimal classification models can be obtained.

13.4 Field Study and Data Collection

To test the feasibility and validity of the genetic algorithm for identifying activity type and trip purpose proposed in the previous section, three types of data are needed. Firstly, GPS tracking data should be collected. Respondents need to be recruited to carry the GPS data logger wherever they go during the survey days. Second, GIS land use data like road network and Point of Interest (e.g. typical buildings such as shopping mall, school, hospital and park etc.) need to be collected. Third, to evaluate the accuracy of prediction by the proposed method, respondents are also required to provide activity diaries as well as their socioeconomic characteristics.

13.4.1 Case and Sample

The field study was conducted in Guangzhou from January 11th to May 30th 2011. Guangzhou, the capital city of Guangdong province of China and a major metropolis in the Pearl River Delta, was chosen for our field study because of the availability of detailed land use data, sufficient complexity of urban environment and convenience for us to collect data. With a limited research budget and number of GPS devices, we recruited in total 21 respondents with different socio-demographic backgrounds such as occupation and age. The respondents participated in the survey for 325 days, resulting in a dataset of 168 person-days in which a total of 859 activities were reported by the respondents.

13.4.2 Collection of GPS Tracking Data and Activity Dairies

As already mentioned above, GPS has been widely used because of its good performance in providing accurate three-dimensional spatial and temporal information in outdoor environments. Recently, cellular phone tracking based on Global System for Mobile Communication (GSM) network has attracted attention due to the popularity of smart phones. Although the GSM network is widespread, its positioning accuracy is only about 50–100 m (Retscher and Kealy 2005; Sun et al. 2009). Despite that some of the operating systems like Android in HTC/Google and iPhone OS in iPhone have embedded Assisted GPS (AGPS), our test of iPhone 4 in Hong Kong shows the accuracy varies a lot from several meters to dozens of meters in different situations. Worse still, the use of AGPS in mobile phone greatly reduced its battery life. On the other hand, further improvements of GPS technology in size, weight and power supply make it more convenient for respondents to bring with wherever they go. For the above reasons, this study chose to use the more reliable portable GPS devices for collecting tracking data. The type of portable GPS devices (size: 90 mm × 44 mm × 10 mm; weight: 50 g) that we used is assisted by GSM signal and can achieve a positioning accuracy of about 2.5 m. In total 11 devices were used. Except that two devices were out of order due to battery problems, all the other devices work well. An on-line monitoring system was developed by the GPS device provider. Once the devices are turned on and begin tracking, they are highlighted on the right panel of the system. These GPS tracking data with WGS-84 Coordinate System can be immediately uploaded to this system and displayed on the digital map when start and end time is provided.

As noted before, each of the participants was required to bring the above GPS device when performing their daily activities and report all their activities and trip making through a paper-based questionnaire for a number of days. Besides some instructions of GPS device in the questionnaire, spaces were also left for participants to fill in some basic data like home, school or work address, places that they often visit for shopping, entertaining, etc. Most importantly, they were asked to record each trip and corresponding details such as date, start/end time, activity type, destination/place name and street address.

To make the data collection process smoothly, participants were encouraged to charge the battery of GPS device timely. As for the potential problems of GPS signal blockage and data loss caused by high-rise building and cold/warm start, they were reminded to pay attention to the indicator light of GPS signal on the device and try to make sure that the device is at work all the time. Respondents were encouraged to report any problem with the GPS device during the data collection process.

In most existing studies on GPS applications, tracking data were recoded for every seconds. For example, Moiseeva et al. (2010) collected data with a time interval of 3 s, while Bohte and Maat (2008) logged data every 6 s. With such detailed data, one is able to reconstruct every single movement or standstill of the object. However, if a GPS receiver logs individuals' positioning information second by second continuously for 24 h, the amount of data would be enormous. These massive data not only cause much trouble in data preparation because of the need to remove a great volume of redundant data (not to mention the tedious work of data filtering for removing noise data), they also post high requirement for the memory capacity and battery life of GPS devices. In fact, if the purpose is to collect activity-travel behavior data, not the exact and detailed trajectories of movement, in other words, if the purpose is to distinguish activity and travel, one does not need the GPS tracking data in such a frequency. For all these reasons, our GPS tracking data were mostly recorded for every 1-min. To support our argument and for comparison purposes, some of them were recorded for every 3 min. The 1-min data involves 100 person-days and 549 reported activities, whilst the 3-min ones 68 person-days and 310 reported activities. As evidenced later, the successful rate of detecting trips and activities from 3-min data is not very much different from that of 1-min data.

13.4.3 Collection of GIS Land Use Data

The following GIS land-use data were collected: Guangzhou administrative boundary, a complete road network, as well as Point of interests (POIs) including restaurants, banks, post offices, medical and sports facilities, places for entertainment, shopping places, schools, government departments, places of welfare and religious activities in both point and polygon patterns.

Among these GIS data, Guangzhou administrative boundary and road network were collected from Statistics Bureau of Guangzhou while the layer of restaurants came from National Geomatics Center of China (NGCC). These data were in WGS 84 (short for World Geodetic System 1984) Coordinate System, which is a commonly used geocentric coordinate system worldwide and is currently the reference system being applied to the Global Positioning System. All the other POI layers were collected from Guangzhou Bureau of Urban Planning in the same but unknown coordinate system. These data were the latest and most complete information we could obtain. Obviously, these data can cover most of the trip purposes we need to classify, including home, work, school, shopping, recreation, personal affairs and welfare/religion activities. Among these seven trip purposes, the first two categories can be identified according to participants' basic data filled in the questionnaire; schools are comprised of elementary and secondary school, universities and colleges; shopping places consist of markets and shopping malls; public places of recreation refer to restaurants, sports courts and parks; facilities for personal affairs includes post offices, banks, governments and medical establishments, while the last category refers to welfare and religion institutions.

13.5 Detecting Activity Types and Trip Purposes

13.5.1 Data Preprocessing

13.5.1.1 Adjusting All Data into the Same Coordinate System

As mentioned earlier, some of the POI layers are referenced in a coordinate system different from the WGS-84 system in which the GPS tracking data and other GIS data are referenced. We first need to transform these POI layers to the WGS-84 coordinate system so that a composite map that includes the base map of Guangzhou boundary, land use layers and GPS track points in the WGS-84 system are developed for further analysis. This transformation is accomplished through spatial adjustment in ArcMap by linking and comparing a number of control points having the same relative positions in the source and destination layers so as to derive the transformation formulas from one coordinate system to another. To minimize the correction errors, a total of eight control points scattered at different corners of the urban area were chosen. The residual errors associated with each link and the overall Root Mean Square (RMS) error were estimated and the results indicate that the transformation meets the accuracy requirement. After the establishment of the transformation formulas, the feature layers can be adjusted accordingly.

13.5.1.2 Preprocessing GPS Tracking Data

The original GPS tracking data contains only the spatio-temporal trajectories of objects, including latitude and longitude coordinates, date, time, speed information, temperature as well as some indicators concerning the status of the device. The original GPS data may contain some noise data (tracking points that go beyond the boundary of the study area or drift far away from other points) resulting from many reasons including the lack of sufficient number of visible satellites because of urban canyon effects. These noise data were first removed through a data filtering process. Information on HDOP value, the number of visible satellites, speed and local conditions were used to evaluate the quality of the GPS data and identify the noise data and decide which tracking points should be filtered. After the GPS data had been cleaned up, map matching was conducted to map the GPS trajectories to the road network and land use on the GIS map. One may use the buffer analysis based on distance from GPS point to road network proposed by Wolf (2000) or the method suggested by Chung and Shalaby (2005) to combine distance and azimuth data to identify the links traveled. Most of the GPS tracking data in this study match well with the GIS map. Not much effort on map matching was needed in this case.

13.5.1.3 Activity and Trip Identification

Since the GPS tracking data are continuous spatial and temporal trajectories, trips and activities need to be identified from these data. Different methods can be used to identify activities in the outdoor and indoor environments. The time interval between two consecutive track points caused by signal loss in an indoor environment is used to identify indoor activities. The departure time and duration of indoor activities can be identified in its next record when satellite signals are received again. The distance between two adjacent points can be used as supplement information to differentiate an indoor activity from traveling by underground transportation. For example, one may consider a case with the time interval of more than 10 min and the distance of more than 2,000 m as underground traveling but not a trip end or activity in the indoor environment. Trip ends for outdoor activities are distinguished by calculating the density of tracking points. The point analyst tools in ArcMap were employed to calculate point density and identify trip ends and outdoor activities. A previous study suggests that tracking points with densities at least twice that of the others have a better chance to be trip ends. Similar principles were adopted to identify outdoor activities in this study.

Altogether 713 activities or trips in 157 person-days have been successfully detected. Comparing to the 859 activities in 168 person-days reported by the 21 participants, an overall successful rate of 83 % was achieved. This is compatible to that of other studies such as the 81 % successful rate reported by Stopher et al. (2006). Our investigation into the discrepancies between what derived from the GPS data and that reported by the respondents revealed that there were 14 activities detected from GPS data but not reported by respondents. These were most likely the short-duration activities under reported by respondents including buying something at convenient shops; posting a letter; etc. These activities could easily ignored or forgotten by respondents. This is the so-called under-reporting problem of the traditional questionnaire-based data collection methods. Nevertheless, some of those activities might be mistakenly identified from the prolonged waiting time at public transport station or in a traffic jam. On the other hand, there were in total 160 activities that were reported by respondents but not identified from the GPS data. This is largely resulted from the signal loss due to the short period of exposure to outdoor environment. The 11 person-days missed in the GPS data were caused by the fact that the respondents were not much exposed to the outdoor environment because they lived and worked very close to subway stations and they traveled only in the subways during those days. In addition, the inappropriate usage of the device (e.g. the GPS device was stored deep in the carrying bag) that make it difficult to receive signals also accounted for some of the discrepancies.

We have also separately calculated the successful rate for the 1-min data and that for the 3-min data. The results show that the successful rate of the 1-min data was 84 %, whilst that of the 3-min data was 82 %. Though that for the 1-min data is higher than that for the 3-min data, but the difference is very marginal. However, the 3-min data saved lots of storage space, battery life and preprocessing efforts.

13.5.1.4 Identifying Possible Trip Purposes

To identify the possible trip purposes associated with trip ends, in addition to the POI information, individuals' basic information like the addresses of home, school or work place, places that they often visit are also used. Based on the addresses that respondents filled in the questionnaire, we made use of Google Earth and ArcMap to obtain the exact positions of these places in terms of latitude and longitude coordinates. A separate layer of the positions of home, workplace and often visited places was established for each participant. Once this was done, we then associated trip ends data with personal basic data and the land-use types nearby to identify possible trip purposes using buffer analysis. In the earlier years of GPS-based travel surveys, when the accuracy of positioning is concerned, Wolf et al. (2004) define that trip ends within 200 m from the home location were assigned with the activity purpose 'home', and other point of interests within the distance of 300 from each ends were derived as possible purposes. Years later, with the improvement of GPS technologies, Bohte and Maat (2008) reduce the area of buffer zone and argued that the area of 'home' and 'work' should be larger since the location of which would be visited frequently (i.e. if distance between trip end and POI <50 m, then category = category POI ('shopping', 'recreation', 'culture', 'medical', 'kids' or 'railway station', if distance between home/work and trip end < 100 m, then category = 'home'/'work'). Similarly, Moiseeva et al. (2010) also set the radius of 100 m for the personal spatial data and 50 m for the land use data in the region of interest. Consequently, taking the previous experience and positioning accuracy into consideration, buffer zones in this case study were defined as within a distance of 100 m or 50 m respectively from home/work address and other frequently visited places/point of interest. Finally, Land use features which fall into the buffer zones of trip ends are regarded as possible purposes.

13.5.2 Operationalizing and Calibrating the Algorithm

Based on the above discussions, five variables including land use types of trip ends (in terms of possible trip purposes), trip durations, the starting time of the trips, age of the respondents and day of the week are used to encode the chromosome.

We consider ‘age’ as a representative of socio-demographic variables because it is a proxy for employment status, life cycle and life style. It is thus an important explanatory variable for activity patterns. Further, activity patterns on weekdays and weekends are usually quite different, thus day of the week in terms of the division between weekdays and weekends is also included.

If all five variables are used to encode the chromosome, the algorithm will become very complicated. Instead we stratify the cases by day of the week and age of respondents into four cases: younger or older than 25 years old and weekday or weekend. The other three variables are used as genes to develop chromosomes specifically for each of the four cases. The Michigan approach is used to encode the chromosomes. In other words, separate classification models are developed for each purpose. The encoded chromosome is illustrated as follows:

$$\text{if } \underbrace{(\ C_j \in) \text{ and } (V_{1L}A_1) \text{ and } (V_{2L}A_2V_{2R})}_{\text{Conditions}}, \text{ then } \underset{\substack{\downarrow \\ \text{Conclusion}}}{C_j} .$$

The first condition ($C_j\in$) is to judge whether C_j is one of the possible trip purposes; A_1 and A2 are the attributes representing start time and duration for stay respectively, V_{iL}, V_{iR} are left and right boundary value of attributes; C_j represents one of the trip purposes. In this study, seven categories of trip purpose are differentiated including home (C1), work (C2), school (C3), shopping (C4), recreation (C5), personal affairs (C6), and volunteering/religion activities (C7).

Following Freitas (2002), the fitness function is defined based on the confidence factor and the completeness factor. In other words, the fitness value of each chromosome depends on how well the attributes match with those existing combination of conditions and conclusion derived from data provided by respondents.

In order to ensure that stronger chromosomes have greater chance of survival than the weaker ones, firstly we use the elite strategy to reserve the best individuals in each generation. In other words, the chromosome with the highest fitness value in the current generation directly survives into the next generation without crossover or mutation. The other chromosomes are selected through the roulette wheel selection process, which ensure that each chromosome has the chance to be chosen and the ones with higher fitness values are more likely to be selected.

Chromosomes selected from the above method are then randomly chosen to be operated by one-point crossover through crossover rate (a threshold between 0 and 1 to judge which chromosomes should conduct crossover operation). In case unreasonable chromosomes are generated during the process, this step would be operated to the boundary values of the same attribute like starting time or duration from different individuals. Similarly, mutation also operates to the chromosomes which are randomly chosen through mutation rate (a threshold between 0 and 1 to judge which individuals should conduct mutation operation), and we can perform this operation by changing boundary value of one attribute to another random number between 0 and 1.

To calibrate the algorithm, we need to decide on the size of the population, maximum number of generations, and the probabilities of crossover and mutation. It should be noted that similar to other machine learning methods, parameters in GA such as population size, maximum generation, as well as crossover and mutation probability (Pc and Pm) need to be adjusted and improved continuously during the evolutionary process. Obviously, if the population size is set unreasonably, it's difficult to find out the optimum solution or the convergence time would be greatly extended. In this study, the population size is set to be between 20 and 30. GA is sensitive to the crossover and mutation rates, a crossover rate that is too high may lead to premature convergence of the genetic algorithm, in contrast, it may slow down the progress of evolution (Kaplan and Hegarty 2005). Similarly, a very small mutation rate may lead to genetic drift while a mutation rate that is too high may result in the loss of some good solutions. Lots of experiments in the previous studies indicate that Pc value should range from 0.5 to 0.9 while Pm value should be set between 0.05 and 0.2 (Şen and Öztopal 2001; Yang et al. 2006).

13.5.3 Results

Figure 13.1 shows part of the algorithm's output file, which includes the input parameters in the top left box as well as the first three and last five generations of the chromosome evolution. As can be seen in the bottom right box, after 80 generations' evolution in this run, we got the best classification model with the fitness value of 0.7347 for the purpose of "Recreation": 0.7793, 0.0127, 0.1432, and 5.0000, which indicates 18:42 as start time, 18 min minimum duration, 3 h and 26 min maximum duration and trip purpose 'recreation' respectively. The whole sentence can be interpreted as: on condition that "5" is included as one of the possible purposes, if activity starts after 18:42 while its time duration falls into the interval between 18 min and 3 h and 26 min, it is regarded to be the best model to identify trip purpose as "Recreation".

After running dozens of times for each trip purpose category, best classification models having the highest fitness value for each category can be identified. Table 13.1 lists the best classification model for all categories of trip purposes specifically for each of the four cases identified earlier. Information on which run of the algorithm and which generation the best model was obtained is also given in the table. No classification models are given for the cases in which the trip purposes are not reported by respondents.

As shown in Table 13.1, classification models for activity types/trip purposes such as "Home", "Work" and "School" have rather high value of fitness, higher than that for "shopping", "Recreation" and "Personal affairs". This may be due to fact that the land use data for home, work and school are more complete and updated than that for other trip purposes. Another possible reason is that activity types of shopping, recreation and personal affairs are more likely performed at places with mixed land use than other types. It is usually more difficult to identify trip purposes

```
Classification of "for Recreation" using Genetic Algorithm:
**************************************************
              Basic Parameters of GA:
        population size(population)  =     25
        maximum generation(maxGen)   =     80
        crossover rate(pcross)       =   0.60
        mutation rate(pmutation)     =   0.10
Run Times =  1
The current generation is:        1
 The average fitness value of this population is:   0.1657
 The maximum fitness value is:        0.4571
 The minimum fitness value is:        0.0455
 The best individual is coded as: 0.7793  0.0183  0.7750  5.0000
Fitness value of the best individual is:   0.4571
The current generation is:        2
 The average fitness value of this population is:   0.1501
 The maximum fitness value is:        0.6125
 The minimum fitness value is:        0.0455
 The best individual is coded as: 0.6964  0.0127  0.1432  5.0000
Fitness value of the best individual is:   0.6125
The current generation is:        3
 The average fitness value of this population is:   0.1775
 The maximum fitness value is:        0.6125
 The minimum fitness value is:        0.0455
 The best individual is coded as: 0.6964  0.0127  0.1432  5.0000
Fitness value of the best individual is:   0.6125
------

The current generation is:        76
 The average fitness value of this population is:   0.2132
 The maximum fitness value is:        0.7347
 The minimum fitness value is:        0.0455
 The best individual is coded as: 0.7793  0.0127  0.1432  5.0000
Fitness value of the best individual is:   0.7347
The current generation is:        77
 The average fitness value of this population is:   0.2071
 The maximum fitness value is:        0.7347
 The minimum fitness value is:        0.0455
 The best individual is coded as: 0.7793  0.0127  0.1432  5.0000
Fitness value of the best individual is:   0.7347
The current generation is:        78
 The average fitness value of this population is:   0.1617
 The maximum fitness value is:        0.7347
 The minimum fitness value is:        0.0455
 The best individual is coded as: 0.7793  0.0127  0.1432  5.0000
Fitness value of the best individual is:   0.7347
The current generation is:        79
 The average fitness value of this population is:   0.1444
 The maximum fitness value is:        0.7347
 The minimum fitness value is:        0.0455
 The best individual is coded as: 0.7793  0.0127  0.1432  5.0000
Fitness value of the best individual is:   0.7347
The current generation is:        80
 The average fitness value of this population is:   0.1965
 The maximum fitness value is:        0.7347
 The minimum fitness value is:        0.0455
 The best individual is coded as: 0.7793  0.0127  0.1432  5.0000
Fitness value of the best individual is:   0.7347
```

Fig. 13.1 Example of model outputs

Table 13.1 Case specific classification models

Starting time	Duration (min(h:min))	Duration (max(h:min))	Fitness value	Purpose	Run	Generations
Age > 25& weekday						
17:26	9:09	14:50	0.8858	Home	42nd	70
7:58	0:50	9:42	0.9025	Work	37th	61
9:13	3:47	7:46	0.6400	School	24th	34
16:02	0:06	3:14	0.5333	Shopping	40th	37
17:53	0:47	2:27	0.6944	Recreation	62nd	60
11:14	0:07	1:58	0.4444	Personal affairs	39th	61
\	\	\	\	Volunteering/ religion		
Age >25 & weekend						
12:46	1:26	20:38	0.9420	Home	40th	33
\	\	\	\	Work		
\	\	\	\	School		
13:40	0:36	6:24	0.4630	Shopping	38th	67
10:46	1:15	4:02	0.7872	Recreation	27th	35
11:18	0:07	1:56	0.6227	Personal affairs	29th	38
\	\	\	\	Volunteering/ religion		
Age < =25 & weekday						
11:57	0:36	18:08	0.9112	Home	35th	62
8:06	2:27	10:22	0.8674	Work	29th	42
8:04	1:30	5:08	0.7347	School	21st	52
19:01	0:06	4:33	0.4444	Shopping	36th	57
15:12	0:53	3:27	0.7347	Recreation	31st	26
11:22	0:05	1:49	0.6392	Personal affairs	27th	41
\	\	\	\	Volunteering/ religion		
Age < =25 & weekend						
11:50	0:11	19:09	0.9481	Home	35th	57
\	\	\	\	Work		
\	\	\	\	School		
12:21	0:09	4:23	0.7103	Shopping	39th	37
11:04	0:14	12:21	0.8048	Recreation	36th	56
9:42	0:10	2:11	0.6227	Personal affairs	37th	29
9:25	0:17	3:35	0.6400	Volunteering/ religion	24th	9

at places with mixed land use. On the other hand, there are no obvious differences between the trip purpose categories regarding at which run of the algorithm or at which generation the best classification models were obtained. This is probably due to the randomness of the genetic algorithm.

The classification models developed can now be applied to detect trip purposes for given data on land use, starting time, duration, age and day of the week. There might be different ways to apply the models for detection. One way is to assign the purpose of a classification model that has the closest distance to the given data. The Euclidean distance can be used to measure the differences between the data to be classified and different classification models. Thus, if the purpose of the classification model falls within the possible purposes of a trip end, we can then calculate the distance by adding the distances from the given data on starting time and duration to those values specified in the classification models of possible purposes. The smaller the gap it is with one classification model, the higher possibility it is classified as the trip purpose of the model. The distance concerning the starting time can be acquired directly by the point-to-point comparison. As for that concerning duration, if the given data lies within the minimum and maximum value of duration in the classification model, the distance for duration would be zero. Otherwise, distances will be calculated respectively between the given duration and the two boundary values in the model. The shorter distance is considered to be the distance from the duration boundaries of the model.

13.5.4 Internal and External Validity

To assess the applicability of a model or algorithm, it is important to evaluate its validity. Cook and Campbell (1979) elaborate the types of validity and define that internal validity addresses whether or not an observed covariation could be considered a causal relationship. In contrast to internal validity, external validity evaluate whether or not an observed causal relationship could be generalized to different measures, persons, settings, and times. For relatively small data sets, internal validation of models may not be sufficient and indicative for their performance in other studies. External validation may therefore be necessary before implementing these models in practice (Bleeker et al. 2003). To evaluate the validity of the algorithm developed in this study, we can compare the trip purposes detected by the algorithm to the real trip purposes reported by respondents. The 1-min data set (i.e., a total of 549 activities) is used to develop the classification models. These models are then applied to detect trip purposes for the 1-min dataset to assess internal validity and for the 3-min dataset (about 310 activities) to evaluate external validity. The results on internal validity are presented in Table 13.2 and that on external validity in Table 13.3.

Table 13.2 shows the cross-tabulation of the trip purposes reported by respondents and that detected by the models. The accuracy rate for each category is also reported. As one may tell from the table, the internal validity of the model is in general very high. The accuracy rates for three of the seven categories are higher than 85 %. As expected, home and school have the highest accuracy rates. The accuracy rates for shopping and recreation are also rather high. On the other hand,

Table 13.2 Internal validity

Trip purposes predicted by models	Trip purposes reported by respondents							Missing land-use data cases	Number of cases	Accuracy rates (%)
	C1	C2	C3	C4	C5	C6	C7			
C1: home	109			9					118	92.4
C2: work		18	1	2	5				26	69.2
C3: school			21		3				24	87.5
C4: shopping				45	3	1	1	7	57	79.0
C5: recreation	1			13	94				108	87.0
C6: personal affairs	4		1	23	19	46		21	114	40.3
C7: Volunteering/ religion					1		2		3	66.7

Table 13.3 External validity

Trip purposes predicted by models	Trip purposes reported by respondents							Missing land-use data cases	Number of cases	Accuracy rates (%)
	C1	C2	C3	C4	C5	C6	C7			
C1: home	64			5	2			1	72	88.9
C2: work		29		4		2			35	82.9
C3: school			8						8	100
C4: shopping				18	2			3	23	78.3
C5: recreation		1	1	10	59				71	83.1
C6: personal affairs			1	5	9	17		8	40	42.5
C7: Volunteering/ religion										

a relative low percentage of personal affairs are correctly detected by the models. As indicated in the table, a major reason is that there are many cases that the land use data for personal affairs are missing. From this point of view, this is caused by availability of data, not by the algorithm. In a way similar to Table 13.2, Table 13.3 reports the external validity. Comparing these two tables, one may easily find out that the external validity is very much comparable to the internal validity. Though for some cases, the external validity is lower than the internal validity, but in some other cases, the external validity is even higher than the internal validity.

Overall, the internal and external validities show that the algorithm and data mining approach proposed in this study can detect activity types and trip purposes with accuracy rates reasonably good and comparable to that of other studies. Though for some trip purposes the accuracy rate of detection are not very high, we believe that it is very much caused by data availability rather than the algorithm.

13.6 Discussions and Conclusions

Over the past few years, although GPS-enabled methods have attracted increasing attention in the collection of individuals' activity-travel behavior data, the identification of activity type and trip purpose is still considered as a bottleneck and keeps passive GPS from large-scale applications in behavior studies. It was against this background that this study was initiated and conducted. We developed a new method based on a genetic algorithm for the detection of trip purpose. We established a set of classification models through self-learning from data on land use type of trip ends, trip duration and timing, socioeconomic characteristics of respondents, and other relevant information. To test the applicability and validity of the proposed method, we conducted a field study and collected data in 2011. The data were used to test the applicability and validity of the method. The evaluation results indicated that more than 80 % of the major trip purposes or activity types have been successfully detected based on the proposed method. Both the internal and external validity tests have demonstrated satisfactory ability of this proposed method for identifying activity type and trip purpose. Most importantly, this learning method has great potentials to achieve better performance when more data are involved.

This study therefore contributes to the improvement of trip purpose detection by developing a self-learning method with less artificial interference and establishing classification models separately according to different characteristics of respondent and situation.

It should be noted that as an exploratory attempt, this study unavoidably has limitations. In particular, accuracy of trip purpose detection varied considerably according to different categories, which was mainly caused by the lack of data for some land-use types. The success rate of trip purpose detection therefore depends heavily on the accuracy and completeness of GIS data, especially for those land-use features. Furthermore, although the overall accuracy already looks promising in this pilot test, this approach to identify trip purpose from GPS data is at the experimental stage which needs to be improved in both algorithm design and sample method. Therefore, several issues may be identified as concerns for future research.

In the short term, participants with wider range of identities and age groups should be recruited so that more data in both quantity and diversity can be collected for training and testing. Input data of the algorithm including the variables and personal information should be more diversified according to different trip purpose. The design and parameters of GA also need to be improved, in particular, respondents' socio-economic characteristic such as sex, age and occupation should be considered as variables within the algorithm so as to participate in the evolution process rather than just establish different classification models for each case as employed in this study. Although the process of introducing more variables into GA would be challenging, to effectively minimize arbitrariness and subjectivity as typically revealed in previous studies and achieve better classification models, the benefit of doing so is worth the efforts. We believe that the development of advanced learning algorithms will further enhance the success for detecting activity travel behavior data while reducing respondents' burden gradually.

In the medium term, the rapid development of positioning technologies, in particular, the use of mobile phone and wireless communication technologies for indoor positioning, is very likely to provide much greater positioning accuracy. Therefore, the post-processing of detecting useful data from GPS loggers can be much easier.

In the long term, it should be noted that logged data presents high demand for memory capacity of GPS or GSM device because it has to record and store huge mass of tracking data. With the help of 3G mobile communication technologies and other high-speed facilities stored with a full set of deduction program, real-time tracking of individuals' behavior can be theoretically an inevitable trend and potentially realized in the future study.

Acknowledgements This research is supported by two General Research Fund (GRF) projects of the Hong Kong Research Grant Council (HKBU245008 and HKBU244610) and a FRG project of Hong Kong Baptist University (FRGI/10-11/049).

References

Andre, J., Siarry, P., & Dognon, T. (2001). An improvement of the standard genetic algorithm fighting premature convergence in continuous optimization. *Advances in Engineering Software, 32*(1), 49–60.

Bleeker, S. E., Moll, H. A., Steyerberg, E. W., Donders, A. R. T., Derksen-Lubsen, G., Grobbee, D. E., & Moons, K. G. M. (2003). External validation is necessary in prediction research: A clinical example. *Journal of Clinical Epidemiology, 56*(9), 826–832.

Bohte, W., & Maat, K. (2008). Deriving and validating trip destinations and modes for multi-day GPS-based travel surveys: A large-scale application in the Netherlands. In 8th international conference on survey methods in transport: Harmonization and data comparability, France.

Buckles, B. P., & Petry, F. E. (1992). Genetic algorithms: Introduction and applications. In IEEE Computer Society Press, Catalog #2935, Los Alamitos, CA, USA.

Buliung, R. N. (2005). *Activity/travel behaviour research: Approaches and findings with. Identification of research themes and emerging methods* (Working Paper Series No. 008). Hamilton: Center for Spatial Analysis, McMaster University.

Ceylan, H., & Bell, M. G. H. (2004). Traffic signal timing optimization based on genetic algorithm approach, including drivers' routing. *Transportation Research Part B, 38*, 329–342.

Chen, C., & Mokhtarian, P. L. (2006). Tradeoffs between time allocations to maintenance activities/travel and discretionary activities/travel. *Transportation, 33*, 223–240.

Chen, C., Gong, H., Lawson, C., & Bialostozky, E. (2010). Evaluating the feasibility of a passive travel survey collection in a complex urban environment: Lessons learned from the New York City case study. *Transportation Research Part A, 44*, 830–840.

Chung, E., & Shalaby, A. (2005). A trip reconstruction tool for GPS-based personal travel surveys. *Transportation Planning and Technology, 28*(5), 381–401.

Cook, T. D., & Campbell, D. T. (1979). *Quasi-experimentation: Design and analysis issues for field settings*. Chicago: Rand McNally.

Curtis, C., & Perkins, T. (2006). *Travel behaviour: A review of recent literature*. http://urbanet.curtin.edu.au/local/pdf/ARC_TOD_Working_Paper_3.pdf

Deng, Z. W., & Ji, M. H. (2010). Deriving rules for trip purpose identification from GPS travel survey data and land use data: A machine learning approach. In B. Mao, Z. Tian, H. Huang, & Z. Gao (Eds.), *Proceedings of the seventh international conference on traffic and transportation studies*, held in Kunming, China, August 3–5. Published by ASCE (American Society of Civil Engineering).

Draijer, G., Kalfs, N., & Perdok, J. (2000). Global positioning system as data collection method for travel research. *Transportation Research Record, 1719*, 147–153.

Forrest, S. (1993). Genetic algorithms: Principles of natural selection applied to computation. *Science, New Series, 261*(5123), 872–878. Aug. 13, 1993.

Forrest, T. L., & Pearson, D. F. (2005). Comparison of trip determination methods in household travel surveys enhanced by a global positioning system. *Transportation Research Record: Journal of the Transportation Research Board, 1917*(2005), 63–71.

Freitas, A. A. (2002). A survey of evolutionary algorithms for data mining and knowledge discovery. In A. Ghosh & S. Tsutsui (Eds.), *Advances in evolutionary computation* (pp. 819–845). New York: Springer.

Gen, M., Li, Y. Z., & Ida, K. (1999). Solving multi-objective transportation problem by spanning tree-based genetic algorithm. *Journal of IEICE Trans Fundamentals, E82A*, 2802–2810.

Goldberg, D. E. (1989). *Genetic algorithms in search, optimization, and machine learning*. Boston: Addison-Wesley Longman.

Greene, D. P., & Smith, S. F. (1993). Competition-based induction of decision models from examples. *Machine Learning, 13*, 220–257.

Griffin, T., & Huang, Y. (2005, November 9–11). A decision tree classification model to automate trip purpose. In S. Dascalu (Ed.), *Derivation: Proceedings of the ISCA 18th international conference on computer applications in industry and engineering* (pp. 44–49). Honolulu: ISCA. ISBN 1-880843-57-9

Holland, J. H. (1992, July). Genetic algorithms. *Scientific American*, pp. 66–72.

Janikow, C. Z. (1993). A knowledge-intensive genetic algorithm for supervised learning. *Machine Learning, 13*, 189–229.

Kaplan, E. D., & Hegarty, C. J. (2005). *Understanding GPS: Principles and applications* (2nd ed., p. 726). Boston: Artech.

Lu, X., & Pas, E. I. (1999). Socio-demographics, activity participation and travel behaviour. *Transportation Research Part A, 33*, 1–18.

Moiseeva, A., Jessurun, J., & Timmermans, H. (2010). Semiautomatic imputation of activity-travel diaries using GPS traces, prompted recall, and context-sensitive learning algorithms. In 89th annual meeting of Transportation Research Board, Washington, DC.

Nicolas, P. G., & Hao, J. K. (2001). A genetic algorithm for the classification of natural corks. In *Proceedings of the genetic and evolutionary computation conference* (pp. 1382–1388). San Francisco: Morgan Kaufmann.

Noda, E., Freitas, A. A., & Lopes, H. S. (1999). Discovering interesting prediction rules with a 805 genetic algorithm. In *Proceedings of the Congress on Evolutionary Computation-1999 (CEC 1999)* (pp. 1322–1329). Washington, DC: IEEE.

Park, B., Messer, C. J., & Urbanik, T. (1999). Traffic signal optimization program for oversaturated conditions: Genetic algorithm approach. *Journal of the Transportation Research Board, 1683*, 133–142.

Pattnaik, S. B., Mohan, S., & Tom, V. M. (1998). Urban bus transit route network design using genetic algorithm. *Journal of Transportation Engineering, 1998*, 124(4).

Punch, W., Goodman, E., Pei, M., Chia-Shun, L., Hovland, P., & Enbody, R. (1993). Further research on feature selection and classification using genetic algorithms. In *Proceedings of the 5th international conference on genetic algorithms* (pp. 557–564). San Francisco: Morgan Kaufmann

Razali, N. M., & Geraghty, J. (2011). Genetic algorithm performance with different selection strategies in solving TSP. In Proceedings of the World Congress on engineering 2011: Vol 2, London, UK.

Retscher, G., & Kealy, A. (2005). Ubiquitous positioning technologies for intelligent navigation systems. *The Journal of Navigation, 59*, 91–103.
Roeva, O. (2008, May 2–4). Improvement of genetic algorithm performance for identification of cultivation process models. In *Proceedings of the 9th WSEAS international conference on evolutionary computing*. Sofia, Bulgaria. ISBN: 978-960-6766-58-9.
Şen, Z., & Öztopal, A. (2001). Genetic algorithms for the classification and prediction of precipitation occurrence. *Hydrological Sciences Journal, 46*(2), 255–267.
Stopher, P., Collins, A. (2005, January). Conducting a GPS prompted recall survey over the internet. In Proceedings of the 84th annual meeting of the Transportation Research Board, Washington, DC.
Stopher, P., FitzGerald, C., & Zhang, J. (2006). *Advances in GPS technology for measuring travel*. Institute of Transport and Logistics Studies, Paper ITLS-WP-06-15, University of Sydney.
Stopher, P., FitzGerald, C., & Zhang, J. (2008). Search for a global positioning system device to measure person travel. *Transportation Research Part C, 16*(3), 350–369.
Sun, B., Wang, D., & Xia, L. (2009, December 10–12). GPS-enabled automated data collection for studying individuals' activity-travel behavior. In *Proceedings of the 14th international conference of Hong Kong society for transportation studies* (pp. 5–15). Hong Kong.
Timmermans, H., Anastasia, M., & Joran, J. (2009). Collecting activity-travel data using modern ICT: Recent developments and prospects. In *Proceedings of 14th international conference of Hong Kong Society for Transportation Studies, Hong Kong* (pp. 5–15).
Wang, D., & Law, F. Y. T. (2007). Impacts of Information and Communication Technologies (ICT) on time use and travel behavior: A structural equations analysis. *Transportation, 34*, 513–527.
Wolf, J. (2000). *Using GPS data loggers to replace travel diaries in the collection of travel data*. Dissertation, School of Civil and Environmental Engineering, Georgia Institute of Technology, Atlanta, GA.
Wolf, J., Schönfelder, S., Samaga, U., Oliveira, M., & Axhausen, K. W. (2004). 80 weeks of GPS-traces: Approaches to enriching the trip information. *Transportation Research Record, 1870*, 46–54.
Yang, Y. F., Lohmann, P., Heipke, C. (2006). Genetic algorithms for the unsupervised classification of satellite images. *Symposium of ISPRS Commission III, Photogrammetric Computer Vision PCV'6, 36*(3), 179–184.

Chapter 14
Quantitative Analysis of Climate Change and Human Crises in History

Harry F. Lee and David D. Zhang

14.1 Introduction

A climate-crisis relationship has long been conceived: Ellsworth Huntington's (1907) *The Pulse of Asia: A Journey in Central Asia Illustrating the Geographic Basis of History* is believed to be the first scholarly work mentioning the climate-crisis relationship.[1] Unfortunately, research in this area stagnated after Huntington. The stagnation was, in part, attributable to the absence of accurate and high-resolution paleo-climate reconstructions. Moreover, most of the related studies about the topic were qualitative and based on selective historical cases (e.g., Utterström 1955; Le Roy Ladurie 1972; Bryson and Murray 1977; Gribbin 1978; Hinsch 1988). Although they provide evidence that climatic factors can contribute to human crises in some parts of the world, the evidence itself is anecdotal and it remains unclear to what extent these case-specific findings can be generalized. There was a lack of compelling evidence to confirm the climate-crisis relationship. Research about the climate-crisis relationship also fell from favor because the topic smacked of environmental determinism. In extreme cases during the early twentieth century, some historians actually ruled out investigating natural phenomena altogether, regarding them as purely accidental facts unrelated to human history (Pfister 2007).

[1]Huntington (1907) suggests that in historical China, the Mongol conquests in the thirteenth century and Manchu conquests in the seventeenth century were primarily triggered by climate change.

H.F. Lee (✉) • D.D. Zhang
Department of Geography and International Centre of China Development Studies, University of Hong Kong, Pokfulam Road, Hong Kong, Hong Kong
e-mail: harrylee@hku.hk; zhangd@hkucc.hku.hk

M.-P. Kwan et al. (eds.), *Space-Time Integration in Geography and GIScience*, DOI 10.1007/978-94-017-9205-9_14

Since the mid-1990s, there has been rapid advancement in high-resolution paleo-climate reconstructions and a growing concern about the possible catastrophic consequences of global warming. Against this background, research about the climate-crisis relationship in human history has been revitalized. Nevertheless, the proof of the relationship in recent studies remains vague, owing to the reliance on individual cases (e.g., deMenocal 2001; Cullen et al. 2000; Polyak and Asmerom 2001; Yancheva et al. 2007). As there is not much quantitative evidence about the relationship, no one can say with certainty how far and to what extent human crises are really caused by climate change. At the same time, for those studies which suggest the climate-crisis relationship, the causal mechanism that describes how climate deterioration is eventually translated into human crises has rarely been investigated. Therefore, the relationship remains in a black box. This probably explains why climate-crisis relationship is one of the most controversial topics in academia.

Driving to the heart of the controversy about the climate-crisis relationship over the past few years, we took a pioneering approach to examine quantitatively the connection between deteriorating climate[2] and human crises. In addition, the causal mechanism responsible for the connection was also scientifically explored. It is worth mentioning that we based our work on high-resolution paleo-climate reconstructions and fine-grained historical socio-economic datasets to verify the climate-crisis relationship, and our findings were derived from all of the known cases, instead of individual examples. By doing so, we provided important scientific evidence about the climate-crisis relationship in recent human history. This book chapter is a brief summary of our recent research findings that have been published in the journals *Chinese Science Bulletin*, *Climatic Change*, *Global Ecology and Biogeography*, *Human Ecology*, and the *Proceedings of the National Academy of Sciences of the United States of America*, respectively.

14.2 Quantitative Analysis of Climate-Crisis Relationship

Since the mid-2000s, we have conducted a series of studies verifying the connection between climate change and human crises (in recent human history) via quantitative and statistical methods (Zhang et al. 2005, 2006, 2007a, b, 2011a, b). Our study areas include China, Europe, and other countries/regions in the Northern Hemisphere during the pre-industrial era. To facilitate our research, we posited a set

[2]Deteriorating climate refers to climatic cooling or warming. Cooling shortens the crop growing season and reduces farmland area (Galloway 1986); warming shortens the duration between sowing and harvesting and increases evapotranspiration (Lobell and Field 2007). Both are detrimental to agricultural productivity, especially to a primarily agricultural economy characterized by a low level of technology and high dependence on climate. Together with the side effects of climate change, such as shifts in rainfall pattern, the carrying capacity of the agro-economy shrinks significantly in a deteriorating climate.

of hypotheses, as follows: in the pre-industrial period, the main source of livelihood was agriculture; traditional agriculture was very much dictated by the whims of climate and weather conditions; any deterioration of climate would trim agricultural production; yield reduction would trigger famine, tax revolt, and a weakening of state power; the deficit in livelihood resources was aggravated by the population expansion accumulated in the previous favorable climate. Thus, human crises were likely to erupt during the period of deteriorating climate. Given the above set of hypotheses, our basic premise in previous studies was that, if empirically observed human crises were correlated with climate change in a statistically significant manner (i.e., human crises became more frequent in a deteriorating climate), then we could confirm the connection between climate change and human crises. Three major types of human crises were covered in our previous studies: population checks, population collapses, and socio-political chaos. Our research findings about the climate-crisis connection could be summarized as follows.

14.2.1 Population Checks

In Thomas Malthus' (1798) classic treatise, *An Essay on Population*, three types of population checks are mentioned: wars, famines, and epidemics. Of these three types of population checks, we put more emphasis on wars because of their catastrophic and long-lasting social consequences. We found that in China in AD 1000–1911, like climate variations[3] (Fig. 14.1a), the war frequency in China[4]

[3]Briffa and Osborn (2002) chose five representative climate series of the last millennium in the Northern Hemisphere (Jones et al. 1998; Mann et al. 1999; Briffa 2000; Crowley and Lowery 2000 and Esper et al. 2002) to discuss differences between the records of various independent paleo-temperature studies. The five recalibrated temperature anomaly series were arithmetically averaged to give the Northern Hemispheric temperature anomaly series (in °C, from the AD 1961–1990 mean) (Fig. 14.1a). Although there are China-wide temperature reconstructions with a millennium year length of record (Wang et al. 2001; Yang et al. 2002), they did not reach the annual scale that is required by our research. Given that the major long-term cooling events revealed by the temperature reconstructions of China and the Northern Hemisphere were basically synchronous in the past millennium, Briffa and Osborn's (2002) 'averaged' Northern Hemisphere temperature anomaly series was chosen as the standard paleo-temperature record to quantitatively delineate the cold and warm phases. The boundaries of the warm and cold phases were delineated at the mean temperature point between minimum and maximum values of two contiguous phases on Briffa and Osborn's (2002) averaged series. A cold or warm phase would be determined if the average temperature change had an amplitude >0.14 °C, in order to obtain an equal aggregate duration of cold and warm periods. Based on the averaged series, six major cycles of warm and cold phases were identified between AD 1000 and 1911. The cold phases spanned AD 1110–1152, 1194–1302, 1334–1359, 1448–1487, 1583–1717, and 1806–1911, while the warm phases spanned AD 1000–1109, 1153–1193, 1303–1333, 1360–1447, 1488–1582, and 1718–1805 (Fig. 14.1a).

[4]Our war data was elicited from a multi-volume compendium which scrupulously records the wars that took place in China from 800 BC to AD 1911 (Editorial Committee of Chinese Military History 1985). There were 1,679 wars during the period AD 1000–1911, including 682 rebellions (Fig. 14.1b).

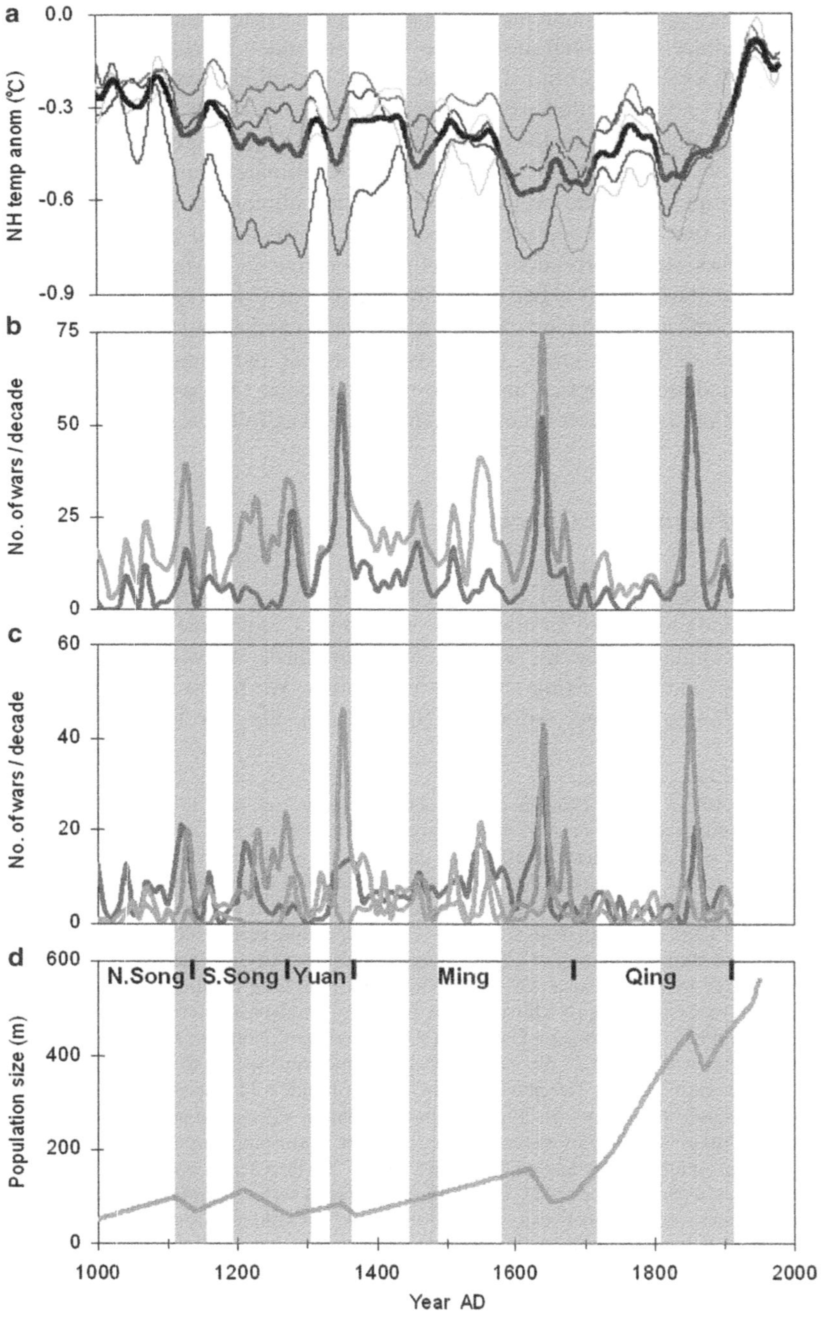
a
0.0
-0.3
-0.6
-0.9
NH temp anom (°C)
b
75
50
25
0
No. of wars / decade
c
60
40
20
0
No. of wars / decade
d
600
400
200
0
Population size (m)
N.Song
S.Song
Yuan
Ming
Qing
1000
1200
1400
1600
1800
2000
Year AD

Table 14.1 Pearson's correlation coefficients between temperature anomaly and wars in China (in different analytical units), AD 1000–1911

Analytical units	Total wars	Rebellions	Others	North	Central	South
Phase	−0.693*	−0.686*	−0.014	−0.511	−0.666*	−0.525
Decadal	−0.179	−0.229*	0.030	−0.090	−0.214*	−0.065
Annual	−0.115**	−0.156**	0.026	−0.060	−0.133**	−0.034

*Significant at 0.05 level (2-tailed) ($P < 0.05$); **Significant at 0.01 level (2-tailed) ($P < 0.01$)

demonstrated a cyclic pattern, with a turbulent period followed by a relatively tranquil one (Fig. 14.1b) (Zhang et al. 2006). Eight of the ten war peaks (>25 wars/decade) coincided with cold phases. Two of the three very high war peaks (>50 wars/decade) occurred in the coldest phases (temperature is below −0.5 °C). All cold phases have one or two war peaks. Rebellion was the dominant war category (Fig. 14.1b, red line) and its variation was highly correlated with climatic changes. Regarding the distribution of wars in different geographic regions[5] (Fig. 14.1c), in warm and humid South China, war frequency variations were less sensitive to temperature changes. In contrast, war outbreaks in North China were closely associated with cold phases. In Central China, six of the seven war peaks occurred in cold phases, and they all closely followed cold phases. We calculated Pearson's correlation coefficients between wars and temperature at three different time scales: phase, decade, and annual (Table 14.1). At the phase scale, total wars, rebellions, and Central China wars were significantly correlated with temperature. At the decadal scale, rebellions and Central China wars were significantly correlated with temperature. At the annual scale, total wars, rebellions, and Central China wars were significantly correlated with temperature.

Fig. 14.1 Climate change, wars, population growth, and dynastic changes in China in AD 1000–1911. (**a**) Normalized temperature change records for the last millennium for land areas in the Northern Hemisphere north of 20° N: Jones et al. (1998, *dark blue line*); Mann et al. (1999, *pink line*); Briffa (2000, *yellow line*); Crowley and Lowery (2000, *turquoise line*); Esper et al. (2002, *violet line*); and the average of these five normalized series (*bold black line*). Cold phases are shaded as *gray strips*. (**b**) Frequency of total wars (*sky blue line*) and rebellions (*red line*). (**c**) Frequency of wars in North China (*red line*), Central China (*sky blue line*), and South China (*bright green line*). (**d**) Dynastic changes and population size (in million) in China (Color figure online)

[5]According to the physical regionalization of China (Zhao 1986), China is divided into three macro regions, namely: (1) North China – with continental humid, semi-humid, semi-arid, and arid temperate climate influenced by both the monsoons and the westerlies. Its average annual temperature ranges from very low to 14 °C. Major agriculture products are spring wheat (northern part) and winter wheat (southern part). Economic activities are mainly pastoral because of the relatively low average annual precipitation of 50–750 mm, with >100 frost days per year; (2) Central China – with a climate dominated by the monsoons, with annual temperature ranging from 14 to 18 °C, and 10–80 frost days per annum. The region has served as China's major rice producing area; and (3) South China – with south sub-tropical and tropical climates and average annual temperature ranging from 19 to 22 °C. The long growing season permits double- or triple-cropping in a year. Frost days are <10 per annum.

Our results reveal that high war frequencies echoed more stringent resource conditions. For instance, the correlation coefficients for total wars, rebellions, and Central wars became stronger from the annual scale to phase scale (Table 14.1), because prolonged cooling exhausted the stored livelihood resources and eventually brought wars. In humid tropical and sub-tropical South China, the influence of cooling in cold phases on agricultural production might have a subdued effect on resource reduction because of a rich endowment in heat and moisture in the coastal region. Even if cooling were severe enough to affect cropping, the more flexible farming system in the South, with a wide range of domesticated species, could adopt alternative crops. Therefore, human reaction to cooling in South China was not so sensitive and severe. In contrast, the climate in Central China is controlled by the monsoons: cold-dry air masses move in from Siberia in the winter and warm-humid marine air masses come in from the southeast and southwest in the summer. It should be noted that many studies of China's paleo-climate indicated that the cold periods were dominated by winter monsoons from Siberia and hence were drier than warm periods (An 2000; Li et al. 2000). Thus, cold periods could have imposed a double jeopardy in terms of coldness and dryness, bringing a highly stressed condition for agriculture in Central China. North China, where the main sustenance was grazing, was sensitive to cooling which could reduce short-term production and trigger long-term loss of ecosystem productivity due to land degradation and desertification. In addition, unlike crop produce, pastoral animal resources could not be stored for a long time. Thus in the north, the onset of a cold period would soon be followed by wars. It should be noted that although most of the war peaks occurred in cold phases, the correlation between temperature and wars was not significant. At the time when the whole of China or part of Central China was ruled by northern nomadic tribes (South Song, Yuan, and Qing Dynasties), a period which covered more than 500 years and over half of the study period, people in the north could freely shift to the south or acquire their livelihood from the south; thus, the war frequency in North China was reduced in the cold phases (respectively in the thirteenth, fourteenth, seventeenth, and nineteenth centuries). At the annual scale correlation analysis, if we disregarded the occupation years, the correlation between temperature and wars in North China (i.e., $r = -0.208$, $P < 0.01$, not shown in Table 14.1) was more significant than those in other areas (cf. Table 14.1).

We extended our exploration of the climate-war relationship from China to different geographic regions in the Northern Hemisphere (Zhang et al. 2007a). We found that in the same manner as the Northern Hemisphere temperature variations[6] (Fig. 14.2a), the incidence of warfare in the Northern Hemisphere, Europe, Asia, and the arid areas of the Northern Hemisphere (i.e., the arid zone from Eurasia to North Africa) in AD 1400–1900[7] occurred in a cyclic pattern, with a turbulent period

---

[6]Our temperature data come from Mann and Jones' (2003) Northern Hemisphere annual temperature trajectory, supplemented by the Southern Hemisphere and global temperature trajectories.

[7]Our war data was elicited from Brecke's (1999) *Conflict Catalog*, which is the most inclusive global war dataset so far, documenting a total of 2,912 wars fought in AD 1400–1900. The

followed by a relatively tranquil one (Fig. 14.2b). Two periods plagued by unrest and warfare and three relatively peaceful periods inversely followed the temperature undulation. This pattern appeared not only at a continental scale, but also in three war databases with different violence thresholds at the global scale[8] (Fig. 14.2c). In short, synchronous periods of relative peace and turbulence during those 500 years were a global phenomenon seemingly linked to temperature change.

At the century time scale, war frequencies and the ratio of wars per year in different centuries were calculated at various geographic scales (global, continental, and country).[9] Those calculations demonstrated that the number and ratio of wars in the mild eighteenth century were the lowest compared with other centuries. The results indicate that the worldwide war ratio during cold centuries was 1.93 times greater than that of the mild eighteenth century, and 1.77, 1.91, 1.50, and 2.24 times greater for the Northern Hemisphere, Asia, the arid areas of the Northern Hemisphere, and Europe, respectively. At the country scale, only 27 of 170 countries and areas around the world had higher war ratios in the eighteenth century than in the fifteenth to seventeenth and nineteenth centuries. Most of the 'exceptional cases' are located near the Equator, where there was no obvious Little Ice Age and the resulting cooling that would have a significant effect on agriculture.

To refine the analysis, a series of Pearson's correlation analyses were carried out at the much lower scale of an annual level time-frame.[10] The results show that the numbers of wars per year, in all war classifications, were significantly negatively correlated with the annual changes of the temperature anomalies at the global level (Table 14.2). The number of wars in the Northern Hemisphere and the Southern Hemisphere also correlated significantly with their temperature variations. Further analysis of the frequency of wars in Europe, Asia, and the arid areas of the Northern Hemisphere also shows significant correlations with the Northern Hemisphere temperature anomaly for their geographical patterns (Table 14.2). The values of the coefficients at the annual scale also reflect environmental vulnerability and population density. As arid regions are the most vulnerable to climate change, the highest coefficient should and does appear in the war frequencies for the arid areas of the Northern Hemisphere. Values are lower for Asia, probably because much of the continent for which there is conflict data is subject to a wet tropical

---

wars include all recorded violent conflicts that meet Richardson's magnitude ≥ 1.5 criterion (≥ 32 deaths). The geographical locations of wars in the dataset are divided by their natural locations.

[8]Apart from Brecke's *Conflict Catalog*, we also included another two global war datasets for comparison: Wright's (1942) dataset, which documents all hostilities involving members of the family of nations, whether international, civil, colonial, or imperial, which were recognized as states of war in the legal sense or which involved over 50,000 troops; and Luard's (1986) dataset, which documents those encounters that involved at least one sovereign state and involved substantial, organized fighting over a significant period (also known as principle wars).

[9]The calculation of the war frequencies and the ratio of wars was based on Brecke's (1999) *Conflict Catalog*.

[10]Data were smoothed by the 40-year Butterworth low-pass filter to remove fluctuations on time-scales <40 year prior to statistical analysis.

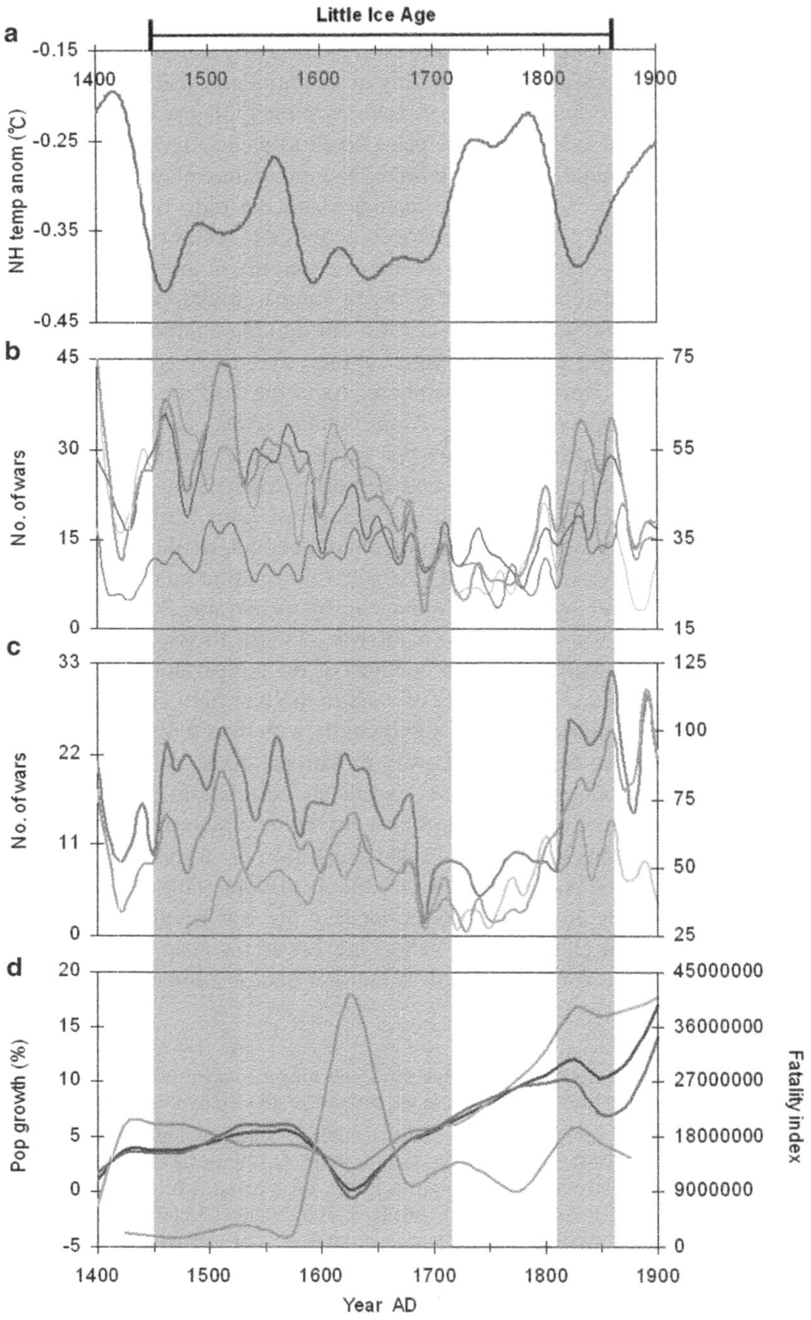
Little Ice Age
a
-0.15
-0.25
-0.35
-0.45
NH temp anom (℃)
1400
1500
1600
1700
1800
1900
b
45
30
15
0
No. of wars
75
55
35
15
c
33
22
11
0
No. of wars
125
100
75
50
25
d
20
15
10
5
0
-5
Pop growth (%)
45000000
36000000
27000000
18000000
9000000
0
Fatality index
Year AD

Table 14.2 Pearson's correlation coefficients between temperature anomaly and wars in different geographic regions in the world in AD 1400–1900. All data have been smoothed by 40-year Butterworth low-pass filter prior to statistical analysis

Variables		
War	Temp anomaly	Correlation coefficient (r)
Worldwide[a]	Global	−0.279***[e]
Worldwide[b]	Global	−0.285***
Worldwide[c]	Global	−0.329***
N. Hemisphere[c]	N. Hemisphere	−0.390***
S. Hemisphere[c]	S. Hemisphere	−0.408***
Asia[c]	N. Hemisphere	−0.235***
Arid areas in the N. Hemisphere[c]	N. Hemisphere	−0.473***
Europe[c]	N. Hemisphere	−0.458***
Western Europe[c]	N. Hemisphere	−0.329***
Eastern Europe[c]	N. Hemisphere	−0.548***
North America[c]	N. Hemisphere	−0.112*
South America[c]	S. Hemisphere	−0.188***
West & Central Africa[c]	S. Hemisphere	−0.404***
East & South Africa[c]	S. Hemisphere	−0.444***
China[d]	N. Hemisphere	−0.264***

*Significant at 0.05 level (2-tailed) ($P < 0.05$); ***Significant at 0.001 level (2-tailed) ($P < 0.001$)
[a]Data source: Wright (1942)
[b]Data source: Luard (1986)
[c]Data source: Brecke (1999)
[d]Data source: Editorial Committee of Chinese Military History (1985)
[e]n = 419 (i.e., AD 1482–1900)

or subtropical environment, where a fall in temperature would have a smaller effect than elsewhere on agricultural production and would not reduce food resources to the same extent because there is more alternative food. The American continents had a very low population density and a large amount of fertile land during this period, and their values are therefore either insignificant or of very low significance. Europe and Africa have either large cold areas or desert areas relatively vulnerable to cooling, and we find that warfare there is highly correlated with the temperature anomalies. At the global scale, the temperature anomaly is simply an arithmetic

Fig. 14.2 Climate change, wars, and population growth in different geographic regions in the world in AD 1400–1900. (**a**) Temperature anomaly (°C) in the Northern Hemisphere that is smoothed by 40-year Butterworth low-pass filter. (**b**) Number of wars in the Northern Hemisphere (*bright green*), Asia (*pink*), Europe (*turquoise*), and the arid areas in the Northern Hemisphere (*orange*). (**c**) Number of wars worldwide as recorded by Wright (1942) (*turquoise*), Luard (1986) (*orange*), and Brecke (1999). (**d**) Population growth rate (in 20-year units) in Europe (*turquoise*), Asia (*pink*), and the Northern Hemisphere (*blue*), together with the Northern Hemisphere fatality index (*bright green*, in 50-year units). Cold phases are shaded as *gray stripes*. All war time series are in decadal units. The *bright green curves* correspond to the right y-axis (Color figure online)

mean of the Northern Hemisphere and Southern Hemisphere temperature anomalies (Mann and Jones 2003). The values of the coefficient for wars at the global scale are thus lower than those of the Northern Hemisphere and Southern Hemisphere alone.

We also examined the major types of population checks (i.e., famines, wars, and epidemics) in the Northern Hemisphere[11] in relation to climate change in AD 800–1900[12] (Fig. 14.3) (Zhang et al. 2011b). Regressions were run to estimate the relative sensitivity of various population checks to temperature change in the Northern Hemisphere.[13] Results show that temperature was negative and highly significant in all regressions (Table 14.3). A 10 % decrease in temperature produces on average 2.2%, 1.2 %, and 1.1 % increases in famines, wars, and epidemics, respectively, in the Northern Hemisphere. Although both warming and cooling can shrink land carrying capacity, the regression results in the previous section and here indicate that cooling was generally more detrimental to human societies at the hemispheric and global scales in history.

14.2.2 Population Collapses

As for the connection between climate change and population collapses[14] in China, five demographic collapses occurred during the last millennium, each with population losses ranging from over 30 million to 80 million[15] (Fig. 14.1d) (Zhang et al. 2006). These collapses happened in cold phases. On the contrary, all warm phases experienced fast population growth.[16] The lag times (within a range of 5–50 years) for population collapses were gradually extended from the first to last cold phases. This could be explained by gradual introduction of new crops, improvement in food production and storage and irrigation technologies.

---

[11]Our famine time series was derived from Walford (1970) and Golkin (1987); our war time series was derived from Kohn (1999); and our epidemics series was derived from Cliff et al. (1998), Kohn (2001), and Xiao and Liu (2005).

[12]The centennial climate variability in the Northern Hemisphere was elicited by arithmetically averaging the 12 most recent and authoritative paleo-temperature reconstructions chosen by the Intergovernmental Panel on Climate Change (in °C, from the AD 1961–1990 mean) (Jansen et al. 2007), then smoothed by the 100-year Butterworth low-pass filter to remove fluctuations on time-scales <100 years (Fig. 14.3a).

[13]The independent variables are time and temperature anomalies in the Northern Hemisphere. Time (t) presumably represents technology and/or capital accumulation. An attempt is made to eliminate the trend from the population, using parabolic (t and t^2), squared (t^2) and cubic (t^3) terms (see Galloway 1986). The regressions were corrected for autoregressive disturbances using the Prais-Winsten estimation method. Results show that the various detrending procedures did not affect the significance of temperature.

[14]The term 'population collapse' refers to negative population growth.

[15]The calculation of population losses was based on Jiang (1993).

[16]See footnote no. 3 for the dates of cold and warm phases in China.

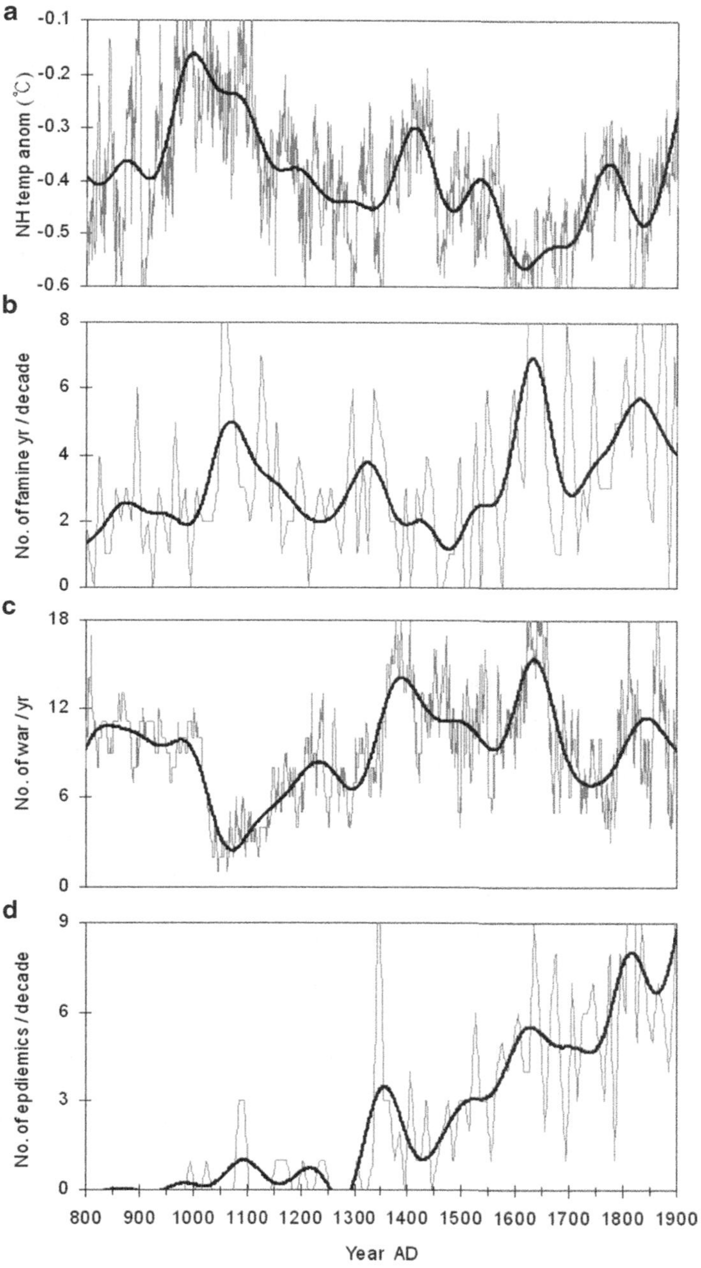

Fig. 14.3 Climate change and populations checks in the Northern Hemisphere in AD 800–1900. (**a**) Northern Hemisphere temperature anomaly (°C) from the AD 1961–1990 mean. (**b**) Years with famine per decade. (**c**) Number of wars per year. (**d**) Number of deadly epidemic events of six diseases (malaria, plague, typhus, measles, smallpox, and dysentery) per decade. Both series have been smoothed by the Butterworth 100-year low-pass filter (described by the *bold curve*)

Table 14.3 Regressions of various population checks in the Northern Hemisphere on time and Northern Hemisphere temperature anomalies in AD 800–1900. Population checks and temperature anomalies data have been smoothed by 100-year Butterworth low-pass filter prior to statistical analysis. The regressions are corrected for autoregressive disturbances using the Prais-Winsten estimation method. Elasticity can be interpreted as the percentage of population check changes in response to a one percent increase in temperature. t = calendar year divided by 10^3. R^2_{adj} = adjusted R^2 calculated for the untransformed variables

Dependent		Independent variable					
variable	Constant	t	t^2	t^3	Temp	Elasticity	R^2_{adj}
Famine	−4.445	6.713***	−1.441***		−3.493***	−0.212	0.309
Famine	−0.383		0.936**		−3.555***	−0.216	0.279
Famine	0.205			0.386**	−3.580***	−0.217	0.298
War	9.333	−2.783	1.164		−3.678***	−0.119	0.226
War	7.651		0.180		−3.653***	−0.118	0.218
War	7.774			0.071	−3.657***	−0.118	0.218
Epidemics	7.643	−18.248***	9.772***		−1.680***	−0.113	0.927
Epidemics	−3.398		3.328***		−1.514***	−0.103	0.886
Epidemics	−2.120			1.559***	−1.670***	−0.113	0.914

*Significant at 0.05 level (2-tailed) ($P < 0.05$); **Significant at 0.01 level (2-tailed) ($P < 0.01$); ***Significant at 0.001 level (2-tailed) ($P < 0.001$)

We also found a very significant correlation between the Northern Hemisphere temperature swings and population growth rates for the world, Northern Hemisphere, Asia, North America, and Europe, as well as for China for the period AD 1400–1900[17] (Table 14.4) (Zhang et al. 2007a). The correlation coefficients for Europe and North America are lower than the others. The reason for this is that Europeans and Americans lived in an ecologically open system that began during the Industrial Revolution, one in which European population pressure could be relieved by cross-continental colonization and migration. When the nineteenth century (the age of Great Migration in Europe) population data were not included in the analysis, the correlations for Europe and North America were almost as high as those for Asia and the whole Northern Hemisphere. Similarly, the large-scale trans-Atlantic migration from Europe to North America in AD 1700–1900, which is supposed to be driven by cooling, can also be substantiated by the significant negative correlation between the Northern Hemisphere temperature anomaly and North America's population growth ($r = 0.341$, $P < 0.001$).

In addition, we explored quantitatively the temporal pattern, spatial pattern, and triggers of population collapses[18] in relation to climate change at the global scale

[17] See footnote no. 6 for the details of Northern Hemisphere temperature change. The calculation of population growth rate was based on McEvedy and Jones (1978). Prior to statistical analysis, all data were smoothed by the 40-year Butterworth low-pass filter to remove fluctuations on time-scales <40 years.

[18] The identification of population collapses was based on McEvedy and Jones (1978).

Table 14.4 Pearson's correlation coefficients between temperature anomaly and population growth rate in different geographic regions in the world in the Little Ice Age. All data have been smoothed by 40-year Butterworth low-pass filter prior to statistical analysis

Region	AD 1400–1900	AD 1500–1900	AD 1400–1800	AD 1500–1800
World	0.387***	0.518***	0.584***	0.754***
N. Hemisphere	0.355***	0.528***	0.523***	0.765***
Asia	0.434***	0.593***	0.550***	0.751***
Europe	0.151***	0.362***	0.228***	0.587***
North America	0.230***	0.326***	0.428***	0.565***
China	0.388***	0.574***	0.471***	0.718***

***Significant at 0.001 level (2-tailed) ($P < 0.001$)

in AD 800–1900[19] (Fig. 14.4) (Zhang et al. 2011b). There were 88 population collapses at the country/regional scale, of which 38 were caused by wars, 33 by epidemics, 6 by famines, 5 by war-epidemics synthesis, 4 by cross-oceanic colonization, 1 by famine-migration synthesis, and 1 by famine-war synthesis. There were 7, 70, and 11 (the average number of population collapses per century was 4, 16, and 2) in the Medieval Warm Period, cold phases (C1–C4), and mild phases, respectively. Most of the collapses happened in C1 (41) and C3 (23) – the two long cold phases. Nearly 70 % of the collapses clustered in AD 1040–1050 (7), AD 1340–1350 (36), and AD 1592–1620 (17) – although the three short climate-deteriorating periods spanned less than 2 % of the study period (Fig. 14.4b, grey bars). In contrast, the mild phases spanned over 50 % of the study period, but the associated number of population collapse incidents was only 11. Furthermore, the average population growth rate and temperature in the Northern Hemisphere moved in the same direction. The long cold phases were associated with population decline; after-cooling mild phases were associated with rapid population growth, while warm phases were associated with diversified population growth rates in different climatic zones[20] (Fig. 14.4b).

[19]See footnote no. 12 for centennial climate variability in the Northern Hemisphere from AD 800 to 1900. Based on the temperature series, we established warming and cooling thresholds according to the averaged temperature anomalies of the starting century of the Medieval Warm Period and Little Ice Age, respectively. A period in which the temperature anomaly was > -0.3 °C was classified as a warm phase; a period in which the temperature anomaly was < -0.42 °C was classified as a cold phase. It should be noted that the average AD 1961–1990 temperature anomaly is 0 °C, which is c. 0.3 °C higher than that of the Medieval Warm Period. In line with these criteria, we identified a warm phase in the Medieval Warm Period (AD 954–1114) and four cold phases in the Little Ice Age [C1 (AD 1236–1359), C2 (AD 1459–1510), C3 (AD 1554–1741), C4 (AD 1804–66)]. These represent periods of climate deterioration. The remaining years were classified as mild phases (Fig. 14.4a).

[20]The delineation of major climatic zones in the Northern Hemisphere is made according to the modified Köppen classification system (see London Times 2007). Four climate zones are delineated as follows: (1) tropical humid – rainy climate with no winter, coolest month above 18 °C; (2) warmer humid – rainy climate with mild winter, coolest month above 0 °C, but below 18 °C,

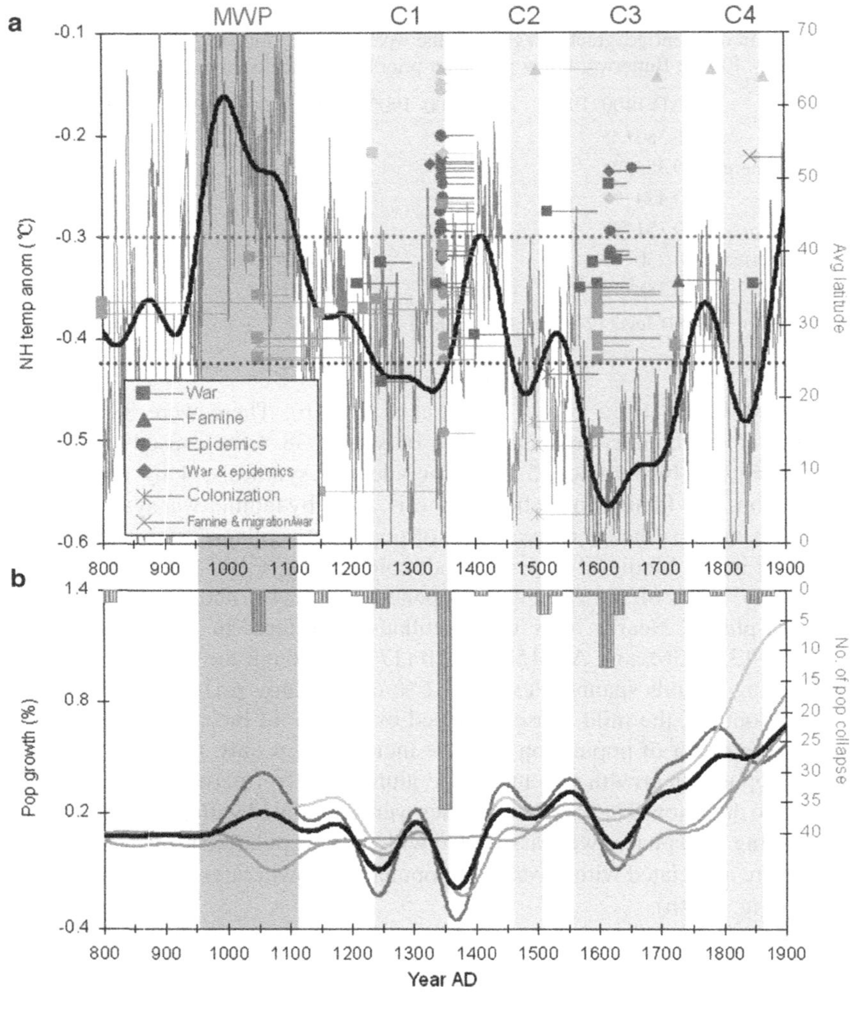

Fig. 14.4 Climate change, population collapse, and population growth in the Northern Hemisphere in AD 800–1900. (**a**) Northern Hemisphere temperature anomaly (°C) from the AD 1961–1990 mean and population collapse. The temperature series (*grey line*) has been smoothed by the Butterworth 100-year low-pass filter to characterize its centennial variability (*bold black line*). *Red* and *blue dotted horizontal lines* denote the warming threshold (> −0.3 °C) and cooling threshold (< −0.42 °C), respectively. Regarding the details of each population collapse; the average latitude of the relevant country/region corresponds to the right Y-axis and the associated climatic zone is identified by colour (*red*, warmer humid zone; *turquoise*, cooler humid zone; *bright green*, tropical humid zone; *orange*, dry zone); cause is represented by a symbol (see figure legend); and duration is revealed by the length of the line. (**b**) Population collapse and population growth rate. The *grey bars* represent the number of population collapses in 20-year units, which corresponds to the right Y-axis (inverted). Population growth rate in different climatic zones is identified by colour (*red*, warmer humid zone; *turquoise*, cooler humid zone; *bright green*, tropical humid zone; orange, dry zone), and *bold black* (Northern Hemisphere). The *red* shaded area represents the warm phase in the Medieval Warm Period (MWP), while the *blue* shaded area represents the cold phases in the Little Ice Age (C1–C4) (Color figure online)

Although population size in the dry zone comprised only 15 % of the Northern Hemisphere total, 37.5 % (33) of the population collapses happened there. The population growth rate in the dry zone fluctuated along with temperature change. Every warming/cooling resulted in population decline before 1800 (Fig. 14.4b, orange line). Twenty-nine population collapses happened in the dry zone in the Medieval Warm Period, C1 and C3. Population grew slowly in mild phases. From AD 1 to 1800, total population increased by only 76.9 %, while population in warmer humid and cooler humid zones increased by 493.1 % and 890.6 %, respectively.

The dry zone was the only zone that suffered from population collapses in the Medieval Warm Period. After the population drop of the Medieval Warm Period, the population sizes of some countries there (Egypt, Libya, Tunis, and Iraq) did not reach their previous high until AD 1800 (McEvedy and Jones 1978). This indicates that the agro-ecosystem in arid and semi-arid regions is highly sensitive to changes in temperature and precipitation (Zhang et al. 2007a). Furthermore, most of the countries and regions in the dry zone were long-civilized and underwent long-term intensive cultivation. The resultant land degradation might have magnified and prolonged the ecological impact of climate change (i.e. shrinkage of land carrying capacity). This explains why the average population collapse duration in the dry zone (102.8 years) was much longer than that in other zones (50.3 years) (Fig. 14.4a). Although the Muslim agricultural revolution spread to all arid areas in northern Africa and western Asia beginning in the eighth century, it could not help dissipate the demographic impact of climate change. All population collapses in the Medieval Warm Period located in the dry zone's major agricultural regions (Algeria, Arabia, Egypt, Libya, Morocco, Tunisia, and Turkey) were primarily triggered by nomadic tribal invasions (Issar 1995). As the zone became hotter and drier in the Medieval Warm Period (von Rad et al. 1999; Enzel et al. 2003), nomadic tribes in pastoral areas suffered most, forcing them to invade adjacent agricultural regions for subsistence. Subject to further warming, farmlands in those regions were degraded to pastoral land (McEvedy and Jones 1978; Issar 1995).

Warming and cooling were detrimental to the dry zone, but in warmer humid and cooler humid zones, in middle and high latitudes, only cooling was detrimental. Forty-five out of the total of 51 population collapses in those zones occurred in cold phases, while warming boosted population growth rates by increasing the local land carrying capacity (Galloway 1986; Lee et al. 2008; Lee and Zhang 2010; Lee et al. 2009). Population size in warmer humid and cooler humid zones made up nearly 70 % of the Northern Hemisphere total, which largely determined the overall trend of Northern Hemisphere population growth (Fig. 14.4b). A disparity of agro-ecological sensitivity between the two zones may be revealed by their respective population growth rates. The cooler humid zone had a gentle fluctuation

warmest month above 10 °C; (3) cooler humid – rainy climate with severe winter, coldest month below 0 °C, warmest month above 10 °C; and 4) dry – dry climate; limits are defined by formulae based on rainfall effectiveness.

of population growth (Fig. 14.4b, turquoise line), while the rate in the warmer humid zone oscillated more (Fig. 14.4b, red line). As agricultural production in the southern part of the warmer humid zone (i.e. southern Europe and China) involves multiple cropping that is only sustainable in a warm climate, the land carrying capacity in the warmer humid zone was more vulnerable to cooling relative to other zones, resulting in greater fluctuations in population growth.

The demographic impact of deteriorating climate was not great in the tropical humid zone. The zone had steady population growth and only four population collapses, owing to the rich biodiversity and large land carrying capacity there. There were two slight drops in population growth (Fig. 14.4b, green line): one in the Medieval Warm Period because of the hot-drought climate in India (Walford 1970) and one from AD 1550 to 1750, caused by the colonization-induced disasters in the New World (McEvedy and Jones 1978).

Given the significant association between temperature change and population collapses, regressions were run to estimate the relative sensitivity of population growth in the Northern Hemisphere and various climatic zones to climate change[21] (Table 14.5) (Zhang et al. 2011b). For the Northern Hemisphere, temperature was positive and highly significant in the regressions in which a 10 % increase in temperature produced on average a 3.1 % increase in population growth rate. At the regional level, the warmer humid zone was shown to be the region most sensitive to temperature change, in which a 10 % increase in temperature produced on average a 2.9 % increase in population growth rate. This is consistent with the above findings regarding temperature change and population collapses in the warmer humid zone. In tropical humid, cooler humid, and dry zones, even though temperature positively correlated with population growth, their association (elasticity) was relatively weak compared with that of the warmer humid zone. This might be attributable either to their relatively low population density (McEvedy and Jones 1978) or to favorable geographic context, which provides some spare land carrying capacity to buffer climate-induced subsistence shortage and its demographic impact (Zhang et al. 2007a). Dry zone is the only region in which population collapses occurred in both warm and cold phases. Such a non-stationary relationship may dampen the regression results.

14.2.3 Socio-political Chaos

In China over the last millennium, the six cold phases could be further divided into long ones (each lasting 105–133 years) and short ones (each lasting 25–43 years).[22]

[21] See footnote no. 13 for the specifications of the regression models.

[22] See footnote no. 3 for the dates of the cold phases in China.

Table 14.5 Regressions of the population growth rates in various climate zones on time and Northern Hemisphere temperature anomalies in AD 800–1900. Population growth and temperature anomalies data have been smoothed by 100-year Butterworth low-pass filter prior to statistical analysis. The regressions are corrected for autoregressive disturbances using the Prais-Winsten estimation method. Elasticity can be interpreted as the percentage of population growth in response to a one percent increase in temperature. t = calendar year divided by 10^3. R^2_{adj} = adjusted R^2 calculated for the untransformed variables

		Independent variable					
Dependent variable	Constant	t	t^2	t^3	Temp	Elasticity	R^2_{adj}
N. Hemisphere	1.175	−1.799***	0.845***		0.499***	0.301	0.781
N. Hemisphere	0.087		0.209***		0.516***	0.312	0.611
N. Hemisphere	0.153			0.101***	0.505***	0.305	0.677
Warmer humid	0.606	−0.516	0.325		0.833***	0.288	0.561
Warmer humid	0.293		0.141**		0.838***	0.290	0.468
Warmer humid	0.362			0.063**	0.832***	0.288	0.507
Cooler humid	2.159	−4.068***	1.890***		0.110*	0.063	0.783
Cooler humid	−0.290		0.456***		0.147**	0.084	0.609
Cooler humid	−0.173			0.227***	0.121*	0.069	0.682
Tropical humid	1.422	−2.446***	1.084***		0.154***	0.200	0.784
Tropical humid	−0.051		0.222***		0.176***	0.229	0.579
Tropical humid	−0.005			0.112***	0.163***	0.210	0.647
Dry	2.242	−4.077***	1.766***		0.130**	0.083	0.606
Dry	−0.214		0.328***		0.168***	0.107	0.407
Dry	−0.160			0.169***	0.147***	0.093	0.470

*Significant at 0.05 level (2-tailed) ($P < 0.05$); **Significant at 0.01 level (2-tailed) ($P < 0.01$); ***Significant at 0.001 level (2-tailed) ($P < 0.001$)

We found that all dynastic changes,[23] which result from nation-wide socio-political chaos, occurred in cold phases (Table 14.6) (Zhang et al. 2006). The only exception was the collapse of Yuan Dynasty and the establishment of Ming Dynasty in AD 1368, which happened only 9 years after the end of the AD 1334–1359 cold phase (Fig. 14.1d and Table 14.6). Of the six cold phases, five experienced dynasty collapse. It can be noted that the three long cold phases in the last millennium (i.e., AD 1194–1302, 1583–1717, and 1806–1911) brought about the collapse of the three longest dynasties: Song (c. AD 960–1279), Ming (c. AD 1368–1644), and Qing (c. AD 1644–1911).

[23]The times of dynastic changes were based on official records published by government bodies and historians. The dynasties included those that ruled most parts of China, and those established by remote tribes that once occupied an area equivalent to over ten provinces of the current Chinese territory.

Table 14.6 Climate change and dynastic changes in China in AD 1000–1911

Year	Average temperature anomaly (°C)	Climatic phases	No. of wars	War ratios	Dates of dynastic changes
AD1000–1109	−0.252	Warm	135	1.23	
AD1110–1152	−0.368	Cold	93	2.16	Establishment of Jin (AD1115), collapses of North Song (AD1127) and Liao (AD1125)
AD1153–1193	−0.315	Warm	41	1.00	
AD1194–1302	−0.419	Cold	252	2.31	Establishment of Great Mongol (AD1206) and Yuan (AD1271), collapses of Jin (AD1234), South Song (AD1279)
AD1303–1333	−0.362	Warm	33	1.07	
AD1334–1359	−0.454	Cold	90	3.46	
AD1360–1447	−0.345	Warm	189	2.15	Establishment of Ming and collapse of Yuan (AD1368)
AD1448–1487	−0.461	Cold	89	2.23	
AD1488–1582	−0.392	Warm	208	2.19	
AD1583–1717	−0.534	Cold	266	1.97	Establishment of Qing (AD1636) and collapse of Ming (AD1645)
AD1718–1805	−0.413	Warm	72	0.82	
AD1806–1911	−0.456	Cold	204	1.93	Establishment of the Republic of China and collapse of Qing (AD1911)

We also found that in Europe, when temperature[24] was below -0.1σ (above -0.1σ), Europe moved into "dark ages" ("golden ages") over the past

[24]The temperature anomaly series in Europe was derived from two authoritative temperature reconstructions at the annual scale. This first one is Luterbacher et al.'s (2004) temperature reconstruction for European land areas (25°W to 40°E and 35°N to 70°N) spanning AD 1500–2003. The second temperature reconstruction is associated with Osborn and Briffa's (2006) temperature dataset spanning AD 800–1995, which contains 14 regional temperature-related proxy records. However, only those regional temperature series nested within Europe were combined to show

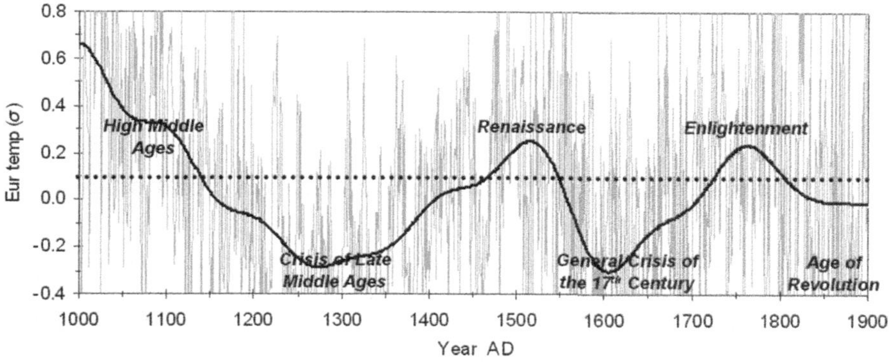

Fig. 14.5 Climate change and the golden and dark ages in Europe in AD 1000–1900. The European temperature anomaly series (σ) has been smoothed by 100-year Butterworth low-pass filter. Horizontal dotted line represents the threshold in defining crisis periods (i.e., temperature $\leq -0.1\sigma$). The periods in which temperature $\geq -0.1\sigma$ stand for golden ages, while the periods in which temperature $\leq -0.1\sigma$ stand for dark ages

millennium[25] (Fig. 14.5) (Zhang et al. 2011a). The periods in which temperature was lower than -0.1σ were AD 1212–1381 (corresponding to the Crisis of Late Middle Ages) and AD 1568–1665 (corresponding to the General Crisis of the Seventeenth Century), whereas the periods in which temperature was higher than -0.1σ were the tenth to twelfth centuries (corresponding to the High Middle Ages), the late-fourteenth to early sixteenth centuries (corresponding to the Renaissance), and the late-seventeenth to eighteenth centuries (corresponding to the Enlightenment). Basically, the alternation of the above warm ($> -0.1\sigma$) and cold ($< -0.1\sigma$) phases coincided with the alternation of "golden" and "dark" ages as defined by historians (Lyon et al. 1969; Roberts 1996). The Age of Revolution in the early nineteenth century (one of the darker ages) is an exceptional case, as the temperature during the time was higher than -0.1σ. Nevertheless, when compared with the other so-called dark ages, the socio-political impact brought by the Age

---

the temperature change in Europe over time. It was done by normalizing each of the above series and then taking their arithmetical average. The above two temperature reconstructions were derived from different proxies and reconstructed by different methods. In order to combine the two reconstructions together, each of them was normalized to homogenize the original variability of all series. It should be noted that this transformation cannot preserve the numerical values of temperature variation, but will provide the relative amplitude of temperature change. Then, the two normalized series were arithmetically averaged and then smoothed by 100-year Butterworth low-pass filter to generate the Europe temperature composite.

[25]According to Lyon et al. (1969) and Roberts (1996), the delineation of golden and darks ages in Europe over the last millennium is as follows: Golden ages – High Middle Ages (eleventh to thirteenth centuries), Renaissance (fifteenth to mid-sixteenth centuries), and Enlightenment (mid-seventeenth to late eighteenth centuries). Dark ages – Crisis of Late Middle Ages (fourteenth century), General Crisis of the Seventeenth Century (mid-sixteenth to mid-seventeenth centuries), and the Age of Revolution (late eighteenth to mid-nineteenth centuries).

of Revolution was not a general one. Even though there was an upsurge in grain price, social disturbance, war, migration, and demographic crises did not happen. This matches with the fact that the cooling during the time was mild.

14.3 Causal Mechanism for the Climate-Crisis Relationship

It has been shown in the previous section that various human crises became more frequent in a period of deteriorating climate. The timing of the peaks of population checks, population collapses, and socio-political chaos coincided, which implies their interrelationship. In addition, the connection between climate change and human crises was evidenced by quantitative and statistical findings. Although our fundamental hypotheses (cf. Quantitative analysis of climate-crisis relationship) are shown to be valid in explaining the climate-crisis relationship, the process that translated climate change into various human crises is far too complicated to be captured by them. To work out the complete set of causal linkages involved in the translation of climate deterioration into human crisis,[26] we proceeded to examine the climate-crisis causal mechanism in a period that contained both periods of harmony and times of crisis (Zhang et al. 2011a). Because the General Crisis of the Seventeenth Century in Europe was marked by widespread economic distress, social unrest, and population decline (Fischer 1996; Goldstone 1991; Parker and Smith 1978; Aston 1966), we systematically collected and tabulated all available historical data about climate, agro-ecology, economy, society, human ecology, and demography in Europe, AD 1500–1800. Sixteen variables[27] were identified (Fig. 14.6) that facilitate our exploration of specific causal mechanisms between climate change and human crisis. To elicit the real association between climate change and the cyclic pattern of different variables, the variables with obvious long-term trends (agricultural production index, grain price, real wages, body height, and population size) were linearly detrended (Chu and Lee 1994; Galloway 1986).

Our study period covers both mild and cold phases of the Little Ice Age in the Northern Hemisphere[28] (Fig. 14.6a), while the fluctuations of all agro-ecological,

[26]Given that we addressed whether climate change is a credible cause for large-scale societal crisis from the macro-historic perspective, macro-historic and aggregate features are privileged over micro-historic and individual ones; general trends are preferred to particular moments or events; and broad distinctions or geographical uniformities take precedence over localized analyses.

[27]Please refer to the SI Appendix, Materials and Methods I–XI of Zhang et al.'s (2011a) study for the specification and data source of each variable.

[28]Based on the Northern Hemisphere (Fig. 14.6a, red line, cf. footnote no. 12) and European (Fig. 14.6a, black line, cf. footnote no. 24) temperature anomaly series, we divided our study period into Mild Phase 1 (AD 1500–1559; average temperature $= 0.43\sigma$), Cold Phase (AD 1560–1660; average temperature $= -0.59\sigma$), and Mild Phase 2 (AD 1661–1800; average temperature $= 0.24\sigma$). The Cold Phase coincided with the General Crisis of the Seventeenth Century. In Mild Phase 2, there was brief cooling in AD 1700 and 1750.

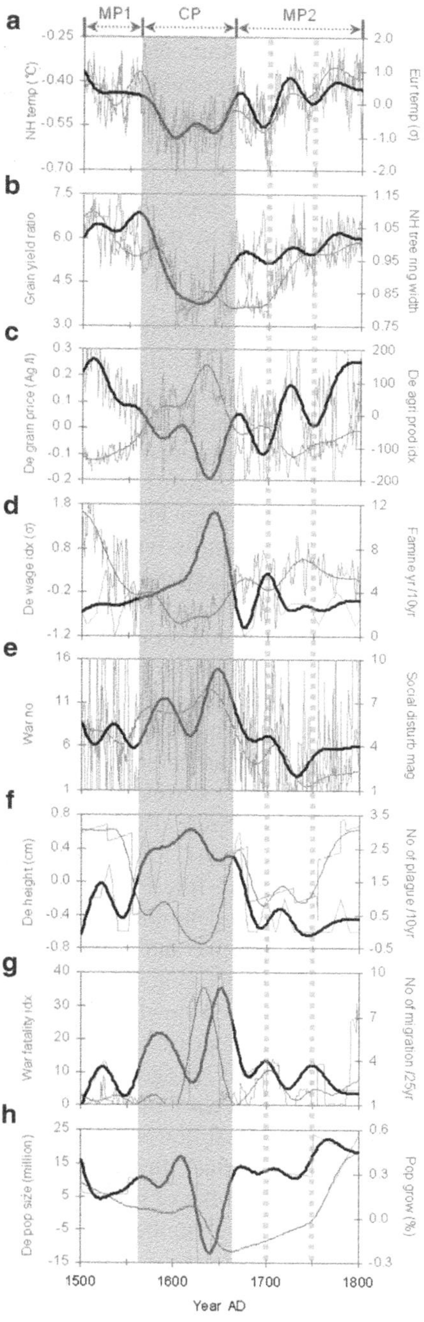

Fig. 14.6 Responses of different variables in human society to climate change in Europe in AD 1500–1800. (**a**) Northern Hemisphere temperature anomaly (°C, *red line*) and Europe temperature anomaly (σ, *black line*). (**b**) Ratio of grain yield to seed (*red line*) and Northern Hemisphere extra-tropical tree-ring widths (*black line*). (**c**) Detrended grain price (Ag/L, *red line*) and detrended agricultural production index (*black line*). (**d**) Detrended wage index (σ, *red line*) and number of famine years per decade (*black line*). (**e**) Number of wars (*red line*) and magnitude of social disturbances (*black line*). (**f**) Detrended human height (in cm, *red line*) and number of plagues per decade (*black line*). (**g**) War fatality index (*red line*) and number of migrations per quarter century (*black line*). (**h**) Detrended population size (in millions, *red line*) and population growth rate (%, *black line*). All data have been smoothed by 40-year Butterworth low-pass filter. The *blue* shading represents the crisis period (Cold Phase), and the *blue dashed line* represents short-term cooling (Color figure online)

socioeconomic, human ecological, and demographic variables corresponded very well with temperature change and were in successive order. The variables of the bio-productivity, agricultural production, and food supply per capita (FSPC) sectors responded to temperature change immediately, whereas the social disturbance, war, migration, nutritional status, epidemics, famine, and population sectors responded to the drop in FSPC with a 5- to 30-year time lag. The adverse effect of the two short-term, minor cooling episodes in Mild Phase 2 (Fig. 14.6, blue dotted lines) also was reflected by the variables' fluctuations in annual and decadal units, such as Northern Hemisphere tree-ring width, grain yield, grain price, and agricultural production index.

Cooling in the Cold Phase dampened agro-ecosystem output by shortening plant growing seasons and shrinking the cultivated land area (Galloway 1986). The ratio of grain yield to seed decreased along with temperature decline (Fig. 14.6b, red line). Tree-ring width (a variable of bio-productivity) also varied in response to temperature change, decreasing rapidly in AD 1560–1650 (Fig. 14.6b, black line). Grain yield links directly to agricultural production, which is represented by the agricultural production index. Although in the long term the agricultural production index moved upwards with population size, it decreased or stagnated in a cold climate and increased rapidly in a mild climate at the multi-decadal time-scale (Fig. 14.6c, black line).

Although agricultural production decreased or stagnated in a cold climate, population size continued to grow. Hence, two variables of FSPC – grain price and real wages of labor – changed considerably, and economic crisis followed. Grain price is determined by both demand and supply and is an important indicator of the boom-and-bust cycle in an agrarian economy. The detrended grain price (Fig. 14.6c, red line) was inversely correlated with every fluctuation of the agricultural production index and temperature. Real wages of labor (Fig. 14.6d, red line) varied inversely with grain price and followed agricultural production and temperature change closely. Given the low FSCP, famine became more frequent (Fig. 14.6d, black line), resulting in deteriorating nutritional status and ultimately in reduced human body height (Koepke and Baten 2005). The average height of Europeans followed temperature closely (Fig. 14.6f, red line) and declined 2 cm in the late sixteenth century. It increased slowly with rising temperatures only after AD 1650.

Inflating grain prices and declining real wages bred unbearable hardship in all walks of life, triggering many social problems and intensifying existing social conflicts. Peaks of social disturbance such as rebellions, revolutions, and political reforms followed every decline of temperature, with a 1–15-year time lag (Fig. 14.6e, black line). Many disturbances eventually developed into armed conflicts. The number of wars increased by 41 % in the Cold Phase (Fig. 14.6e, red line). Although the number of wars decreased in the interval AD 1620–1650, these wars were comparatively more lethal and longer lasting (e.g., the Thirty Years War) (Brecke 1999). Annual war fatalities from AD 1620–1650 were >12 times those in the period AD 1500–1619 (Fig. 14.6g, red line).

More frequent and severe economic chaos, famine, social disturbance, and war pushed people to emigrate. In Europe, migration (Fig. 14.6g, black line) peaked during AD 1580–1650, overlapping exactly with the peak of social disturbance. This correlation indicates that social conditions are imperative in driving migration (Wrigley and Schofield 1981). Migration, coupled with individuals' deteriorating health caused by poor nutrition, facilitated the spread of epidemics (Zhang et al. 2007a). The number of plagues peaked during AD 1550–1670 (Fig. 14.6f, black line), reaching the highest level throughout the study period (i.e., AD 1500–1800). Population growth rate, which is codetermined by famine, epidemics, and war, fluctuated in a complex manner. When peaks in war fatalities and famine occurred during AD 1620–1650, the annual population growth rate (Fig. 14.1h, black line) dropped dramatically from 0.4 to −0.3 %. Population collapse occurred (Fig. 14.1h, red line), and the European population dropped to its lowest point (105 million people) in AD 1650.

In general, variables in European societies (except population) reacted linearly to temperature change at the multi-decadal time scale (Fig. 14.6). Some variables, however, responded exponentially to cooling in AD 1620–1650. We further examined the time-series of those variables and found that, after cooling, population pressure rose after AD 1560 (the agricultural production index declined, and annual population growth was 0.4 %) to the point that a significant reduction of population size was necessary to ease food strain in Europe. The triggers of population collapse were war and famine. After AD 1618, many large-scale wars and famines occurred in Europe. The war fatality index was 20 times higher than the AD 1500–1617 average and persisted at a similar level for the next 32 years (Fig. 14.6g, red line). During AD 1618–1649, 10 million people perished in wars (Brecke 1999).

Humans have served as both producers and consumers in Earth's ecosystem since the Agricultural Revolution. During great wars and famines, death rates exceeded birth rates, causing substantial reduction of the agricultural production workforce. Also, collapse of agricultural production infrastructure caused by wars left behind massive damage to carrying capacity and sustainability (Zhang et al. 2006). Consequently the role of the human population as a producer became less significant. Although the temperature and grain yield in AD 1600–1620 and 1620–1650 were similar, in AD 1621 the feedback effect of population collapse brought about a 13 % reduction in agricultural production, which had stagnated for 50 years. Such a huge decrease caused an exponential increase in grain price (+200 %), famine (+250 %), war fatality (+1,350 %), social disturbance (+100 %), migration (+250 %), and other population checks. On the other hand, real wages, body height, and epidemics remained at the same level, and the number of wars dropped slightly (Fig. 14.6). This complex relationship between agricultural production and population size continued until AD 1650, when temperature and thus agricultural production increased.

At the end of the Cold Phase there was an augmentation of agricultural production that, together with a population slump, led to a rise in FSPC and the recuperation of most of European society after AD 1660. This date marks the end of the General Crisis of the Seventeenth Century and the start of the Enlightenment era. The mild climate in the eighteenth century created human ecological harmony, leading to a speedy economic and population recovery in Europe. Although the short cooling in AD 1700 and 1750 caused minor fluctuations in grain yield, real wages, grain price, famine, war, social disturbance, and migration, its impact was not strong enough to cause general crisis and population collapse (Fig. 14.6).

Cooling triggered a chain of responses in variables pertaining to European physical and human systems. All 16 of the variables we identified are categorized into 11 sectors according to the response time of variables to cooling, together with natural laws and social theories related to different variables. Five of these sectors contain two variables with the same properties (e.g., the variables ‘NH temperature’ and ‘European temperature’ belong to the climate change sector). We then identified a set of causal linkages among the 11 sectors, demonstrating how climate change brings about various human crises (Fig. 14.7).

In the set of causal linkages, climate change and associated bio-productivity fluctuation are revealed as the ultimate cause of economic, social, human ecological, and demographic problems. If the climate change and bio-productivity sectors are disregarded, various linkages within the human system seem to be driven endogenously by population growth. The concept of a population-driven human system is prevalent among social scientists, demographers, and economists (Lee and Anderson 2002; Turchin 2003), but ignoring the impact of climate forces on human systems may lead to false conclusions. Although the causal linkages in Fig. 14.7 are theoretically reasonable, the time sequence and predictability of the linkages should be verified statistically before any definite conclusions are drawn.

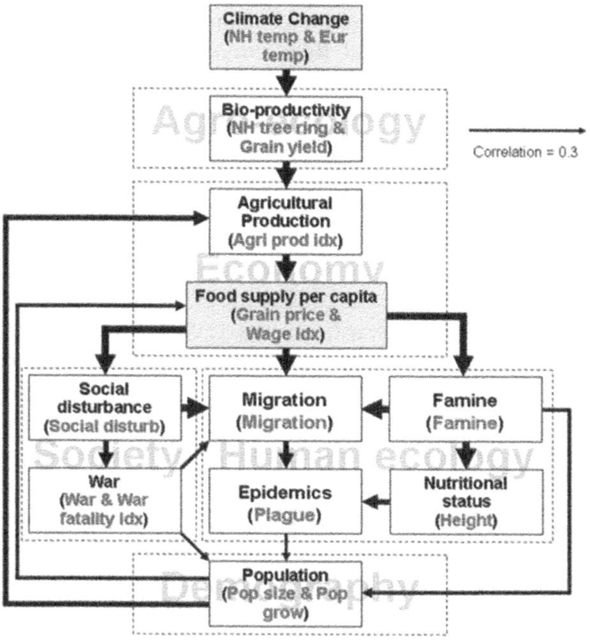

Fig. 14.7 Set of causal linkages from climate change to human crises in pre-industrial Europe. The terms in *bold black* type are sectors, and terms in *red* type within parentheses are variables that represent the sector. The thickness of the *arrow* indicates the degree of average correlation (Color figure online)

We further validated the time sequence and predictability of the causal linkages in Fig. 14.7 by using Granger Causality Analysis[29] (Zhang et al. 2011a). Via the Granger Causality Analysis, the causal relationship between variables is confirmed only if the cause precedes the effect in time and the causal series contains special information that could better explain and forecast the series being caused (Granger 1988). The causal linkages in Fig. 14.7 boiled down to these relationships: Climate change → bio-productivity → agricultural production → FSPC; FSPC → social disturbance → war; FSPC → famine → nutritional status; FSPC, social disturbance, war, and famine → migration; nutritional status and migration → epidemics; war, famine, and epidemics → population; population → agricultural production; and population → FSPC. Our Granger Causality Analysis results show that all null hypotheses of these linkages were rejected (13 linkages with $P < 0.01$ and 4 linkages with $P < 0.05$), implying that causal relationships between climate change and various human crises are statistically valid (Table 14.7).

14.4 Discussions

In our studies, we scientifically demonstrated the connection and also the causal mechanisms between climate change and various human crises in pre-industrial human societies. In addition, we showed that the societal impact brought about by global climate change was geographically diversified. In regions where population pressure and/or agricultural dependency on climate were higher, the climate-crisis relationship was also stronger and more apparent.

Since 2005, the year in which our first quantitative study about the climate-war relationship in historical China (Zhang et al. 2005) was published, there has been a surge in quantitative empirical large-N studies that explore whether and how climate change affects human crises, which allows more scientifically robust and generalizable arguments about the climate-crisis nexus to be generated. For those large-N quantitative studies that span several centuries or even longer period (Zhang et al. 2010; Tol and Wagner 2010; Wang et al. 2010), all of their findings concur with our conclusion about the importance of climate change in affecting human societies. However, for those large-N studies which use empirical data for the

[29] If one factor (*A*) causes another factor (*B*), then changes of A happen first, followed by changes of *B*, that is the cause precedes the effect. Under this circumstance, Granger Causality Analysis could be used to test whether *A* which precedes *B* could better help predict *B* by adding *A*. Then, we consider *A* is the Granger cause of *B* (i.e., the causal relationship between *A* and *B*). Granger Causality Analysis has been used widely in business, economics, sociology, psychology, politics, biology, and medicine. It also is regarded as an effective method to verify causal relationships in the social sciences. Before Granger Causality Analysis, an Augmented Dickey–Fuller test was adopted to check the stationarity of data. Any nonstationary data were subjected to first- or second-level differencing. Then regressions were run (by controlling the number of lags) to identify the causal relation.

Table 14.7 Granger causality analysis for each of the linkages as shown in Fig. 14.7

Causal linkage (null hypothesis)	*F*	*P*
Climate change does not *Granger-cause* bio-productivity	207.485	0.000***
Bio-productivity does not *Granger-cause* agricultural production	7.440	0.007**
Agricultural production does not *Granger-cause* food supply per capita	9.834	0.002**
War does not *Granger-cause* population	391.805	0.000***
Epidemics do not *Granger-cause* population	103.054	0.000***
Famine does not *Granger-cause* population	155.736	0.000***
Population does not *Granger-cause* agricultural production	5.731	0.017*
Population does not *Granger-cause* food supply per capita	67.664	0.000***
Food supply per capita does not *Granger-cause* famine	10.307	0.000***
Famine does not *Granger-cause* nutritional status	2.139	0.009**
Nutritional status does not *Granger-cause* epidemics	2.345	0.004**
Food supply per capita does not *Granger-cause* social disturbance	1.971	0.024*
Social disturbance does not *Granger-cause* war	3.256	0.000***
Social disturbance does not *Granger-cause* migration	1.786	0.037*
War does not *Granger-cause* migration	2.250	0.006**
Food supply per capita does not *Granger-cause* migration	2.164	0.008**
Migration does not *Granger-cause* epidemics	1.835	0.031*

*Significant at 0.05 level (2-tailed) ($P < 0.05$); **Significant at 0.01 level (2-tailed) ($P < 0.01$); ***Significant at 0.001 level (2-tailed) ($P < 0.001$)

past few decades only (i.e., especially AD 1950 and forward) or even shorter time periods (AD 1980 and forward) (Burke et al. 2010; Hsiang et al. 2011; Hendrix and Glaser 2007; Hendrix and Salehyan 2012; Jensen and Gleditsch 2009; Meier et al. 2007; Miguel et al. 2004; Raleigh and Kniveton 2012; Koubi et al. 2012; Theisen 2012), their results for the climate-crisis relationship are diverse. Indeed, in those studies, there is some confusion whether it is the effect of 'climate' or 'weather' that is being tested, because climate is usually defined as 'average weather' over a 30-year period (see Gleditsch 2012). It should be emphasized that short-term variations in climate may not have a significant effect on human crises, but the long-term variations do, because worsening 'climate' conditions contribute to a gradual build-up of tension. Crises occur only if the tension exceeds the buffering capacity of human societies (in terms of migration, economic change, innovation, trade, peaceful resource redistribution, and so on) for a significantly long period (Zhang et al. 2007a, 2011a).

On the other hand, there are also some recent studies based on selective individual cases that question the role of climate change in triggering human crises. For instance, Butzer (2012) argues that between the fourteenth and eighteenth centuries, Western Europe responded to extreme climatic perturbations by innovation and intensification; such modernization was decentralized, protracted, flexible, and broadly based. Under extreme stress, people try new solutions. After overcoming initial, ideological dissonance, people can indeed come together to support change.

Human crises should be attributable to socio-economic factors instead of climate change. Such a viewpoint and approach are also adopted by other scholars (Salehyan 2008; Fan 2010; Tertrais 2011). To facilitate a fruitful discussion about the climate-crisis relationship, we try to clarify some issues here.

First, the crucial issue linking scale to explanation is whether the variables used to explain a phenomenon are themselves located at the same scale. Causal explanation and generalization relevant to one scale regime are unlikely to be appropriate at others (Gibson et al. 2000). When employing variables at a different level to explain phenomenon at a particular level, we must be aware of some explanatory fallacies. Individualistic fallacies may occur when research imputes the cause of higher-level (or macro) patterns to be the same as that causing lower-level (or micro) patterns. Ecological fallacies are those that impute the cause of lower-level (or micro) patterns to be the same as those operating at a higher (or macro) level. We based our research upon aggregate statistics to conclude the climate-crisis relationship. While our findings may not be applicable to explain every individual historical incident, it would be conceptually wrong to simply explore individual cases to support/reject aggregate statistical results.

Second, although social factors may explain some site-specific short-term crises in a specific time slot, they cannot explain the widespread and simultaneous occurrence of long-term crises across different countries which have different social settings (i.e., characterized different stages of civilization, culture, economic development, and resources), at least in the pre-industrial era. In fact, climate-induced societal change can be measured at different scales, whereas the magnitude of change depends upon the economic impact of climate deterioration. We established the connection and underlying causal mechanisms between climate change and various human crises at hemispheric, continental, and macro-regional scales by scientific means. We concluded that climate change was the ultimate cause of human crisis in pre-industrial societies at the macro-historical level.

Our research findings challenge a widely accepted basic concept of Malthusian theory in ecological and population research, which states that subsistence level increases arithmetically while population increases geometrically (Malthus 1798) (Fig. 14.8a). According to the theory, land carrying capacity is assumed to be essentially constant or possibly increasing monotonically. Population checks (i.e., famines, wars, and epidemics) and the associated population collapses/ socio-political chaos occur when faster population growth overshoots the subsistence level. Since the land carrying capacity was understood as relatively fixed, the root of historical human crises has been attributed simply to excessive population growth. In fact, Malthus (1798) is insightful in pointing out that the root cause of human crises is the ecological imbalance between population size and land carrying capacity, which is attributable to divergent rates of population growth and subsistence increase. However, our studies emphasize that, given the technological limitations of agricultural production during the time in question, long-term climate changes would have a significant impact on food supplies, and that the constant or possibly monotonic increase in land carrying capacity (assumed by Malthus 1798,

Fig. 14.8 Comparison of population size and subsistence level in Europe in AD 1500–1800. (**a**) Malthusian scenario, in which population size and subsistence level are purely hypothetical. (**b**) Real world situation. The *black line* represents population size (in millions, corresponds to the left Y-axis), while the *grey line* denotes subsistence level (corresponds to the right Y-axis). The *grey* shaded area represents a cold period with the Northern Hemisphere temperature anomaly below −0.5 °C (Color figure online)

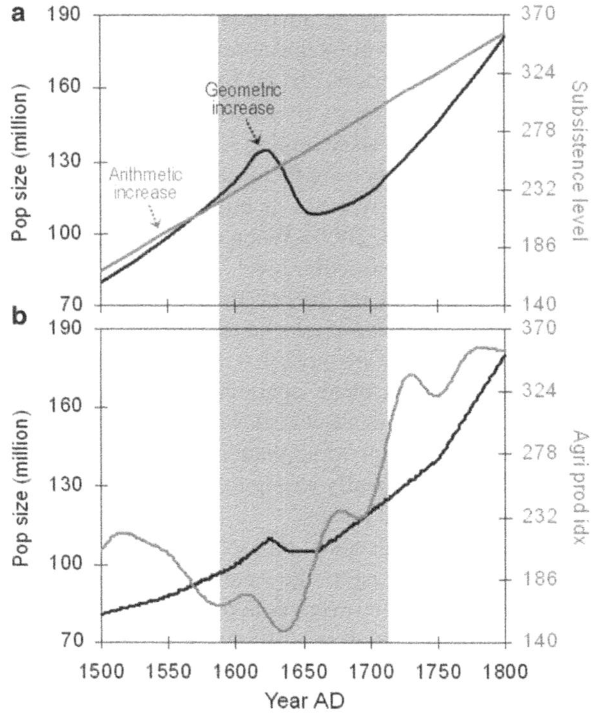

and many other scholars) may not be true, at least in the pre-industrial era. The upward trend of land carrying capacity is, in fact, characterized by shorter-term recurring oscillations (ascending wave pattern) in accordance with the alternation of cold and warm climate (Zhang et al. 2007a; Lee et al. 2008, 2009; Lee and Zhang 2010). On the other hand, in agrarian societies, population expands to the land carrying capacity (demographic saturation), which reduces the natural buffering capacity that would provide for societal resilience (e.g. tracts of non-cultivated arable land) (Wood 1998; Fagan 2000). This renders the population vulnerable to any shocks in food supplies. Such circumstances interact with the climate-induced decline of land carrying capacity to produce human crises.

By examining the climate-crisis connection in Europe from AD 1500 to 1800, we found that the subsistence level in Europe increased in an ascending but oscillating manner, and its ups and downs matched closely with the alternation of warm and cold climate (Fig. 14.8b) (Zhang et al. 2011b). We also discovered that the subsistence level fluctuated more deeply than population size did, and that agricultural shrinkage brought by cooling since the late sixteenth century preceded the population collapse in the mid-seventeenth century. Such a sequential relationship (or cause-and-effect relationship) further supports our studies.

14.5 Conclusions

Previous research about the climate-crisis relationship was constrained by the availability of paleo-climate reconstructions (cf. Introduction). In recent years, the situation has been completely changed. There has been a blossoming of paleo-climate reconstructions, which are not only in high temporal resolution, but also in fine spatial resolution. Subject to the increasing availability and precision of paleo-climate data, climate-crisis research in pre-industrial societies was further facilitated. On the other hand, the linkages and feedback loops for the climate-population relationship may be non-stationary. Two specific methods are recommended to tackle the possible non-stationary components of the relationship: cross-correlation function and wavelet analysis. Cross-correlation function measures overall associations for the whole frequency-time domain. The overall cross-correlation function associations are often determined by predominant associations of some frequencies, which often mask weak associations of other frequencies or time periods (Bloom et al. 2007). Wavelet analysis is used to examine non-stationary time series, particularly the causal links of multiple non-stationary time series (Cazelles et al. 2008; Zhang et al. 2009). The continuous wavelet transform decomposes the time series into both time and frequency components, the calculation of the wavelet power spectrum quantifies in the time-frequency domain the distribution of the variance of the time series (Cazelles et al. 2008; Grinsted et al. 2004). Those who are interested in the details and application of the above methods can refer to Zhang et al.'s (2010) study. Hopefully, with the application of more sophisticated methods, the role played by climate change in human history can be fully unveiled in the near future.

Acknowledgements This research was supported by the Hui Oi-Chow Trust Fund (201302172003 and 201205172003), HKU Seed Funding Programme for Basic Research (201109159014), and Research Grants Council of The Government of the Hong Kong Special Administrative Region of the People's Republic of China (HKU745113H, HKU758712H, HKU7055/08H, and HKU705-HSS-12).

References

An, Z. S. (2000). The history and variability of the East Asian paleomonsoon climate. *Quaternary Science Reviews, 19*(1–5), 171–187.

Aston, T. H. (1966). *Crisis in Europe: 1560–1660*. London: Routledge & Kegan Paul.

Bloom, R. M., Buckeridge, D. L., & Cheng, K. E. (2007). Finding leading indicators for disease outbreaks: Filtering, cross-correlation, and caveats. *Journal of the American Medical Informatics Association, 14*(1), 76–85.

Brecke, P. (1999). Violent conflicts 1400 A.D. to the present in different regions of the world. In 1999 meeting of the Peace Science Society (International), Ann Arbor, MI.

Briffa, K. R. (2000). Annual climate variability in the Holocene: Interpreting the message of ancient trees. *Quaternary Science Reviews, 19*(1–5), 87–105.

Briffa, K. R., & Osborn, T. J. (2002). Paleoclimate – Blowing hot and cold. *Science, 295*(5563), 2227–2228.
Bryson, R. A., & Murray, T. J. (1977). *Climates of hunger: Mankind and the world's changing weather*. Madison: University of Wisconsin Press.
Burke, M. B., Miguel, E., Satyanath, S., Dykema, J. A., & Lobell, D. B. (2010). Climate robustly linked to African civil war. *Proceedings of the National Academy of Sciences of the United States of America, 107*(51), E185.
Butzer, K. W. (2012). Collapse, environment, and society. *Proceedings of the National Academy of Sciences of the United States of America, 109*(10), 3632–3639.
Cazelles, B., Chavez, M., Berteaux, D., Ménard, F., Vik, J. O., Jenouvrier, S., et al. (2008). Wavelet analysis of ecological time series. *Oecologia, 156*(2), 287–304.
Chu, C. Y. C., & Lee, R. D. (1994). Famine, revolt, and the dynastic cycle – Population dynamics in historical China. *Journal of Population Economics, 7*(4), 351–378.
Cliff, A., Haggett, P., & Smallman-Raynor, M. (1998). *Deciphering global epidemics: Analytical approaches to the disease records of world cities, 1888–1912*. Cambridge: Cambridge University Press.
Crowley, T. J., & Lowery, T. S. (2000). How warm was the medieval warm period? *Ambio, 29*(1), 51–54.
Cullen, H. M., De Menocal, P. B., Hemming, S., Brown, F. H., Guilderson, T., & Sirocko, F. (2000). Climate change and the collapse of the Akkadian empire: Evidence from the deep sea. *Geology, 28*(4), 379–382.
deMenocal, P. B. (2001). Cultural responses to climate change during the Late Holocene. *Science, 292*(5517), 667–673.
Editorial Committee of Chinese Military History. (1985). *Zhongguo junshishi* (Vol. II). Beijing: Jiefangjun Chubanshe.
Enzel, Y., Bookman, R., Sharon, D., Gvirtzman, H., Dayan, U., Ziv, B., & Stein, M. (2003). Late holocene climates of the near East deduced from dead sea level variations and modern regional winter rainfall. *Quaternary Research, 60*(3), 263–273.
Esper, J., Cook, E. R., & Schweingruber, F. H. (2002). Low-frequency signals in long tree-ring chronologies for reconstructing past temperature variability. *Science, 295*(5563), 2250–2253.
Fagan, B. (2000). *The little ice age: How climate made history 1300–1850*. New York: Basic Books.
Fan, K.-W. (2010). Climatic change and dynastic cycles in Chinese history: A review essay. *Climatic Change, 101*(3–4), 565–573.
Fischer, D. H. (1996). *The great wave: Price revolutions and the rhythm of history*. New York: Oxford University Press.
Galloway, P. R. (1986). Long-term fluctuations in climate and population in the preindustrial era. *Population and Development Review, 12*(1), 1–24.
Gibson, C. C., Ostrom, E., & Ahn, T. K. (2000). The concept of scale and the human dimensions of global change: A survey. *Ecological Economics, 32*(2), 217–239.
Gleditsch, N. P. (2012). Whither the weather? Climate change and conflict. *Journal of Peace Research, 49*(1), 3–9.
Goldstone, J. A. (1991). *Revolution and rebellion in the early modern world*. Berkeley: University of California Press.
Golkin, A. T. (1987). *Famine: A heritage of hunger*. Claremont: Regina Books.
Granger, C. W. J. (1988). Some recent development in a concept of causality. *Journal of Econometrics, 39*(1–2), 199–211.
Gribbin, J. (1978). *Climatic change*. Cambridge: Cambridge University Press.
Grinsted, A., Moore, J. C., & Jevrejeva, S. (2004). Application of the cross wavelet transform and wavelet coherence to geophysical time series. *Nonlinear Processes in Geophysics, 11*(5/6), 561–566.
Hendrix, C. S., & Glaser, S. M. (2007). Trends and triggers: Climate, climate change and civil conflict in sub-Saharan Africa. *Political Geography, 26*(6), 695–715.

Hendrix, C. S., & Salehyan, I. (2012). Climate change, rainfall, and social conflict in Africa. *Journal of Peace Research, 49*(1), 35–50.

Hinsch, B. (1988). Climatic change and history in China. *Journal of Asian History, 22*(2), 131–159.

Hsiang, S. M., Meng, K. C., & Cane, M. A. (2011). Civil conflicts are associated with the global climate. *Nature, 476*(7361), 438–441.

Huntington, E. (1907). *The pulse of Asia: A journey in Central Asia illustrating the geographic basis of history*. London: Houghton Mifflin.

Issar, A. S. (1995). Climate change and the history of the Middle East. *American Scientist, 83*(4), 350–355.

Jansen, E., Overpeck, J., Briffa, K. R., Duplessy, J.-C., Joos, F., & Masson-Delmotte, V. (2007). Palaeoclimate. In S. Solomon, D. Qin, M. Manning, Z. Chen, M. Marquis, K. B. Averyt, M. Tignor, & H. L. Miller (Eds.), *Climate change 2007: The physical science basis. Contribution of Working Group I to the fourth assessment report of the intergovernmental panel on climate change* (pp. 433–498). Cambridge: Cambridge University Press.

Jensen, P. S., & Gleditsch, K. S. (2009). Rain, growth, and civil war: The importance of location. *Defence and Peace Economics, 20*(5), 359–372.

Jiang, T. (1993). *Zhongguo jindai renkou shi*. Hangzhou: Hangzhou University Press.

Jones, P. D., Briffa, K. R., Barnett, T. P., & Tett, S. F. B. (1998). High-resolution palaeoclimatic records for the last millennium: Interpretation, integration and comparison with General Circulation Model control-run temperatures. *Holocene, 8*(4), 455–471.

Koepke, N., & Baten, J. (2005). The biological standard of living in Europe during the last two millennia. *European Review of Economic History, 9*(1), 61–95.

Kohn, G. C. (1999). *Dictionary of wars*. New York: Facts on File.

Kohn, G. C. (2001). *Encyclopedia of plague and pestilence* (Rev. ed.). New York: Facts on File.

Koubi, V., Bernauer, T., Kalbhenn, A., & Spilker, G. (2012). Climate variability, economic growth, and civil conflict. *Journal of Peace Research, 49*(1), 113–127.

Le Roy Ladurie, E. (1972). *Times of feast, times of famine: A history of climate since the year 1000*. London: Allen & Unwin.

Lee, R., & Anderson, M. (2002). Malthus in state space: Macroeconomic-demographic relations in English history, 1540 to 1870. *Journal of Population Economics, 15*(2), 195–220.

Lee, H. F., & Zhang, D. D. (2010). Changes in climate and secular population cycles in China, 1000 CE to 1911. *Climate Research, 42*(3), 235–246.

Lee, H. F., Fok, L., & Zhang, D. D. (2008). Climatic change and Chinese population growth dynamics over the last millennium. *Climatic Change, 88*(2), 131–156.

Lee, H. F., Zhang, D. D., & Fok, L. (2009). Temperature, aridity thresholds, and population growth dynamics in China over the last millennium. *Climate Research, 39*(2), 131–147.

Li, B. S., Zhang, D. D., Jin, H. L., Wu, Z., Yan, M. C., Wu, S., et al. (2000). Paleo-monsoon activities of Mu Us Desert, China since 150 ka BP – A study of the stratigraphic sequences of the Milanggouwan Section, Salawusu River area. *Palaeogeography Palaeoclimatology Palaeoecology, 162*(1–2), 1–16.

Lobell, D. B., & Field, C. B. (2007). Global scale climate-crop yield relationships and the impacts of recent warming. *Environmental Research Letters, 2*(1), art. no. 014002.

Luard, E. (1986). *War in international society: A study in international sociology*. London: Tauris.

Luterbacher, J., Dietrich, D., Xoplaki, E., Grosjean, M., & Wanner, H. (2004). European seasonal and annual temperature variability, trends, and extremes since 1500. *Science, 303*(5663), 1499–1503.

Lyon, B., Rowen, H. H., & Hamerow, T. S. (1969). *A history of the western world*. Chicago: Rand McNally.

Malthus, T. R. (1798). *An essay on the principle of population*. London: printed for J. Johnson.

Mann, M. E., & Jones, P. D. (2003). Global surface temperatures over the past two millennia. *Geophysical Research Letters, 30*(15), Art. No. 1820.

Mann, M. E., Bradley, R. S., & Hughes, M. K. (1999). Northern hemisphere temperatures during the past millennium: Inferences, uncertainties, and limitations. *Geophysical Research Letters, 26*(6), 759–762.

McEvedy, C., & Jones, R. (1978). *Atlas of world population history*. London: Allen Lane.

Meier, P., Bond, D., & Bond, J. (2007). Environmental influences on pastoral conflict in the Horn of Africa. *Political Geography, 26*(6), 716–735.

Miguel, E., Satyanath, S., & Sergenti, E. (2004). Economic shocks and civil conflict: An instrumental variables approach. *Journal of Political Economy, 112*(4), 725–753.

Osborn, T., & Briffa, K. R. (2006). The spatial extent of 20th-century warmth in the context of the past 1200 years. *Science, 311*(5762), 841–844.

Parker, G., & Smith, L. M. (1978). Introduction. In G. Parker & L. M. Smith (Eds.), *The general crisis of the seventeenth century* (pp. 1–25). London: Routledge & Kegan Paul.

Pfister, C. (2007). Climatic extremes, recurrent crises and witch hunts: Strategies of European societies in coping with exogenous shocks in the late sixteenth and early seventeenth centuries. *Medieval History Journal, 10*(1–2), 33–73.

Polyak, V. J., & Asmerom, Y. (2001). Late Holocene climate and cultural changes in the southwestern United States. *Science, 294*(5540), 148–151.

Raleigh, C., & Kniveton, D. (2012). Come rain or shine: An analysis of conflict and climate variability in East Africa. *Journal of Peace Research, 49*(1), 51–64.

Roberts, J. M. (1996). *A history of Europe*. Oxford: Helicon.

Salehyan, I. (2008). From climate change to conflict? No consensus yet. *Journal of Peace Research, 45*(3), 315–326.

Tertrais, B. (2011). The climate wars myth. *Washington Quarterly, 34*(3), 17–29.

Theisen, O. M. (2012). Climate clashes? Weather variability, land pressure, and organized violence in Kenya, 1989–2004. *Journal of Peace Research, 49*(1), 81–96.

Times, L. (2007). *Times atlas of the world* (4th ed.). London: Times Books.

Tol, R. S. J., & Wagner, S. (2010). Climate change and violent conflict in Europe over the last millennium. *Climatic Change, 99*(1–2), 65–79.

Turchin, P. (2003). *Historical dynamics*. Princeton: Princeton University Press.

Utterström, F. L. G. (1955). Climatic fluctuations and population problems in early modern history. *The Scandinavian Economic History Review, 3*(1), 3–47.

von Rad, U., Schaaf, M., Michels, K. H., Schulz, H., Berger, W. H., & Sirocko, F. (1999). A 5000-yr record of climate change in varved sediments from the oxygen minimum zone off Pakistan, northeastern Arabian Sea. *Quaternary Research, 51*(1), 39–53.

Walford, C. (1970). *The famines of the world: Past and present*. New York: Burt Franklin.

Wang, S. W., Gong, D. Y., & Zhu, J. H. (2001). Twentieth-century climatic warming in China in the context of the Holocene. *Holocene, 11*(3), 313–321.

Wang, X., Chen, F., Zhang, J., Yang, Y., Li, J., Hasi, E., et al. (2010). Climate, desertification, and the rise and collapse of China's historical dynasties. *Human Ecology, 38*(1), 157–172.

Wood, J. W. (1998). A theory of preindustrial population dynamics – Demography, economy, and well-being in Malthusian systems. *Current Anthropology, 39*(1), 99–135.

Wright, Q. (1942). *Study of war* (Vol. 1). Chicago: University of Chicago Press.

Wrigley, E. A., & Schofield, R. S. (1981). *The population history of England, 1541–1871: A reconstruction*. London: Arnold.

Xiao, S., & Liu, A. (2005). *History of pestilence*. Taipei: Zhiqing Pindao.

Yancheva, G., Nowaczyk, N. R., Mingram, J., Dulski, P., Schettler, G., Negendank, J. F. W., et al. (2007). Influence of the intertropical convergence zone on the East Asian monsoon. *Nature, 445*(7123), 74–77.

Yang, B., Braeuning, A., Johnson, K. R., & Shi, Y. (2002). General characteristics of temperature variation in China during the last two millennia. *Geophysical Research Letters, 29*(9), Art. No. 1324.

Zhang, D., Jim, C. Y., Lin, C. S., He, Y. Q., & Lee, F. (2005). Climate change, social unrest and dynastic transition in ancient China. *Chinese Science Bulletin, 50*(2), 137–144.

Zhang, D. D., Jim, C. Y., Lin, G. C. S., He, Y. Q., Wang, J. J., & Lee, H. F. (2006). Climatic change, wars and dynastic cycles in China over the last millennium. *Climatic Change, 76*(3–4), 459–477.

Zhang, D. D., Brecke, P., Lee, H. F., He, Y. Q., & Zhang, J. (2007a). Global climate change, war, and population decline in recent human history. *Proceedings of the National Academy of Sciences of the United States of America, 104*(49), 19214–19219.

Zhang, D. D., Zhang, J., Lee, H. F., & He, Y. Q. (2007b). Climate change and war frequency in Eastern China over the last millennium. *Human Ecology, 35*(4), 403–414.

Zhang, Z., Cazelles, B., Tian, H., Stige, L. C., Bräuning, A., & Stenseth, N. C. (2009). Periodic temperature-associated drought/flood drives locust plagues in China. *Proceedings of the Royal Society B: Biological Sciences, 276*(1658), 823–831.

Zhang, Z., Tian, H., Cazelles, B., Kausrud, K. L., Bräuning, A., Guo, F., et al. (2010). Periodic climate cooling enhanced natural disasters and wars in China during AD 10–1900. *Proceedings of the Royal Society B: Biological Sciences, 277*(1701), 3745–3753.

Zhang, D. D., Lee, H. F., Wang, C., Li, B., Pei, Q., Zhang, J., et al. (2011a). The causality analysis of climate change and large-scale human crisis. *Proceedings of the National Academy of Sciences of the United States of America, 108*(42), 17296–17301.

Zhang, D. D., Lee, H. F., Wang, C., Li, B., Zhang, J., Pei, Q., et al. (2011b). Climate change and large scale human population collapses in the pre-industrial era. *Global Ecology and Biogeography, 20*(4), 520–531.

Zhao, S. (1986). *Physical geography of China*. Beijing: Science Press.

Chapter 15
Analyzing the Patterns of Space-Time Distances for Tracking the Diffusion of an Epidemic

Tzai-Hung Wen and Yu-Shiuan Tsai

15.1 Introduction

Understanding the dynamics of how infectious diseases spread in time and space is the primary concern of epidemic control and prevention. Most spatial epidemiological studies use innovative spatial statistical methods to identify spatial or spatial-temporal clusters in an epidemic and their associations with environmental risk factors (Carpenter 2001; Cowled et al. 2009; Wen et al. 2006). Epidemic risk areas could be identified as abnormal clusters in which the observed number of cases exceeds the expected number of cases in a given time and location. A geographic information system (GIS) is a computational tool used to visualize and analyze spatial-temporal clustering of diseases; however, complex human behaviors, such as contacts and movement, are difficult to incorporate into clustering analyses.

The pattern of interpersonal contacts is one important risk factor in modeling epidemic dynamics. Contact networks can be formulated such that individuals are modeled as nodes, with contacts modeled as links. Therefore, social-network analysis (SNA) is used to model individual contact behaviors with respect to disease transmission (Firestone et al. 2011; Jones and Handcock 2003; Tsai et al. 2011). Network topological characteristics are then used as epidemiological risk indices to measure patterns of spread within a group of susceptible populations. For example, the degree of network clustering could indicate the infection risk of an individual. A higher network-clustering coefficient indicates a closer connection to his/her neighbors and, therefore, a higher probability of infection. Degrees of

T.-H. Wen (✉) • Y.-S. Tsai
Department of Geography, National Taiwan University, Room #501, Geography Building, National Taiwan University, No. 1, Sec. 4, Roosevelt Road, Taipei 10617, Taiwan
e-mail: wenthung@ntu.edu.tw; mathoowind@gmail.com

M.-P. Kwan et al. (eds.), *Space-Time Integration in Geography and GIScience*, DOI 10.1007/978-94-017-9205-9_15

separation could indicate epidemic spread via long-range interactions. A higher degree of separation between two persons means that transmission of infection between them would take longer. Finally, the distribution of degree centrality could identify whether super-spreaders, who have higher degree centrality values, exist (Huang et al. 2010); however, the diffusion of epidemic diseases is not a single event, and the capture of the temporal process of social contacts must rely on intensive surveys and methodological improvements in social-network analysis.

Additionally, geo-computational models, such as agent-based modeling (ABM), have also been used to detect early warnings of critical transitions and assess interventions against epidemics in a system that emphasizes the importance of spatial-temporal dynamic processes (Barrett et al. 2005). Epidemic dynamics can be modeled as interactions between different categories of populations, including susceptible, exposed, infectious and recovered populations. Model parameters, such as interpersonal contact rate, infection probability, transmission rate and recovery rate, must be derived from mathematical models or estimated from empirical evidence, and these parameters are sensitive to predicted results (Pastor-Satorras and Vespignani 2002). Recently, geo-computational models have been used to simulate the complex interactions of the host, agent and environment in models assessing the impact of climate warming on epidemic spread (Gong et al. 2011; Morin and Comrie 2010).

Although spatial-temporal clusters and diffusion processes of epidemics have been studied extensively in recent decades (Cowled et al. 2009; Wen et al. 2012; Wilesmith et al. 2003), spatial methods of tracking the possible sources of infection have not been comprehensively developed. Current methods of tracking sources of infection and routes of transmission are time-consuming, difficult and intensive, involving epidemiological investigations and laboratory diagnosis with genotyping technology. The objective of this study is to propose an innovative methodology that considers only the spatial-temporal relationships of illness onsets, combining exploratory spatial analysis and network topological indicators, degree centrality and network clustering coefficient, to identify space-time clusters, track possible sources of an epidemic and measure transmission risk of an individual. Analyzing temporal progression of these space-time clusters of an epidemic, two geographic diffusion patterns, contagious and relocation processes, can be further differentiated. It will be beneficial when establishing appropriate interventions for specific epidemic period efficiently.

The structure of this paper is as follows: in the next section we describe the methodology for tracking epidemic diffusion in time and space. Section 15.3 presents a hypothetical example to explain how the proposed analytical procedure works. The methodology is further applied to the dengue fever epidemic in Kaohsiung City, Taiwan in 2009–2010 as the case study. The background of the case study and major findings are reported in Sect. 15.4. Final section states our conclusions.

15.2 Methodology

The proposed methodological framework is composed of two procedures for tracking epidemic diffusion. In the first stage, GIS is used to analyze the week of onset and the residential location of each confirmed case to establish case-to-case distances across space and time. Cumulative pairs of points calculated at a given space-distance and time-distance are counted. The diagram of space-time distances allows the identification of the space-time clustering of cases. Then, the space-time pairs of these clustering cases are selected and mapped. In the second stage, the link-node network based on these pairs is further analyzed. In this paper, we use the link-node topology of space-time distances to determine the network's topological indicators, such as the degree centrality scores of each case, which can be used to identify possible sources of infection (individuals with higher values of the indicators). Each procedure is described below in detail.

15.2.1 Definitions of Space-Time Distances

Basing our model on the Knox statistic (Knox and Bartlett 1964; Kulldorff and Hjalmars 1999), we used the illness onset time and residential locations of each infected case to determine case-to-case distances in space and time. The individual's temporal distance is defined as the difference in the onset week between any two infected cases, designated as u and v, as in Eq. 15.1:

$$\text{Time} - \text{distance}_{u,v} = \left|\text{InfectTime}_u - \text{InfectTime}_v\right|, \tag{15.1}$$

where InfectTime_u is the onset time of infected individual u. Moreover, when we use aggregated areal data (e.g., district or county) as the spatial unit for analyzing diffusion, we can define the district's temporal distance as the difference in the time of first onset between any two districts, denoted as U and V, as in Eq. 15.2:

$$\text{District's Time} - \text{distance}_{U,V} = \left|\min_{u\in U}\left(\text{InfectTime}_u\right) - \min_{v\in V}\left(\text{InfectTime}_v\right)\right|. \tag{15.2}$$

where $u \in U$ such that $\text{InfectTime}_{u_0} = \min_{u\in U}\left(\text{InfectTime}_u\right)$ and $v \in V$ such that $\text{InfectTime}_{v_0} = \min_{v\in V}\left(\text{InfectTime}_v\right)$.

Space-distance is defined as the distance between the locations of any two infected individuals, denoted as u and v, as in Eq. 15.3:

$$\text{Space-distance}_{u,v} = \left|\text{location}_u - \text{location}_v\right|. \tag{15.3}$$

Similarly, the district's space-distance can be defined as the distance between any centroids of two districts, denoted as *U* and *V*, as in Eq. 15.4

$$\text{District's Space-distance}_{U,V} = |\text{location}_u - \text{location}_v|, \text{ where } u \in U \text{ and } v \in V \tag{15.4}$$

15.2.2 Stage 1: Diagram of Space-Time Distances

The diagram of space-time distances can be established by counting the cumulative pairs of points calculated at a given space–distance and time–distance. The diagram is a two-dimensional contour plot; the x-axis is the time-distance, and the y-axis is the space-distance. The values of the contour are the cumulative values of the space-time pairs *DR(i, j)*, which are defined as in Eq. 15.5

$$DR(i,j) = \left| \left[\bigcup_{RSD(m,n)=i} (m,n) \right] \cap \left[\bigcup_{RTD(p,q)=j} (p,q) \right] \right|, \tag{15.5}$$

where *(m,n)* is a pair, with *m* and *n* ranging from 1 to the total number of cases. RSD *(m,n)* is the rounded spatial distance at row *m* and column *n*. *RTD (m,n)* is the rounded temporal distance at row m and column n. Therefore, DR (i,j) is the number of space-time pairs when $RSD = \text{i}$ and $RTD = \text{j}$.

Higher values of the *DR (i,j)* indicate that more cumulative pairs of points were found at a given space- and time-distance, showing space-time clustering of cases (Kulldorff and Hjalmars 1999). The outlier-detection technique is used to determine the number of cumulative pairs of points that is significant. The threshold in this study is defined as the third quartile plus 3 times the inter-quartile range (IQR) (Kvanli et al. 2005). When the number of cumulative pairs of points was greater than the "upper fence" value, the pairs of space-time distances were selected and the spatial distribution of these pairs was mapped.

15.2.3 Stage 2: Analyzing Link-Node Structure of Space-Time Pairs

We regarded infected cases as nodes and selected space-time pairs of these cases as links. A link-node network of space-time pairs was established. The network topology of space-time distances could thus be analyzed further.

In this study, we used two network indicators to measure the risk of epidemic diffusion. If one node that exists in many pairs represents a person who has many connections with others in time and space, then he/she could be regarded as a possible infection source. This concept is identical to degree-based centrality (the number of links that a node has), which is a commonly used measure in social-network analysis (Freeman 1978). The average nodal degree centrality < k > is defined as follows:

$$\langle k \rangle = \frac{1}{N} \sum_{v} k_v \quad (15.6)$$

where N is the total number of nodes and k_v is the number of links (neighbors) that node v has.

If node v has k_v neighbors and a total of Ev edges among these k_v nodes, the degree of network clustering coefficient < C_v > can be defined as the density of connection of the neighbors of node v (as shown in Eq. 15.8). A larger value indicates that a given set of neighbors are well connected with each other, which means that a person with higher network clustering could have a higher probability of infection if one of his or her neighbors were to be infected. This indicator can measure transmission risk at a node in the network of space-time distances.

$$\langle C_v \rangle = \left\langle \frac{2E_v}{k_v (k_v - 1)} \right\rangle = \frac{1}{N} \sum_{v} \frac{2E_v}{k_v (k_v - 1)} \quad (15.7)$$

15.3 A Hypothetical Example

To explain how the proposed analytical method works in our study, we will give a hypothetical example of how to use the spatiotemporal relationships of infected cases to track the spread of an epidemic. Suppose the data D are obtained by spatial-temporal data acquisition and that the data contain the identification (ID), the residential location, and the onset time for each individual (as shown in Table 15.1). Figure 15.1 shows the spatial distribution of these cases and nodes. Darker colors indicate onset of illness at an earlier stage of the epidemic.

Table 15.1 Residential locations and illness onset time of the hypothetical epidemic

	Residential locations		
ID	X-coordinate	Y-coordinate	Illness onset time
1	1	2	6
2	2	1	7
3	3	4	1
4	3.5	3	0
5	4	4.5	2
6	4.5	3.5	1
7	6	5.5	2
8	5.5	1.5	5
9	6	2.5	3
10	7	1	4

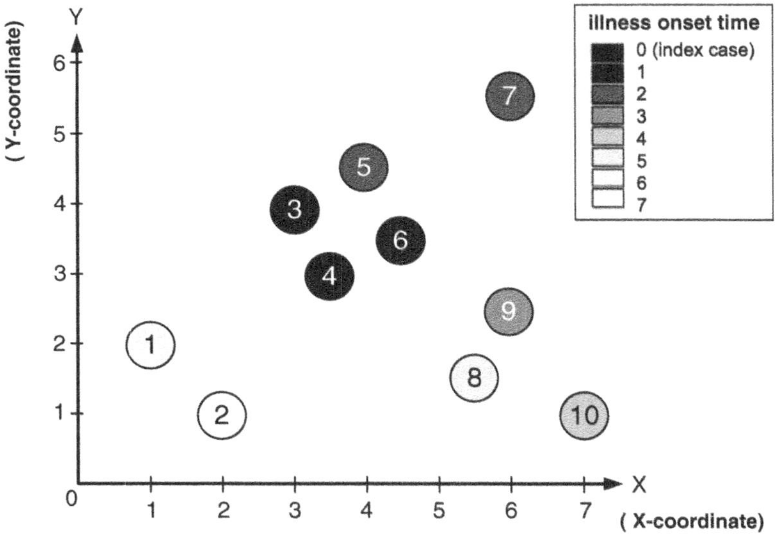

Fig. 15.1 Spatial locations and onset illness of the hypothetical epidemic

Node	1	2	3	4	5	6	7	8	9	10
1		1	3	3	4	4	6	5	5	6
2	1		3	3	4	4	6	4	4	5
3	5	6		1	1	2	3	4	3	5
4	6	7	1		2	1	4	3	3	4
5	4	5	1	2		1	2	3	3	5
6	5	6	0	1	1		3	2	2	4
7	4	5	1	2	0	1		4	3	5
8	1	2	4	5	3	4	3		1	2
9	3	4	2	3	1	2	1	2		2
10	2	3	3	4	2	3	2	1	1	

Rounded spatial distance

Rounded temporal distance

Fig. 15.2 Rounded time-distance (RTD) and space-distance (RSD) two-dimensional matrix

The first step is to establish space-time distances. The time-distance (TD) is the difference in illness onset time to detection between any two individuals (see Eq. 15.1), and the space-distance (SD) is the geographic distance between any two individuals' residences (see Eq. 15.3). The data *D* can be transformed into the SD-and-TD two-dimensional matrix. Because the elements in the SD-and-TD matrix are continuous values and the values cannot be accumulated, the element values should be made discrete and grouped in intervals. In the example, the element is integerized by rounding. The transformed rounded time-distance (RTD) and space-distance (RSD) two-dimensional matrix is shown in Fig. 15.2.

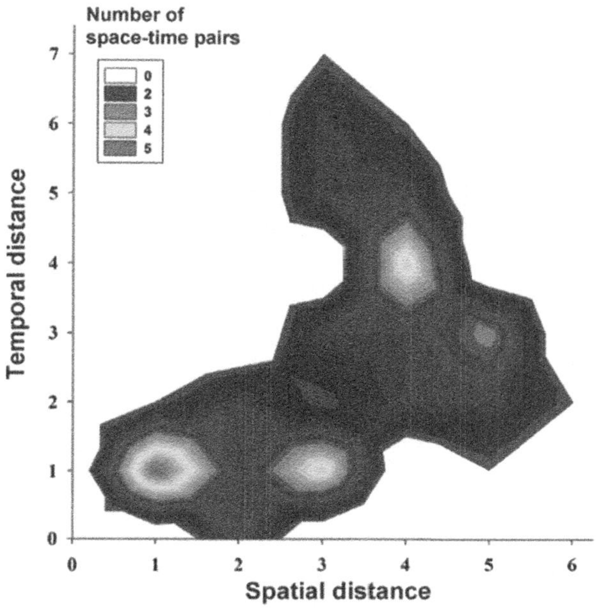

Fig. 15.3 Space-time distances of the hypothetical epidemic

The diagram of space-time distances, derived by counting RSD and RTD (Eq. 15.6) from the two-dimensional matrix, can be constructed as shown in Fig. 15.3. A higher value in the diagram indicates a greater number of cumulative pairs of space-time points. In this example, we used outlier detection to find the significant large number of cumulative space-time pairs of points to identify space-time clustering. We then selected these space-time pairs for further analysis.

Figure 15.4 shows the structure of a network constructed from selected space-time pairs. There are two components; therefore, the spread of this epidemic could be characterized as relocation-dependent. In the larger component, nodes with higher degree centrality values (nodes 3, 4, 5 and 6) could be the infection sources; the infection then diffused to nodes 8, 9 and 10. Node 4 has the highest network-clustering coefficient, indicating that this area has the highest transmission risk (Table 15.2). Thus, the spread of infection could be identified within the component.

Through the methods we proposed, the diffusion of a hypothetical epidemic can be depicted clearly. The areas with higher degree centrality represent probable sources of infection, and the nodes with higher network clustering represent areas with the greatest transmission risk. Later in time in the model, the epidemic spread to neighboring areas (contagious diffusion). Late in the epidemic, the disease spread to areas farther away, leaving the area that had been the source of infection (relocation diffusion).

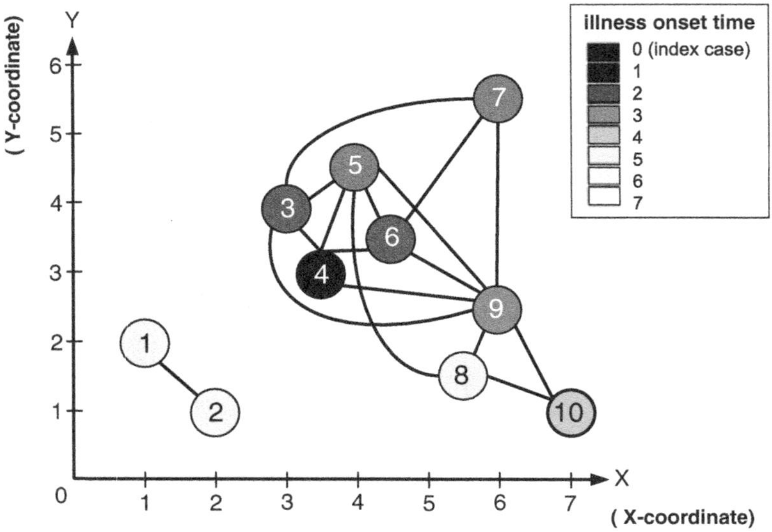

Fig. 15.4 The structure of a network constructed from selected space-time pairs

Table 15.2 Degree centrality and network clustering coefficients of each infected case in the hypothetical epidemic

ID	Degree centrality	Network clustering coefficient
1	1	–
2	1	–
3	4	0.67
4	4	**0.83**
5	**5**	0.60
6	4	0.67
7	3	0.67
8	3	0.67
9	7	0.43
10	2	1

Note: The largest number of each network indicator is shown in *bold*

15.4 A Dengue Epidemic in Kaohsiung as a Case Study

This study focused on a dengue epidemic in Kaohsiung as a case study to assess the feasibility of our proposed methodology. Kaohsiung City, located in southern Taiwan, is the second largest metropolitan city, the largest commercial harbor and an important land-, marine- and air-transportation hub in Taiwan. Figure 15.5 shows the location of Kaohsiung City and one of its satellite cities, Fengshan City. This two-city area was used as our study area because it was the site of a major dengue epidemic in southern Taiwan.

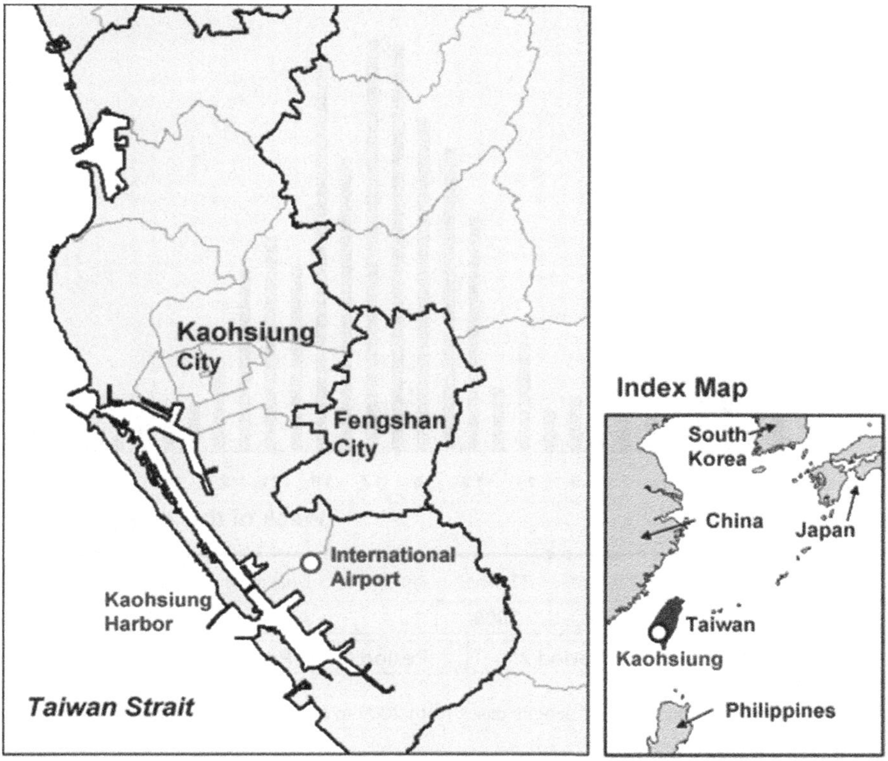

Fig. 15.5 Locations of Kaohsiung and Fengshan, Taiwan

The data on dengue cases was collected through the national surveillance system of Taiwan Centers for Disease Control (Taiwan CDC), which reported 785 confirmed dengue cases (DEN) in the two-city area from July 2009 to February 2010. The epidemic curve in Fig. 15.6 shows two significant epidemic waves. The period of the initial wave was from July to late August. The second wave (the major epidemic) lasted from early September to February of the next year. To understand the progression of the epidemic, we divided this epidemic into five periods: the initial wave (period 1, July-late August), pre-epidemic of the second wave (period 2, early September-October), major epidemic of the second wave (period 3, November), post-epidemic of the second wave (period 4, December), and January- February of the next year (period 5).

Figure 15.7 shows the cumulative pairs of space–time distances for the dengue epidemic. Deeper red coloring indicates the presence of more cumulative pairs in a given space–time distance interval, which also indicates spatial–temporal clustering or interaction. The space–time clusters were also identified by outlier detection and are represented as red dots. Four space–time clustering patterns were identified.

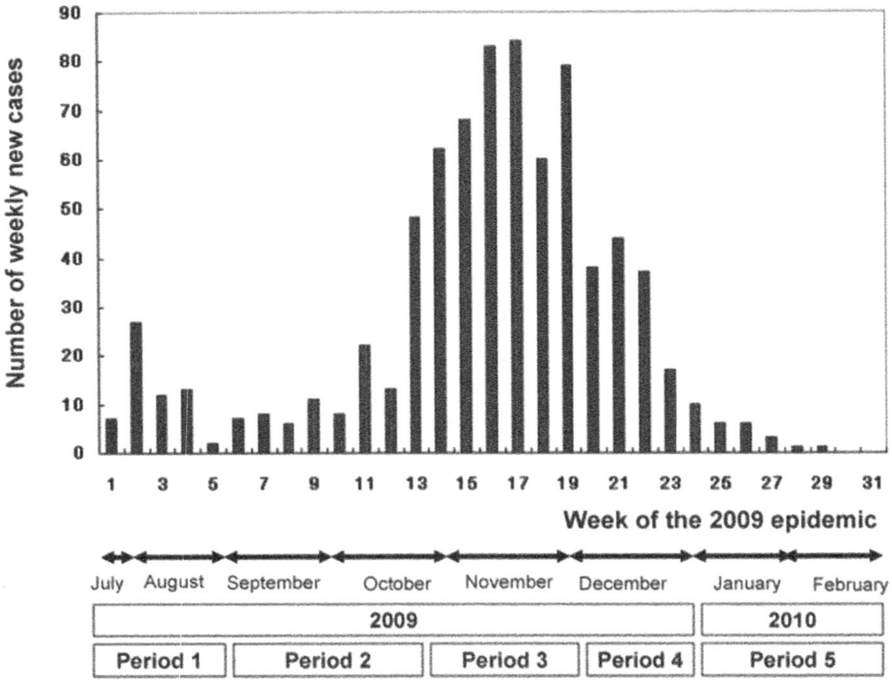

Fig. 15.6 The epidemic curve of dengue cases from 2009 to early 2010, Kaohsiung

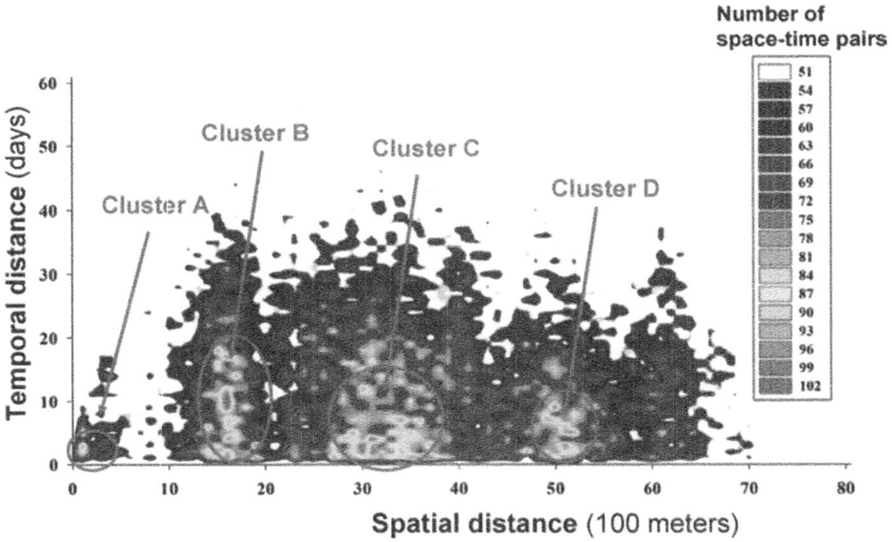

Fig. 15.7 Space-time distances of the 2009 dengue epidemic in Kaohsiung

The first space–time pairs clustered within 100–300 m and 1–4 days; the second clustered within 1.3–1.8 km and 1–20 days; the third clustered within 2.6–3.9 km and 1–20 days; and the last clustered within 4.6–5.5 km and 1–20 days. In other words, in the first two space–time clusters, the average interval between the onset of illness for any two cases was less than 20 days and the residential distance between the two cases was less than 2 km. Thus, most cases could have been infected due to neighborhood characteristics. On the other hand, the residential distance between the last two space–time clusters was in the range 2.6–5.5 km. The spread of infection across this distance could be attributed to population mobility.

We then mapped four major clusters of space–time pairs. The network density of these pairs is shown in Fig. 15.8. The intense spots in Cluster A and Cluster B, which has smaller space-time distances, are located in south Kaohsiung and at the border between the two cities, respectively. Cluster C and Cluster D, which has a larger range of space-time distances, were scattered across the study area, but intense spots were also centered on the border between the two cities. The results suggest that the infected cases in the border have intense spatial-temporal interactions.

Two network indicators to measure diffusion risk are summarized in Table 15.3. The average degree centrality reflects the connections of one infected person. The results showed that Cluster C has the highest degree centrality. Table 15.4 also shows that most of the cases in Cluster C occurred in Period 3 of the epidemic. These results suggest that Period 3 is the transition stage from local to large-scale transmission and that infected cases with higher centrality in Cluster C could be the sources of large-scale transmission. The other indicator, the network-clustering coefficient, reflects the risk of transmission. Cluster A has the highest values of this indicator and occurred in the first period of the epidemic (Table 15.5). The results suggest that areas of Cluster A could be where infected persons initiated local transmission.

In summary, the diffusion patterns show that the epidemic originated in south Kaohsiung (Cluster A). Infections in the initial stage resulted from neighborhood characteristics, such as vacant lots or houses with poor garbage disposal. The epidemic followed the "contiguous diffusion" pattern and gradually spread to the border between the two cities in Period 2 (Cluster B). Possibly due to demographic factors, such as population mobility, the scale of epidemic followed the "relocation diffusion" pattern and spread across a larger range of space-time distances in Periods 3 and 4 (Cluster C and D).

15.5 Conclusions

Space and time are two important dimensions in describing epidemic dynamics and risk characteristics. This study applied space-time distances to determine network indicators that could differentiate among contagious and relocation diffusion patterns to model epidemic progression. Using the dates of onset and residential locations of infected persons, the proposed method also identified possible sources

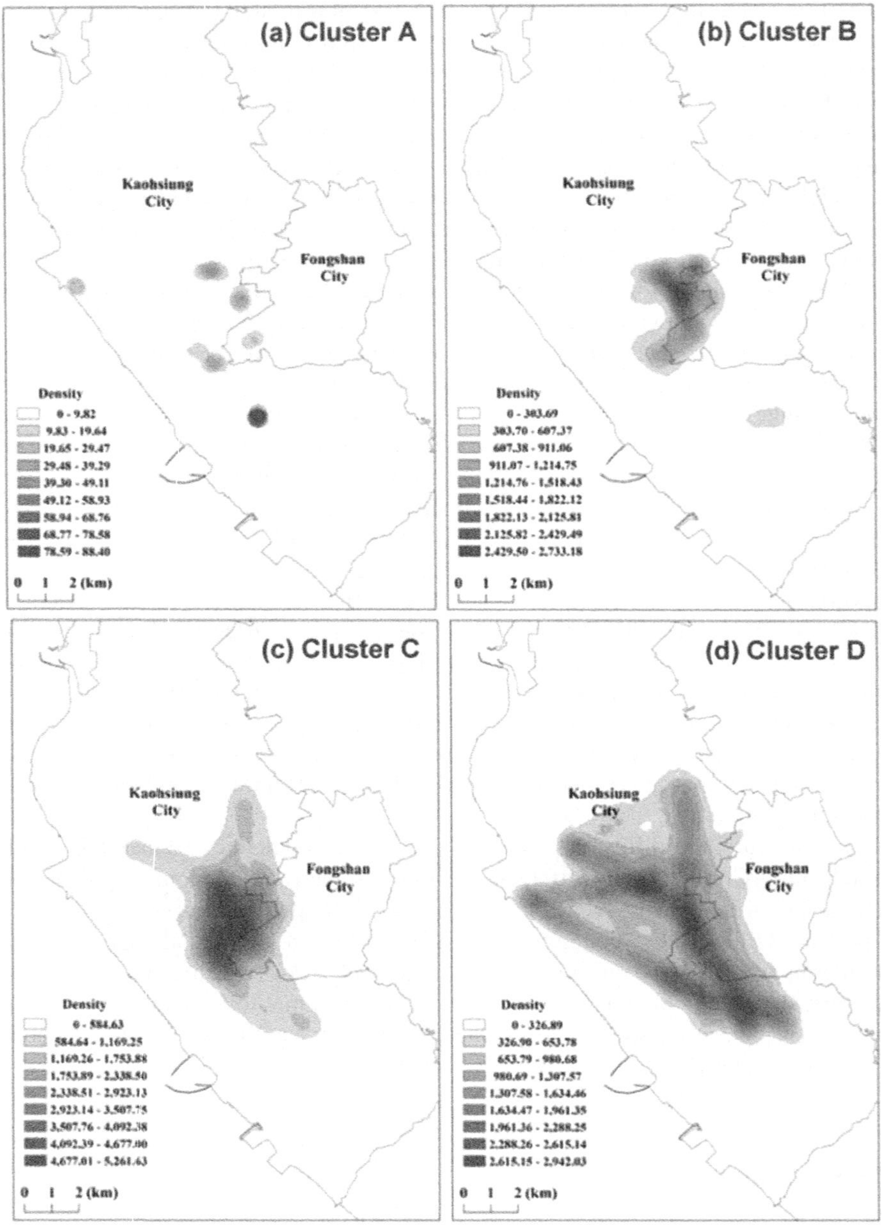

Fig. 15.8 The network density of the selected space-time pairs of the 2009 dengue epidemic in Kaohsiung

Table 15.3 Average degree centrality and network clustering coefficient of each space-time cluster of the dengue epidemic in Kaohsiung, 2009

	Cluster A	Cluster B	Cluster C	Cluster D
Average degree centrality	3.51	25.60	**51.93**	36.27
Average network clustering coefficient	**0.36**	0.09	0.10	0.04

Note: The largest number of each network indicator is shown in *bold*

Table 15.4 Temporal progression of the infected cases with top 10 % of the degree centrality in each space-time cluster

	Period 1	Period 2	Period 3	Period 4	Period 5
Cluster A	**37**	18	19	6	0
Cluster B	0	1	**52**	27	0
Cluster C	0	0	**69**	11	0
Cluster D	0	3	**55**	22	0

Note: The largest number in each cluster is shown in *bold*

Table 15.5 Temporal progression of the infected cases with top 10 % of the network clustering coefficient in each space-time cluster

	Period 1	Period 2	Period 3	Period 4	Period 5
Cluster A	**39**	13	30	26	0
Cluster B	0	9	27	**39**	5
Cluster C	0	21	**36**	14	7
Cluster D	0	8	27	**36**	6

Note: The largest number in each cluster is shown in *bold*

of infection and areas with high transmission risk. This method also provides broader insights into modeling spatial-temporal epidemic dynamics.

However, this study had some limitations. Not only spatial-temporal relationships of infected persons, but molecular surveillance with genotyping technology and known epidemiological links could confirm disease transmission occurred. Although the methodology proposed in this paper does not provide biological evidences of tracking disease transmission, it could be the complementary procedure of improving geographic targeting of interventions, rather than replacing current epidemiological investigation practices. Social relationships are also other important clues to tracking disease transmission. The procedure of measuring geospatial proximity between infected persons would extend social behaviors/distances (such as population mobility, contact patterns, and so on) in the further study.

Acknowledgements The research was supported by the grants of National Science Council in Taiwan (NSC 98-2410-H-002-168-MY2, NSC 101-2119-M-002-020). The authors also acknowledge the financial support provided by Infectious Diseases Research and Education Center, Department of Health and National Taiwan University. The funder had no role in study design, data collection and analysis, or preparation of the manuscript.

References

Barrett, C. L., Eubank, S. G., & Smith, J. P. (2005). If smallpox strikes Portland.... *Scientific American, 292*(3), 54–61. doi:10.1038/scientificamerican0305-54.
Carpenter, T. E. (2001). Methods to investigate spatial and temporal clustering in veterinary epidemiology. *Preventive Veterinary Medicine, 48*(4), 303–320. doi:10.1016/S0167-5877(00)00199-9.
Cowled, B., Ward, M. P., Hamilton, S., & Garner, G. (2009). The equine influenza epidemic in Australia: Spatial and temporal descriptive analyses of a large propagating epidemic. *Preventive Veterinary Medicine, 92*(1–2), 60–70. doi:10.1016/j.prevetmed.2009.08.006.
Firestone, S. M., Ward, M. P., Christley, R. M., & Dhand, N. K. (2011). The importance of location in contact networks: Describing early epidemic spread using spatial social network analysis. *Preventive Veterinary Medicine, 102*(3), 185–195. doi:10.1016/j.prevetmed.2011.07.006.
Freeman, L. C. (1978). Centrality in social networks conceptual clarification. *Social Networks, 1*(3), 215–239. doi:10.1016/0378-8733(78)90021-7.
Gong, H., DeGaetano, A. T., & Harrington, L. C. (2011). Climate-based models for West Nile Culex mosquito vectors in the Northeastern US. *International Journal of Biometeorology, 55*(3), 435–446. doi:10.1007/s00484-010-0354-9.
Huang, C. Y., Tsai, Y. S., & Sun, C. T. (2010). Effects of friend-making resources costs and remembering on acquaintance networks. *PHYSICA A: Statistical Mechanics and Its Applications, 389*(3), 604–622. doi:10.1016/j.physa.2009.09.038.
Jones, J. H., & Handcock, M. S. (2003). Social networks (communication arising): Sexual contacts and epidemic thresholds. *Nature, 423*(6940), 605–606.
Knox, E., & Bartlett, M. (1964). The detection of space-time interactions. *Journal of the Royal Statistical Society, Series C, 13*(1), 25–30.
Kulldorff, M., & Hjalmars, U. (1999). The Knox method and other tests for space-time interaction. *Biometrics, 55*(2), 544–552. doi:10.1111/j.0006-341X.1999.00544.x.
Kvanli, A. H., Pavur, R. J., & Keeling, K. B. (2005). *Concise managerial statistics* (1st ed., p. 656). Cincinnati: South-Western College Publishing.
Morin, C. W., & Comrie, A. C. (2010). Modeled response of the West Nile virus vector Culex quinquefasciatus to changing climate using the dynamic mosquito simulation model. *International Journal of Biometeorology, 54*(5), 517–529. doi:10.1007/s00484-010-0349-6.
Pastor-Satorras, R., & Vespignani, A. (2002). Immunization of complex networks. *Physical Review E, 65*(3), 36104. doi:10.1103/PhysRevE.65.036104.
Tsai, Y.-S., Huang, C.-Y., Wen, T.-H., Sun, C.-T., & Yen, M.-Y. (2011). Integrating epidemic dynamics with daily commuting networks: Building a multilayer framework to assess influenza A (H1N1) intervention policies. *Simulation, 87*(5), 385–405. doi:10.1177/0037549710379481.
Wen, T.-H., Lin, N. H., Lin, C.-H., King, C.-C., & Su, M.-D. (2006). Spatial mapping of temporal risk characteristics to improve environmental health risk identification: A case study of a dengue epidemic in Taiwan. *Science of the Total Environment, 367*(2–3), 631–640. doi:10.1016/j.scitotenv.2006.02.009.
Wen, T.-H., Lin, M.-H., & Fang, C.-T. (2012). Population movement and vector-borne disease transmission: Differentiating spatial-temporal diffusion patterns of commuting and non-commuting dengue cases. *Annals of the Association of American Geographers, 102*(5), 1026–1037. doi:10.1080/00045608.2012.671130.
Wilesmith, J. W., Stevenson, M. A., King, C. B., & Morris, R. S. (2003). Spatio-temporal epidemiology of foot-and-mouth disease in two counties of Great Britain in 2001. *Preventive Veterinary Medicine, 61*(3), 157–170. doi:10.1016/j.prevetmed.2003.08.002.

Chapter 16
Assessing Dynamic Exposure to Air Pollution

Mei-Po Kwan, Desheng Liu, and Jaclyn Vogliano

16.1 Introduction

Accurate assessment of human exposures to air pollution (e.g., nitrogen dioxide and fine particulate matter) is critical to a wide range of geographic and health studies, such as modeling the association between air pollution exposure and health outcomes (e.g. asthma, premature mortality, and cancer), evaluating health impacts or disease burdens in different population groups, and determining environmental injustice that may be occurring with respect to exposures (Brunekreef and Holgate 2002). Methodologies of exposure assessment thus have significant implications for enhancing our understanding of and shedding light on how certain social issues can be addressed (e.g., how air pollution differentially affect different social groups). Health studies in the past typically assess air pollution exposure by assigning temporally averaged (e.g. annual mean) pollution concentration at people's residential locations (Jerrett et al. 2005). However, air pollution exposure is much more complicated as the exposure process is determined by the interaction of two complex geographic processes: air pollution concentrations and people's activity-travel patterns, both vary in space and time in highly complex ways.

M.-P. Kwan (✉)
Department of Geography and Geographic Information Science, University of Illinois at Urbana-Champaign, Champaign, IL 61820, USA

Department of Human Geography and Spatial Planning, Utrecht University, Utrecht, The Netherlands
e-mail: mpk654@gmail.com

D. Liu
Department of Geography, The Ohio State University, Columbus, OH 43210-1361, USA
e-mail: liu.738@osu.edu

J. Vogliano
NextEra Energy Resources, LLC, P.O. Box 14000 Juno Beach, FL 33408-0420, USA
e-mail: Jaclyn.Vogliano@nexteraenergy.com

M.-P. Kwan et al. (eds.), *Space-Time Integration in Geography and GIScience*, DOI 10.1007/978-94-017-9205-9_16

As air pollution sources are closely associated with particular types of productive activities (e.g., manufacturing) and traffic, exposure tends to be low during the night when people stay at home. In contrast, peak exposure to air pollution occurs in the daytime when people are at work or traveling to and from them (Briggs 2005). Consequently, individual air pollution exposures exhibit significant spatial-temporal variability – since people spend different amounts of time at different locations throughout the day. To evaluate individual air pollution exposure more accurately, it is necessary to take the temporal variation of both air pollution concentration and people's daily movement into account (Briggs 2005; Hoek et al. 2008). Yet, there has been limited research to date on exposure assessment that takes into account the complex interaction between temporal variations in air pollution concentrations and people's activity-travel patterns (with Gulliver and Briggs [2005] as an important exception).

This chapter presents a method for improving air pollution exposure assessment through integrating the spatial-temporal dynamics of pollution concentration and detailed space-time activity patterns of individuals. Specifically, we focus on the assessment of exposure to particulate matter with aerodynamic diameter of 10 μm or less (PM_{10}) in Franklin County, Ohio (USA). Particulate matter concentrations were first estimated with an air dispersion model using data of industrial point source emissions in the study area. These estimates were then adjusted for background concentrations based on data from three local monitors. Using these estimates and an activity-travel diary dataset, two notions of exposure to air pollution were implemented: (a) static exposure evaluated at an individual's residential location; and (b) dynamic exposure assessed based on the locations an individual undertake out-of-home activities and trips, including locations en route from one location to another.

The chapter then analyzed whether there are significant differences between static and dynamic exposures to particulate matter for individuals of various gender, income, education, and age subgroups. It also examine whether individuals of these subgroups experienced different levels of dynamic exposure to air pollution. Results from the analyses did not reveal any significant difference in static and dynamic exposure between different gender, income, education, and age subgroups. There were also no significant differences in dynamic exposure among different gender, education and age subgroups. But dynamic exposures for different income subgroups were different. The chapter then provides interpretations of these results in the specific context of the study area and limitations in the data available for modeling the geographic patterns of air pollution.

16.2 Assessment of Individual Exposure to Air Pollution

Geographers and health researchers have long been concerned with the health effects of people's exposure to air pollution. Essential tasks in this kind of studies include determining the geographic distribution of pollution concentrations across the study area and determining which portion of the population is at greatest risk. They seek to address four important questions: (a) What are the primary sources of specific pollutants? (b) What environmental factors influence the dispersion of air borne particles? (c) How can one accurately portray relevant pollution concentrations across a particular study area? (d) How can one assess the exposure to air pollution of different individuals who live and undertake daily activities in different parts of the study area?

16.2.1 Estimating the Spatial and Temporal Patterns of Air pollution

In efforts to understand people's exposure to air pollution, past studies have used different methods to estimate the spatial and temporal distribution of pollution concentrations relative to the geographic distribution of the affected population. Six broad categories of methods have been used to date: proximity-based assessments, geostatistical interpolation, land use regression models, dispersion models, integrated models, and hybrid models (Jerrett et al. 2005). Each method entails varying levels of complexity, data requirement, sophistication, and time commitment. Several researchers have conducted health-related studies using various combinations of these six air pollution assessment methods. A brief critical overview of proximity-based assessments, geostatistical interpolation, land use regression models, and dispersion models is provided as follows.

Proximity-based assessments establish relationships between air pollution and exposure by assuming that individuals face greater risk if they are located near major emission sources (Jerrett et al. 2005). A common approach of this method is to relate distance to heavily traveled roadways, traffic density, and surrounding populations. Venn et al. (2001), for instance, adopted this approach and applied a buffer of 150 m around major roadways that intersected the study area to capture school children's exposure to traffic-related air pollution. Janssen et al. (2001), on the other hand, gathered actual measurements of traffic-related air pollutants outside and inside of schools located 400 m from major highways. The study found an association between $PM_{2.5}$ concentrations for both indoor and outdoor air with an increase in truck traffic density and also a decrease with increasing distance from the roadways.

Proximity assessment offers a simple approach for exposure analysis. The method, however, has several limitations. First, the dynamic nature of human activities and travel is ignored. As English et al. (1999) argued, assessment of potential exposure in this approach focuses narrowly on people's residence, work, and/or school locations while other daily activities are overlooked. Clearly, human activities are not limited to two or three isolated localities throughout a typical 24-h day (Kwan 2009, 2012a, b). As Gulliver and Briggs (2005) highlighted, although most people spend a relatively small amount of their time traveling, journey-time exposures make up a disproportionately large proportion of individual exposure to ambient (and especially traffic-related) air pollution. A static analysis of air pollution exposure may thus lead to serious underestimation of people's actual exposures.

Additionally, proximity-based assessment is heavily based on traffic emissions that are generated by a combination of various types of vehicles. But levels of emissions may vary considerably with different mix of cars and trucks that occupy the roadways (Kanaroglou et al. 2000; Gertler 2003). Since most previous studies did not distinguish vehicle type when making emission estimates, assessments of people's air pollution exposure may be inaccurate. Further, dispersion characteristics of air pollution may be influenced by topographical and meteorological factors. The isotropic dispersion assumption that air borne particles experience the same dispersion pattern in all directions surrounding the study area can be invalid in most situations (Jerrett et al. 2005). It is unrealistic to assume that wind speed, wind direction, weather patterns, atmospheric stability, and land use patterns have no effect on air pollution dispersal. Thus, proximity-based assessment tends to oversimplify the analysis of air pollution exposure. Reality encompasses much more complex scenarios – for example, human activities take place at many locations throughout the day and external factors influence the dispersal of particulate matter concentrations.

Geostatistical interpolation is another approach to pollution assessment. It uses deterministic or stochastic geostatistical techniques to estimate pollution levels at unsampled locations across a study area, and uses these values to interpolate a continuous surface of pollution concentrations for the entire area (e.g., Farhang 1983; Pengelly et al. 1984; Finkelstein et al. 2003). A critical component of this method is a dense network of monitoring stations that is distributed throughout the study area. Several interpolation techniques were used to estimate pollution concentrations, including inverse distance weighting, splines, Theissen triangulation, and kriging. Among them kriging is the most commonly used method in air pollution studies (Jerrett et al. 2001). An advantage of kriging interpolation over the other methods is the establishment of both predicted values at unsampled locations and their standard errors. Having access to error estimates helps inform the researchers about where predicted values may be more or less reliable.

Implementing interpolation techniques in air pollution studies is a more sophisticated approach when compared to proximity-based assessments. But this approach also has its limitations. While interpolation methods utilize real pollution measurements for computing exposure estimates, data availability may be an issue. Jerrett et al. (2005), for instance, estimated that a dense network of sampling sites ranging from 10 to 100 stations is required for meaningful interpolation results depending on the size of the study area. But it is difficult to obtain an adequate geographic coverage of pollution concentration readings for certain areas since government monitoring stations are mostly located in heavily populated areas or near major roadways or industrialized areas and are sparsely scattered in rural areas (Bell 2006). Further, the temporal resolution of monitored data is often low (e.g., on 6-day intervals or at particular hours of the day). As a result, estimates for days or times when concentrations were not sampled can be very crude. Finally, when individual exposures are assessed based on these crude air pollution data without consideration of people's activity patterns (i.e. time spent at home, work, and school), the actual levels of individual exposure can be significantly underestimated.

The third method for air pollution assessment is land use regression modeling, which utilizes land use attributes and traffic characteristics as the basis for making pollution estimates (e.g., Briggs et al. 1997). More specifically, these models predict the level of pollution concentrations for unsampled locations in a study area using data on traffic volume, land cover, and topography. The output from a land use regression model is a pollution surface created based on the data collected at the monitor sites. Similar to the previous two methods, land use regression analysis requires measurement data for the target pollutants. Once again, the availability of a dense monitoring network plays a crucial role in the reliability of land use regression pollution estimations, where an increase in monitor density helps to reduce prediction errors. Studies have shown that data derived from a single monitoring location can only represent a small area surrounding a particular site (e.g., Briggs et al. 1997). Reliance on a sparse monitoring network for estimating exposure levels can thus lead to considerable uncertainty and error.

The fourth approach to modeling spatial variations of air pollution concentrations is the utilization of dispersion models. These models conceptualize pollution dispersion in terms of deterministic processes, which are implemented through Gaussian plume equations (e.g., Gualtieri and Tartaglia 1998; Clench-Aas et al. 1999b; Bellander et al. 2001). The core components that formulate these processes include data on emissions, meteorological conditions, and topography. Data on emissions, also referred to as background concentrations, are usually obtained from government monitoring stations and serve as a model calibrator. Data on wind speed, wind direction, outside surface temperature, solar radiation, and atmospheric stability comprise the meteorological component. Furthermore, topographical information relating to the elevation of the study area is essential along with emission data that can be subdivided into two main categories: stationary and mobile. These three broad types of factors are used in dispersion models to estimate spatial variations of air pollution concentrations across a specified region. Once all parameters are obtained, dispersion calculations are performed to represent hourly variations or yearly averages in pollution concentrations.

Air pollution exposure studies that utilize dispersion models illustrate the advantages of integrating both spatial and temporal disparities of pollutant levels across a given region (e.g., Bartonova et al. 1999; Clench-Aas et al. 1999a). This method overcomes the need for extensive monitoring networks that are common with previously discussed techniques such as proximity-based assessment, geostatistical interpolation, and land use regression. Further, external parameters which diffuse air pollutants such as wind, atmospheric stability, topography, and traffic flows are accounted for in dispersion frameworks (Jerrett et al. 2005). Dispersion models can be applied to various spatial and temporal scales, such as short term air pollution episodes in urban areas or pollution transfer for large regions over space and time.

Dispersion models, however, have several drawbacks. Multiple parameters must be entered for each individual source and hence a substantial amount of time is needed for data preparation. Gaussian assumptions about dispersion patterns may be unrealistic (Jerrett et al. 2005). Estimation error may be present when there are temporal mismatches between emissions and meteorological data. For instance, this can happen when meteorological data are collected by the hour while emissions data are provided as an annual rate. As a result, exposure estimations have the potential to contain substantial error.

16.2.2 Assessing Population or Individual Exposure to Pollution

Another important component of exposure assessment is to determine which portion of the population is exposed to elevated levels of air pollution. Here, pollution exposure can be analyzed in terms of distinct population groups or at the individual level. For an aggregate assessment, the total number of people living in each square kilometer can be examined in relation to varying hourly concentrations for a well-defined area in the study area (e.g., Slørdal 1998). On the other hand, point estimations can be made for all homes and buildings within a given area (Clench-Aas et al. 1999a). For individual assessment, long and short term exposure can be determined using different approaches. When concerned with long term exposure, dispersion models can be used to estimate concentrations surrounding the home, work place, or school for a specific individual based on yearly, monthly, or seasonal averages (e.g., Bartonova et al. 1999). Conversely, short term exposure can be analyzed using a diary method (Duan 1982; Gulliver and Briggs 2005). Diary entries provide information regarding location and time of targeted individuals, which in turn can be related to the spatial and temporal variations of estimated air pollutants (Bartonova et al. 1999; Clench-Aas et al. 1999b).

A traffic exposure study in Oslo, Norway provides an example of individual assessment that used the diary method (Bartonova et al. 1999). In this research, dispersal patterns of particulate matter and nitrogen oxides related to traffic and spatial heating sources were estimated using a dispersion model. Estimations were provided for a 1-km resolution grid on an hourly basis which reflected air pollution patterns for an urban region. A subgroup of the participants provided activity diaries that contained a detailed description of their daily movement. Participant activities were recorded by the hour for a 2–3 week period (Bartonova et al. 1999). For the study area, concentration estimations for each receptor location were provided for each hour during a 3 month period. These calculations were used to represent annual average concentrations for a 3-year time frame. On the other hand, individual exposure levels were provided for each location that was visited on an hourly basis. Here, receptor values at each location were assigned to the individual when they were situated within the study area while grid values were assigned outside of the study area (Bartonova et al. 1999). If multiple locations were visited during a 1 h time frame, an average was taken of the outdoor concentrations at each stop.

While air pollution exposure is determined by the interaction of air pollution concentrations and people's activity-travel patterns, few past studies have taken into account the fact that people are at different locations at different times of the day (Kwan 2013; Richardson et al. 2013). To evaluate people's exposures to air pollution more accurately, it is important to also consider their daily movement (Kwan 2012a, b). To address this issue, we present in what follows a method for integrating the spatial-temporal dynamics of pollution concentrations and people's space-time activity patterns. We develop and explore two notions of exposure: (a) static exposure that evaluates exposure based on an individual's residential location; and (b) dynamic exposure that assesses exposure based on various locations that an individual travels to throughout the day, including the routes taken to travel between these locations.

16.3 Study Area and Data

The study area for this research is Franklin County located at the heart of the Columbus Metropolitan Area in Ohio, USA. It is a mid-sized metropolitan area with a dispersed urban structure and a population of about 1.8 million in 2010. Suburbanization in the past three decades has led to the spatial expansion of its urban area, a hollowing out of the inner city, and strong auto dependence.

To evaluate individual exposure to air pollution, two main types of data were used in this study: estimates of 24-h PM_{10} concentrations and activity-travel diary data for Franklin County residents. The 24-h particulate matter concentrations were estimated with an air dispersion model that utilizes three types of data: emission data, meteorological data, and topographic data. Characteristics of these data are described as follows, while the estimation procedures are provided in the next section.

First, emission data of the 119 major point sources of industrial pollutants in the study area were obtained from the Ohio Environmental Protection Agency's 1999 emissions inventory (Fig. 16.1). Variables from these data include latitude, longitude, base elevation, stack height, stack temperature, exit flow rate, and the total amount of pollution emitted (tons per year) of each source. Second, meteorological data of both surface and upper air used to determine the dispersal patterns of pollutants were obtained from the Columbus International Airport and the Dayton Wright Patterson Air Force Base. From both datasets, variables such as wind direction, wind speed, and temperature were used to determine atmospheric turbulence characteristics, which in turn influence the overall dispersal patterns of the emitted pollutants. Finally, data about the topographic features in the study area were drawn from USGS 7.5 min digital elevation model (DEM) quads of the study area. Typically, the elevation of the study area will have a stronger influence on

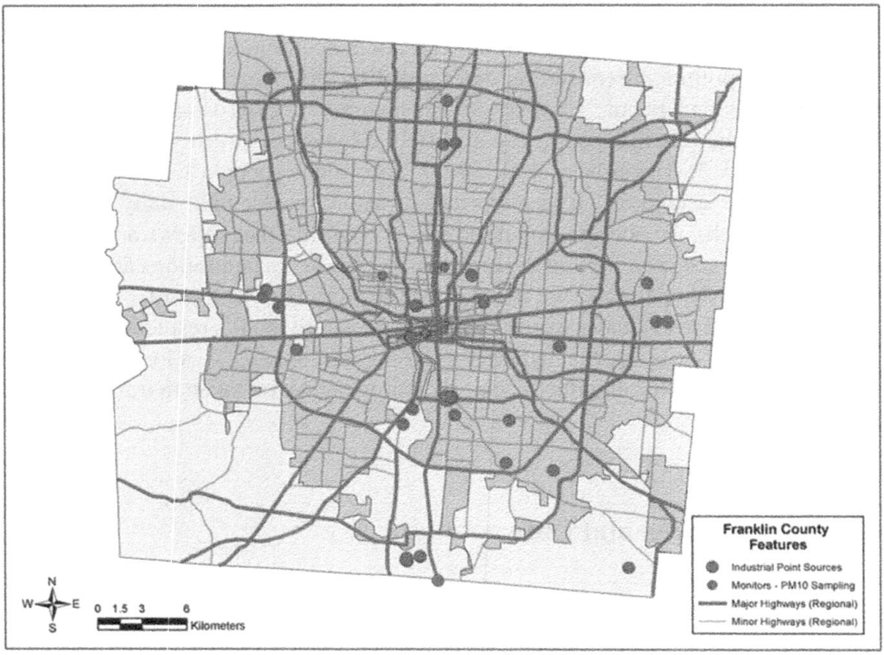

Fig. 16.1 The three local monitor sites and the 119 major industrial point sources in Franklin County, Ohio

the dispersal patterns if mountainous or hilly regions are present. However, since Franklin County is relatively flat, topographic influences on pollution dispersion are minimal in this study although they were incorporated in the model.

Another component that was added to the pollution estimates generated by the dispersion model was the level of background concentrations, which accounted for the pollution from all other sources that were not considered by the dispersion model. In this study, these other sources included major industrial sources located outside of Franklin County, minor sources that do not have emission inventories (restaurants, auto body shops, gas stations, and dry cleaners), mobile sources, fugitive emissions (land fills, storage piles), dust from unpaved roadways, and environmental tobacco smoke. These concentration values were obtained from the three local monitoring sites, all located within 6 km from the city center of Columbus. These values were then extrapolated to create a continuous surface covering the entire study area and added onto the results of the estimated concentrations from the air dispersion model.

The second type of data used in this study was an activity-travel diary dataset collected in the 1999 Mid-Ohio Area Household Travel Survey. The data were collected by the Mid-Ohio Regional Planning Commission (MORPC) over a 5-month period from February to June 1999 from 5,555 households located in Franklin County and several adjacent countries. About 13,500 individuals and

52,000 trips were represented in this dataset. Among these individuals, only those who resided in Franklin County and did not undertake any travel outside Franklin County were included in this study. Data items in this dataset include personal and household information such as age, gender, education, and race of each individual, household income, vehicle ownership, and number of household members. Many characteristics of the activities and trips participants undertook in an assigned survey day are also included in the dataset, such as activity location (latitude and longitude) and timing of activities and trips.

16.4 Method

Assessment of both static and dynamic exposure in this study was performed through three procedures: (1) estimation of air pollution concentrations using an air pollution dispersion model; (2) extrapolation and addition of monitored concentrations; and (3) evaluation and comparison of individual exposures using both static and dynamic measures. These procedures were implemented using an air dispersion modeling software package (BEEST Suite) and ArcGIS 9.2.

16.4.1 Estimation of Particulate Concentrations

To estimate air pollution dispersion, this study used a Gaussian plume model in which air pollution is represented as a plume coming from the top of a stack (point source) characterized by a height and diameter. The vertical displacement of the plume is determined from the stack dimensions along with stack gas exit velocity and temperature. After the plume has reached the stack height, the pollution is dispersed in three dimensions. This model was implemented using the Industrial Source Complex Short Term 3 (ISCST3) model in the BEEST Suite and the following Gaussian equation:

$$X = \frac{QKVD}{2\pi U_s \sigma_y \sigma_z} exp\left[-0.5\left(\frac{y}{\sigma_y}\right)^2\right]$$

where:

Q = pollutant emission rate (mass per unit time)
K = a scaling coefficient to convert calculated concentrations to desired units (default value of 1×10^6 for Q in g/s and concentration in $\mu g/m^3$
V = vertical term for source elevation, receptor elevation, plume rise, etc.
D = decay term
σ_y, σ_z = standard deviation of lateral and vertical concentration distribution (m)
U_s = mean wind speed (m/s) at release height

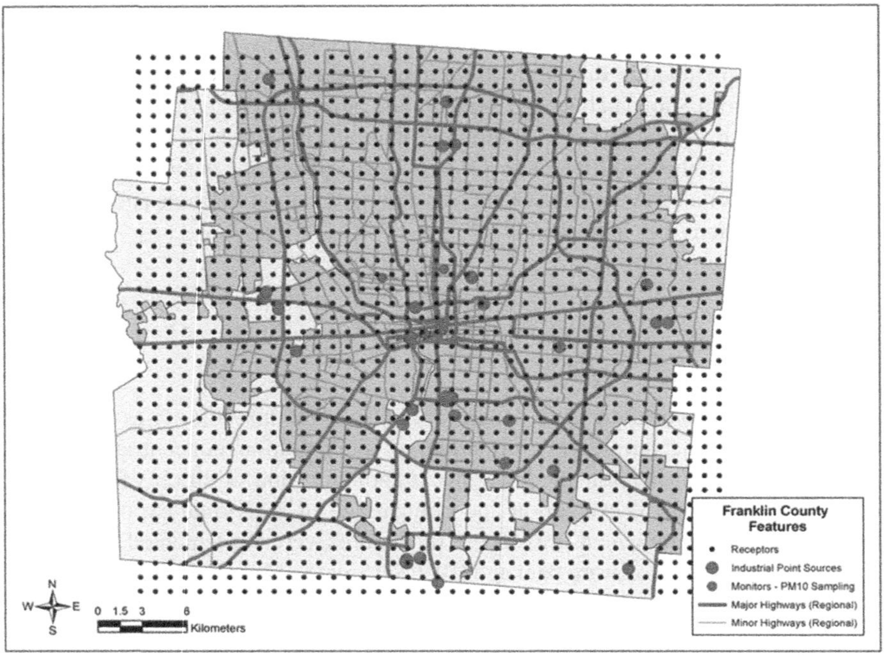

Fig. 16.2 The 1,520 receptor locations used in the dispersion modeling

Based on this conceptualization and implementation of the dispersion process, pollution concentrations were estimated for every hour at every receptor location for eight selected days in March, April and May of 1999: March 2nd, 4th, 5th; April 5th, 7th, 9th; May 6th, 7th. These days were selected to coincide with the survey days of the activity-travel diary dataset used in the study, and this allows for an analysis of 24-h particulate matter exposure for survey participants. Receptor locations are user-defined sites at which the model will estimate concentration values from multiple pollution sources. For this study, a 38 by 40 uniform grid with one kilometer of separation between each receptor was used – thus creating 1,520 receptor locations to cover the extent of Franklin County (Fig. 16.2).

Based on data of the 119 industrial point sources, pollution concentration estimates were generated by the dispersion model for the 1,520 receptor locations for a total of 192 h (8 days × 24 h). Using these concentration values and the kriging method, 192 surfaces of particulate matter concentrations were then created. These surfaces were then used to assess individual exposure for selected participants in the activity-travel diary survey. Maps generated by the air dispersion model indicate that certain hot spots with elevated levels of particulate matter exist across Franklin County (Fig. 16.3). These areas of high pollution concentrations are located mainly in the southeast of downtown Columbus and around the intersection of two major

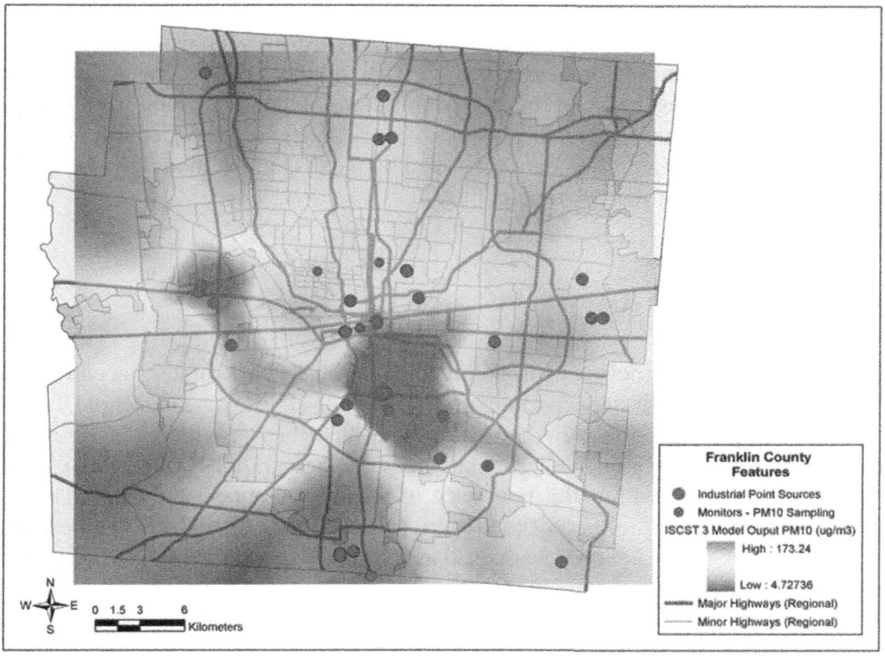

Fig. 16.3 Eight-day pollution estimates indicating areas of high PM_{10} Concentrations ($\mu g/m^3$) in the study area

highways on the west of Franklin County. This suggests that different individuals may experience different levels of exposure due to the locations of their activities and trips on the selected survey days.

16.4.2 Background Concentrations

Besides pollution estimates generated with data on the 119 industrial point sources, background concentrations from other sources were obtained from the three local monitoring sites and added onto the results of the estimated concentrations from the air dispersion model. These other sources of particulate matter include industrial processes, transportation, fuel combustion, indoor sources, environmental tobacco smoke, wood burning, cooking, and solid waste. The contribution from each source varies by region and density (i.e., road networks, industrial plants, etc.).

The process of incorporating background concentrations consisted of three main steps. First, hourly concentration readings from the three local monitors were obtained. Second, particulate concentrations generated by the dispersion model at each monitor site for every hour were then subtracted from the monitored value. The result represents the background concentration at a particular monitor site from all contributing sources except major industrial sources. A background concentration value was then assigned to each of the 1,520 receptor locations through extrapolating the background concentrations of the three monitor sites using a nearest neighbor function. Finally, values of these background concentrations were then added to the air pollution data generated by the dispersion model for the 1,520 receptor locations for the 192 time points.

16.4.3 Exposure Assessment

To assess individual exposure to air pollution based on the particulate concentrations generated by the above steps, participants of the Mid-Ohio Area Household Travel Survey whose assigned survey days (days on which they provided activity-travel data) matched the days of the particulate matter concentrations, whose trips were undertaken entirely within the boundary of Franklin County, and whose travel locations were successfully geocoded were selected from the dataset. There were 292 such participants and they were used as the subsample for assessing individual exposure to particulate matters in this study.

Two exposure measures were evaluated in this study: static exposure and dynamic exposure. Static exposure assumes that an individual's exposure is determined only by the level of pollution concentrations at the residential location and thus assesses individual exposure based only on the home location. This value was obtained by a simple summation of particulate concentrations at the home location of each selected participant for the 192 time points (24 h for the 8 survey days). Dynamic exposure, on the other hand, takes into account the effects of people's travel on their exposure. To obtain this value, people's location at different times of a day was first identified and their exposure at a particular time point was then determined based on where they were at that time. Finally all exposure values thus identified for the 192 time points were summed up to obtain the person's dynamic exposure.

Since particulate concentrations in the study area were estimated on an hourly basis, locations of the 292 selected participants at the beginning of each hour for the entire 24-h period of the eight survey days were first identified. If an individual was undertaking an activity at a stationary location (i.e., home, workplace, etc.) at the start of the hour, then the concentration value associated with that location was applied to his or her 24-h summation of particulate matter exposure. On the other hand, an individual may be traveling at the start of a given hour. For this kind of cases, location of the individual along the route at the beginning of a given hour was estimated using information on the origin and destination locations, the departure and arrival times of each trip, as well as the travel speeds of the road segments traversed. Because the activity-travel diary dataset does not provide information about the actual routes taken, this estimation was based on the assumption that the shortest path between an original and destination location was used for all the trips by the participants.

16.5 Results

To examine whether there are significant differences between static and dynamic exposures for the subsample as a whole and for various subgroups, the 292 individuals in the subsample were grouped into different gender, income, education, and age groups. First, participants' age was categorized into three groups: young (19–40), middle (41–61), and old (62–83) age groups. These age categories were derived based on the range and distribution of the youngest and oldest individuals in the subsample. The nine income groups in the original dataset were re-classified into three categories: low, middle, and high income groups. Lastly, the six categories of education in the original dataset were reclassified into three groups: less than high school/high school, vocational/some college, and undergraduate/graduate.

Using these groupings, several paired t-tests were performed to examine whether there are significant differences between static and dynamic exposures to particulate matter. Paired t-tests were used because two different exposure measures for the same individual were compared and they were thus not independent of each other. As shown in Table 16.1, the results indicate that, for the entire subsample of 292 individuals, static exposures (231.05 $\mu g/m^3$) were basically the same as dynamic exposures (229.89 $\mu g/m^3$). The small difference of 1.16 $\mu g/m^3$ was not statistically significantly ($p = 0.57$).

As Table 16.1 also shows, there are no significant differences between static and dynamic exposures for all gender, income, education, and age subgroups in the subsample. These differences range from 0.59 $\mu g/m^3$ (high school or below) to 9.94 μg/m3 (low income). These results suggest that taking into account of people's out-of-home activities and trips did not seem to make a difference in the level of exposure they experienced – as also indicated by the high Pearson correlation coefficients (from 0.82 to 0.96) between the static and dynamic exposures for

Table 16.1 Static and dynamic exposures to air pollution

	Static exposure ($\mu g/m^3$)	Dynamic exposure ($\mu g/m^3$)	Mean difference	Paired t-test p-value	Pearson correlation
All individuals (292)	231.05	229.89	1.16	0.57	0.88
Men	231.05	229.30	1.75	0.55	0.87
Women	231.05	230.49	0.56	0.84	0.89
Low income	237.81	247.74	9.94	0.21	0.86
Middle income	242.21	239.37	3.84	0.21	0.89
High income	223.34	221.55	1.79	0.56	0.88
High school or below	232.34	232.93	0.59	0.84	0.96
Vocational/some college	223.64	224.56	0.92	0.76	0.93
Undergraduate/graduate	235.75	233.08	2.67	0.44	0.83
Young	237.56	237.71	0.15	0.97	0.89
Middle age	226.26	223.41	2.85	0.37	0.82
Old	235.18	234.40	0.78	0.84	0.94

Table 16.2 ANOVA test of differences in dynamic exposure within subgroups

	df	SS	MS	F	Significance
Gender					
Regression	1	103.59	103.59	0.02	0.88
Residual	290	1399,392.73	4,825.49		
Total	291	1399,496.32			
Income					
Regression	2	28,856.61	14,428.30	2.95	0.05
Residual	267	1304,492.48	4,885.74		
Total	269	1333,349.09			
Education					
Regression	2	3,834.64	1,917.32	0.39	0.68
Residual	282	1376,755.56	4,882.11		
Total	284	1380,590.20			
Age					
Regression	2	14,630.09	7,315.05	1.50	0.22
Residual	284	1383,769.88	4,872.43		
Total	286	1398,399.97			

each subgroup in the subsample in Table 16.1. Among all of the subgroups, only the dynamic exposures for different income subgroups show a pattern: dynamic exposures increase as one move from high- to middle-income group and from middle- to low-income group. This suggests that ignoring the location of low-income individuals' daily activities and trips may lead to an underestimation of their actual exposures to air pollution. This in turn has significant implications for any examination of environmental justice.

To explore the participants' dynamic exposures to air pollution, differences within various gender, income, education, and age subgroups of individuals in the subsample were examined using analysis of variance (ANOVA). As Table 16.2 shows, results of the ANOVA tests indicate that there are no significant differences in exposures to air pollution within different gender, education, and age subgroups in the subsample. The two-tail tests for these three categories yield significance levels that are greater than 0.05. Exposures, however, are different among the income subgroups. Income was thus the only factor identified in this analysis that had a significant effect on the participants' exposure to air pollution. But since income is closely related to education, it is not clear why education was not a significant factor influencing individual exposures. In sum, analysis of within-group differences for the gender, income, education, and age subgroups did not yield a clear or consistent pattern of statistically significant differences in dynamic exposures.

Finally, we used multiple linear regression analysis to examine the effects of gender, income, education, and age (as independent variables) on individual dynamic exposures (as the dependent variable). Due to the categorical nature of the independent variables, dummy variables were used to model the effect of different categories in each subgroup (e.g., age was modeled by three variables: young, middle, and old age). A regression coefficient of 0.18 and an R^2 value of 0.03 were obtained for the regression model, suggesting that the independent variables only accounts for 3 % of the dynamic exposures of the individuals in the subsample. Among the independent variables, only income has a significant but weak influence on individual exposure to air pollution ($p = 0.03$).

16.6 Discussion and Conclusion

This study explored a method for integrating dynamic pollution concentrations and detailed space-time activity patterns of individuals in air pollution assessment. It implemented two notions of exposure to air pollution: static exposure evaluated at an individual's residential location, and dynamic exposure assessed based on various locations that individual travels to throughout the day. Results from the analysis do not support the expectation that individual exposures as evaluated by the static and dynamic measures differ significantly. In addition, gender, education, and age do not seem to have any influence on people's dynamic exposure to air pollution. In sum, the analysis did not yield a clear or consistent pattern of statistically significant differences between static and dynamic exposures, as well as among various population subgroups using the dynamic measure. In other words, results from this study do not seem to be supportive of the argument that taking people's daily movement into account in exposure assessment is crucial for the accuracy of such assessment.

These results, however, need to be interpreted in the specific context of the study area and limitations in the data available for modeling the geographic patterns of air pollution. First, instead of suggesting that people's daily movement does not affect their exposure to air pollution in significant ways, the results are likely caused by the specific geography of urban opportunities in the study area and the activity-travel patterns of the survey participants. As shown in Fig. 16.3, the highest levels of pollution concentrations are found mainly in the southeast of downtown Columbus and in the west at the intersection of two major highways. But the highest concentrations of urban opportunities (e.g., shopping and recreational facilities) are located largely in northern suburbs where pollution concentrations were low. Thus even as survey participants undertook activities and traveled outside of their home, their activities locations would tend not to coincide geographically with areas of highest pollution concentrations.

Further, the Gaussian formulation of the dispersion model used in the study is such that pollution concentrations are subject to a strong distance decay effect, where the highest levels of particulate matter concentrations are located close to the sources but quickly decline as distance increases from these pollution sources. Thus, unless an individual's activities or trips were located close to these pollution sources, the levels of air pollution they experienced might not be high enough for revealing significant differences between their static and dynamic exposures. Because of these factors, results of the study do not necessarily mean that people's daily movement does not affect their exposure to air pollution in general.

Also important in leading to the results is the lack of an adequate geographic coverage of the mobile and minor pollution sources of particulate matter in the study area. These sources include vehicular traffic, restaurants, auto body shops, gas stations, dry cleaners and home heating systems. Generally speaking, the only way to account for pollution generated by these sources is to rely on local monitors that collect samples at varying temporal resolutions. Ideally, a dense monitoring network that represents urban, suburban and rural areas within the study area would capture the impact of these sources on the quality of surrounding air. This in turn would allow people's dynamic exposure to be assessment more accurately. There are, however, only three local monitors in the study area which are all located in the north of and within 6 km from downtown Columbus. This extreme lack of monitored data made it difficult to accurately represent particulate matter emissions from the mobile and minor pollution sources in the study area. Further, location of these monitors also means that the pollution sampling was only representative of areas fairly close to the city center. As a result, although emission data from the three local monitors were incorporated into the pollution estimates in the study, they did not seem to have effectively captured people's dynamic exposures (in light of the fact that many trips were made to the urban opportunities located in the northern suburbs which have no local pollution monitors). Thus to more accurately assessed people's dynamic exposure to air pollution, many more monitoring sites would be necessary for this type of analysis in the future.

While the results are less promising than what were expected, this study still contributes to air pollution assessment methods in important ways. It proposed the notion of dynamic exposure, and developed and implemented a GIS-based method for evaluating people's exposure to air pollution that takes into account of the location of their out-of-home activities and trips. The method will be especially relevant for future studies on environmental health because high-resolution space-time data on human movement are becoming increasingly available through the use of new survey methodologies and advance geospatial technologies (e.g., accurate route data can now be collected using GPS). To be able to assess people's dynamic exposures more accurately, however, not only a dense monitoring network is needed to capture particulate concentrations, local monitors should also provide pollution data at higher temporal resolution (e.g., for every hour or 30-min intervals).

References

Bartonova, A., Clench-Aas, J., Gram, F., Gronskei, K. E., Guerreiro, C., & Larssen, S. (1999). Air pollution exposure monitoring and estimation. Part V. Traffic exposure in adults. *Journal of Environmental Monitoring, 1*, 337–340.

Bell, M. L. (2006). The use of ambient air quality modeling to estimate individual and population exposure for human health research: A case study of ozone in the Northern Georgia Region of the United States. *Environmental International, 32*, 586–593.

Bellander, T., Berglind, N., Gustavsson, P., Jonson, T., Nyberg, F., & Pershagen, G. (2001). Using geographical information systems to assess individual historical exposure to air pollution from traffic and house heating in Stockholm. *Environmental Health Perspectives, 109*, 633–639.

Briggs, D. (2005). The role of GIS: Coping with space (and time) in air pollution exposure assessment. *Journal of Toxicology and Environmental Health, Part A, 68*, 1243–1261.

Briggs, D. J., Collins, S., Elliott, P., Fischer, P., Kingham, S., & Lebret, E. (1997). Mapping urban air pollution using a GIS: A regression-based approach. *International Journal of Geographical Information Science, 11*, 699–718.

Brunekreef, B., & Holgate, S. T. (2002). Air pollution and health. *Lancet, 360*, 1233–1242.

Clench-Aas, J., Bartonova, A., Bohler, T., Gronskei, K. E., Sivertsen, B., & Larssen, S. (1999a). Air pollution exposure monitoring and estimating. Part I. Integrated air quality monitoring system. *Journal of Environmental Monitoring, 1*, 313–319.

Clench-Aas, J., Bartonova, A., Gronskei, K. E., & Walker, S. E. (1999b). Air pollution exposure monitoring and estimation. Part IV. Urban exposure in children. *Journal of Environmental Monitoring, 1*, 333–336.

Duan, N. (1982). Models for human exposure to air pollution. *Environment International, 8*, 305–309.

English, P., Neutra, R., Scalf, R., Sullivan, M., Waller, L., & Zhu, L. (1999). Examining associations between childhood asthma and traffic flow using a geographic information system. *Environmental Health Perspectives, 107*, 761–767.

Farhang, A. C. (1983). *The effects of wind on air quality in Hamilton, Ontario Report AQRB-83-001-M Atmospheric Environment Service*, Environment Canada, 4905 Dufferin Street, Downsview, Toronto.

Finkelstein, M. M., Jerrett, M., DeLuca, P., Finkelstein, N., Verma, D. K., & Chapman, K. (2003). Relation between income, air pollution and mortality: A cohort study. *Journal of Canadian Medical Association, 169*, 397–402.

Gertler, A. W. (2003). Diesel vs. gasoline emissions: Does PM from diesel or gasoline vehicles dominate in the U.S. In R. Joumard (Ed.), *Proceedings of the 12th international symposium: Transport and air pollution* (pp. 185–192). Arcueil: INRETS.

Gualtieri, G., & Tartaglia, M. (1998). Predicting urban traffic air pollution: A GIS framework. *Transportation Research Part D, 3*, 329–336.

Gulliver, J., & Briggs, D. J. (2005). Time-space modeling of journey-time exposure to traffic-related air pollution using GIS. *Environmental Research, 97*, 10–25.

Hoek, G., Beelen, R., de Hoogh, K., Vienneau, D., Gulliver, J., Fischer, P., et al. (2008). A review of land-use regression models to assess spatial variation of outdoor air pollution. *Atmospheric Environment, 42*, 7561–7578.

Janssen, N. A. H., van Vliet, P. H. N., Aarts, F., Harssema, H., & Brunekreef, B. (2001). Assessment of exposure to traffic related air pollution of children attending schools near motorways. *Atmospheric Environment, 35*, 3875–3884.

Jerrett, M., Burnett, R. T., Kanaroglou, S., Eyles, J., Brook, J. R., & Giovis, C. (2001). A GIS-environmental justice analysis of particulate air pollution in Hamilton, Canada. *Environment and Planning, A, 33*, 955–973.

Jerrett, M., Arain, A., Kanaroglou, P., Beckerman, B., Potoglou, D., & Sahsuvaroglu, T. (2005). A review and evaluation of intraurban air pollution exposure models. *Journal of Exposure Analysis and Environmental Epidemiology, 15*, 185–204.

Kanaroglou, P. S., Buliung, R. N., & Taylor, B. W. (2000). *Mobile source emissions in urban areas: The relative contribution of trucks*. Paper presented at the fifth international conference on protection and restoration of the environment, Thassos Island, Greece.

Kwan, M.-P. (2009). From place-based to people-based exposure measures. *Social Science and Medicine, 69*, 1311–1313.

Kwan, M.-P. (2012a). How GIS can help address the uncertain geographic context problem in social science research. *Annals of GIS, 18*(4), 245–255.

Kwan, M.-P. (2012b). The uncertain geographic context problem. *Annals of the Association of American Geographers, 102*(5), 958–968.

Kwan, M.-P. (2013). Beyond space (as we knew it): toward temporally integrated geographies of segregation, health, and accessibility. *Annals of the Association of American Geographers, 103*(5), 1078–1086.

Pengelly, L. D., Kerigan, A. T., Goldsman, C. H., & Inman, E. M. (1984). The Hamilton study: Distribution of factors confounding the relationships between air quality and respiratory health. *Journal of the Air Pollution Control Association, 34*, 1039–1043.

Richardson, D. B., Volkow, N. D., Kwan, M.-P., Kaplan, R. M., Goodchild, M. F., & Croyle, R. T. (2013). Spatial turn in health research. *Science, 339*(6126), 1390–1392.

Slørdal, L. H. (1998). *Calculation of exposure to NO2, PM10, and PM2.5 for Oslo, Drammen, Bergen and Trondheim* (NILU OR 38/98). Kjeller: Norwegian Institute for Air Research (in Norwegian).

Venn, A. J., Lewis, S. A., Cooper, M., Hubbard, R., & Britton, J. (2001). Living near a main road and the risk of wheezing illness in children. *American Journal of Respiratory and Critical Medicine, 164*, 2177–2182.

Chapter 17
Spatiotemporal Change Pattern Mining: A Multi-disciplinary Perspective

Xun Zhou, Shashi Shekhar, and Pradeep Mohan

17.1 Introduction

Given a definition of change and a dataset about a spatiotemporal (ST) phenomenon, ST change pattern mining is the process of identifying the location and/or time frame of shifts in phenomenon. Detecting patterns of change over space and/or time is an increasingly important activity in application domains ranging from climate science to public health. Researchers have developed numerous techniques to facilitate the mining of such patterns. Addressing domain specific challenges, they have often worked in distinct research settings, most notably time series analysis, remote sensing, and spatial statistics. Although they tend to target different aspects of the change pattern mining problem, there is much researchers could learn from one another to advance their respective areas.

Related literature includes discipline-specific surveys. For example, literature on time series analysis (Basseville and Nikiforov 1993; Shaban 1980; Zacks 1983) focused on methods for the change point detection (a.k.a., "abrupt change" detection) problem. In remote sensing, some survey papers (Radke et al. 2005; Coppin et al. 2002; Singh 1989; Im and Jensen 2005) mainly focus on change detection techniques using bitemporal or multi-temporal image sequences. A tutorial by Wong and Neill (2009) discusses the problem of "event detection" which aims to discover abnormal behaviors of data. The authors cover time series change detection and

X. Zhou (✉) • S. Shekhar • P. Mohan
Department of Computer Science and Engineering, University of Minnesota, Twin Cities, Minneapolis, MN, USA
e-mail: xun@cs.umn.edu; shekhar@cs.umn.edu; mohan@cs.umn.edu

M.-P. Kwan et al. (eds.), *Space-Time Integration in Geography and GIScience*, DOI 10.1007/978-94-017-9205-9_17

spatial cluster detection but not spatial wombling (Barbujani et al. 1989), which is another important change pattern in spatial statistics. Comprehensive reviews that compare techniques across disciplines are lacking.

The use of discipline-specific survey to understand change pattern mining techniques is challenging because different research areas employ similar terminology to describe different phenomena. For example, the term abrupt change in time series analysis refers to a change in the statistical distribution of data at a certain time point (Basseville and Nikiforov 1993) (i.e., temporal non-stationarity). However, in spatial wombling, it describes a significant difference of value across boundaries separating different spatial areas (Barbujani et al. 1989; Fagan et al. 2003). By contrast, the patterns gradual change (Chapin et al. 2004; Verbesselt et al. 2010) and interval of abrupt change (Zhou et al. 2011) in their respective disciplines share a similar footprint in time, which is an interval.

In this chapter, we define an interdisciplinary framework for change pattern mining. Specifically, we make the following contributions. We propose a taxonomy classifying the output patterns of different techniques according to their change footprint, i.e., points, intervals, regions, etc. We compare concepts across disciplines to identify synonyms and homonyms from a change pattern footprint perspective. We show opportunities for cross-fertilization of concepts, theories, and techniques across disciplines, such as the scan statistics (Kulldorff 1997) for public health being applied in climate change research, etc.

Figure 17.1 shows the four main components of a pattern mining process: data types describing the ST phenomenon, a definition of a change pattern, an approach for mining the pattern, and the change pattern discovered (ST footprint). The rest of the chapter is organized according to this framework: The next section surveys different spatiotemporal data types and statistical data models for change pattern mining. The section that follows presents definitions of change patterns from calculus and statistic perspectives. Then a description of the change pattern footprint taxonomy is provided, follows by a summary of the major techniques used by each discipline. Finally, we conclude the chapter with future research plans.

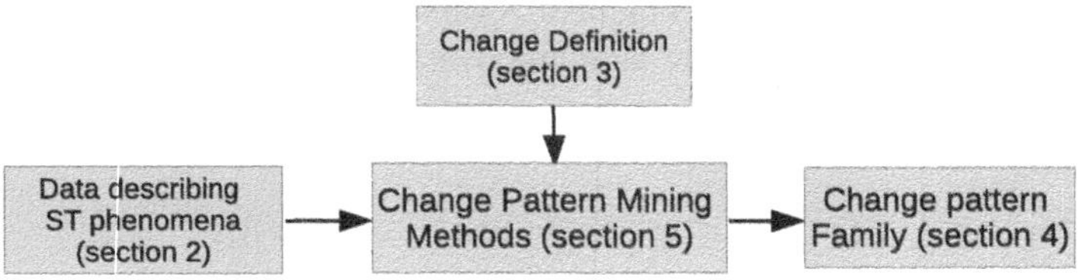

Fig. 17.1 A flow chart showing ST change pattern mining process

17.2 Data Describing ST Phenomena

In this section we identify the various data types for describing ST change phenomena that can be found in literature and statistical data models. Table 17.1 shows a summary of input data types for change pattern mining.

17.2.1 Temporal Data

Figure 17.2a shows a time series from climate science. It is a plot of the Sahel rainfall index (JISAO 2011), a normalized rainfall from June to October over the Sahel region in Africa, over the past 111 years. Change pattern analysis on such datasets is of interest to climate change researchers.

17.2.1.1 Event Reports

Event reports are collections of time-stamped data instances. Unlike time series, event report data may not be uniformly distributed in time, and usually have spatial dimensions such as the locations of the events. Disease reports in public health and crime reports in public safety are typical event report data. (Event reports are discussed further in Sect. 17.2.3)

17.2.2 Spatial Data

17.2.2.1 Point Data

Point data on two dimensional space usually represents the locations of a collection of events. Each point is associated with a pair of coordinates. In a public health

Table 17.1 Input data types for change pattern mining and corresponding research area

	Data type	Spatial statistics	Time series analysis	Image processing
Temporal	Time series		x	
Spatial	Point data	x		
	Polygon data	x		
	Lines and paths	x		
	Spatial field	x		
Spatio-temporal	ST event reports	x		
	Image sequences			x
	ST raster data	x	x	

dataset, points may represent instances of disease. In a public safety dataset, points may represent crime incidents. Figure 17.2b shows an example of spatial point data in crime analysis. This is a map of drunk driving locations in Nebraska where the two coordinates represent the longitude and latitude of the incidents (Mohan et al. 2011).

17.2.2.2 Lines and Paths

Spatial paths are sequences of values that are taken at locations along a curve in a spatial framework. For example, in Fig. 17.3, the data curve on the right shows the vegetation cover observed in Africa (Tucker et al. 2004) measured in Normalized Difference Vegetation Index (NDVI) (shown on left) along 18.5 E degree longitude. Value change pattern along the path may reflect environmental change in different latitudes.

17.2.2.3 Polygons

Polygon data on two dimensional space usually represent two dimensional spatial objects such as lakes, or spatial regions in space such as political jurisdictions. Some

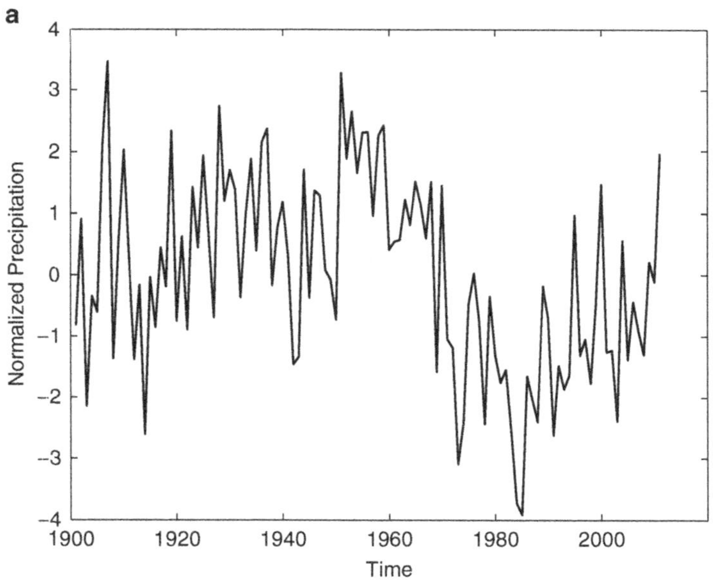

Fig. 17.2 Examples of temporal and spatial input data (**a**) A time series (**b**) Spatial point data (**c**) Spatial polygon data (Adapted from Lu and Carlin (2005) with permission)

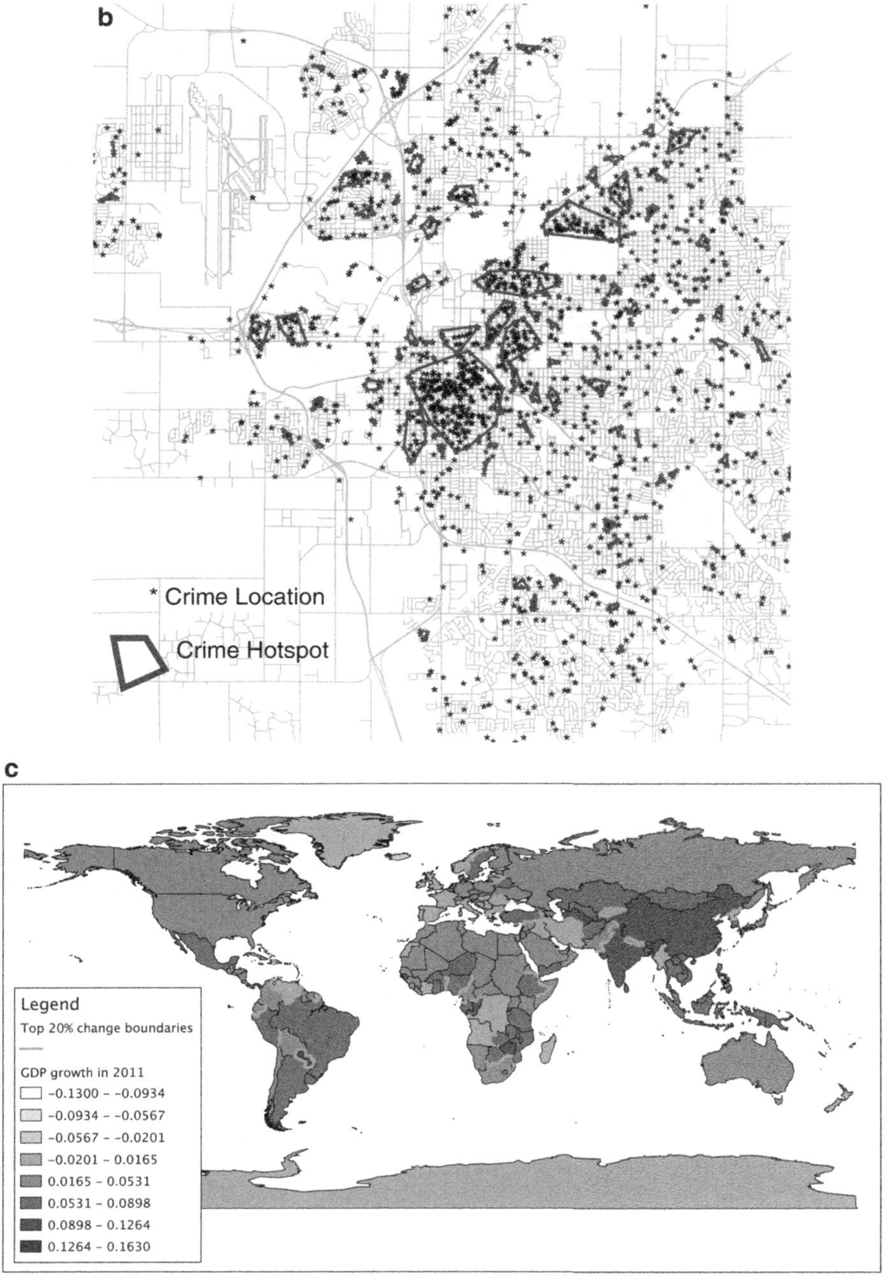

Fig. 17.2 (continued)

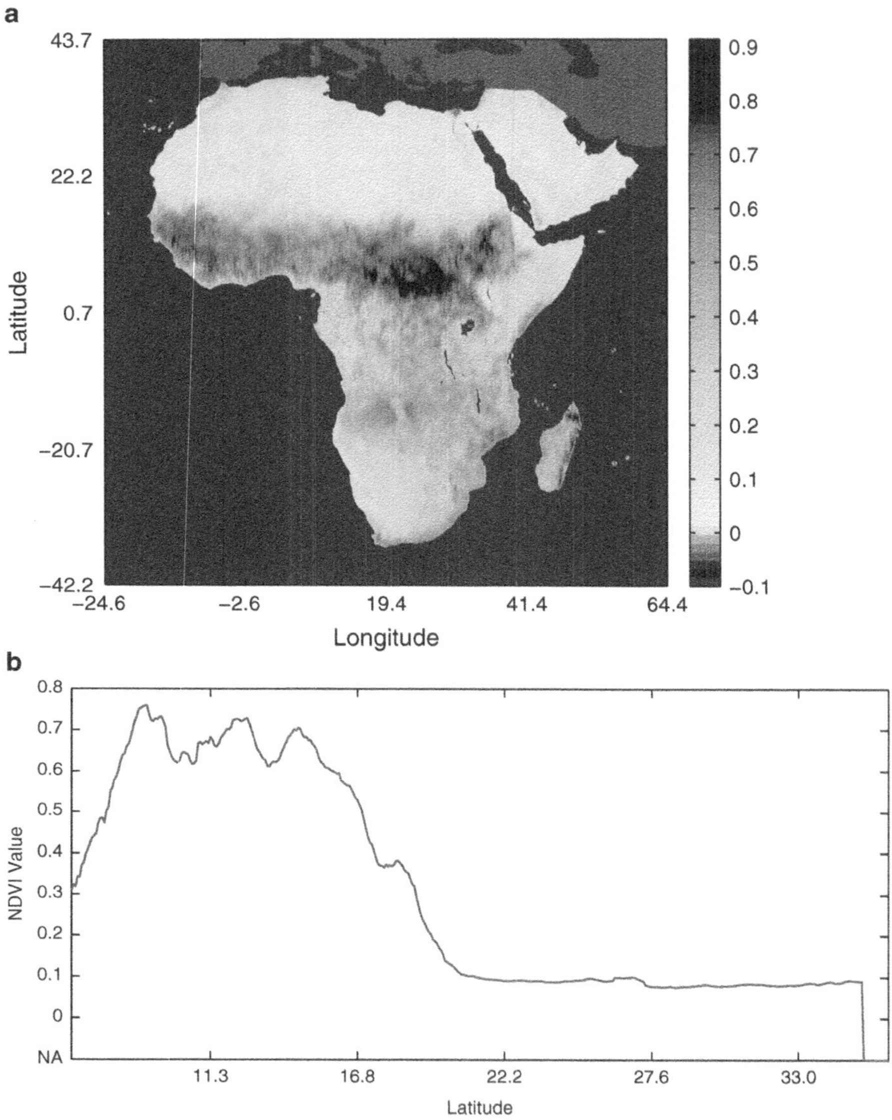

Fig. 17.3 Spatial path data (**a**) Vegetation cover in Africa (**b**) Value observed along a longitudinal path

non-spatial features such as population in each region may be associated with each of the polygons. For example, Fig. 17.2c shows an example of global GDP growth data where each polygon represents a country and the color code represents the GDP growth of each country in 2011 (CIA 2013).

17.2.2.4 Spatial Raster Data

Spatial framework is a finite grid imposed on the underlying space (Worboys and Duckham 2004), such as the latitude-longitude system, etc. A finite set of n computable functions or simple fields $\{f_i,\ 1 \le i \le n\}$, where f_i: Spatial framework → Attribute domain (A_i), map the spatial framework F onto different attribute domains A_i (Shekhar and Chawla 2003). For example, Fig. 17.4a shows the global precipitation distribution represented as spatial raster dataset. The dataset describes the reanalysis results of rainfall amount all over the world in Jan 1, 1948, where

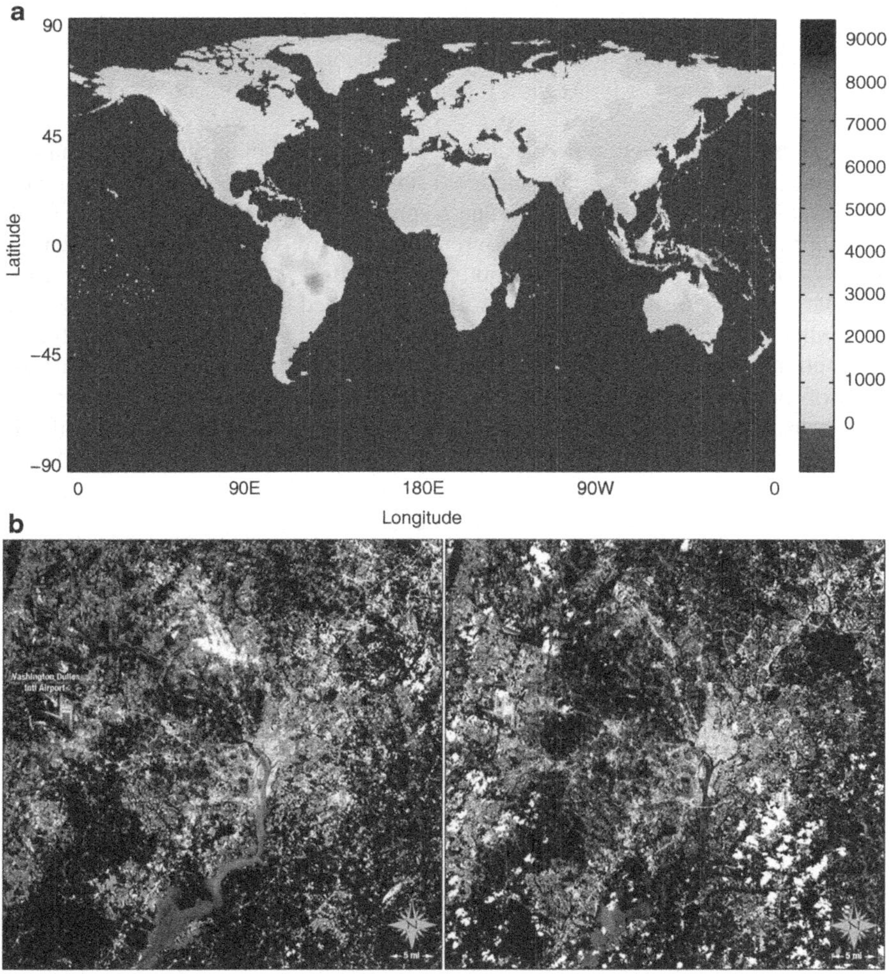

Fig. 17.4 More examples of spatial data (**a**) Image sequence data (Landsat imagery courtesy of NASA Goddard Space Flight Center and U.S. Geological Survey) (**b**) A spatial raster dataset from climate science

each pixel represents an area of 0.5 by 0.5 degrees (Jones and Harris 2008). This data type is commonly seen in climate and earth science. Finding spatial areas with value changes may help identify patterns of environmental change.

17.2.3 Spatiotemporal Data

17.2.3.1 ST Event Reports

Discrete events may be represented as points in space and time. Each event has a time stamp, a pair of coordinates, and possibly other information. For example, in public health report data, each point rep- resents one disease case, along with its location, time, and other information such as patients information. In public safety data, each crime report is also a ST point with information such as crime type, loss, etc.

Furthermore, ST event reports can also be represented in an aggregated manner given a spatial zoning (e.g., jurisdictions) schema. The number of reports in each zone can be totaled so that each zone (represented as a polygon in space) will have a series of report totals at different time. These datasets may be used for the detection of disease or crime outbreak patterns. Spatiotemporal raster data: Similar to spatial raster data, STraster data are gridded space-time framework with a number of functions defined over it. It is usually used to represent the distribution of continuous variables over space and time. For example, the daily precipitation data of the world shown in Fig. 17.4a is available for a period of 106 years. The entire dataset thus can be represented as ST raster dataset where each spatial grid represent a fixed area (e.g., 0.5 by 0.5 degrees in latitude and longitude), and each grid in time represent a day. Finding change pattern on such data (e.g., decline in rainfall) can be used to confirm theories of climate changes.

17.2.3.2 Image Sequences

An image sequence contain images or snapshots of a spatial region at two time instances (e.g., satellite image). Such datasets are commonly used in the area of remote sensing to identify the changes that occur between the two time points. Figure 17.4b shows an example of image sequences where the two snapshots of Washington D.C. were taken in 1984 and 2011 respectively (NASA 2011).

17.2.4 Statistical Models of Data

17.2.4.1 Temporal Statistical Model

Time series are usually modeled as i.i.d. samples drawn from a statistical distribution. However, it is always the case that time-referenced attributes are auto-correlated. More complicated models of time series have been applied to model temporal data, such as Markov chain where the probability distribution of a value at time t depends only on the value at t-1 (Gilks et al. 1995).

17.2.4.2 Spatial Data Models

Three statistical models are used to model spatial data (Banerjee et al. 2003). The geostatistical model deals with continuous spatial surface with discrete sampled locations. Tools such as Kriging are used to interpolate unsampled values. The lattice model (a.k.a. areal model) deals with continuous space partitioned into regular grids or irregular spatial regions. Interactions between partitions are characterized by the spatial neighborhood relationship (e.g., topological connectivity). Data models such as the spatial autoregressive regression model (SAR) and Markov random field (MRF) can be applied on such datasets. Finally, point process is used to model discrete spatial point data, such as locations of event reports. Typically, point patterns include completely random, clustered or de-clustered.

Traditional statistical analysis assume that data samples are independently and identically-distributed. This assumption does not hold for spatial data which exhibits special features due to its unique spatial structure. However, unique spatial structure exist in spatial data. First of all, spatial data is highly self-correlated. A fundamental observation is that "Everything is related to everything, but nearby things are more related than distant things", a.k.a. the first law of geography (Tobler 1970). This feature of spatial data is called spatial autocorrelation. For example, precipitation observed at nearby ground stations should be similar. A second feature of spatial data is the underlying process differ from place to place. For example, in a spatial point process, data may be clustered in certain regions instead of uniformly distributed over space.

17.3 Definitions of Change

Intuitively, we find it helpful to classify definitions of "change" in ST data as either calculus-based or statistics-based. In the calculus-based definitions, data value in neighboring time instances or spatial locations are compared. By contrast, in statistics, data is often viewed as samples from the underlying distribution. A parameter of data distribution (e.g., mean) is employed to measure the difference between subsets of the data. From these two ways of defining change, we can derive the change patterns and mining techniques.

17.3.1 Calculus-Based Definitions

The concept of "change" was originally formalized in calculus. When traversing along a one dimensional function $y = f(x)$ where y is a dependent output on an independent input x, the change of y near a given locations x_0 is computed as the difference between the function values at x_0 and at its very near location $x_0 + \Delta x$: $\Delta Y = f(x_0 + \Delta x) - f(x_0)$, where Δx is the distance between the two locations.

When computing the change rate of y with respect to x, differentiation (Wikipedia 2012-D) is often used. It is the change in y against the change in x: $m = \Delta y/\Delta x$. This rate of change is defined as the derivative of y with respect to x. For a non-vertical line, the derivative is also the slope. Slope is a measure of the steepness of the line and is constant between any pair of locations. For an arbitrary curve, the derivative may be different from place to place. The derivative at any location is then computed as the limit when choosing an extremely small Δx : $\lim_{\Delta x \to 0} \frac{\Delta y}{\Delta x}$. Figure 17.6 shows examples of calculus-based change definitions of change. Figure 17.5a shows three functions f_1, f_2 and f_3, where f_1 and f_2 are linear function and f_3 is a polynomial function. Their derivative functions are shown in Fig. 17.5b where f_1', and f_2' are constant and f_3' is a decreasing line.

Based on the above definition, the sign of the derivative indicates the change trend (i.e., positive for increase and vice versa). The value of derivative indicates the rate of change. Figure 17.5c shows a function with a change of value occurred between 4 and 6. The derivative function (shown in Fig. 17.5d) is negative in this part, indicating the change. For two or more dimensional data, change between adjacent locations can occur along any direction. In calculus, a vector representation is used to model a change of value in space. Gradient and divergence are the measurements of such changes. Change pattern that are defined based on calculus include spatial boundary analysis (wombling), where a change is defined as locations with high gradient value and a collection of such locations forms the boundary area (Barbujani et al. 1989; Fitzpatrick et al. 2010). The problem of change interval discovery (Zhou et al. 2011) employs a similar idea.

17.3.2 Statistical-Based Definitions

Statistical models assume that the data are samples drawn from an underlying generative process using certain known or unknown value distributions. Instead of looking at value changes in the dataset, statistical approaches often model changes in terms of statistical parameters such as shifts in mean, standard deviation, quantile, percentile, etc.

For example, Fig. 17.6a shows the normalized Sahel precipitation index (July-October precipitation in Sahel region, Africa) from 1900 to 2011 (JISAO 2011).

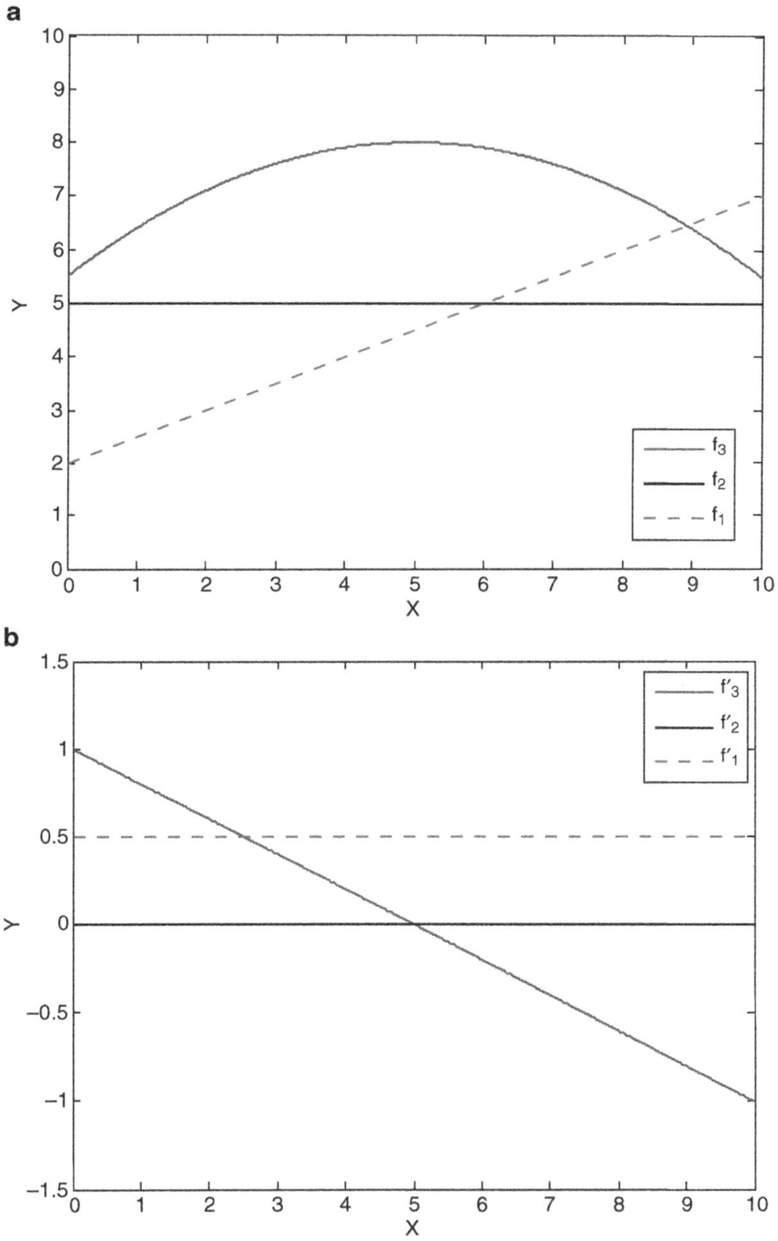

Fig. 17.5 An example of 1-d function and derivatives (**a**) Three functions with a change in value (**b**) The first derivative functions of f1, f2, and f3 (**c**) A function with a change in value (**d**) The first derivative of the function

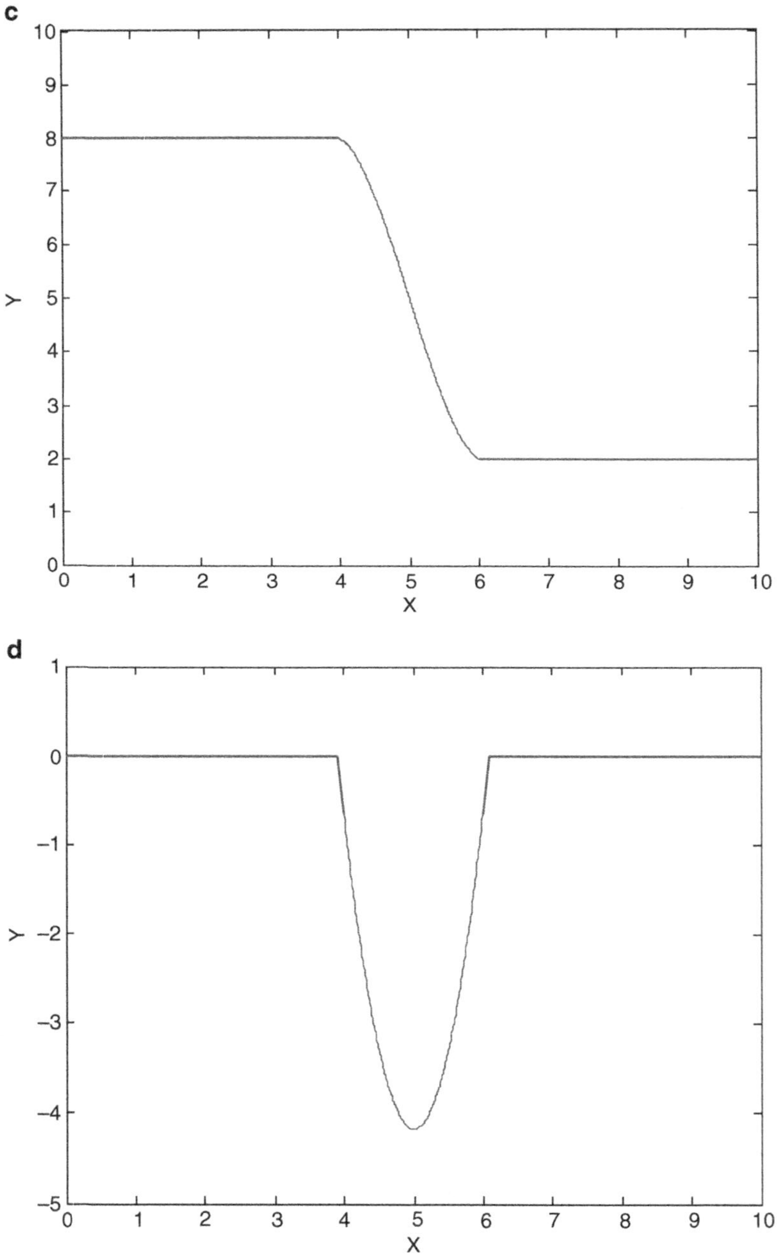

Fig. 17.5 (continued)

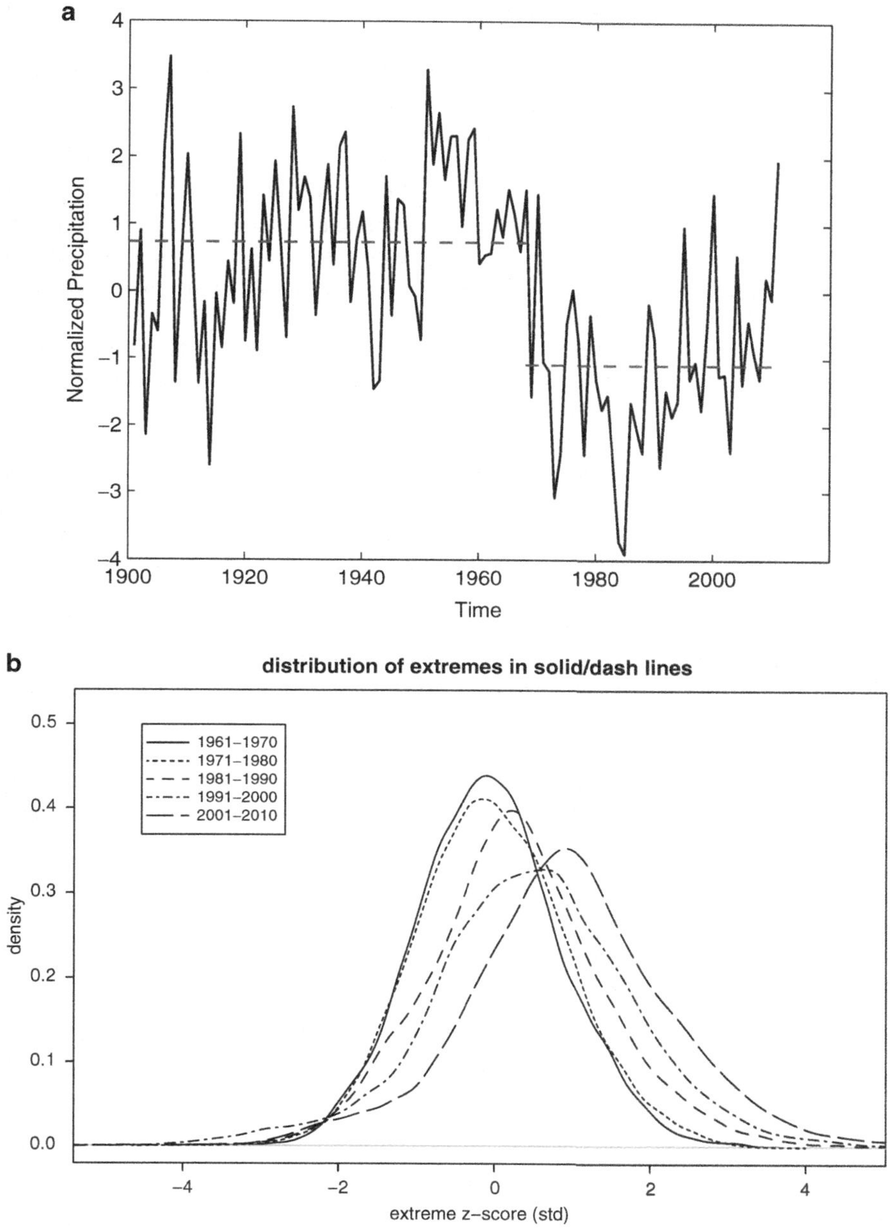

Fig. 17.6 An example of 1-d function and derivatives (**a**) A dataset with a change in mean (**b**) Change in distribution of global temperature anomalies

The data mean shifted to a lower level around the year 1968 (marked as dashed line) which reflects the documented drought of rainfall in this area (Dai et al. 2004) Fig. 17.6b shows an example of distribution change. The curves are the distributions of global land temperature anomalies (with respect to the baseline in 1951–1980) in different decades (Hansen et al. 2012) which are approximately normally distributed. As can be seen, the curves are shifting towards the right where more high temperature are observed. This is considered as evidence of global climate change (Hansen et al. 2012). In spatial applications such as disease mapping, the number of points (reports) in a spatial region is usually assumed to follow a Poisson distribution. A change in the distribution indicates a higher risk of disease (i.e., outbreak) in this area (Kulldorff 1997).

As noted in Sect. 17.2, different applications may make different assumptions on the underlying data distribution. Based on the assumption of the underlying data distribution, statistical models can be classified as either parametric or the non-parametric models. A parametric model may assumes the underlying distribution is known. A non-parametric model may assume that the distribution is known without the parameters being specified, or even that the distribution is not known. For example, the top 5 % of the global temperature anomaly may be considered as extremes. To identify a change in extreme level, a parametric model which assumes that the data follows a normal distribution (e.g., like the curves shown in Fig. 17.6b) would simply compute, where μ and σ are the mean and standard deviation of the normal distribution. However, a non-parametric approach would compute the 5 % percentile of the dataset since the normality assumption does not hold in the data.

Change patterns whose definitions are statistics-based include abrupt change in time series (Basseville and Nikiforov 1993), spatial clusters (e.g., outbreak of disease or crimes) in spatial statistics (Kulldorff 1997; Kulldorff and Nagarwalla 1995), etc.

17.4 Change Pattern Families

In this section, we provide a taxonomy classifying different change patterns based on their spatiotemporal footprints in the embedded ST framework. These ST footprints includes points, intervals/curves, regions, and sub-spaces. A decision tree classifying the output patterns and related techniques is given in Fig. 17.7. These change pattern can also be viewed from space and time perspectives respectively. A temporal change pattern may include time points or intervals. A spatial change pattern may include spatial location, paths and boundaries, and spatial regions. The combinations of these two dimensions creates spatiotemporal output patterns, such as spatiotemporal sub-spaces. Figure 17.8 shows a chart from this perspective.

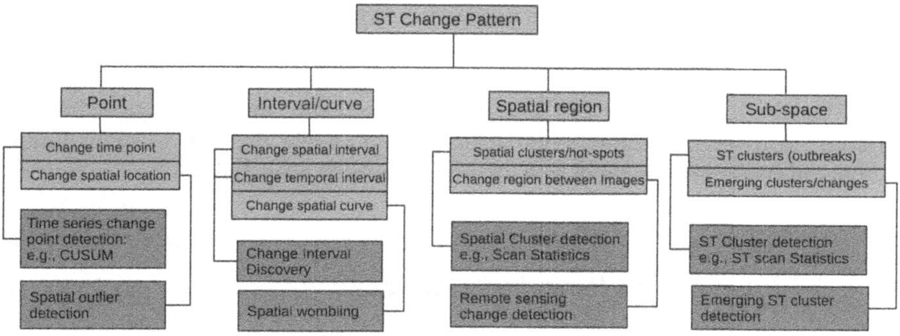

Fig. 17.7 A classification of ST change pattern mining output

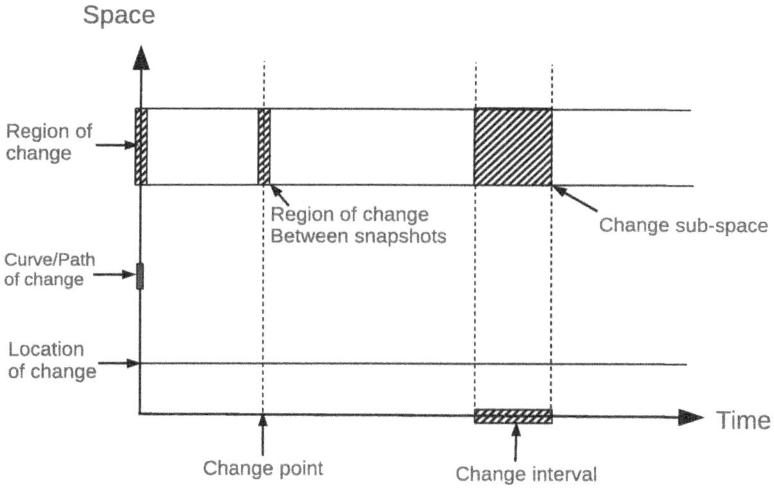

Fig. 17.8 A space-time view of the change pattern families

17.4.1 Point Changes

The first type of change footprint is a "point", that is a single frame in time or a location in space. The point pattern in time series analysis is defined as the time frame before and after which the data distributions are different (Basseville and Nikiforov 1993) (statistical change). This change may be led by the non-stationarity of the underlying process, and is also referred to as "abrupt change". For example, in the Sahel rainfall index time series (previously shown in Fig. 17.6a), the mean of the data changed at the time frame of the year 1968.

A similar pattern can be found in spatial data (e.g., raster data) where the discontinuity occurs at a location. Due to spatial autocorrelation in most geographical attributes, the value observed at a location is expected to be similar to that of its neighboring locations. The spatial locations where attribute value are significantly different from surroundings are called spatial outliers. For example, a huge new "McMansion" (Wikipedia 2012-M) built in an older modest neighborhood is considered a spatial outlier. Finding outliers in spatial datasets is an important activity in application domains that use image analysis or spatial statistic techniques.

17.4.2 Interval and Boundary Changes

A change pattern can also appear as a linear collection of locations/time frames. Two specific patterns of this type are change interval and boundary.

17.4.2.1 Interval Changes

On the temporal dimension of a dataset, an increase of value may occur within an interval of time where a persistent trend can be observed. A large fraction of time frames in this period may experience a significant shift of value. Figure 17.9a, shows intervals of change in the smoothed Sahel rainfall index time series, where the precipitation decreased rapidly between 1966 and 1971.

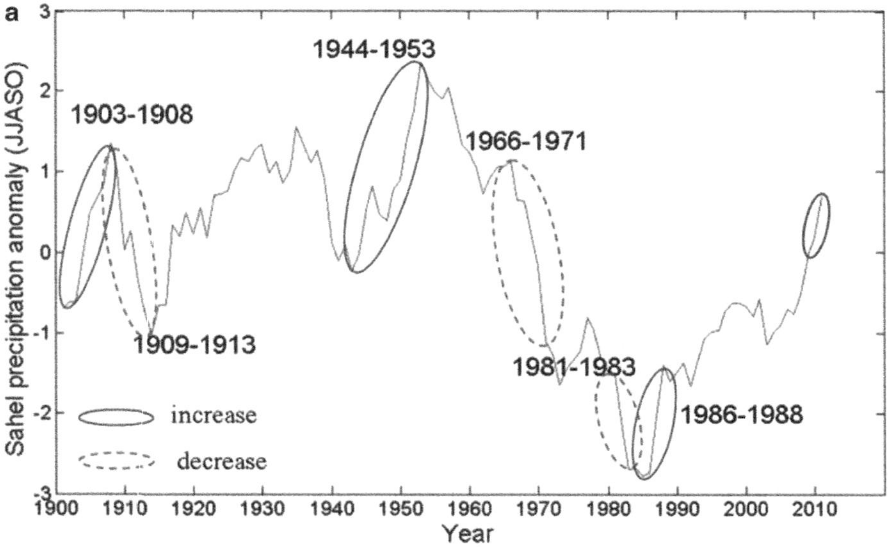

Fig. 17.9 An example of ST change interval pattern (**a**) An example of change interval in a time series (**b**) An example of change interval in a spatial path

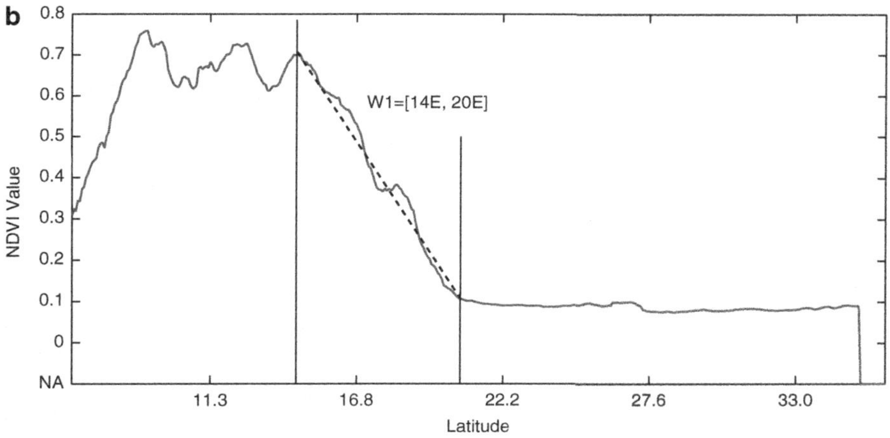

Fig. 17.9 (continued)

Change interval may also be observed along a path taken in a spatial field. For example, on the spatial path shown in Fig. 17.3, the value decreases rapidly from high level (rain forest and savanna) to very low (desert), as shown in Fig. 17.9b. This helps researchers understand the spatial footprint of the transitional zone (a.k.a. ecotone) which may shift in response to climate change. A formal definitions of this pattern has been proposed as an instance of the interesting sub-path/interval discovery problem (Zhou et al. 2011) and a data mining approach to the problem was provided.

Besides interval of change, terms describing this kind of change include gradual change in boundary analysis in ecology (Fitzpatrick et al. 2010) and in time series analysis for climate science (Chapin III 2004).

17.4.2.2 Spatial Boundaries

A number of applications aim to find a curve on two dimensional space that separate different regions. Given a spatial point, polygon or raster dataset, the output pattern consists of multiple curves where value difference on two sides of the curve is significantly higher than what would be expected in a randomly generated dataset. Figure 17.2c shows the top 20 % change boundaries (highlighted) found on the global GDP growth data (CIA 2013). This problem is also referred to as spatial wombling (Womble 1951), where techniques have been proposed to discover boundaries on point, polygon and raster data. Terms such as rapid change (Liang et al. 2009), abrupt change (Lu and Carlin 2005) are also used in the definitions of a similar pattern, whose footprint is also referred to as "Edge" or "Zones of Abrupt Change" (Liang et al. 2009), etc.

17.4.3 Spatial Regions of Change

Spatial regions of change refer to areas in the spatial framework that have significantly different data in comparison to the rest of the data or the expected value in the same area. Specifically, there are two patterns in this category, namely, spatial clusters, and changes between images.

17.4.3.1 Spatial Clusters

Spatial clusters refers to the gathering pattern of point data in space. In real world datasets, such as disease reports and crime reports, data points are not uniformly distributed over space (a.k.a. heterogeneity). The area with dense reports are considered as having a higher risk of disease or crime. Figure 17.2b shows the footprint (highlighted polygons) of spatial cluster in the given point (crime) dataset. In some scenarios, the expected numbers of reports is available from historical records or theoretical models. For example, the number of disease reports is usually assumed to follow a Poisson distribution (Banerjee et al. 2003). Areas with a statistically significant higher number of reports are also considered as spatial clusters, which may indicate an ongoing outbreak of disease or crimes (Kulldorff and Nagarwalla 1995). On areal-aggregated or raster datasets, a cluster may also refer to a number of polygonal areas or a collection of spatial grids where the intensity (e.g., the number of reports) inside is much higher than outside. Such clusters are also referred to as "spatial hot-spots" in spatial statistics (Chainey et al. 2008).

A number of techniques have been proposed to detect spatial clusters and hotspots. Among them, scan statistics (Kulldorff 1997) is a representative method that has been widely used in epidemiology. We discuss scan statistics in more detail in the next section.

17.4.3.2 Change Between Images

In remote sensing, it is of interest to identify spatial regions of change between two or among a series of raster images of the same spatial area. This problem is referred to as "change detection" in remote sensing and image processing (Coppin et al. 2002; Radke et al. 2005). Change detection on satellite images help identify new buildings or suspicious object movements in a region. For example, in the example in Fig. 17.4b, the area in the red box is one of the identified change regions (near Washington Dulles International Airport) where land use has significantly changed during the years between the two snapshots. The change region can be identified as whole, or on a pixel-by- pixel basis.

17.4.4 ST Subspace of Change

A ST subspace of change refers to the footprint of a change of phenomena (e.g., disease outbreak) occurring in both space and time. We further divide it into two types, namely, ST clusters and ST emerging clusters.

17.4.4.1 ST Clusters

The spatial cluster pattern may be generalized to the temporal dimension. In a ST event report dataset, a ST cluster refers to (region, time-interval) tuples where the density of points is higher than the rest of the data, or compared to the expected number. A ST cluster implies that a spatial region has a higher risk of disease or crime during a certain period of time. On an aggregated or raster ST dataset, a ST hotspot pattern can be defined in a similar manner.

17.4.4.2 Emerging ST Clusters

An emerging ST clusters is a special type of ST cluster, where the magnitude of change (e.g., difference from expected value) is monotonically changing over time. Though they have the same ST footprint as ordinary ST clusters, emerging clusters characterize the evolution of ST clusters over time. They thus resemble change interval on the temporal dimension.

A spatiotemporal version of scan statistics (Kulldorff et al. 1998) has been proposed to discover ST clusters. A model, proposed by Neill et al. (2005) employs similar ideas for discovering emerging ST clusters.

17.5 Change Pattern Mining Approaches

This section introduces a few representative techniques from various change pattern categories. Table 17.2 listed these techniques with their respective pattern footprint. Due to space limitations, readers are directed to more detailed reviews of further explorations.

17.5.1 Change Point Detection Techniques

Many change point detection techniques have been proposed. One is the Cumulative Sum (CUSUM), first proposed by Page in 1954 (Page 1954), CUSUM is an on-line approach to process time series such as system monitoring signals. The method

Table 17.2 List of techniques and related categories

Techniques	Area of research	Pattern name used	Change pattern footprint	Primary applications
CUSUM	Time series analysis	Abrupt change, change point	Time point	System control
S-Outlier detection	Spatial statistics	Spatial outlier	Spatial location	Image analysis, transportation
Interesting interval discovery	Time series analysis, spatial statistics	Sub-path/interval of abrupt change	Interval	Climate science
Statistical spatial Wombling	Spatial statistics	Wombling, rapid change, abrupt change	Spatial curves	Public health, ecology
Scan Statistics	Spatial statistics, cluster detection	Spatial cluster	Spatial regions	Public health
Change detection	Image Processing	Change	Spatial regions/pixels	Remote sensing
ST cluster detection	Spatial statistics	ST cluster, outbreak	ST subspace	Public health
Emerging ST cluster detection	Spatial statistics	Emerging cluster	ST subspace	Public health

keeps a cumulative sum of a score S_0 based on a statistical parameter θ of the data stream. When the sum exceeds a certain threshold, a change can be flagged. Formally, the score (for detecting positive change) is defined as $S_0 = 0$, and $S_{n+1} = \max(0, S_n + x_n - \theta_n)$ where x_n is the nth data value, and θ_n is the parameter. A number of works under the CUSUM framework have been proposed, defining θ as the mean, standard deviation, or likelihood function of the data (Kawahara and Sugiyama 2009; Kucera et al. 2007).

Other change point detection techniques may use likelihood ratio tests, maximum-likelihood estimation (MLE), and model learning to find the change point. Detailed reviews of techniques can be found in the time series analysis literature (Basseville and Nikiforov 1993; Shaban 1980; Zacks 1983).

17.5.2 Spatial Outlier Detection Techniques

Spatial outlier detection techniques have been discussed in the literature of spatial statistics. One example is the Moran Scatterplot (Anselin 1995) approach. The z-score (normalized value) at a location is plotted against the average z-score at its neighboring locations. Points falling into the second and the fourth quadrants are considered as spatial outliers. Other statistical approaches include variogram cloud, scatterplot, and spatial statistic $Z_{s(x)}$, etc. A unified definition of spatial outlier detection along with a computational framework can be found in the literature of data mining (Shekhar et al. 2003).

17.5.3 Change Interval Discovery

The change interval discovery problem was proposed to model an enduring change of value that takes place during a time period or within a spatial path. Such change patterns may be found in precipitation time series where a multi-decadal drought occurs, or on a spatial path across different ecological zones where vegetation cover shifts quickly from forest to desert. An algebraic interest measure named "sameness degree" was proposed based on the differences between neighboring values

$$\text{Sameness Degree} = \frac{\text{AVG } \{\Delta x\}}{\text{AVG} \geq \theta \{\Delta x\}}$$

(Zhou et al. 2011). An interval with sameness degree exceeding a threshold is considered as an interval of change. The authors also provided a computational approach for enumerating and filtering sub-intervals.

17.5.4 Spatial Wombling Techniques

Spatial wombling origins from ecology where boundaries between ecological zones are investigated (Womble 1951). Statistical approaches have been designed to find statistically significant borders

Between different regions or contour maps. In a hierarchical Bayesian approach (Lu and Carlin 2005), value differences between neighboring locations or areas that exceed a threshold are selected as candidates, and a boundary likelihood value (BLV) is estimated using Markov Chain Monte Carlo (MCMC) simulations to find statistically significant boundaries. Other approaches have been proposed on geostatistical (Liang 2009), areal (Lu and Carlin 2005), and point process (Liang et al. 2009) data models.

17.5.5 Spatial/ST Scan Statistics and Emerging Cluster Detection

Spatial scan statistics (Kulldorff and Nagarwalla 1995) employs a hypothesis testing, where the null hypothesis H_0 is: the probability of disease inside any region is the same as that outside the region, and the alternative hypothesis H_1 is that in a certain region Z, there is a higher probability (q_{in}) of each disease inside than outside (q_{out}). A likelihood ratio score is then maximized among all the possible spatial regions to find the most likely cluster of disease. The original scan statistic assumes that the events follow a Bernoulli or a Poisson model. Later extensions on the same ideas explored assumptions of normal (Kulldorff et al. 2009), exponential (Huang et al. 2007), ordinal (Jung et al 2007), and non-parametric models (Kulldorff et al.

2005). Scan statistic is also generalized to space-time to handle ST event report datasets (Kulldorff et al. 1998; Kulldorff 2001).

The computational efficiency was further optimized by Neill and Moore (2004) by employing a top-down pruning of the search space. A Bayesian version of scan statistics is also proposed where an inference is used instead of the frequentist hypothesis testing (Neill et al. 2006). A further step was made to model the emergence of an outbreak inside a spatiotemporal subspace (Neill et al. 2005). The null and alternative hypotheses were modified as follows: H_0: the probability that one instance is at risk in the cluster for each day (t_{min}, t_{min+1}, ..., W) is the expected value p, and H_1: The above probability in days t_{min}, t_{min+1}, ..., T are $q \times p$ where $1 \leq q_1 \leq q_2 \leq \ldots \leq q_T$. Computational methods were also provided for the efficient computation of this pattern.

17.5.6 Remote Sensing Image Change Detection

Techniques of change detection in image processing for remote sensing have been extensively discussed in the literature. Typical methods on bi-temporal change detection include simple differencing (Rosin 1998; Rosin and Ioannidis 2003), where a value difference between pixels at the same location from two images are compared with a threshold for change determination, hypothesis testing, where the null hypothesis of no change is tested against the alternative hypothesis of change using significance test or likelihood ratio test (Barkat 2005; Aach and Kaup 1995; Thoma and Bierling 1989; Dempster et al. 1977; Fukunaga and Hayes 1989). Another approach for image change detection is predictive modeling (Hsu et al. 1984; Yakimovsky 1976; Elfishawy et al. 1991; Jain and Chau 1995).

17.5.7 Opportunities for Cross-Fertilization

ST cluster detection techniques are mostly applied on disease mapping and crime analyses. We argue that other application domains can also benefit from these techniques as they may also face the problem of cluster/hotspot detection. For example, a pattern of extreme rainfall change at regional scales due to global climate change is of interest to climate science research. To do this, the underlying statistical models may need to switch from Binomial or Poisson distributions to those modeling tail behaviors of continuous variables, such as the Generalized Extreme Value (GEV) distributions (Coles 2001), and the spatial dependencies among locations may need to be addressed.

The concept of return period (Wikipedia 2012-R) is used to characterize the expected years before seeing the next climate events (rainfall, temperature) of the same intensity. A return period of a rainfall event is 100 years means that the probability that the intensity of this rainfall is exceeded in any year is $1/100 = 0.01$.

This concept is used in statistical analysis for climate extreme change detection. However, a similar concept can be introduced to ST cluster detection where a return period of disease outbreak can be estimated based on the probability that an outbreak of the same extent in each time step (e.g., month). This may help the understanding the spreading pattern of epidemic diseases.

17.6 Conclusion and Future Work

This chapter presents an interdisciplinary review of change pattern mining techniques in time series analysis, spatial statistics, and remote sensing to facilitate cross-fertilization of techniques and ideas across disciplines. A taxonomy based on ST footprint was proposed to better address the challenges posed by terminology differences across these disciplines. Cross-fertilization ideas such as the application of ST cluster detection techniques on climate change study has been discussed. More ideas might be generated when more interdisciplinary comparison of techniques are made. In the future, we wish to integrate techniques from a broader range of disciplines to enrich cross-fertilization ideas. We also plan to explore new input data types and patterns of change pattern mining. Finally, we plan to investigate other taxonomies to classify change pattern mining techniques.

Acknowledgement This material is based upon work supported by the National Science Foundation under Grant No. 1029711, IIS-1320580, 0940818 and IIS-1218168, the USDOD under Grant No. HM1582-08-1-0017 and HM0210-13-1-0005, and the University of Minnesota under the OVPR U-Spatial.

References

Aach, T., & Kaup, A. (1995). Bayesian algorithms for adaptive change detection in image sequences using Markov random fields. *Signal Processing: Image Communication, 7*(2), 147–160.

Anselin, L. (1995). Local indicators of spatial association-Lisa. *Geographical Analysis, 27*(2), 93–115.

Banerjee, S., Carlin, B. P., & Gelfand, A. E. (2003). *Hierarchical modeling and analysis for spatial data* (Vol. 101). London: Chapman & Hall/CRC.

Barbujani, G., Oden, N. L., & Sokal, R. R. (1989). Detecting regions of abrupt change in maps of biological variables. *Systematic Biology, 38*(4), 376–389.

Barkat, M. (2005). *Signal detection and estimation*. London: Artech House Publishers.

Basseville, M., & Nikiforov, I. V. (1993). *Detection of abrupt changes: Theory and application* (Vol. 15). Englewood Cliffs: Prentice Hall.

Chainey, S., Tompson, L., & Uhlig, S. (2008). The utility of hotspot mapping for predicting spatial patterns of crime. *Security Journal, 21*(1), 4–28.

Chapin, F. S., III, Callaghan, T. V., Bergeron, Y., Fukuda, M., Johnstone, J. F., Juday, G., & Zimov, S. A. (2004). Global change and the boreal forest: Thresholds, shifting states or gradual change? *AMBIO: A Journal of the Human Environment, 33*(6), 361–365. BioOne.

CIA. (2013). *CIA World Factbook*. Available at: http://www.cia.gov/library/publications/the-world-factbook/. Accessed 13 Jan 2013.

Coles, S. (2001). *An introduction to statistical modeling of extreme values*. New York: Springer.

Coppin, P., Lambin, E., Jonckheere, I., & Muys, B. (2002). Digital change detection methods in natural ecosystem monitoring: A review. In *Series in Remote Sensing-Proceedings of the First International Workshop on the Analysis of Multi-Temporal Remote Sensing Images-MULTITEMP, 2*, 3–36.

Dai, A., Lamb, P. J., Trenberth, K. E., Hulme, M., Jones, P. D., & Xie, P. (2004). The recent Sahel drought is real. *International Journal of Climatology, 24*(11), 1323–1331.

Dempster, A. P., Laird, N. M., & Rubin, D. B. (1977). Maximum likelihood from incomplete data via the EM algorithm. *Journal of the Royal Statistical Society: Series B Methodological, 39*, 1–38.

Elfishawy, A. S., Kesler, S. B., & Abutaleb, A. S. (1991). Adaptive algorithms for change detection in image sequence. *Signal Processing, 23*(2), 179–191.

Fagan, W. F., Fortin, M. J., & Soykan, C. (2003). Integrating edge detection and dynamic modeling in quantitative analyses of ecological boundaries. *BioScience, 53*(8), 730–738.

Fitzpatrick, M. C., Preisser, E. L., Porter, A., Elkinton, J., Waller, L. A., Carlin, B. P., & Ellison, A. M. (2010). Ecological boundary detection using Bayesian areal wombling. *Ecology, 91*(12), 3448–3455.

Fukunaga, K., & Hayes, R. R. (1989). The reduced Parzen classifier. *IEEE Transactions on Pattern Analysis and Machine Intelligence, 11*(4), 423–425.

Gilks, W. R., Richardson, S., & Spiegelhalter, D. (Eds.). (1995). *Markov chain Monte Carlo in practice: Interdisciplinary statistics*. London: Chapman & Hall/CRC.

Hansen, J., Sato, M., & Ruedy, R. (2012). Perception of climate change. *Proceedings of the National Academy of Sciences, 109*(37), E2415–E2423.

Hsu, Y. Z., Nagel, H. H., & Rekers, G. (1984). New likelihood test methods for change detection in image sequences. *Computer Vision, Graphics, and Image Processing, 26*(1), 73–106.

Huang, L., Kulldorff, M., & Gregorio, D. (2007). A spatial scan statistic for survival data. *Biometrics, 63*(1), 109–118.

Im, J., & Jensen, J. R. (2005). A change detection model based on neighborhood correlation image analysis and decision tree classification. *Remote Sensing of Environment, 99*(3), 326–340.

Jain, Z. S., & Chau, Y. A. (1995). Optimum multisensor data fusion for image change detection. *IEEE Transactions on Systems, Man, and Cybernetics, 25*(9), 1340–1347.

JISAO (2011). *Sahel rainfall index data*. Joint Institute for the Study of the Atmosphere and Ocean (JISAO). http://jisao.washington.edu/data/sahel/. Accessed Sept 1 2011.

Jones, P., & Harris, I. (2008). *CRU Time Series (TS) high resolution gridded datasets*. University of East Anglia Climatic Research Unit (CRU), NCAS British Atmospheric Data Centre, http://badc.nerc.ac.uk/view/badc.nerc.ac.uk__ATOM__dataent_1256223773328276

Jung, I., Kulldorff, M., & Klassen, A. C. (2007). A spatial scan statistic for ordinal data. *Statistics in Medicine, 26*(7), 1594–1607.

Kawahara, Y., & Sugiyama, M. (2009). Change-point detection in time-series data by direct density-ratio estimation. In *Proceedings of 2009 SIAM international conference on data mining (SDM2009)* (pp. 389–400). Society for Industrial and Applied Mathematics (SIAM), Philadelphia, PA.

Kucera, J., Barbosa, P., & Strobl, P. (2007). Cumulative sum charts-a novel technique for processing daily time series of MODIS data for burnt area mapping in Portugal. In *International workshop on the analysis of multi-temporal remote sensing images, 2007. MultiTemp 2007* (pp. 1–6). Washington, DC: IEEE.

Kulldorff, M. (1997). A spatial scan statistic. *Communications in Statistics: Theory and Methods, 26*(6), 1481–1496.

Kulldorff, M. (2001). Prospective time periodic geographical disease surveillance using a scan statistic. *Journal of the Royal Statistical Society: Series A (Statistics in Society), 164*(1), 61–72.

Kulldorff, M., & Nagarwalla, N. (1995). Spatial disease clusters: Detection and inference. *Statistics in Medicine, 14*(8), 799–810.

Kulldorff, M., Athas, W. F., Feurer, E. J., Miller, B. A., & Key, C. R. (1998). Evaluating cluster alarms: A space-time scan statistic and brain cancer in Los Alamos, New Mexico. *American Journal of Public Health, 88*(9), 1377–1380.

Kulldorff, M., Heffernan, R., Hartman, J., Assunção, R., & Mostashari, F. (2005). A space–time permutation scan statistic for disease outbreak detection. *PLoS Medicine, 2*(3), e59.

Kulldorff, M., Huang, L., & Konty, K. (2009). A scan statistic for continuous data based on the normal probability model. *International Journal of Health Geographics, 8*(1), 58.

Liang, S. (2009). *Bayesian hierarchical modeling of geostatistical and spatial point process data.* PhD thesis, University of Minnesota.

Liang, S., Banerjee, S., & Carlin, B. P. (2009). Bayesian wombling for spatial point processes. *Biometrics, 65*(4), 1243–1253.

Lu, H., & Carlin, B. P. (2005). Bayesian areal wombling for geographical boundary analysis. *Geographical Analysis, 37*(3), 265–285.

Mohan, P., Shekhar, S., Shine, J. A., Rogers, J. P., Jiang, Z., & Wayant, N. (2011). A neighborhood graph based approach to regional co-location pattern discovery: A summary of results. In *Proceedings of the 19th ACM SIGSPATIAL international conference on advances in geographic information systems* (pp. 122–132). New York, NY, USA: ACM.

NASA (2011). *Landsat imagery archive*. Washington DC, 1984–2011. http://landsat.gsfc.nasa.gov/images/archive/c0042.html. Accessed 25 Jan 2013.

Neill, D. B., & Moore, A. W. (2004). Rapid detection of significant spatial clusters. In *Proceedings of the tenth ACM SIGKDD international conference on Knowledge discovery and data mining* (pp. 256–265). New York, NY, USA: ACM.

Neill, D. B., Moore, A. W., Sabhnani, M., & Daniel, K. (2005). Detection of emerging space-time clusters. In *Proceedings of the eleventh ACM SIGKDD international conference on knowledge discovery in data mining* (pp. 218–227).

Neill, D. B., Moore, A. W., & Cooper, G. (2006). A Bayesian spatial scan statistic. *Advances in Neural Information Processing Systems, 18*, 1003.

Page, E. S. (1954). Continuous inspection schemes. *Biometrika, 41*(1/2), 100–115.

Radke, R. J., Andra, S., Al-Kofahi, O., & Roysam, B. (2005). Image change detection algorithms: A systematic survey. *IEEE Transactions on Image Processing, 14*(3), 294–307.

Rosin, P. (1998). Thresholding for change detection. In *Proceedings of sixth international conference on computer vision* (pp. 274–279). Washington, DC: IEEE.

Rosin, P. L., & Ioannidis, E. (2003). Evaluation of global image thresholding for change detection. *Pattern Recognition Letters, 24*(14), 2345–2356.

Shaban, S. A. (1980). Change point problem and two-phase regression: An annotated bibliography. *International Statistical Review, 48*(1), 83–93.

Shekhar, S., & Chawla, S. (2003). *Spatial databases: A tour*. Upper Saddle River: Prentice Hall.

Shekhar, S., Lu, C. T., & Zhang, P. (2003). A unified approach to detecting spatial outliers. *GeoInformatica, 7*(2), 139–166.

Singh, A. (1989). Review article digital change detection techniques using remotely-sensed data. *International Journal of Remote Sensing, 10*(6), 989–1003.

Thoma, R., & Bierling, M. (1989). Motion compensating interpolation considering covered and uncovered background. *Signal Processing: Image Communication, 1*(2), 191–212.

Tobler, W. R. (1970). A computer movie simulating urban growth in the Detroit region. *Economic Geography, 46*, 234–240.

Tucker, C. J., Pinzon, J. E., & Brown, M. E. (2004). *Global inventory modeling and mapping studies* (pp. 1981–2006). College Park: Global Land Cover Facility, University of Maryland.

Verbesselt, J., Hyndman, R., Zeileis, A., & Culvenor, D. (2010). Phenological change detection while accounting for abrupt and gradual trends in satellite image time series. *Remote Sensing of Environment, 114*(12), 2970–2980.

Wikipedia (2012-D). Derivative. http://en.wikipedia.org/wiki/Derivative. Accessed 6 Sept 2012.

Wikipedia (2012-M). Mcmansion. http://en.wikipedia.org/wiki/McMansion. Accessed 6 Sept 2012.

Wikipedia (2012-R). Return period. http://en.wikipedia.org/wiki/Return_period. Accessed 6 Sept 2012.

Womble, W. H. (1951). Differential systematics. *Science, 114*(2961), 315.

Wong, W. K., & Neill, D. B. (2009). *Tutorial on event detection, ACM SIGKDD 2009*. http://www.pptsearch.net/download.php?fid=109746. Accessed 1 Sept 2012.

Worboys, M., & Duckham, M. (2004). *GIS: A computing perspective*. Boca Raton: CRC.

Yakimovsky, Y. (1976). Boundary and object detection in real world images. *Journal of the ACM, 23*(4), 599–618.

Zacks, S. (1983). *Survey of classical and Bayesian approaches to the change-point problem: Fixed sample and sequential procedures of testing and estimation*. Recent advances in statistics: Papers in honor of Herman Chernoff on his sixtieth birthday, p. 245. Waltham, MA: Academic Press.

Zhou, X., Shekhar, S., Mohan, P., Liess, S., & Snyder, P. K. (2011). Discovering interesting sub-paths in spatiotemporal datasets: A summary of results. In *Proceedings of the 19th ACM SIGSPATIAL international conference on advances in geographic information systems* (pp. 44–53). New York, NY, USA: ACM.

Chapter 18
Contemporary Computing Technologies for Processing Big Spatiotemporal Data

Chaowei Yang, Min Sun, Kai Liu, Qunying Huang, Zhenlong Li, Zhipeng Gui, Yunfeng Jiang, Jizhe Xia, Manzhu Yu, Chen Xu, Peter Lostritto, and Nanying Zhou

18.1 Introduction

To better understand physical and social phenomena and the impact of human activities on them, scientists have collected data about the Earth's surface for hundreds of years (Jacobson et al. 2000). For example, the Earth's surface temperature (Kusuda and Achenbach 1965), volcanic eruptions (Robock 2000), hurricanes (Goldenberg et al. 2001), and other parameters of geographic phenomena have been recorded. In the past century, the invention of computers for information processing, launching of satellites and the evolution of remote sensing technologies have greatly improved our capabilities to observe and record various physical parameters of the Earth's surface (Christian 2005). This is reflected in at least five aspects.

First, our ability to record physical phenomenon is growing from one dimension, to two, three, four and even higher dimensions (Akyildiz et al. 2002). In the past, we were only able to quantify the physical phenomena like temperature or snowfall for a specific representative point. For example, Washington D.C. had a historical snowfall of over 2 ft in February 2010. Recent developments have allowed us to use satellite technology to observe the Earth and capture 2D/3D values for a continuous surface on the planet, such as elevation data (Rodríguez et al. 2006).

C. Yang (✉) • M. Sun • K. Liu • Q. Huang • Z. Li • Z. Gui • Y. Jiang • J. Xia • M. Yu • C. Xu
P. Lostritto • N. Zhou
Center for Intelligent Spatial Computing, George Mason University, Fairfax, VA 22030-4444, USA
e-mail: cyang3@gmu.edu; msun@gmu.edu; kliu4@gmu.edu; qhuang1@gmu.edu; zli1@masonlive.gmu.edu; zgui@gmu.edu; yjiang7@gmu.edu; jxia3@gmu.edu; myu7@gmu.edu; cxu3@gmu.edu; plostr@gmu.edu; nzhou@gmu.edu

M.-P. Kwan et al. (eds.), *Space-Time Integration in Geography and GIScience*, DOI 10.1007/978-94-017-9205-9_18

Second, we can now collect important information more frequently using active and passive sensors working at a frequency of years, months, days, or even seconds and milliseconds to detect different parameters that change at different speeds according to their physical nature. For example, biological cells change in milliseconds, temperature will change hourly and the Earth's elevation changes at a rate of years to millions of years without human intervention.

Third, we have sensors of different spatial resolution to detect various phenomena for different usage. In weather forecasting, the resolution generally is at the kilometer level in order to study the global circulation of wind, heat flux, and water, and to assist individual decision making. On the other hand, the study of water contamination for public health purposes requires microscopic observation (Hart and Martinez 2006).

Fourth, sensors with different spectral capabilities were developed to collect a wide range of spectral resolution and scope coverage from nanometer to millimeter. Studies were conducted to better characterize different species in ecological studies and minerals in geological studies based on data collected using hyper-spectral instruments (Hart and Martinez 2006).

Fifth, the growth of sensors in resolution of space, time, spectral, and biological scope has greatly increased our capabilities to observe the Earth surface and provided greater support to enhance understanding and, eventually, the prediction of different geographic phenomena (Akyildiz et al. 2002). We have increased weather forecasting scope/resolution from several hours to 10 days and from state/provinces to cities and zip code level with good accuracy on a global extent. Citizens as sensors have also provided significant amount of data at petabytes per day level through social media and other contemporary technologies, such as smart phones.

These observational or simulation datasets are of great value to different domains of human scientific quests and application developments because they provide baseline snapshot data about the Earth at specific times. For example, the datasets can be (1) managed and utilized to build a Digital Earth (Yang et al. 2008), (2) integrated at different levels for emergency response, such as for hurricane Sandy, and (3) integrated or fed into complex models for better understanding of geographic phenomena which will lead to make better informed predictions regarding issues such as long term climate change, which can better aid policy making (Yang et al. 2010). Further, the analyses of social media data can provide critical information that may not be available through other methods – for example, people twit breakouts of disease, social events, and plan for activities.

Recognizing the demand for big data and their dispersal, various domains and communities, such as the Global Earth Observation (GEO), have been established to share, process, and utilize big data for extracting valuable information and knowledge. Big data will help us gain new insights about the geographic phenomena, and further our understanding of geographic phenomena for making better decisions. However, handling big data poses grand challenges for computing, this chapter introduces through several projects the computing challenges of big data handling and potential solutions for these challenges.

18.2 Computing Challenges

The growth of big data has outpaced the increasing of computing capability and the advancement of data processing technologies and tools in recent years, and poses the challenges of heterogeneity, performance, visualization, on-demand processing, privacy, and social collaborations.

18.2.1 Heterogeneity

Due to the diverse application usages and technologies for data collecting, processing, storage and access, the heterogeneities include the following aspects: (1) Diversified content organization and representation for data models, formats, specifications and encoding methods. For example, the data can be either structured or unstructured, such as text, sensor data, images, audio, video, clickstreams, and log files. (2) The application specific technologies and strategies on storage/access methods including data access and retrieval protocols, authorizations and priorities. (3) The geographically dispersed storage and management of big data that is due to the globalization calling for the integrating of data collected, hosted (stored) and shared by dispersed providers, organizations and governmental agencies all over the world. The heterogeneity has posed the challenges on data discovery (search, identification and indexing) and integration management (Li et al. 2011b; Gui et al. 2012). The major problem is that current data structures/models, algorithms and systems are usually designed to support limited criteria based on particular application requirements. Detailed challenges (Agrawel et al. 2012) include how to rapidly store and retrieve a large volume of data on storage? How to build efficient index and data structure to organize them in file systems? How to automatically generate the right metadata to describe what data are recorded and how they are recorded and evaluated?

18.2.2 Performance

Timeliness is a vital factor in many time-critical applications. For example, we need real-time data analysis in e-business, emergence response and decision-making for natural and man-made disasters. Hence, in the big data era, the capacity of the rapid data acquisition, discovering, integration, processing, modeling and visualization has become an important measurement of system maturity.

18.2.3 Computing Intensity

Big data processing involves executing complex algorithms to extract information from massive raw data. These complex algorithmic processes, such as data aggregation and subset, make the information extraction extremely computational intensive. For example, the computational and storage requirements for deriving regional and global water, energy, and carbon conditions from multi-sensor and multi-temporal datasets far exceed the capacity of a single computer (Kumar et al. 2006). Usually, high performance computing (HPC), such as parallel computing, grid computing and cloud computing, is needed to support the massive volume of big data handling including increasingly sophisticated and complex geophysical algorithms and models (Guan 2008; Huang and Yang 2011; Huang et al. 2012).

18.2.4 Visualization

Visualizing 3D/4D environmental big data is critical to understand and predict geographic phenomena for relevant decision making. Efforts of developing visualization tools to support the analysis of dynamic geographic phenomena started approximately a few decades ago (Hibbard et al. 1994). Despite the developments of geovisualization tools such as virtual globes, displaying the dynamic phenomena represented by 3D/4D data remains a challenge due to the complexity and volume of the data. As online visualization becomes a generic requirement for big data in many occasions, it introduces more challenges on integrating, processing and visualizing geospatial data in a web-based environment. Effective visualization relies on efficient data organizations to facilitate fast access of these 3D/4D data, advanced visualization algorithms to display the phenomena described by such data, and state-of-the-art computing techniques to address the computational complexity (Li et al. 2011a). The advances in Graphics Processing Units (GPUs) have yielded better rendering capacity as well as computing power, serving as ubiquitous and affordable computing devices for big data applications (Stone et al. 2007).

18.2.5 Applications

Geographic applications with massive data involved are complex. More than often, the computing demands exceed the computing capacity and become a driver for computing science advancements (Yang et al. 2011b). Different computing strategies should be adopted to speed up the computing process for different applications (Huang et al. 2012). For example, climate change analysis poses grand challenges for computing sciences when we perform climate modeling and simulation over global area for a long time interval, such as 10–200 years. In addition, the initialization of such models requires different types of data from different organizations and in different formats. To better address twenty-first century geographic science and application challenges, cross cutting and interoperability solutions are required to integrate interdisciplinary data, information, and knowledge for science advancements.

18.2.6 Privacy

Privacy is another critical concern, and one that becomes increasingly important as many big data collected daily are personal related (which sensitively reflects the personal information, activities and behaviors in both physical and virtual world). Therefore, there is increasing public concern regarding the inappropriate use of personal data, particularly through converging of data from multiple sources. Managing privacy is both a technical and a sociological problem, which must be addressed jointly from both perspectives to achieve the promise of big data.

18.2.7 Massive Collaboration

Many analytics for big data will not only involve the computational processes automatically conducted by machines, but will also involve human interventions. For example, in web x.0 era, end users are involved as the sensors or data providers. Many data are collected and contributed by the users. In big data processing procedures, it is important to support the input and collaboration from multiple human experts in different domains to have better understanding and wise decisions.

A systematic solution for big data should not only tackle the physical and technical challenges but also consider and coordinate the policies and social aspects. This chapter will focus on the computing technology.

18.3 Big Data Management

The first challenge about big data is discoverability and accessibility. This section describes with an example how metadata, catalogs, index, and database are used to facilitate big data discovery and access using database archived metadata and index for component, service and datasets (Fig. 18.1).

18.3.1 Metadata and GEOSS Clearinghouse

Metadata is structured information that describes, explains, locates, and makes it easier to retrieve, use, or manage data or an information resource. It is often defined as data about data. In general, there are two kind of metadata: one is descriptive data, which describes the content and serves as a brief abstract of the data; and another is structural metadata, which describes the design and specification of data structures. Metadata is critical for big data management and discovery. Well designed and described metadata mechanism could better support management and discovery

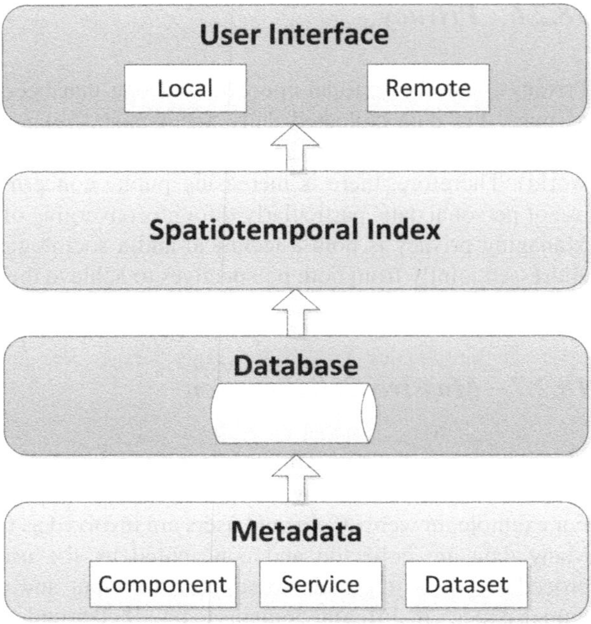

Fig. 18.1 Big data discovery flow with GEOSS clearinghouse as an example

process by providing basic description of data so that we can handle relative small scale data (e.g. metadata with 10–20 elements) before getting to real big data.

Global Earth Observation System of Systems (GEOSS) clearinghouse is one of such components (GEO 2012). GEOSS clearinghouse facilitates better integration and access to geospatial resources for GEOSS and other global communities by providing metadata/data management, and discovery capability. In GEOSS clearinghouse, metadata, data content and online access are standardized through different metadata schema, which includes: (1) FGDC CSDGM Metadata Standards (FGDC 2012), (2) Dublin core (Dublin 2012), (3) ISO-19115, and (4) ISO-19139. These metadata standards can be extended to different data series and datasets including geographic features and feature properties. For example, ISO standards precisely define the description of geographic information and provide the mandatory and conditional metadata sections, metadata entities and metadata elements. These standardized metadata could be archived in various catalogs that provide searching capabilities for different domain users (Huang et al. 2011).

18.3.2 Database and Index

The GEOSS clearinghouse supports MySQL, Postgresql, Oracle, MckoiDB, and SQL server as the database according to different requirements. It also support

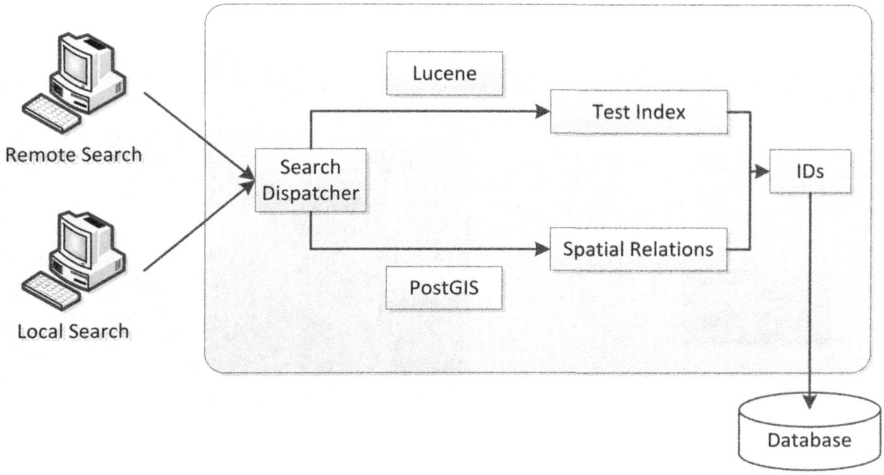

Fig. 18.2 GEOSS index kernel

SQL Azure, which is a cloud-based database service provided by Microsoft. The Postgresql version of the GEOSS clearinghouse is persistently maintained and hosted on server and cloud computing platform. Postgresql database with its PostGIS spatial extension provides high performance search capabilities for both spatial information and text information, management of big data exceeding 4 terabytes, and support to concurrent user accesses.

A properly designed indexing mechanism can significantly improve big data discovery by supporting high-speed data retrieval process (Theodoridis et al. 1996). In general, GEOSS clearinghouse indexes text description of the features (such as natural resources and national hydrology) and geographic location and temporal extension of features and boundaries. Figure 18.2 shows the database and index framework of GEOSS clearinghouse. Białecki et al. (2012) is used to index one dimension text-based data. Spatial index mechanism is supported by Geotools or Postgis spatial extension. By adopting the index mechanism, GEOSS clearinghouse can support high-speed data retrieval process for data discovery.

Maintaining a consistently quick response time is critical, especially when GEOSS clearinghouse needs to store millions to billions of metadata. The spatial and temporal dimensions of the data cause this problem. Research shows that if the response time is longer than 3 s, the users will become frustrated (Nah 2004). Besides adopting index mechanisms, GEOSS clearinghouse also employed buffering and multi-threading mechanisms. Based on the above strategies, the GEOSS clearinghouse provides consistently quick responses (1–2 s) for concurrent searches from different domain users.

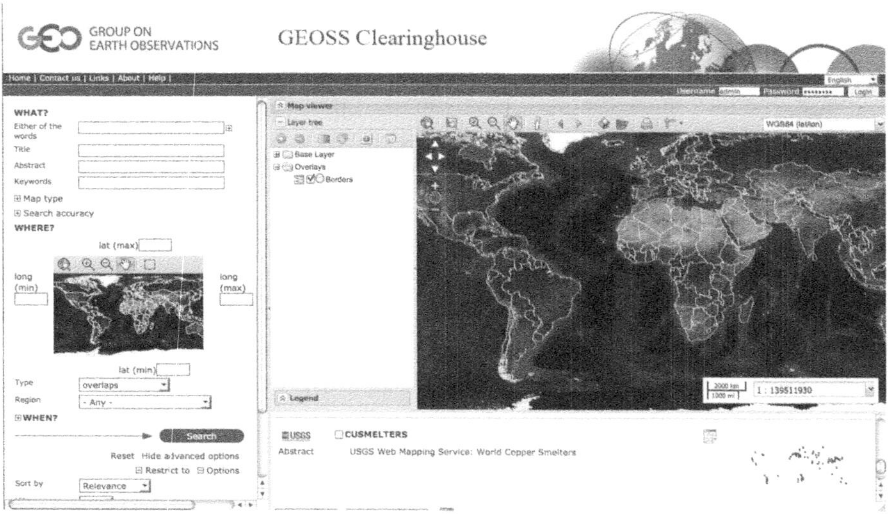

Fig. 18.3 Web GUI of GEOSS clearinghouse

18.3.3 Data Access and User Interaction

The GEOSS clearinghouse provides geospatial datasets and resources through both local (Fig. 18.3) and remote interfaces. For local interface, users access big data through the GEOSS clearinghouse portal, which provides a simple and easy-understanding search interface. Users can search metadata based on the title, keywords, any text, spatial bounding box and so on. In addition, users can access data according to the data category. For remote interface, the GEOSS clearinghouse supports various protocols for domain users to search geospatial datasets through CSW, SRU, RSS and other standard interfaces. After the search process, discovered results will be shown on a result page. The result page supports different sorting rules (relevance, Date, title alphabetical order, popularity, and rate). Icons are used to tag provider through a thumbnail. The buttons for access tools are dynamically added to each record panel according to resource types. The original metadata file can be accessed through the supper link for the domain user. Users can also interact with integrated visualization tools for viewing data services. This integrated tool is supported by OpenLayer (2012).

18.3.4 Challenges and Computing Demands

To address the spatiotemporal data management challenges, metadata is used in GEOSS clearinghouse to describe data; a spatiotemporal index with high performance is built based on PostGIS database to support data discovery; both local and remote access functions are provided to users to search, present and use geospatial data.

GEOSS clearinghouse is the engine of the GEOSS and provides discovery and access functions to worldwide users with big data; hence, GEOSS clearinghouse needs a well-designed scalability function to support reasonable concurrent search requests. To meet this requirement, GEOSS clearinghouse is deployed on Amazon EC2 cloud. Using the cloud resources, GEOSS clearinghouse can provide higher performance and more concurrent requests by improving the flexibility and scalability (Huang et al. 2010).

18.4 Big Data Processing

Due to the complexity and large volume of big data, single computing resource cannot produce and handle the data sufficiently. After a scientific model is developed, the model needs to be sent to different computing platforms for simulations. Results or observations will be collected and stored into distributed data storages. Big data processing is related to further operations on these data, which includes data preprocessing, data management of intermediate productions and data usage (analysis and visualization, Fig. 18.4). We will use the processing of ModelE data of climate@home project as an example to explain how data processing is handled.

18.4.1 Data Preprocessing

Raw data returned into storages are usually not well organized, fragment, and unusable. They may be in different formats, projections, spatial or temporal coverage. Therefore, data preprocessing such as reformatting, re-projecting, scaling,

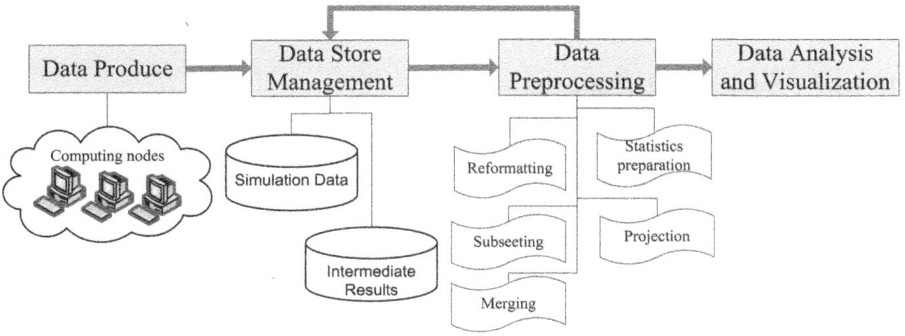

Fig. 18.4 Typical big spatiotemporal data processing workflow

and intermediate result pre-calculating is required before data are delivered or used in analysis and visualization. Data preprocessing are not performed arbitrarily. In contrary, the process is application-dependent.

18.4.1.1 Reformat

Many scientific model simulations generate results in binary format. But in order to facilitate efficient and interoperable data access and utilization, many standard data formats, such as Hierarchical Data Format (HDF) and Network Common Data Form (NetCDF), are developed. These formats are commonly featured with the following characteristics: (1) self-describing, (2) machine independence, (3) supporting array-oriented scientific data, and (4) efficient access to massive data with build-in index. To archive ModelE simulation, we transferred results from binary file into NetCDF format immediately after results are produced. Running the model once will generate hundreds of variables. These variables are separated into 11 different NetCDF files depending on the dimension of the variables. And some variables with the same dimensions are stored in the same file in order to share common metadata description.

18.4.1.2 Rescale

ModelE data are concatenated and subset to change both temporal and spatial coverage. For example, 10-year data that contain 120 monthly data over global scale are concatenated in a time series. Accordingly, a time dimension is added to NetCDF file structure. The resultant data can avoid the time consumption on repeatedly reading metadata for the monthly data. Meanwhile, files in regional scale can be extracted from the original global scale data. For example, if a variable like ozone is requested quite frequently in polar areas, then the data array in polar areas of this variable can be subset from the global dataset. Scientists can save time by reading only a small subset of the entire data. However, redundant data storage is sacrificed to obtain a better performance.

18.4.1.3 Intermediate Statistics

Since climate analysis is based on averaged values of quarterly or monthly data over a long time period, statistics like annual mean must be calculated before other advanced analyses to avoid wasting time on repeated calculations for statistics. According to previous studies (Potter 2009; Schmidt et al. 2006; James et al. 2003), annual mean is one of the most popular mean values used in further analysis. Calculating 100 1-year mean of data produced by ModelE may take up to 25 s (Sun et al. 2012). In order to save the time, annual mean is the necessary statistic which should be prepared in the data-preprocessing stage. The annual mean data are permanently stored in data storage. Most analyses will start from this intermediate data. The other intermediate statistics are calculated depending on specific application demands.

18.4.2 Issues About Data Management

Data preprocessing changes data properties and generates intermediate results. These changes are sent back to data storage for permanent archive. This proportion of intermediate production also requires well organization in the database. Usually, intermediate productions take more space in the data storage. Database designers need to assess trade-off between better performance and compact data storage. Visiting frequency and retrieving volume from users should be considered to assess the trade-off. Therefore, an efficient design of scientific data management should consider the emergence of intermediate productions and the possible demand from data preprocessing and data usage at the very beginning.

18.4.3 Data Analysis

After preprocessing, data are ready for analysis. Data analysis is the process to discover patterns and useful principles. The typical data analysis for climate studies using ModelE data includes: mean values, correlation between variables, stationary over time series, quality of forecasting, and spatiotemporal patterns (von Storch and Zwiers 1999). Although scientists have already developed mature statistical methods to perform these analyses, they are various challenges for dealing with big data.

Performance is a big issue, especially for on-demand analysis request through Internet. For example, producing the Taylor diagram, which compares the 5-years global means of simulated data from 300 model scenarios to the observation data for detecting simulation quality, takes up to 900 s including the time for calculating the mean values (Sun et al. 2012). Techniques such as generating intermediate statistics discussed in the last section can speed up the computing performance. R[1] has developed multi-thread technology to enable statistical analysis on big data for pursuing higher performance. GPU is of great potential to execute parallel computing.

Discovering knowledge from massive and multiple dimensional numerical data is another problem. In the context of geospatial sciences, geo-visualization (e. g., maps) has been proved to be an efficient method for a prompt understanding of complex geospatial data (Roth et al. 2008). Therefore, geovisual analytics integrating multiple interactive tools, dynamic graphs and live-linked views of data representation has the potential to provide an intuitive and heuristic method for analyzing big data.

[1]http://www.revolutionanalytics.com/products/enterprise-big-data.php

In addition, interoperability can also be an obstacle for analyses involving multiple data resources and multiple functions. This problem surfaces up when we perform analysis of evaluating simulation quality which may involve observation and simulation data in different formats.

18.5 Big Data Visualization

Data visualization aims to convert data and information/pattern derived from the data into graphical representations to be easily understood by the users. The visualization pipeline consists of three main parts (Fig. 18.5): (1) Data acquisition is to quickly access the raw data from distributed storage before visualizing; (2) Filtering process is to extract the visualization data from raw data that can't be directly used to visualize in both 2D and 3D; (3) Rendering and visualization are categorized into two main parts: 2D analytical visualization and 3D analytical visualization. In 2D visualization, five visualizing components are needed and include pixel-based visualization, multiple window display, animation display, subset dataset display and visual analytics. In 3D visualization, volume rendering and iso-surfaces are most regular rendering algorithms used for 3D visualization (Levoy 1988; Robert et al. 1988a, b). In addition, there are many optimizations in the two algorithms for speeding up the rendering performance and improving the capability of real-time rendering.

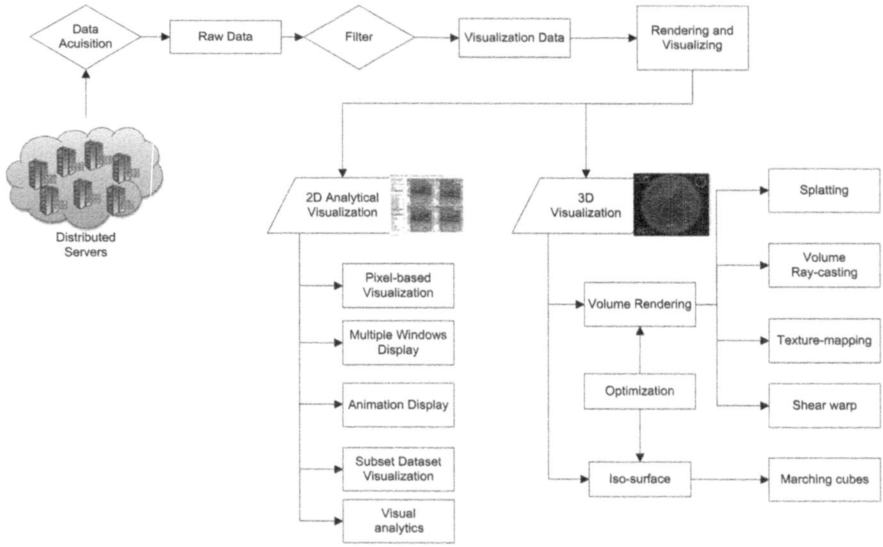

Fig. 18.5 Visualization pipe

18.5.1 2D Analytical Visualization

18.5.1.1 Pixel-Based Visualization Technique for Space-Aware Images

Pixel-based visualization uses 1 pixel to represent a data record, the whole dataset is visualized using a flat two dimensional image. In this image, the pixel color represents the data value, and the pixel location (sequence) in this image represents the two dimensions of this dataset. A legend image is generated accordingly to the value-color mapping. This technique is extremely useful for visualizing space patterns of spatiotemporal data. For 2D data, each dataset is visualized by an image with latitude, longitude and data value mapping to pixel row, column and color (Eq. 18.1).

Equation 18.1: Transformation of 2D dataset to image

$$\mathbf{data\,(latitude,\ longitude) = value \rightarrow pixel\,(row, col) = color\,(RGB)}$$

While the generated images could be simply displayed in a separated window, overlaying them to a geo-referenced map provides more intuitiveness. This is generally achieved by overlaying the images to an existing map application, such as OpenLayers, Google Maps, Bing Maps and Others. To improve the performance, the images of frequently accessed datasets and parameters could be pre-generated. Figure 18.6 shows the monthly average net radiation of Earth in December 1956 simulated by ModelE visualized as an RGB image. This image is overlapped on Microsoft Bing Maps to provide geospatial reference.

18.5.1.2 Multiple Windows to Display Multiple Parameters Simultaneously

For data with many parameters, displaying multiple parameters in multiple windows (one parameter each window) provides an effective way for end users to easily compare different parameters. For example, displaying net thermal radiation and net solar radiation in two map windows for the same study area and same time (period) enables scientists efficiently examine the relationship between the two climatic parameters.

This technique should allow the users to customize how many parameters they want to compare simultaneously and select the parameters, time periods and study areas they want to compare. How to organize the data to optimize the parameters, time periods and area selections in different datasets is a great challenge due to the intrinsic characteristics of big data. Figure 18.7 is a screenshot showing four climatic variables overlapped in four Google Earth windows. When users interact with one map window, such as zoom in, the other three map windows will simultaneously zoom to the same point for better comparing selected variables.

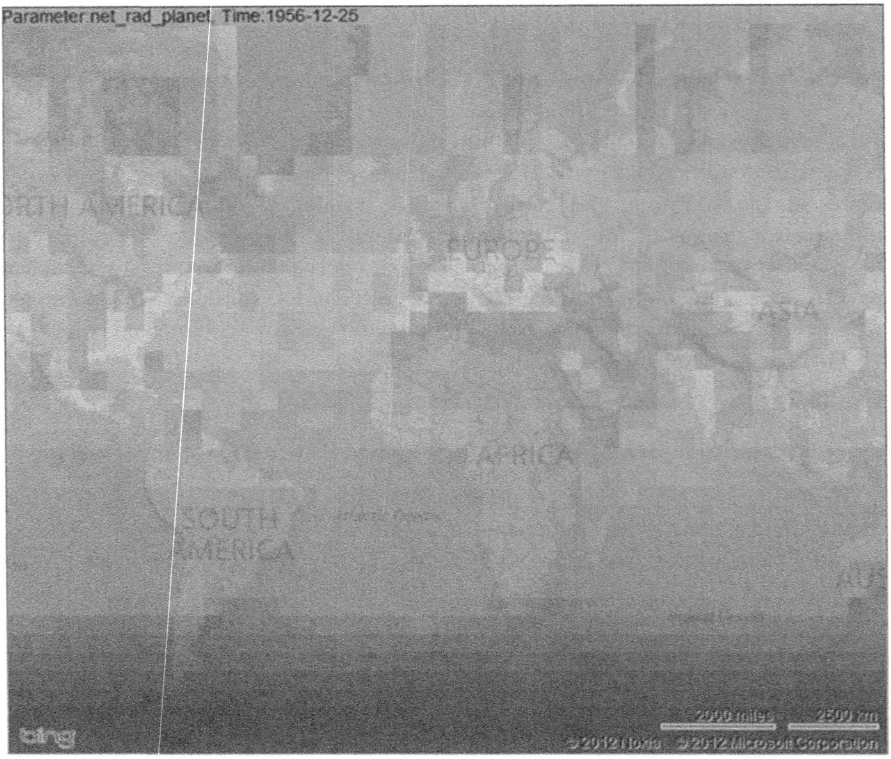

Fig. 18.6 Data visualized as an image and overlapped in Bing Maps

18.5.1.3 Animation to Display Time Series

Animated maps can help detect simple pattern changes of values through time. Combining multiple parameters display with animation allows users to simultaneously compare and analyze spatiotemporal data in both temporal dimension and space dimension. Users will then obtain the generic knowledge how different spatial patterns of multiple variables change over the same time period. For example, from 1960s to 1990s, land surface temperature increases and the largest change appear around polar areas.

18.5.1.4 Subset and Visualize Dataset

While displaying the entire geographic region of a dataset is useful for general overview of the dataset, it is usually more useful to subset the dataset based on an Area of Interested (AOI) selected by a user and then visualize and analyze

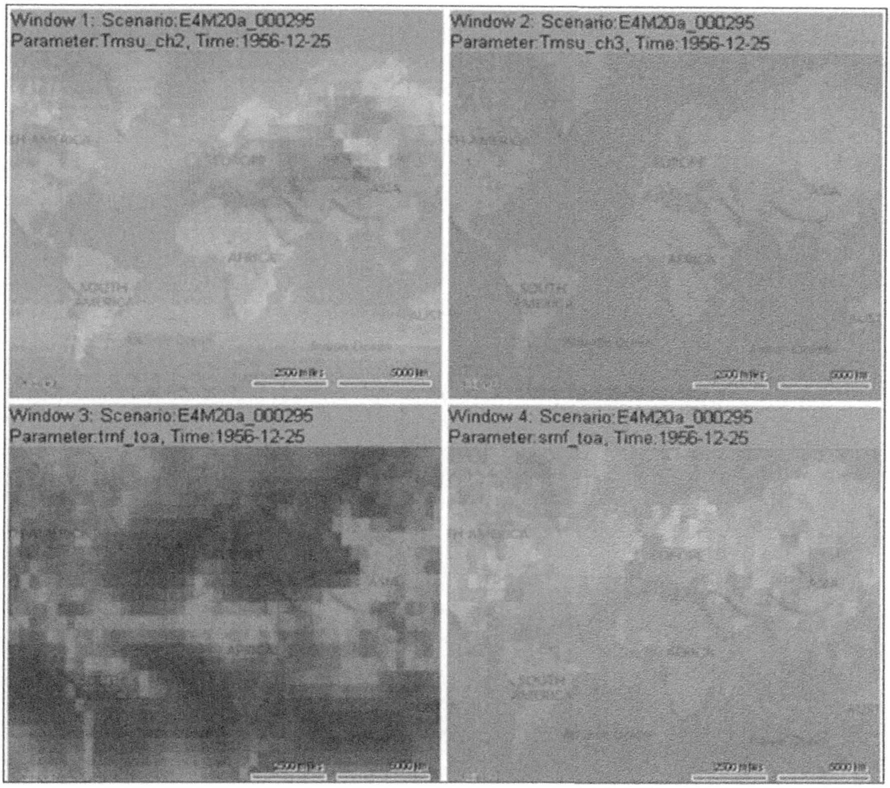

Fig. 18.7 Four climate variables displayed in four windows

the extracted data. To address the challenge brought by this demand, we need an intuitive interface to select the AOI(s), a high performance algorithm to extract the required dataset from the big data repository, on-the-fly visualization algorithm to render the extracted dataset as an image.

It is relatively easy to implement the previous four steps in a desktop application environment. In the Web-based environment, which is more prevalent in the WWW era, more research are needed for improving efficiency of performance, user interactivity and the data communication between the server and client side.

18.5.1.5 Visual Analytics

Rather than visualizing the general patterns hidden in data, scientists may need visual analytics which allow them to make data comparison in a intuitive way. Usually, analyses of results are represented as numerical data. However, humans

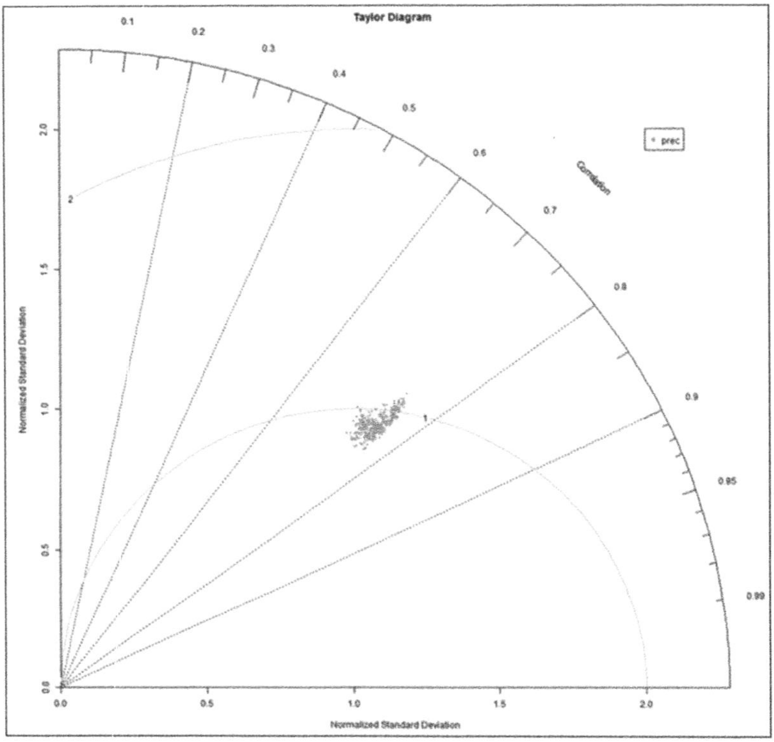

Fig. 18.8 A Taylor diagram generated in our climate@home project

have a relatively weak ability of vision and cognition on identifying underlying principles from the overwhelming amounts of spatiotemporal information presented as textual and numerical data. In this case, approaches that represent information in visual products such as diagrams are required to help researchers comprehend the information (Tufte 1997). For example, the Taylor diagrams may be used to assess the model accuracy that is an important task in climate modeling (Fig. 18.8). Scatter plot may be used to demonstrate the relationship between input parameters and output simulations which is useful for selecting the most appropriate model settings (Fig. 18.9).

However, the design of plot for big data is much harder than small and simple data. Several principles help plot designers to make effective graphics (Kosslyn 2006): (1) neither present too much nor too little information on one static plot or one frame of dynamic plot; (2) communicate using understandable symbols; (3) draw readers' attention by large perceptible differences; (4) group information; (5) make symbols compatible with their meanings; (6) enable accurate comparisons; (7) enable accurate comparisons; and (8) provide context for interpretation.

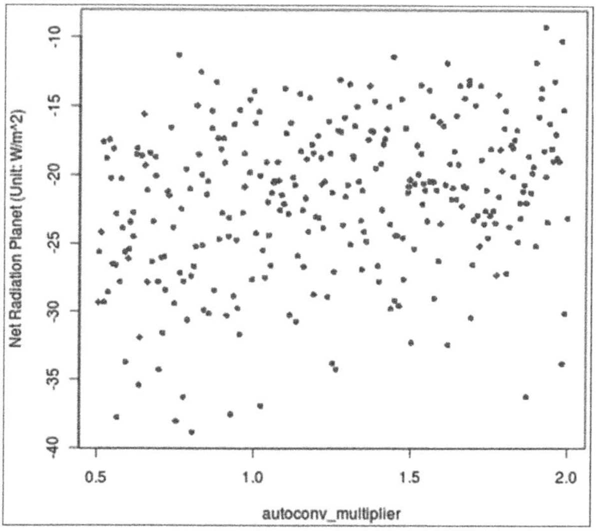

Fig. 18.9 A scatter plot generated in our climate@home project

18.5.2 Multidimensional Rendering and Visualization

Recent decades have generated big data in three dimensional space, time, and multiple principle component variables. Exploring and analyzing valuable information hidden in the large volume of data is a difficult task for scientists (Keim 2002). Visualizing these data with 3D intuitive effect can help us explore potential patterns in data, understand complex phenomena and predict the trend for making decision.

With the development of 3D visualization technologies, traditional 2D visualization like map, plot, bar charts, can be complemented with new technology to visualize and analyze 4D data. There are a large number of 3D visualization techniques which can be used for visualizing 4D big data. Generally speaking, these techniques are divided into two categories: 3D isosurface and Volume rendering. 3D isosurface is a 3D extension of isoline. It is the surface that represents points of a constant value (e.g., pressure, temperature, density) within a 3D space. Marching cubes algorithms (Lorensen and Cline 1987) is a popular method of constructing such isosurfaces from high resolution 3D datasets by creating triangle shaped surfaces with the same vertices density as 3D data. Volume rendering is a technique for visualizing 3D arrays of sampled data (Robert et al. 1988a, b), which is used for the applications in medical imaging and scientific visualization. The classic direct volume rendering algorithms are categorized into volume ray casting, splatting, shear warp and texture mapping. Sometimes, a combination of these techniques is a better way to solve specific visualization problems.

In many cases, regular visualization algorithms cannot satisfy the requirement for rendering with intensive computing requirements. To improve efficiency of visualization, many scientists have focused on optimization techniques. Take volume rendering as example, its use has been limited by its high computational expense.

There are two directions we can solve the intensive computing problem. Firstly, graphics hardware-accelerating is a good way to increase the computing capability. It mainly depends on powerful GPU parallel computing which can provide better rendering capacity and serve high performance application, such as CUDA platform published by NVIDIA. In addition, Distributed visualization strategy and hybrid computing strategy with CPUs and GPUs are also effective ways to tremendously improve the performance of computing resources to achieve the intensive-computing task. Secondly, we can reduce as much volume as possible in current rendering pipe. A typical atmospheric data set can be more than Terabytes in size. To render it fluently require a fast memory. Generally, these optimization techniques are categorized as empty space skipping, early ray termination, Octree and BSP space subdivision, multiple and adaptive resolution representation, and user layers of data for rendering. For example, Li et al. (2011a) gave an Octree-based view dependent LOD strategy to help efficiently visualize large volume data.

As an example, in climate studies, we developed a set of advanced rendering solutions to visualize the dynamic phenomena with virtual globe(World Wind) to handle big data by: (1) Designing a fresh volume render algorithm which use sphere as basic splatting unit to represent each cell from 4D data array, (2) Designing and implementing data processing and render techniques to deal with large amount of rendering volume data, such as LOD, (3) Making full use of machine's computing resources of both CPU and GPU to accelerate the speed of rendering pipeline. As a result, the system (Fig. 18.10) provided more intuitive visual effects and convenient interactivity operations in visualizing dynamic environmental phenomena and helps scientists explore the potential patterns and principles in scientific research.

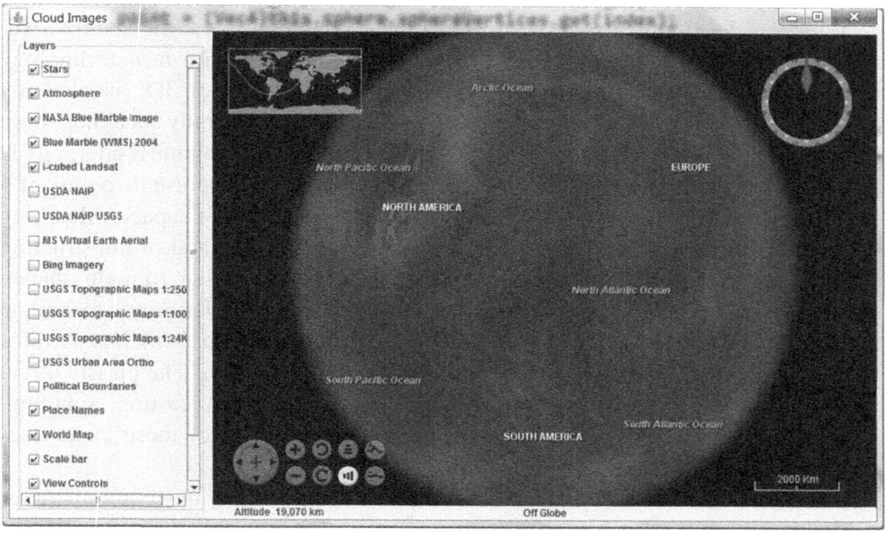

Fig. 18.10 World Wind based 4-dimensional visualization is used to explore climate patterns

18.6 Computing Technologies

To address the computing requirements from model simulations, data discovery by spiking numbers of users, and visualization pipelines, different computing accelerating technologies can be utilized. This section demonstrates how grid computing, cloud computing, and GPU accelerating technologies are utilized to solve the computing challenges.

18.6.1 Cloud Computing for Spike User Access of Clearinghouse

To tackle the data intensity challenges, a series of techniques and services are developed to enable the operations of GEOSS clearinghouse (Huang et al. 2010). Currently, the GEOSS clearinghouse is hosted by the Amazon EC2 to handle potential massive users and data requests. For example, given the elasticity of the computing platform for the clearinghouse, we can autoscale one, two, and five instances to respond to different spike number of user access (Yang et al. 2011a). Figure 18.11 illustrates how the cloud responds to massive concurrent user requests by spinning off new EC2 cloud instances. The example illustrates a varying number of requests to the GEOSS clearinghouse. All instances are running from beginning except the autoscaling case, which has one instance running at beginning and elastically adds instances when needed by concurrent requests. It is observed that when more computing instances are utilized, higher gains in performance can be obtained. The elastic automated provision and release of computing resources greatly prepared us to respond to concurrent access spikes while at the same time reducing cost by sharing computing resources with other applications when there are no concurrent access spikes.

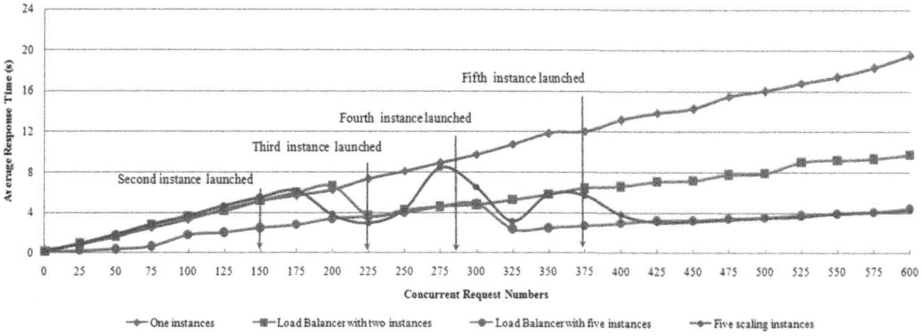

Fig. 18.11 Server response performance comparison by single, two, five, and five autoscaling instances (Yang et al. 2011a)

18.6.2 Grid Computing for Finishing 300 Climate Model Runs

We were requested to run the climate change model ModelE for climate@home project 300 configurations of 10 years within half a month. In average, a 10-year simulation takes approximately 4 days and 15 h to complete on a single computer with a 2.88 GHz CPU speed. It is impossible for us to finish the job with one computer or request HPC resources which would cost us the half month for setting up. We used the Eucalyptus cloud platform to spin off 80 and 160 server instances and used grid computing to coordinate the distribution of tasks and collect results.

Figure 18.12 shows the percentage of computing resource usage, number of CPU cores for computing, and the number of ModelE runs at different time period. On April 11st, 2012, 80 virtual instances were used to run the 300 tasks. At any time, each virtual instance was able to handle one task. Around April 16th, the first 80 runs finished. Around 18th April, another group of 80 instances were able to join

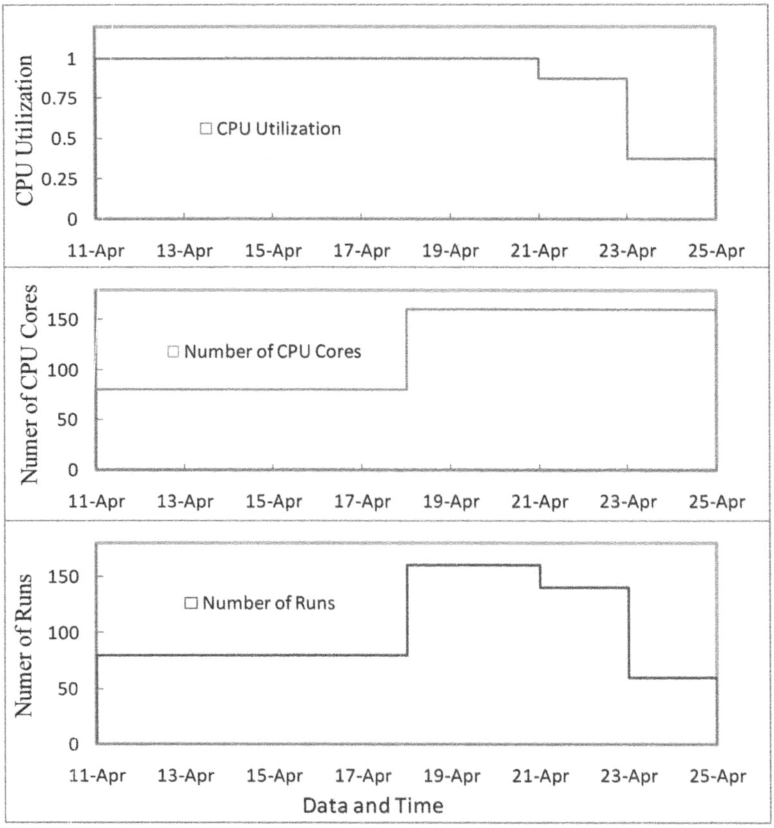

Fig. 18.12 The percentage of computing resource usage, number of CPU cores for computing, and the number of ModelE runs

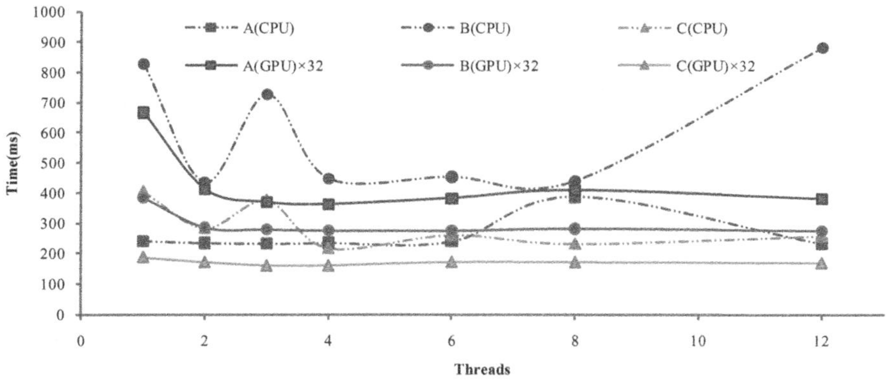

Fig. 18.13 Rendering performance with CPU and GPU computing on three different machines

the computation. Since then, 160 instances were available for the model runs. April 21st, 2012, the 80 tasks starting from April 16th were completed, and the released 80 CPU cores were arranged to run the remaining 60 tasks. Since that time, the total model running tasks by the system were 140. The 80 runs starting from April 18 were finished around Aril 23. With the support of the virtualized resources, the 300 runs were successfully completed within 14 days, rather than 1,387 days, the time required by using one computer.

18.6.3 GPU and CPU for Visualization Pipeline

As a cost-efficient computing technique, GPU-based computing has been employed in various application domains. We investigated the potentials of using GPUs to improve the performance for visualizing of 3D/4D environmental data, which involves multidimensional data handling, mathematical calculations and volumetric rendering. Figure 18.13 shows the rendering performance with CPU and GPU computing on three different machines. We found that, in general, GPU-based rendering is four to five times faster than CPU-based rendering and more stable than CPUs.

18.7 Conclusion

Spatiotemporal big data pose grand challenges in that they are collected and not well structured in heterogeneous formats, storage, archives, managements, processes, and models. Contemporary computing technologies provide enabling capabilities to process the data and extract information/knowledge embedded within the large

Table 18.1 Contemporary technologies for supporting the life cycle of big data

Techs\BD processing	Collection	Storage	Management	Processing	Simulation	visualization
Cloud computing		x	x	x	x	
Spatiotemporal	x	x	x	x	x	x
Data mining		x		x		x
Crowd-sourcing/grid computing	x				x	
GPU computing						x

volume and unstructured system during the lifecycle of data collection/generation, storage/archive, discovery, processing/modeling, analyses, and visualization. Contemporary computing technologies (such as grid computing, cloud computing, and GPU computing) and advanced methodologies (such as spatiotemporal statistics and spatiotemporal data visualization) provide new tools and software to enable handling big data. This chapter uses several projects as examples to illustrate how big data are handled with the computing and advanced technologies. Table 18.1 summarizes how different technologies are used to support the lifecycle of big data handling. The boxes checked means that the technology is used to address the challenges within relevant data handling stage. For example, cloud computing can be used to help manage storage for convenient storage access.

While the processing of big data starts as a new research and in the development direction, many new issues remain to be researched. To name a few: security, large/fast data storage, sufficient computing power, and smart and fast data processing.

- Security is a big concern for both sensitive data and the information systems that handles the big data when the data is archived across the global environment and shared among global users. Relevant policies and regulations should be developed to address this problem.
- Storage is another concern in terms of both volume and speed of storage as well as the manageability of storage for each data access including writing, reading, selecting, and updating.
- Computing power is short for big data handling. How to reduce the computing requirements and best leverage different types of computing resources in combination, such as using both cloud computing and grid computing for data processing.
- Smart and fast data processing is critical for real time applications, such as emergency response and decision relevant event detection.

Acknowledgements Research reported is supported by Microsoft Research, NSF (IIP-1160979), and NASA (NNX12AF89G).

References

Agrawel D., et al. (2012). *Challenges and opportunities with big data*. Big data white paper (Computing Research Association). http://cra.org/ccc/docs/init/bigdatawhitepaper.pdf. Accessed 23 Oct 2012.

Akyildiz, I. F., Su, W. L., Sankarasubramaniam, Y., & Cayirci, E. (2002). A survey on sensor networks. *IEEE Communications Magazine, 8*(40), 102–114.

Białecki, A., Muir, R., & Ingersoll, G. (2012, August 12–16). Apache lucene 4. In A. Trotman, et al. (Ed.), *SIGIR 2012 workshop on open source information retrieval* (pp. 17–24). Portland: Department of Computing Science, University of Otago.

Christian, E. (2005). Planning for the Global Earth Observation System of Systems (GEOSS). *Space Policy, 2*(21), 105–109.

Dublin. (2012). *Dublin core metadata standards*. http://dublincore.org. Accessed 2 Nov 2012.

FGDC (The Federal Geographic Data Committee). (2012). *Geospatial metadata standards*. http://www.fgdc.gov/metadata/geospatial-metadata-standards. Accessed 2 Nov 2012.

GEO (2012). *The GEOSS common infrastructure*. http://www.Earthobservations.org/gci_gci.shtml. Accessed 2 Nov 2012.

Goldenberg, S. B., Landsea, C. W., Mestas-Nuñez, A. M., & Gray, W. M. (2001). The recent increase in Atlantic hurricane activity: Causes and implications. *Science, 5529*(293), 474–479.

Guan, Q. (2008). *Parallel algorithms for geographic processing*. PhD dissertation, University of California at Santa Barbara, Santa Barbara, CA, USA. Advisor(s) Keith Clarke. AAI3319802.

Gui, Z., Yang, C., Xia, J., Liu, K., Xu, C., Li, J., & Lostritto, P. (2012). A performance, semantic and service quality enhanced distributed search engine for improving geospatial resource discovery. *International Journal of Geographical Information Science, 27*, 1109–1132. doi:10.1080/13658816.2012.739692.

Hart, J. K., & Martinez, K. (2006). Environmental sensor networks: A revolution in the earth system science? *Earth-Science Reviews, 78*(3–4), 177–191.

Hibbard, W., Paul, B., Battaiola, A., Santek, D., Voidrot-Martinez, M., & Dyer, C. (1994). Interactive visualization of earth and space science computations. *Computer, 27*(7), 65–72.

Huang, Q., & Yang, C. (2011). Optimizing grid configuration to support geospatial processing – An example with DEM interpolation. *Computers & Geosciences, 37*(2), 165–176.

Huang, Q., Yang, C., Nebert, D., Liu, K., & Wu, H. (2010, November 2). Cloud computing for geosciences: Deployment of GEOSS clearinghouse on Amazon's EC2. In Proceedings of ACM SIGSPATIAL international workshop on high performance and distributed geographic information systems (HPDGIS) (2010), San Jose, CA.

Huang, M., Maidment, D. R., & Tian, Y. (2011). Using SOA and RIAs for water data discovery and retrieval. *Environmental Modelling and Software, 26*(11), 1309–1324.

Huang, Q., Yang, C., Benedict, K., Rezgui, A., Xie, J., Xia, J., & Chen, S. (2012). Using adaptively coupled models and high-performance computing for enabling the computability of dust storm forecasting. *International Journal of Geographical Information Science, 27*, 765–784. doi:10.1080/13658816.2012.715650.

Jacobson, M., Charlson, R. J., Rodhe, H., & Orians, G. H. (2000). *Earth system science: From biogeochemical cycles to global changes* (International geophysics, Vol. 72, p. 527). San Diego: Academic.

James, P., Stohl, A., Forster, C., Eckhardt, S., Seibert, P., & Frank, A. (2003). A 15-year climatology of stratosphere-troposphere exchange with a Lagrangian particle dispersion model: 2. Mean climate and seasonal variability. *Journal of Geophysical Research, 108*, 8522. doi:10.1029/2002JD002639.

Keim, D. A. (2002). Information visualization and visual data mining. *IEEE Transactions on Visualization and Computer Graphics, 8*, 1–8.

Kosslyn, S. M. (2006). *Graph design for the eye and mind*. Oxford: Oxford University Press.

Kumar, S. V., Peters-Lidard, C. D., Tian, Y., Houser, P. R., Geiger, J., Olden, S., Lighty, L., Eastman, J. L., Doty, B., Dirmeyer, P., Adams, J., Mitchell, K., Wood, E. F., & Sheffield, J. (2006). Land information system: An interoperable framework for high resolution land surface modeling. *Environmental Modelling & Software, 21*(10), 1402–1415.

Kusuda, T., & Achenbach, P. R. (1965). *Earth temperature and thermal diffusivity at selected stations in the United States* (No. NBS-8972). Gaithersburg: National Bureau of Standards.

Levoy, M. (1988). Display of surfaces from volume data. *IEEE Computer Graphics and Applications, 8*(3), 29–37.

Li, J., Wu, H., Yang, C., Wong, D., & Xie, J. (2011a). Visualizing dynamic geosciences phenomena using an octree-based view-dependent LOD strategy within virtual globes. *Computers & Geosciences, 37*, 1295–1302.

Li, Z., Yang, C., Wu, H., Li, W., & Miao, L. (2011b). An optimized framework for OGC web service seamlessly integration to support geospatial sciences. *International Journal of Geographic Information Sciences, 25*(4), 595–613.

Lorensen, W. E., & Cline, H. E. (1987). Marching cubes: A high resolution 3D surface construction algorithm. *Computer Graphics (SIGGRAPH'87 Proceedings), 21*, 163–169.

Nah, F. (2004). A study on tolerable waiting time: How long are Web users willing to wait? *Behaviour & Information Technology, 23*(3), 153–163.

OpenLayer. (2012). http://openlayers.org. Accessed 2 Nov 2012.

Potter, K. (2009). Visualization of uncertainty and ensemble data: Exploration of climate modeling and weather data with integrated ViSUS-CDAT systems. *Journal of Physics: Conference Series, 2009*(180), 1–5.

Robert, A. D., Loren, C., & Pat, H. (1988a). Volume rendering. *Computer Graphics, 22*(4), 65–74.

Robert, A. D., Loren, C., & Pat, H. (1988b). Volume rendering. *ACM SIGGRAPH Computer Graphics, 22*(4), 65–74.

Robock, A. (2000). Volcanic eruptions and climate. *Reviews of Geophysics, 38*, 191–219.

Rodríguez, E., Morris, C. S., & Belz, J. E. (2006). A global assessment of the SRTM performance. *Photogrammetric Engineering & Remote Sensing, 72*(3), 249–260.

Roth, R. E., Robbinson, A., Stryker, M., Maceschren, A. M., Lengerich, E. J., & Koua, E. (2008). *Web-based geovisualization and geocollaboration: Applications to public health*. Joint Statistical Meeting. Invited Session on Web Mapping, Denver, CO, USA, 2008.

Schmidt, G. A., Ruedy, R., Hansen, J. E., Aleinov, I., Bell, N., Bauer, M., Bauer, S., Cairns, B., Canuto, V., & Cheng, Y. (2006). Present day atmospheric simulations using GISS ModelE: Comparison to in-situ, satellite and reanalysis data. *Journal of Climate, 19*, 153–192.

Stone, J. E., Phillips, J. C., Freddolino, P. L., Hardy, D. J., Trabuco, L. G., & Schulten, K. (2007). Accelerating molecular modeling applications with graphics processors. *Journal of Computational Chemistry, 28*(16), 2618–2640.

Sun, M., Li, J., Yang, C., Schmidt, G. A., Bambacus, M., Cahalan, R., Huang, Q., Xu, C., & Noble, E. U. (2012). A web-based geovisual analytical system for climate studies. *Future Internet, 4*, 1069–1085. doi:10.3390/fi40x000x.

Theodoridis, Y., Vazirgiannis, M., & Sellis, T. (1996). Spatio-temporal indexing for large multimedia applications. In Proc. of the IEEE conference on multimedia computing and systems, ICMCS.

Tufte, E. (1997). *Visual explanations: Images and quantities, evidence and narrative*. Cheshire: Graphics Press.

von Storch, H., & Zwiers, F. W. (1999). *Statistical analysis in climate research*. Cambridge: Cambridge University Press.

Yang, C., Li, W., Xie, J., & Zhou, B. (2008). Distributed geospatial information processing: Sharing distributed geospatial resources to support digital earth. *International Journal of Digital Earth, 1*(3), 259–278.

Yang, C., Wong, D., Miao, Q., & Yang, R. (2010). *Advanced geoinformation science* (485p.). Boca Raton: CRC Press.

Yang, C., Goodchild, M., Huang, Q., Nebert, D., Raskin, R., Xu, Y., Bambacus, M., & Fay, D. (2011a). Spatial cloud computing: How could geospatial sciences use and help to shape cloud computing. *International Journal on Digital Earth, 4*(4), 305–329.

Yang, C., Wu, H., Huang, Q., Li, Z., & Li, J. (2011b). Using spatial principles to optimize distributed computing for enabling the physical science discoveries. *Proceedings of the National Academy of Sciences of the United States of America, 108*(14), 5498–5503.

Chapter 19
Short-Interval Monitoring of Land Use and Land Cover Change Using a Time Series of RADARSAT-2 Polarimetric SAR Images

Anthony Gar-On Yeh and Zhixin Qi

19.1 Introduction

With rapid growth in economy in the last two decades, cities in some rapidly developing regions in China, such as the Pearl River Delta, have expanded considerably. The demand for land for industrial and residential uses imposed increasing pressure on the management of agricultural and environmental sensitive land. Although a great deal of agricultural land was expropriated to meet the increasing land demand for residence and business, insufficiency of land supply still restricted industrial development in these regions. The insufficiency of land resource caused land prices to soar. As a result, many illegal land uses emerged in some rapidly developing places, such as urban fringes, in these regions. Due to the lack of effective monitoring tools, some illegal land uses have caused irreversible environmental problems, such as loss of farmland, forest degradation, soil erosion, and adverse effects on species diversity (Yeh and Li 1996). Serious environmental deterioration caused by urban sprawl poses a threat to healthy urban development. Short-interval, such as monthly, land use and land cover (LULC) change information is important for the government to control and prevent illegal land development at its early stage. Furthermore, it is also important in analyzing and simulating urban land development and assessing the ecological effect in different scales. Therefore, short-interval monitoring of LULC change in these rapidly developing regions is much needed.

Traditional field-based approaches are limited by their spatial coverage and temporal frequency. Satellite remote sensing has been employed prominently in LULC change monitoring because of its capability to observe land surface in

A.G.-O. Yeh (✉) • Z. Qi
Department of Urban Planning and Design, The University of Hong Kong, Hong Kong, Hong Kong
e-mail: Anthony.Yeh@hku.hk; qizhixin@connect.hku.hk

M.-P. Kwan et al. (eds.), *Space-Time Integration in Geography and GIScience*, DOI 10.1007/978-94-017-9205-9_19

repetitive and consistent manner over large areas. Remote sensing data obtained from different optical sensors are commonly used in characterizing and quantifying LULC change information (Fung 1990; Stefanov et al. 2001; Seto et al. 2002). Numerous studies have been carried out on the use of the visible to shortwave infrared (VSWIR) bands of Landsat Multispectral Scanner (MSS) and Thematic Mapper (TM) data for forestry and agricultural land cover analysis and urban development monitoring (Jensen and Toll 1982; Howarth and Boasson 1983; Coppin and Bauer 1994; Collins and Woodcock 1996; PaxLenney and Woodcock 1997; Li and Yeh 1998). New sensors with higher spatial resolutions, such as SPOT, the Advanced Spaceborne Thermal Emission and Reflection Radiometer (ASTER), the Enhanced TM plus (ETM+), QuickBird, and IKONOS, have also been used for urban studies and land cover change detection (Quarmby and Cushnie 1989; Lupo et al. 2001; Forster et al. 2008; Sonmez et al. 2009).

A number of change detection techniques have been developed in past studies. Reviews of existing change detection methods can be found in many papers (Singh 1989; Coppin et al. 2004; Lu et al. 2004). The current change detection methods can be summarized into two categories: the unsupervised approach and the supervised approach. Among the most widely used unsupervised approaches are image differencing (Weismiller et al. 1977), image rationing (Howarth and Wickware 1981), and change vector analysis (Lambin and Strahler 1994). The unsupervised approach is relatively simple, straightforward, and easy to implement and interpret. However, this approach cannot provide information on the types of LULC changes. The supervised approach can provide information on both changed areas and the types of change these areas undergo because it implements change detection based on separate supervised classification of multi-temporal images. Post-classification comparison (PCC) is a widely used supervised method for change detection (Foody 2001; Wilson and Sader 2002). However, the accuracy of PCC is limited by the accuracy of the classification. Methods combing the unsupervised and the supervised approach were also proposed for change detection (Petit and Lambin 2001; Silapaswan et al. 2001). The unsupervised approach can be used to detect changed areas and then PCC is used to classify detected changed areas to determine types of changes. The hybrid method can reduce the impact of each individual classification on the change detection result and then provide information on types of change.

With these change detection methods, optical remote sensing images have been successfully used in monitoring change for a myriad of application. However, optical remote sensing is limited by weather conditions. Difficulties are encountered in collecting timely LULC information in tropical and sub-tropical regions, such as the Pearl River Delta, characterized by frequent cloud cover. LULC change detection that uses optical remote sensing images in these regions usually covers a long time period, such as several years, due to the unavailability of enough images. Long-interval LULC change information will not be adequate for controlling illegal land use sites in these regions. Many irreversible damages may have already occurred to the environment because of the lack of proper tools in monitoring illegal land development at a more frequent interval. It is important to monitor LULC changes on a regular short-interval, such as monthly, basis to control and prevent illegal land development at its early stage.

Radar remote sensing, which is not affected by clouds, is therefore an effective tool in extracting timely LULC information in regions frequently covered by clouds. Early studies that employed radar remote sensing to investigate LULC information were mainly performed using the space shuttle SIR-C/X-SAR (Saatchi et al. 1997; Pierce et al. 1998). Although the results of these studies are positive, airborne radar imagery systems are only occasionally launched to collect experimental data within a very short period. Regular investigation of timely LULC information using radar remote sensing has become practical only after some operational orbital radar systems with SAR, such as ERS-1 and ERS-2, JERS-1, and RADARSAT-1, were made available for regular data collection. However, most of the existing orbital SAR systems are single-frequency types and might create confusion during the separation and mapping of LULC classes. This confusion can stem from limited information obtained by single-frequency systems (Ulaby et al. 1986; Li and Yeh 2004). To overcome the difficulty presented by single-frequency SAR data, some researchers utilized polarimetric SAR (PolSAR) data to study LULC information (Pierce et al. 1994; Du and Lee 1996; Lee et al. 2001; Freitas et al. 2008). The results show that PolSAR measurements achieve much better classification results than single-polarization SAR. The investigation of LULC information using PolSAR images has become an important research topic since PolSAR images were made available through ENVISAT ASAR, ALOS PALSAR, and RADARSAT-2. Therefore, PolSAR should be a promising tool for short-interval monitoring of LULC change. However, at present, there are still limited studies on the use of PolSAR images to detect short-interval LULC changes. The purpose of this chapter is to explore the potential use of PolSAR data in short-interval monitoring of LULC change.

19.2 Study Area and Data

The study area is located in Panyu District of Guangzhou City in southern China. Panyu lies at the heart of the Pearl River Delta (PRD) and has a total land area of 1,314 km^2 as well as a population of 926,542. This district was an agricultural area before the economic reform in 1978, but has been quickly transformed recently into a rapidly urbanized area. Since Panyu became a district of Guangzhou in July 2000, intensive land development has been implemented to provide housing to the residents of Guangzhou City. Huge profits have been generated through property development, which resulted in the increase in land speculation activities and illegal land use cases.

A time series of RADARSAT-2 PolSAR images with a repeat cycle of 24 days was used in this study (Fig. 19.1). The acquisition parameters of the images are shown in Table 19.1. RADARSAT-2 is the world's most advanced commercial C-band SAR satellite. It is designed with significant and powerful technical advancements, one of which is multi-polarization. RADARSAT-2 can transmit

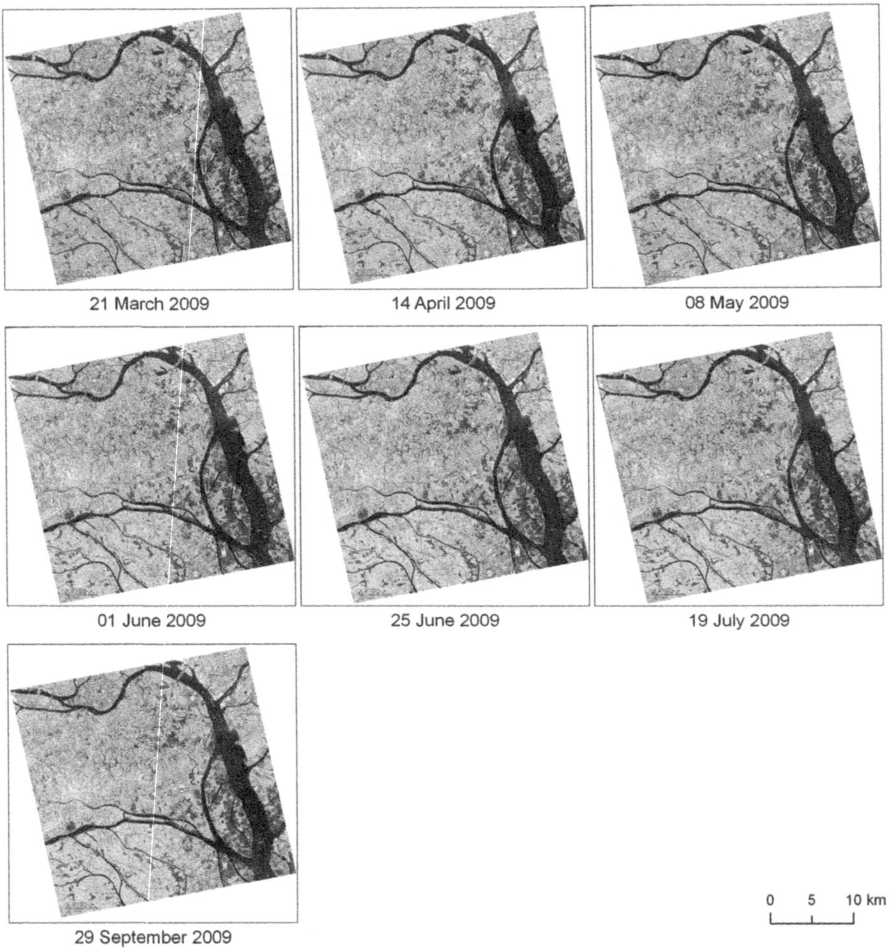

Fig. 19.1 The time series of RADARSAT-2 PolSAR images for short-interval monitoring of LULC change

Table 19.1 Acquisition parameters of the RADARSAT-2 PolSAR images

No.	Acquisition date	Beam	Incident angle	Resolution (range × azimuth)	Orbit
1	2009/03/21	FQ12	31.50	5.2 m × 7.6 m	18-205A
2	2009/04/14	FQ12	31.50	5.2 m × 7.6 m	19-205A
3	2009/05/08	FQ12	31.50	5.2 m × 7.6 m	20-205A
4	2009/06/01	FQ12	31.50	5.2 m × 7.6 m	21-205A
5	2009/06/25	FQ12	31.50	5.2 m × 7.6 m	22-205A
6	2009/07/19	FQ12	31.50	5.2 m × 7.6 m	23-205A
7	2009/09/29	FQ12	31.50	5.2 m × 7.6 m	26-205A

Table 19.2 Number of pixels selected for different types of LULC change

	03/21–04/14	04/14–05/08	05/08–06/01	06/01–06/25	06/25–07/19	07/19–09/29
BL-UB	4,075	1,543	7,514	1,202	1,443	8,134
BL-V	0	1,484	5,301	4,059	1,337	9,009
BL-W	5,090	1,331	10,227	2,538	5,035	545
BL-P	6,526	5,365	333	0	0	10,946
V-BL	10,296	2,709	0	2,988	9,624	7,250
V-UB	256	549	360	2,037	0	3,290
V-W	4,546	1,434	2,407	494	437	1,995
W-BL	5,729	7,748	6,344	11,510	2,655	3,986
W-V	1,018	472	1,585	415	228	5,362
W-P	0	3,890	435	0	0	5,619

horizontal (H) and vertical (V) polarizations depending on the selected mode. Each scattering element (HH, VV, HV, and VH) has varying sensitivities to different surface characteristics and properties, thereby helping improve the discrimination among LULC types.

Field investigations were carried out simultaneously with the acquisitions of the images to collect ground truth. In the field investigations, field plots were selected across the typical LULC classes using a clustered sampling approach (McCoy 2005). In a terrain with poor access, this sampling approach enables the use of most of the accessible sites. GPS was used to record the coordinates of these field plots. LULC classes in the study area can be summarized into five categories: urban/built-up (UB), vegetation (V), paddy (P), water (W), and barren land (BL). On the basis of the experience with multinomial distribution (Congalton and Green 2009), we collected a minimum of 50 samples for each LULC category. The sampling size per field plot in the images ranged from 39 to 603 pixels. This range was determined using the ground coverage in the photos taken during the fieldwork. To evaluate change detection and change type determination results, samples of different types of LULC change were also selected through visual interpretation and field investigations. Visual interpretation was first conducted to identify changed areas from the entire coverage of images. The selected changed areas were then validated in field investigations. The number of plots of different types of LULC change was determined in the light of actual change. More samples were selected for types with more changes, whereas fewer samples were selected for types with fewer changes. Table 19.2 shows the number of pixels selected for each type of LULC change.

19.3 Methodology

This study utilized the time series of RADARSAT-2 PolSAR images to extract short-interval LULC changes. The repeat cycle of RADARSAT-2 is 24 days and thus monthly LULC changes were extracted by using two successive images.

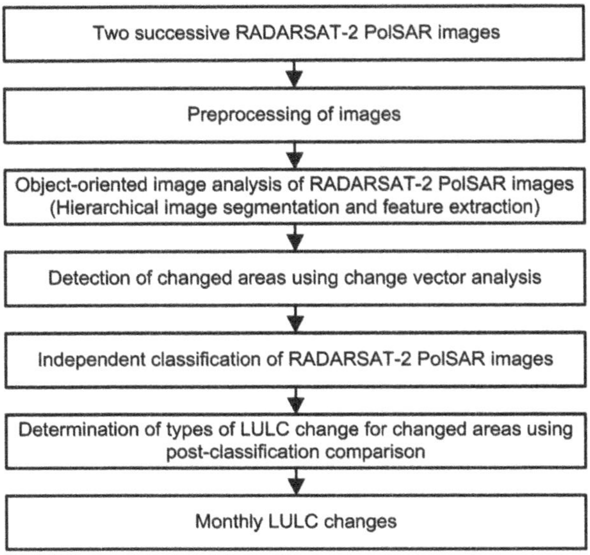

Fig. 19.2 Methodology of extracting monthly LULC changes using two successive RADARSAT-2 PolSAR images

For example, images acquired on March 21 and April 14 were used to extract LULC changes between March 21 and April 14, and then images acquired on April 14 and May 08 were used to extract LULC changes between April 14 and May 08. The methodology of detecting monthly LULC changes using two successive images based on our previous study (Qi and Yeh 2013) is shown in Fig. 19.2.

19.3.1 Preprocessing of RADARSAT-2 PolSAR Images

Image preprocessing, which is critical to change detection, included radiometric correction, speckle filtering, and image registration. Radiometric calibration for the RADARSAT-2 images was performed using PolSARPro_v4.1.5 software (López-Martínez et al. 2005) and applying the sigma look-up table provided in the product. After radiometric correction, the pixel values of the images could be directly related to the radar backscatter of the scene. This is necessary for the comparison of PolSAR images acquired from the same sensor but at different times. A J.S. Lee Sigma filter with a window size of 7×7 was applied on the images to reduce speckles. Compared with other filters, this one effectively retains subtle details and preserves the shape of small land parcels while reducing the speckle effect (Lee et al. 2009). The advantage of this filter will benefit the accurate delineation of tiny land parcels in object-oriented image analysis. Image registration was based on the geometric rectification of the RADARSAT-2 images. The PCI Geomatica software was used to implement the geometric rectification of the images. The RADARSAT-2 image package provides a total of 180 tie points evenly distributed across the entire image.

These tie points tie the line/pixel positions in image coordinates to geographical latitude/longitude and can be used as ground control points (GCP) to register an image to a geocoded target image. This work first created a master geocoded image with the same resolution as the RADARSAT-2 images and then registered the RADARSAT-2 images to this geocoded master image using PCI Geomatica based on the tie points. Visual inspection indicates that the RADARSAT-2 images were registered perfectly.

19.3.2 Object-Oriented Image Analysis of RADARSAT-2 PolSAR Images

Most change detection and classification methods for remote sensing images are pixel-based. When applied to PolSAR images, pixel-based methods have two disadvantages. First, they are prone to be affected by speckles in PolSAR images and produce errors (Qi et al. 2012). Second, they are difficult to use to extract and utilize spatial and textural information, which is helpful in improving classification accuracy of remote sensing data (Gao et al. 2006). Object-oriented image analysis can be used to reduce the speckle effect and extract textural and spatial information to support PolSAR image classification. In object-oriented image analysis, image objects (groups of pixels) are first delineated by using image segmentation techniques, and then change detection and classification are implemented on the object level. In comparison with pixels, image objects are little affected by speckles. Moreover, object-oriented image analysis enables the acquisition of a variety of textural and spatial features for improving classification accuracy. A feature is an attribute that represents certain information concerning objects of interest. Given that regions in an image provide considerably more information than do pixels, many different features for measuring color, shape, and texture of the associated regions can be used (Benz et al. 2004).

In this study, the object-oriented package Definiens Developer 7.0 (Baatz et al. 2004) was used to implement the object-oriented image analysis of RADARSAT-2 images. The multi-resolution segmentation module provided by Definiens Developer 7.0 was used to perform image segmentation based on shape and color homogeneity. For change detection, a straightforward approach for image segmentation is to segment images acquired at different times separately and then overlay them together. However, this method will produce inconsistencies in the delineation of boundaries of objects and result in a large number of fragmental patches in the final segmented image (Li et al. 2009). Such excessive fragmentation can lead to difficulties in change detection. A hierarchical image segmentation procedure was proposed to minimize the inconsistency in delineating objects from two successive images. In considering two co-registered images, image (t_1) and image (t_2), acquired over the same area at different times t_1 and t_2, the procedure of hierarchical segmentation can be summarized as follows: (1) the initial segmentation is applied to image (t_1) with a fixed scale parameter; and (2) the same segmentation

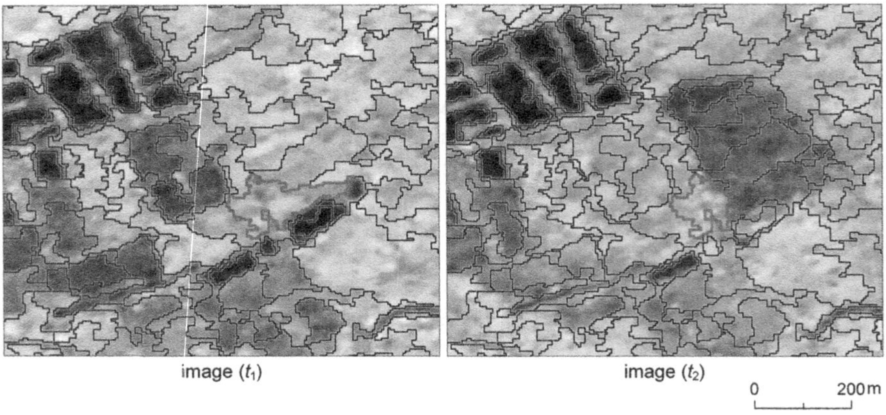

Fig. 19.3 Hierarchical segmentation for delineating image objects from two successive RADARSAT-2 PolSAR images

process is implemented again on image (t_2) while the segmentation result of image (t_1) is taken as the thematic layer for constraint. This procedure causes all object-merging to take place within the boundaries of the segmentation of image (t_1). As shown in Fig. 19.3, new objects are only created in places where the two images are significantly different (e.g., the red polygon in Fig. 19.3). The hierarchical segmentation technique could eliminate inconsistencies in delineating image objects from two successive images. After the delineation of image objects, change detection and classification could be conducted on the object level, and different features could be extracted to support classification.

19.3.3 Combing Change Vector Analysis and Post-classification Comparison for Detecting LULC Changes

Post-classification comparison (PCC) is a commonly used change detection method because it can provide information on both changed areas and the type of change these areas undergo. However, the accuracy of PCC is limited by the accuracy of the classification. It is still difficult to achieve high accuracy for LULC classification using PolSAR images due to the speckle effect and limited spectral information of PolSAR data. Combing unsupervised change detection methods and PCC can reduce the impact of each individual classification on the change detection result and provide information on types of change (Petit and Lambin 2001). The unsupervised approach can be used to detect changed areas and then PCC is used to determine types of changes. Since PCC is only performed on detected change objects, the hybrid method can greatly reduce false alarms caused by the error of independent classification.

This study combined change vector analysis (CVA) and PCC to detect LULC changes based on the object level. CVA is a widely used unsupervised change detection method that uses multichannel images (Malila 1980). CVA can process any number of image channels and can produce detailed change detection information based on the channel change vector obtained by subtracting corresponding image channels of two images acquired at different times. In this study, CVA was used to detect changed objects based on selected features instead of pixel values of image channels. Given that object-oriented CVA can process any number of features, it was suitable for object-oriented change detection using PolSAR images. In object-oriented CVA, change detection is based on feature change vectors (FCVs), which are obtained by subtracting corresponding feature vectors of an image object in two images acquired at different dates. Two co-registered images, image (t_1) and image (t_2), are assumed to be acquired over the same area at different times t_1 and t_2. If k features are extracted from an image object, the feature vectors of the image object in the two images are given by $X = (x_1, x_2, \ldots, x_k)^T$ and $Y = (y_1, y_2, \ldots, y_k)^T$ respectively, the FCVs is defined as

$$\Delta G = X - Y = \begin{pmatrix} x_1 - y_1 \\ x_2 - y_2 \\ \cdots \\ x_k - y_k \end{pmatrix} \tag{19.1}$$

where ΔG includes all the change information between the two images for a given image object, and the change magnitude $\|\Delta G\|$ is computed with

$$\|\Delta G\| = \sqrt{(x_1 - y_1)^2 + (x_2 - y_2)^2 + \cdots + (x_k - y_k)^2} \tag{19.2}$$

The higher the $\|\Delta G\|$ is, the more likely that changes take place. Unsupervised classifiers or threshold methods are commonly applied on the change magnitude to identify changes.

Two aspects are essential for object-oriented CVA: one is the selection of appropriate features to calculate FCVs, and the other is the determination of a suitable threshold to identify changed objects. CVA has been applied on the backscattering matrix of PolSAR data to detect the extent of change caused by an inundation (Shen et al. 2007). Given that the coherency matrix provides more information than the backscattering matrix, this study applied CVA on the coherency matrix to detect LULC changes. A widely accepted assumption is that the statistical distribution of the pixels of change and no-change areas in the change magnitude can be approximated as a mixture of Gaussian distributions (Bovolo and Bruzzone 2007; Camps-Valls et al. 2008). Therefore, the expectation-maximization (EM) algorithm was applied on the change magnitude to identify changed objects. EM is frequently used for data clustering in machine learning and computer vision because it finds clusters by determining a mixture of Gaussians that fit a given data set (Moon 1996). The Weka 3.6 software (Witten et al. 2011) was used to implement EM algorithms to determine the threshold to identify changed objects.

After changed objects were detected using CVA, PCC was performed on the changed objects to determine the type of changes. Many classification methods for PolSAR data have been explored (Chen et al. 1996; Barnes and Burki 2006; Lee et al. 1999; Shimoni et al. 2009). However, so far most of the classification methods are pixel-based. These methods are prone to be affected by speckles in PolSAR images and are hard to utilize textural and spatial information of PolSAR images. Moreover, they cannot take fully use of polarimetric information of PolSAR data for LULC classification. Qi et al. (2012) proposed a new classification method, which integrates polarimetric decomposition, PolSAR interferometry, object-oriented image analysis, and decision tree algorithms, for the classification of PolSAR images. The results show that the proposed method can achieve much higher accuracy than conventional pixel-based classification methods. Polarimetric information has significant implications for identifying different vegetation types and distinguishing between vegetation and urban/built-up. Object-oriented image analysis is helpful in reducing the effect of speckle in PolSAR images by implementing classification based on image objects, and the textural information extracted from image objects is helpful in distinguishing between water and barren land. The decision tree algorithm is efficient in selecting features and implementing classification. In this study, the classification of RADARSAT-2 images was performed using the method proposed by Qi et al. (2012). First, different polarimetric techniques were used to extract polarimetric parameters, and the extracted polarimetric parameters were then combined with the elements of the coherency matrix to form a multichannel image. Second, different features were extracted from the multichannel image for image objects to support LULC classification. Third, a decision tree algorithm was used to select features and create a decision tree for LULC classification. Finally, the final LULC classification was implemented using the constructed decision tree. After the independent classification of RADARSAT-2 PolSAR images, PCC was performed on the changed objects to determine the type of change.

19.4 Results and Discussion

Monthly LULC changes detected using the time series of RADARSAT-2 PolSAR images are shown in Fig. 19.4. Detection accuracy, false alarm rate, and overall error rate are commonly used statistics for evaluating change detection results. The detection accuracy is the percentage of correctly labeled "change" samples. The false-alarm rate is the percentage of erroneously labeled "no-change" samples. The overall error rate is the percentage of erroneously labeled validation samples. These three statistics were calculated using the confusion matrix that was determined using the validation samples of change and no-change. As shown in Fig. 19.5, high accuracies were achieved for monthly LULC change detection by using the time series of RADARSAT-2 PolSAR images. The average overall error rate, average detection accuracy, and average false alarm rate were 1.97 %, 91.29 %, and 1.37 % respectively.

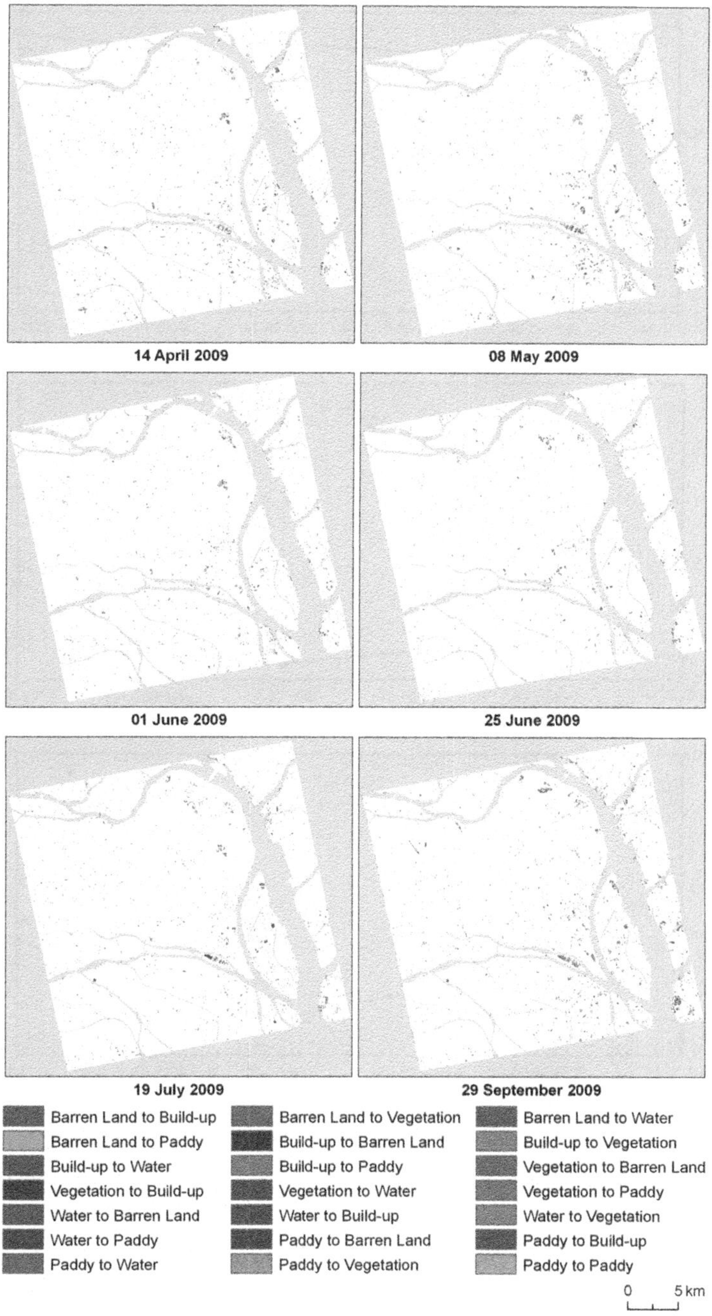

Fig. 19.4 Monthly LULC changes detected using the time series of RADARSAT-2 PolSAR images

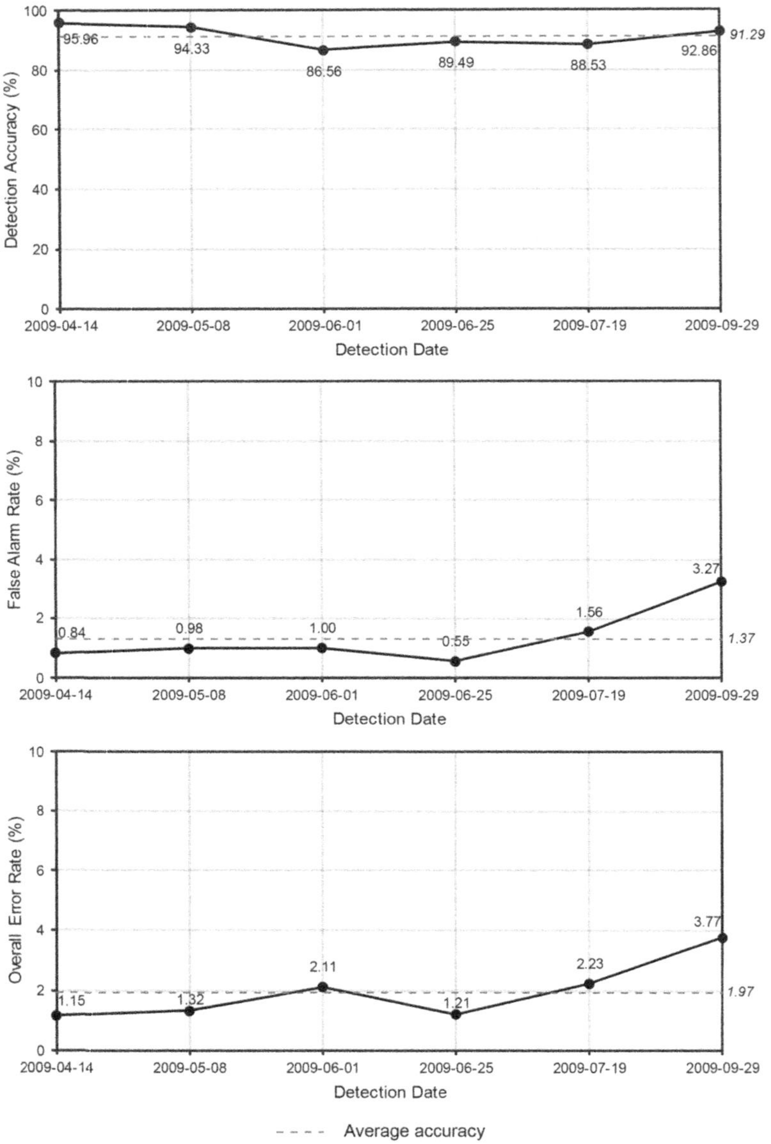

Fig. 19.5 Accuracy of monthly LULC change detection using the time series of RADARSAT-2 PolSAR images

Although the high average accuracies were achieved for change detection, accuracies were different at different times. As show in Fig. 19.5, the overall error rates on September 29 and July 19 were higher than that on the other dates. Many false alarms were found in paddy fields on July 17 and September 29. These false alarms were mainly caused by the growth of paddies. Paddies in the study area have two growth cycles a year. The first growth cycle is usually from early April to late July, and the second growth cycle is from early August to early October. Paddy fields usually turn into barren land covered with stubble and weeds during the other time periods. The study found that paddy fields showed very different backscattering characteristics in PolSAR images with the growth of paddies. The main radar return from paddy fields was paddy-water backscatter (double-bounce scattering) on June 25, but the double-bounce scattering from the paddy fields significantly reduced on July 19 because the water beneath the paddies was almost dried up. Meanwhile, with the growth of paddies, crowns of the paddies became dense, and volume scattering from paddy crowns increased. The main radar return from the paddy fields was dominated by the volume scattering on July 19, and it was hard to distinguish between paddies and other vegetation on that time. Therefore, the change of backscattering characteristics of the paddies fields caused many false alarms in the change detection on July 19. The second growth cycle of paddies started as soon as the end of the first growth cycle. A similar growth process was found in the second growth cycle. The double-bounce scattering from the paddy fields increased again. For the same reason, many false alarms were found in the paddy fields in the change detection on September 29.

The accuracy evaluation for determining types of changes was also carried out based on the confusion matrix that was determined using the validation samples of different types of LULC change. We calculated four statistics for the validation: overall accuracy (OA), estimate of kappa (Kappa), producer's accuracy (PA), and user's accuracy (UA) (Story and Congalton 1986; Congalton and Green 2009). As shown in Fig. 19.6, the average overall accuracy and average kappa value were 72.44 % and 0.68 % respectively. The accuracies were different at different dates, and the overall accuracies and kappa values on September 29 and July 19 were the lowest. Paddy fields were easily confused with vegetation on July 19 because of the similar scattering mechanism, but they became easier to be distinguished from vegetation on September 29. Therefore, many false alarms caused by the growth of paddies were mistakenly identified as changes from vegetation to paddies on September 29. The growth of paddies also caused many false alarms that were misclassified as changes from paddy to vegetation on July 19. Moreover, some paddies were harvested on July 19 and then planted again. These paddies were almost ripe and prone to be confused with vegetation on September 29. This caused some false alarms to changes from barren land to vegetation. The overall accuracies and kappa values on May 08 and June 01 were also low. Some paddies were planted in early April and grew fast from April 14 to May 08. The paddy fields were easily confused with urban/built-up areas due to the strong double-bounce scattering from paddy-water surface on May 08. The fast growth of paddies produced many false alarms to changes from paddy fields to urban/built-up areas in the change detection on May 08. There were also some paddies planted in late April, which were easily confused with barren land or water because of sparse vegetation and irrigation. This created many false alarms to changes from barren land to paddy fields and that from water to paddy fields.

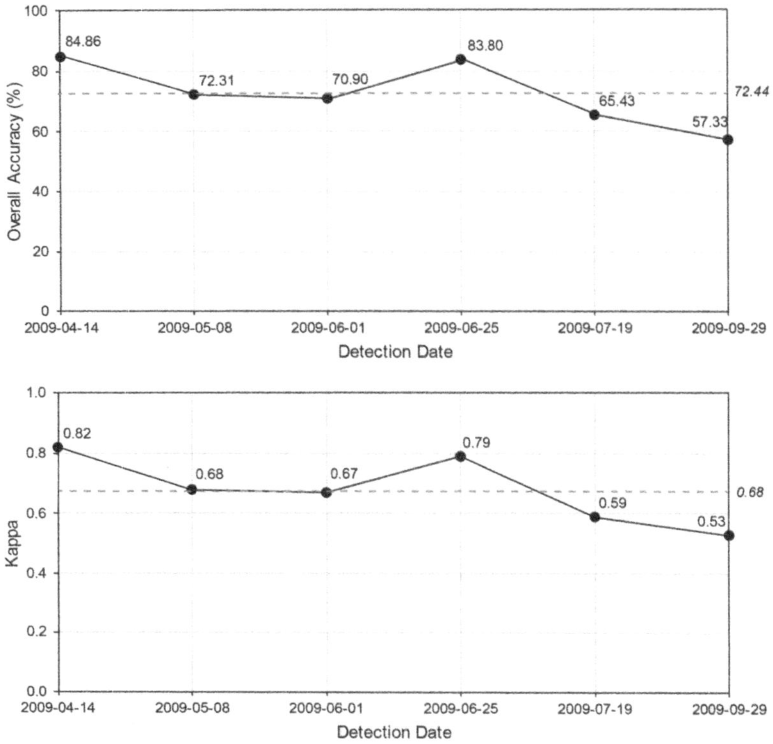

Fig. 19.6 Overall accuracy and kappa value for determining types of LULC change

Illegal land development is commonly related to LULC changes from barren land to urban/built-up areas, that from vegetation to barren land, and that from vegetation to urban/built-up areas. Accurate identification of these types of LULC change are important for detecting potential illegal land use development. The user's and producer's accuracies for determining these types of LULC change were calculated. As shown in Fig. 19.7, the user's and producer's accuracies for changes from vegetation to urban/built-up areas were unstable and very low at some times, and the user's and producer's accuracies for changes from barren land to urban/built-up areas also changed with time. The study found that this was mainly because of the confusion between these two types of LULC change. It is commonly easy to distinguish between barren land and vegetation in PolSAR images because they have different scattering mechanisms. Vegetation has strong volume scattering, while barren land has strong single-bounce scattering because of their relatively flat surface. However, in the study area, barren lands in many construction sites had rough surface because they were dug or covered with mounds. These barren lands were prone to be confused with vegetation because of their rough surface, which causes strong volume scattering. Fortunately, both changes from barren land to

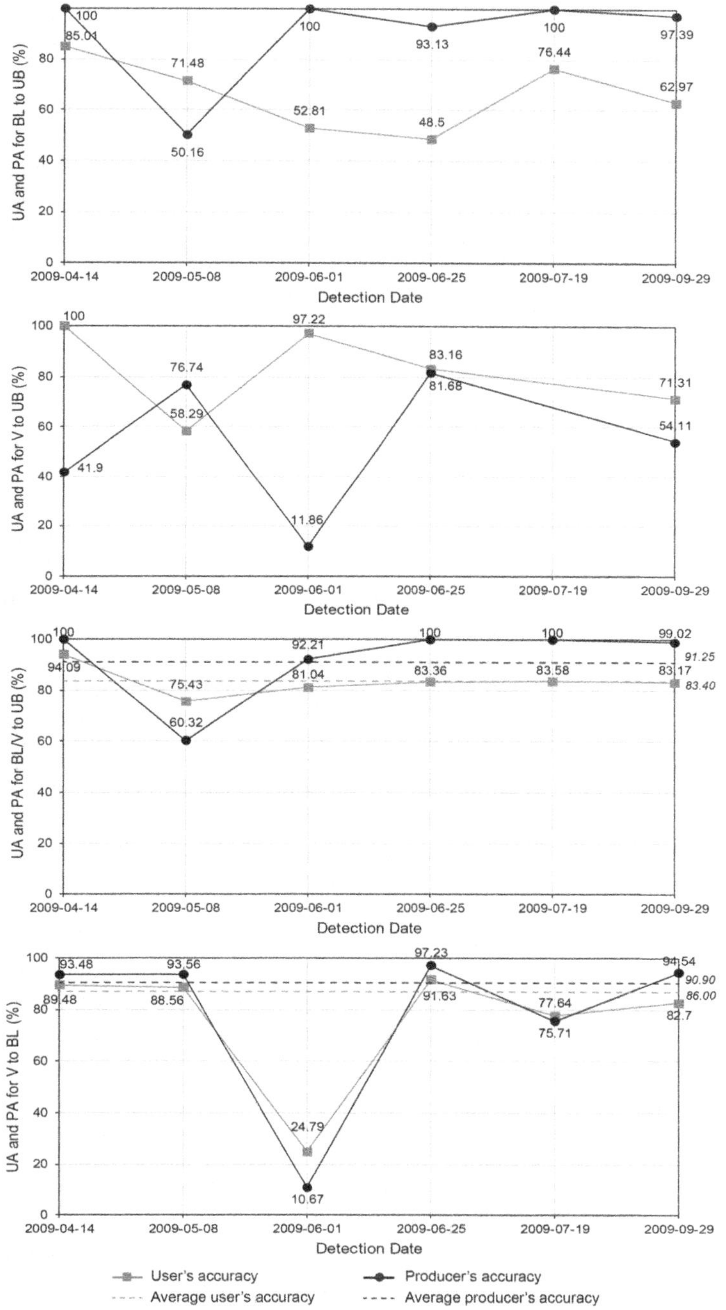

Fig. 19.7 User's and producer's accuracies for determining LULC change types that are important for detecting potential illegal land development

urban/built-up areas and that from vegetation to urban/built-up areas are important for detecting potential illegal land development, and they can be combined together in practical application. The user' and producer's accuracies for changes from barren land or vegetation to urban/built-up areas were also calculated. Figure 19.7 shows that the average user's and producer's accuracies for barren land or vegetation to urban/built-up areas were 83.40 % and 91.25 % respectively. However, the user's and producer's accuracies on May 08 were lower than that on the other dates because paddy growth caused some false alarms to changes from barren land to urban/built-up areas. Figure 19.7 also shows the user's and producer's accuracies were stable and high for determining changes from vegetation to barren land. The average user's and producer's accuracies for changes from vegetation to barren land were 86.00 % and 90.90 % respectively.

19.5 Conclusions

The primary objective of the present study was to explore the potential use of RADARSAT-2 PolSAR images in short-interval monitoring of LULC change. A time series of RADARSAT PolSAR images was used to extract monthly LULC changes. High accuracies were achieved for monthly LULC change detection by using RADARSAT-2 time series images. The average detection accuracy, average false alarm rate, and average overall error rate for detecting changed areas were 91.29 %, 1.37 %, and 1.97 % respectively. The average overall accuracy and average kappa value for determining types of changes were 72.44 % and 0.68 % respectively. Given that illegal land development is commonly related to LULC changes from barren land to urban/built-up areas, that from vegetation to barren land, and that from vegetation to urban/built-up areas, the producer's and user's accuracies for these types of LULC change were also calculated. The average user's accuracy and average producer's accuracy for changes from barren land or vegetation to urban/built-up areas were 83.40 % and 91.25 % respectively. The average user's accuracy and average producer's accuracy for changes from vegetation to barren land were 86.00 % and 90.90 % respectively. The results indicate that it is effective in using a time series of RADARSAT-2 PolSAR images in short-interval monitoring of LULC change, especially in monitoring potential illegal land development.

Paddy cultivation and growth had a great influence on the use of RADARSAT-2 PolSAR images in the monthly monitoring of LULC change. Many false alarms were found in paddy fields due to the growth, planting, and harvesting of paddies, and many of them were mistakenly identified as other LULC change types. This degraded the accuracy of monthly LULC change detection using RADARSAT-2 PolSAR images. Future studies will be carried out on utilizing more multi-temporal PolSAR images, such as three or four images, instead of two images to extract short-interval LULC changes. Compared with LULC changes caused by human activities, changes caused by paddy growth and cultivation change regularly according to season and have a regular seasonal temporal pattern. Therefore, multi-temporal PolSAR images should have a potential application in distinguishing between changes caused by human activities which are irregular and those caused by paddy growth and cultivation which have a regular seasonal temporal pattern.

References

Baatz, M., Benz, U., Dehghani, S., Heynen, M., Höltje, A., Hofmann, P., et al. (2004). *eCognition professional user guide 4*. Munich: Definiens Imaging.

Barnes, C. F., & Burki, J. (2006). Late-season rural land-cover estimation with polarimetric-SAR intensity pixel blocks and sigma-tree-structured near-neighbor classifiers. *IEEE Transactions on Geoscience and Remote Sensing, 44*(9), 2384–2392.

Benz, U. C., Hofmann, P., Willhauck, G., Lingenfelder, I., & Heynen, M. (2004). Multi-resolution, object-oriented fuzzy analysis of remote sensing data for GIS-ready information. *ISPRS Journal of Photogrammetry and Remote Sensing, 58*(3), 239–258.

Bovolo, F., & Bruzzone, L. (2007). A theoretical framework for unsupervised change detection based on change vector analysis in the polar domain. *IEEE Transactions on Geoscience and Remote Sensing, 45*(6), 218–236.

Camps-Valls, G., Gomez-Chova, L., Munoz-Mari, J., Rojo-Alvarez, J. L., & Martinez-Ramon, M. (2008). Kernel-based framework for multitemporal and multisource remote sensing data classification and change detection. *IEEE Transactions on Geoscience and Remote Sensing, 46*(6), 1822–1835.

Chen, K. S., Huang, W. P., Tsay, D. H., & Amar, F. (1996). Classification of multifrequency polarimetric SAR imagery using a dynamic learning neural network. *IEEE Transactions on Geoscience and Remote Sensing, 34*(3), 814–820.

Collins, J. B., & Woodcock, C. E. (1996). An assessment of several linear change detection techniques for mapping forest mortality using multitemporal Landsat TM data. *Remote Sensing of Environment, 56*(1), 66–77.

Congalton, R., & Green, K. (2009). *Assessing the accuracy of remotely sensed data: Principles and practices*. Boca Raton: CRC Press.

Coppin, P. R., & Bauer, M. E. (1994). Processing of multitemporal Landsat TM imagery to optimize extraction of forest cover change features. *IEEE Transactions on Geoscience and Remote Sensing, 32*(4), 918–927.

Coppin, P., Jonckheere, I., Nackaerts, K., Muys, B., & Lambin, E. (2004). Digital change detection methods in ecosystem monitoring: A review. *International Journal of Remote Sensing, 25*(9), 1565–1596.

Du, L., & Lee, J. S. (1996). Fuzzy classification of earth terrain covers using complex polarimetric SAR data. *International Journal of Remote Sensing, 17*(4), 809–826.

Foody, G. M. (2001). Monitoring the magnitude of land-cover change around the southern limits of the Sahara. *Photogrammetric Engineering and Remote Sensing, 67*(7), 841–847.

Forster, M., Frick, A., Walentowski, H., & Kleinschmit, B. (2008). Approaches to utilising QuickBird data for the monitoring of NATURA 2000 habitats. *Community Ecology, 9*(2), 155–168.

Freitas, C. D., Soler, L. D., Anna, S. J. S. S., Dutra, L. V., dos Santos, J. R., Mura, J. C., & Correia, A. H. (2008). Land use and land cover mapping in the Brazilian Amazon using polarimetric airborne p-band SAR data. *IEEE Transactions on Geoscience and Remote Sensing, 46*(10), 2956–2970.

Fung, T. (1990). An assessment of TM imagery for land-cover change detection. *IEEE Transactions on Geoscience and Remote Sensing, 28*(4), 681–684.

Gao, Y., Mas, J. F., Maathuis, B. H. P., Zhang, X. M., & Van Dijk, P. M. (2006). Comparison of pixel-based and object-oriented image classification approaches – A case study in a coal fire area, Wuda, Inner Mongolia, China. *International Journal of Remote Sensing, 27*(18), 4039–4055.

Howarth, P. J., & Boasson, E. (1983). Landsat digital enhancements for change detection in urban environments. *Remote Sensing of Environment, 13*(2), 149–160.

Howarth, P. J., & Wickware, G. M. (1981). Procedures for change detection using Landsat digital data. *International Journal of Remote Sensing, 2*(3), 277–291.

Jensen, J. R., & Toll, D. L. (1982). Detecting residential land-use development at the urban fringe. *Photogrammetric Engineering and Remote Sensing, 48*(4), 629–643.

Lambin, E. F., & Strahler, A. H. (1994). Change-vector analysis in multitemporal space – A tool to detect and categorize land-cover change processes using high temporal-resolution satellite data. *Remote Sensing of Environment, 48*(2), 231–244.

Lee, J. S., Grunes, M. R., Ainsworth, T. L., Du, L. J., Schuler, D. L., & Cloude, S. R. (1999). Unsupervised classification using polarimetric decomposition and the complex Wishart classifier. *IEEE Transactions on Geoscience and Remote Sensing, 37*(5), 2249–2258.

Lee, J. S., Grunes, M. R., & Pottier, E. (2001). Quantitative comparison of classification capability: Fully polarimetric versus dual and single-polarization SAR. *IEEE Transactions on Geoscience and Remote Sensing, 39*(11), 2343–2351.

Lee, J. S., Wen, J. H., Ainsworth, T. L., Chen, K. S., & Chen, A. J. (2009). Improved sigma filter for speckle filtering of SAR imagery. *IEEE Transactions on Geoscience and Remote Sensing, 47*(1), 202–213.

Li, X., & Yeh, A. G. O. (1998). Principal component analysis of stacked multi-temporal images for the monitoring of rapid urban expansion in the Pearl River Delta. *International Journal of Remote Sensing, 19*(8), 1501–1518.

Li, X., & Yeh, A. G. (2004). Multitemporal SAR images for monitoring cultivation systems using case-based reasoning. *Remote Sensing of Environment, 90*(4), 524–534.

Li, X., Yeh, A. G. O., Qian, J. P., Ai, B., & Qi, Z. X. (2009). A matching algorithm for detecting land use changes using case-based reasoning. *Photogrammetric Engineering and Remote Sensing, 75*(11), 1319–1332.

López-Martínez, C., Ferro-Famil, L., & Pottier, E. (2005). *PolSARpro v4.0 polarimetry tutorial.* Paris: European Space Agency. http://earth.esa.int/polsarpro/tutorial.html

Lu, D., Mausel, P., Brondizio, E., & Moran, E. (2004). Change detection techniques. *International Journal of Remote Sensing, 25*(12), 2365–2407.

Lupo, F., Reginster, I., & Lambin, E. F. (2001). Monitoring land-cover changes in West Africa with SPOT vegetation: Impact of natural disasters in 1998–1999. *International Journal of Remote Sensing, 22*(13), 2633–2639.

Malila, W. A. (1980). Change vector analysis: An approach for detecting forest changes with Landsat. In *Proceedings of the 6th annual symposium on machine processing of remotely sensed data* (pp. 326–336). Purdue University, West Lafayette, IN, USA.

McCoy, R. M. (2005). *Field methods in remote sensing.* New York: The Guilford Press.

Moon, T. K. (1996). The expectation-maximization algorithm. *IEEE Signal Processing Magazine, 13*(6), 47–60.

PaxLenney, M., & Woodcock, C. E. (1997). The effect of spatial resolution on the ability on to monitor the status of agricultural lands. *Remote Sensing of Environment, 61*(2), 210–220.

Petit, C. C., & Lambin, E. F. (2001). Integration of multi-source remote sensing data for land cover change detection. *International Journal of Geographical Information Science, 15*(8), 785–803.

Pierce, L. E., Ulaby, F. T., Sarabandi, K., & Dobson, M. C. (1994). Knowledge-based classification of polarimetric SAR images. *IEEE Transactions on Geoscience and Remote Sensing, 32*(5), 1081–1086.

Pierce, L. E., Bergen, K. M., Dobson, M. C., & Ulaby, F. T. (1998). Multitemporal land-cover classification using SIR-C/X-SAR imagery. *Remote Sensing of Environment, 64*(1), 20–33.

Qi, Z., Yeh, A. G.-O., Li, X., & Lin, Z. (2012). A novel algorithm for land use and land cover classification using RADARSAT-2 polarimetric SAR data. *Remote Sensing of Environment, 118*, 21–39.

Qi, Z., & Yeh, A. G. O. (2013). Integrating change vector analysis, post-classification comparison, and object-oriented image analysis for land use and land cover change detection using RADARSAT-2 polarimetric SAR images. In S. Timpf & P. Laube (Eds.), *Advances in spatial data handling* (pp. 107–123). Berlin: Springer.

Quarmby, N. A., & Cushnie, J. L. (1989). Monitoring urban land cover changes at the urban fringe from spot Hrv imagery in southeast England. *International Journal of Remote Sensing, 10*(6), 953–963.

Saatchi, S. S., Soares, J. V., & Alves, D. S. (1997). Mapping deforestation and land use in Amazon rainforest by using SIR-C imagery. *Remote Sensing of Environment, 59*(2), 191–202.

Seto, K. C., Woodcock, C. E., Song, C., Huang, X., Lu, J., & Kaufmann, R. K. (2002). Monitoring land-use change in the Pearl River Delta using Landsat TM. *International Journal of Remote Sensing, 23*(10), 1985–2004.

Shen, G., Guo, H., & Liao, J. (2007). Change vector analysis method for inundation change detection using multi-temporal multi-polarized SAR images. In Proceedings of SPIE, Wuhan, China.

Shimoni, M., Borghys, D., Heremans, R., Perneel, C., & Acheroy, M. (2009). Fusion of PolSAR and PolInSAR data for land cover classification. *International Journal of Applied Earth Observations and Geoinformation, 11*(3), 169–180.

Silapaswan, C. S., Verbyla, D. L., & McGuire, A. D. (2001). Land cover change on the Seward Peninsula: The use of remote sensing to evaluate the potential influences of climate warming on historical vegetation dynamics. *Canadian Journal of Remote Sensing, 27*(5), 542–554.

Singh, A. (1989). Digital change detection techniques using remotely-sensed data. *International Journal of Remote Sensing, 10*(6), 989–1003.

Sonmez, N. K., Onur, I., Sari, M., & Maktav, D. (2009). Monitoring changes in land cover/use by CORINE methodology using aerial photographs and IKONOS satellite images: A case study for Kemer, Antalya, Turkey. *International Journal of Remote Sensing, 30*(7), 1771–1778.

Stefanov, W. L., Ramsey, M. S., & Christensen, P. R. (2001). Monitoring urban land cover change: An expert system approach to land cover classification of semiarid to arid urban centers. *Remote Sensing of Environment, 77*(2), 173–185.

Story, M., & Congalton, R. G. (1986). Accuracy assessment – A user's perspective. *Photogrammetric Engineering and Remote Sensing, 52*(3), 397–399.

Ulaby, F. T., Kouyate, F., Brisco, B., & Williams, T. H. L. (1986). Textural information in SAR images. *IEEE Transactions on Geoscience and Remote Sensing, 24*(2), 235–245.

Weismiller, R. A., Kristof, S. J., Scholz, D. K., Anuta, P. E., & Momin, S. A. (1977). Change detection in coastal zone environments. *Photogrammetric Engineering and Remote Sensing, 43*(12), 1533–1539.

Wilson, E. H., & Sader, S. A. (2002). Detection of forest harvest type using multiple dates of Landsat TM imagery. *Remote Sensing of Environment, 80*(3), 385–396.

Witten, I. H., Frank, E., & Hall, M. A. (2011). *Data mining: Practical machine learning tools and techniques*. Burlington: Morgan Kaufmann.

Yeh, A. G. O., & Li, X. (1996). Urban growth management in the Pearl River Delta – An integrated remote sensing and GIS approach. *ITC Journal, 1*, 77–86.

Chapter 20
Space-Time Analytics of Tracks for the Understanding of Patterns of Life

May Yuan and Atsushi Nara

Location-aware devices have enabled the recording of personal whereabouts at fine spatial and temporal resolutions. These temporal sequences of personal locations provide unprecedented opportunities to explore patterns of life through space-time analytics of movement and stops of individuals. At a disaggregated level, patterns of life reveal the activities and places as well as the development of routines for individuals. At an aggregate level, patterns of life suggest potential social networks and social hot spots for interactions. Moreover, the concept of "neighborhood" can become personalized and dynamic with space-time analytics to identify the spatial extent to which an individual operates and how the extent varies with temporal granularity. This chapter starts with an overview of space-time track analysis. While time geography has proven useful for analysis of space-time paths and space-time constraints on human activities, its scalability to large data sets is questionable. This chapter provides a conceptual framework and methodology for conducting space-time analysis with a massive number of space-time tracks including over a million points of moves and stops over the course of a year. The examples demonstrate the usefulness of the proposed conceptual framework and methodology to distill complex patterns of life at both disaggregate and aggregate levels that can lead to research opportunities for space-time integration in GIScience for an improved understanding of geography.

M. Yuan (✉)
Geospatial Information Science, University of Texas at Dallas, Richardson, TX, USA
e-mail: myuan@utdallas.edu

A. Nara
Geography, San Diego State University, San Diego, CA, USA
e-mail: anara@mail.sdsu.edu

M.-P. Kwan et al. (eds.), *Space-Time Integration in Geography and GIScience*, DOI 10.1007/978-94-017-9205-9_20

20.1 Time Geography as the Conceptual Foundation

Time geography has long been the de facto framework for geographers to analyze space-time activities of individuals and the everyday workings of society at locales of operation (Berry and Linoff 1997). Central to this analytical framework is an emphasis on modeling individual behavior, not just empirical analysis of large groups (Morrill 2005). Hägerstrand's seminal work on *What about People in Regional Science* advocated for attention to individual human beings in an increasingly complicated environment to examine the quality of life for of individuals and collectives. To support such research, he proposed the basic framework of time geography, including the concepts of lifelines, stations, bundles, domains and prisms (Hägerstraand 1970). Advances in geospatial information analysis and visualization have led to significant research developments in time geography, such as computing activity-travel patterns of masses of individuals (Kwan 2004), activities and interactions in a hybrid physical-virtual space (Shaw and Yu 2009), accessibility with varying travel velocity (Miller and Bridwell 2009), and many other improvements. This chapter offers yet another improvement to time geography with a new approach to examining daily space-time paths of individuals with intensive observations through location-aware devices. We have developed conceptual and computational frameworks for space-time analytics of daily movements of individuals to project their patterns of life and potential social networks. The basic concepts of time geography are used to recognize lifelines, stations, bundles, and domains (Fig. 20.1), but we introduce patterns of life, locales and social networks to construct geospatial narratives of their routine activities, stages in life, and opportunities for social interactions.

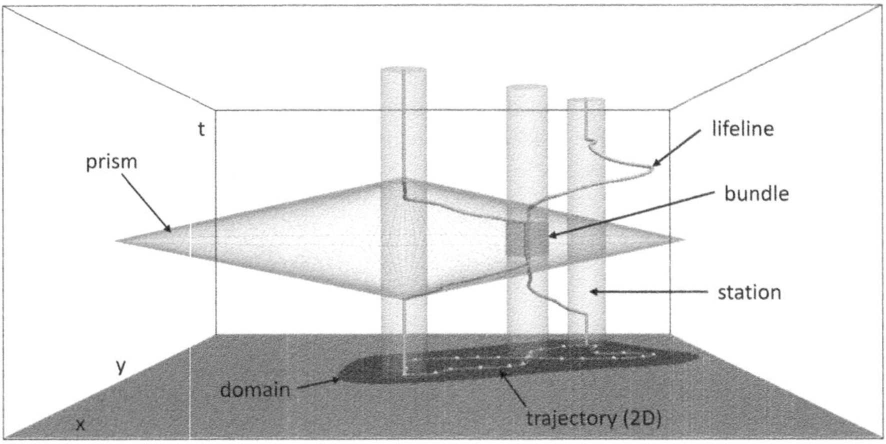

Fig. 20.1 Basic conceptual elements of time geography (Adapted from Hägerstraand 1970)

The proliferation of location-aware devices has led to detailed accounts of entity-location interactions. What are the usual paths that an individual commutes from residence to work? What are common stops along these paths? What are the routine activities that an individual participates in over a period of time? How have the routine activities evolved over time? When and what are occasional outings or routes that one took previously and for what purposes? What are the potential social interactions among individuals? What kinds of and where are places that individuals like to hang out and when? These questions seek to identify patterns of life, socially significant places, and changes in routine activities in space and time. In time geography, lifelines are comparable to the space-time paths taken by individuals, stations correspond to locations visited by individuals, bundles capture locations where multiple individuals visited within space-time proximity, and domains correspond to the spatial extent in which individuals operate. While the conceptual mappings work well, time geography offers no specific quantitative movement analysis of empirical data (Miller and Bridwell 2009).

20.2 From Time Geography to Trajectory Analysis

Along with the growing use of location-aware devices, there are increasingly ubiquitous surveillance sensors. With the availability of location data and sensor data for moving objects, trajectory analysis has quickly gained popularity in GIScience and related disciplines. The often massive, intensive trajectories recorded by location-aware devices or geo-sensor networks at fine spatial and temporal resolutions challenge management, query and analysis of trajectory data (Spaccapietra et al. 2008). In contrast to time geography, trajectory studies have required the development of quantitative methods for movement analysis. In time geography, multiple modes of transportation influence an individual's accessibility in a space-time aquarium with conceptual elements of lifelines, prisms, bundles, etc. that represent human activities. Innovative tools are being developed to visualize such space-time aquaria with functions to query and analyze these conceptual elements (Shaw and Yu 2009). Complementarily, trajectory analysis and trajectory mining, emphasizing the ability to handle massive data, explore effective ways for data aggregation and generalization (Andrienko and Andrienko 2011).

While location-aware devices and geo-sensor networks both provide space-time data on moving objects, the two approaches observe movements in different ways. Surveillance sensors monitor movements in an area, such as severe storms or hurricanes in a region, traffic flows on a highway, vehicles in a parking lot, or patrons at a building entrance. Each object has one trajectory with one activity (e.g. entering a building), and each trajectory is assumed tangent and continuous between beginning and ending points. Location-aware devices or personal diaries (such as geo-tagged tweets or photographs), on the other hand, record the locations of individuals over space and time. One space-time path taken by an individual may

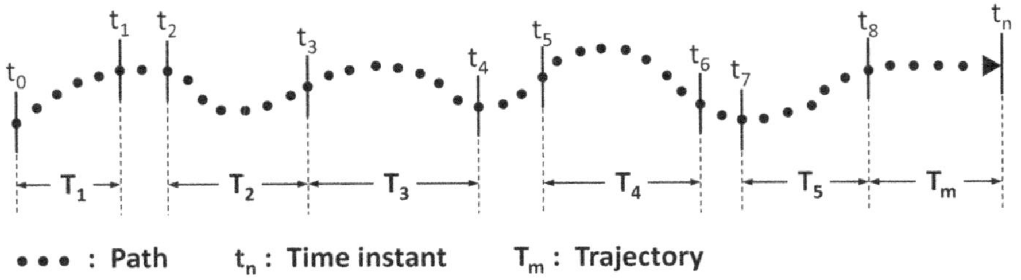

Fig. 20.2 Path and trajectories: Trajectories are defined through the semantic segmentation of a path. Some parts of the path are irrelevant to the application of interest and do not belong to any trajectory (Adapted from Spaccapietra et al. 2008)

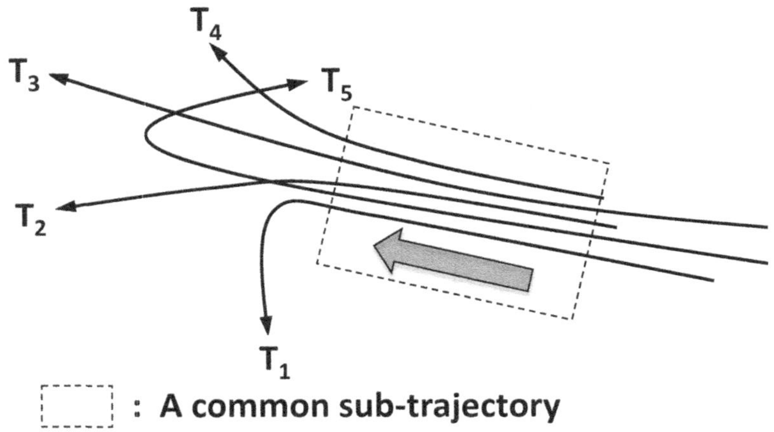

Fig. 20.3 Trajectories and sub-trajectories (Adapted from Lee et al. 2007)

result from multiple activities (e.g. driving, working, shopping), and therefore, may consist of multiple trajectories (Fig. 20.2). Trajectory analysis on both data sources has been built upon descriptive and predictive analyses of trajectory descriptors, similarity indices, clustering patterns, and individual-group dynamics (Long and Nelson 2012). A suite of parameters and derivatives is being built to characterize primitive and compound movement patterns (Dodge et al. 2008), which can serve the basis for a systematic development of quantitative methods for trajectory analysis.

Units of trajectory analysis usually are trajectories, trajectory segments (a.k.a. sub-trajectories, Fig. 20.3), and stops. Measures of trajectory geometry or density are common identifiers for clusters or outliers. Gaffney and Smyth (1999) proposed a model-based clustering algorithm with a probabilistic mixture regression model and the Expectation-Maximization algorithm to analyze hand movement. A similar approach is used to analyze tracks of extratropical cyclones (Gaffney et al. 2007).

By segmenting a trajectory into trajectory segments, movement parameters, such as duration and speed (Dodge et al. 2012) and semantic behavior, such as shopping and work (Popa et al. 2012), can be associated with these features along trajectories for analysis. This approach enables us to extract local behavioral patterns of mobile objects rather than global patterns for the entire trajectories. Several algorithms are designed for trajectory segmentation, such as density-based clustering TRACLUS (Lee et al. 2007), Geometry-based Representativeness (Panagiotakis et al. 2012), and Velocity-based Trajectory Structure (Yan et al. 2012).

In contrast to segmentation, aggregation and connectivity among trajectory segments provide insights into the global trajectory patterns. Andrienko and Andrienko (2011) introduced a trajectory aggregation technique by partitioning the space into compartments, transforming raw trajectory data into moves between the compartments, and aggregating the transformed moves with common origins and common destinations. Guo et al. (2010) proposed a graph-based partition method incorporating the use of trajectory topological relationships to find spatial structures and general patterns of trajectories that were visualized in 2D trajectory density maps at several temporal snapshots. These spatial structures or clusters often lead to interesting semantic implications.

Semantic enrichment contextualizes trajectory segments with behavioral characteristics. Eagle and Pentland (2006, 2009) applied the principle components of movement data, termed eigen-behaviors, to capture the structure of behavioral contexts of individuals, such as staying at home, work, or elsewhere. Availability of a context-rich dataset is a critical factor for success for empirically based research of semantic tracks (Giannotti and Pedreschi 2007). Using bus routes and bus stops data and manually labeled modes of transportation (e.g. foot, bus, or car) to trajectory segments, Patterson et al. (2003) obtained 84 % accuracy in predicting modes of transportation with GPS data sampled at 2–10 s intervals over 3 months during outside activities of the individuals being studied.

The combination of geometric and semantic trajectory analyses deepens the level of behavioral knowledge that can be discovered from trajectories, such as trajectory clusters (grouping), trajectory categories (typing), trajectory sequences (transitioning) and trajectory aggregates (flocking). Laube et al. (2007) introduced a methodology for lifeline context operators and standardizations, and explored the spatio-temporal behaviors of homing pigeons using the sinuosity, rate of change of trajectory sinuosity, navigational displacement, relation between distance to loft and flight sinuosity. Dodge et al. (2009) identified local motion descriptors (i.e., motorcycle, car, bicycle, pedestrian) and global and local motion descriptors (e.g., velocity, acceleration, turning angle, straightness index) as movement signatures to differentiate trajectories from different types of mobile objects. Similarly, Willems et al. (2009) applied Kernel Density Estimation (KDE) to visualize movement patterns of seafaring vessels. The density based visualization has shown to be useful in identifying movement types (e.g. walking or driving; by vessels or speed boats) and gathering places.

Besides mining trajectories from mobile collectives, attention to episodes along individual trajectories offers new insights into behavioral changes in mobile objects, such as one person taking multiple modes of transportation (e.g. bike to a bus stop, ride a bus, bike to building bike rake, and walk into a building), and when mining data streams, the movement patterns of individuals can be revealed in real time (Yan et al. 2011). MobiVis is but one tool developed to visualize social-spatial-temporal patterns of mobile data (Shen et al. 2006; Shen and Ma 2008). The tool incorporates heterogeneous network and semantic filtering techniques based on associated ontology graphs, and the visualization technique of behavior rings that reveal periodic behavioral patterns of individuals and groups.

20.3 From Trajectory Analysis to Space-Time Analytics

Generally speaking, time geography or trajectory analysis approaches explore individual tracks over time or collective tracks within a confined area. Most studies examine geometry, semantics, clusters, classifications, and entity-location interactions of discrete tracks or attempt to generalize a collection of tracks. Whether tracks are taken for the same mobile object is seldom considered, and therefore, methods to reveal routine and incongruent movement patterns are missing in the literature. Yet, temporal patterns that depict transitions between routine and incongruent movements reflect shifts in patterns of life and are essential to many domain applications, in which geospatial distributions of patterns of life and movement are indicative of individual's spatial social adaption or as a measure of a population's collective spatial-social pulses.

Under the assumption that people develop routines and follow these routines over time, their patterns of life form through settling down to the regular activities that take place in space and time. Movement is the process that people engage in to get to the target place at the right time, and congruent movement patterns suggest the formation of space-time routines and therefore, spatial patterns of life. For a population, the summative characteristics for the progression in forming patterns of life and spatial differentials of the process show how a society (or a community) may foster faster or slower adjustments for newcomers and where and when people are gathered in the society.

Analytics remains an ambiguous term that often serves as an umbrella term for statistics, data mining, modeling, simulation, and computational methods to discover and communicate meaningful patterns in data (Shmueli and Koppius 2011). Yet, analytics commonly emphasizes data-centric and data-driven approaches to tackle massive, streaming, heterogeneous and/or unstructured data, especially in business management and marketing (Lavalle et al. 2011), education (Wagner 2012), website use (Marek 2011), and many other applications. In this chapter, we define space-time analytics as methods to simplify the complexity of space-time data into elements and structures that capture the useful information embedded in the data. We emphasize elements and structures that together reveal useful space-time information. Tracks, track segments, stops, and moves are examples of elements in track analysis. Aggregates or disaggregates of these elements may also be useful elements to decipher movement patterns for individuals or collectives. Structures are proximity, connectivity, ordering, or relationships in space, time, and semantics. Here, semantics simply refers to the element's characteristics and meanings of interest. Kinship, for example, is a semantic relationship, and so is management in an operation and participation in a social network or organized/unorganized activity. With this emphasis, space-time analytics focuses on what space-time elements and structures pertain to the research questions of interest and how to elicit these elements and structures from complex data.

20.4 A Conceptual Framework for Space-Time Analytics of Tracks

The conceptual framework of space-time analytics of tracks proposed below, extending the conceptual frameworks of time geography and trajectory analysis, aims to discern space-time patterns of multiple tracks repetitively taken by given individuals, and from the space-time patterns to interpret spatial patterns of life for individuals, their potential interactions, and the spatial patterns of social pulses in the related population. Clarification of terminology is important to communicate concepts precisely. While many terms seem have inconsistent meanings in previous studies, great attention is given to avoid incompatibility of the terms used in the proposed conceptual framework:

- **Path**: the actual route that one took. A path usually follows transportation networks or existing pathways except when travelling in an open space such as a yard, plaza, air or sea.
- **Track**: the line formed by connected observed locations. A track is based on observations taken by GPS points, geo-tagged media, or journey narratives. With a track we know temporally ordered locations visited by a mobile object, but we don't know exactly the path that the object took to travel from one location to the next.
- **Trajectory**: the expected route that one took from one location to the next. A trajectory is a movement estimate based on space-time continuum and space-time constraints. For a car and two consecutive observations of its locations, its trajectory is expected to follow one of the road options between the two locations, except for accidents such as driving into a lake. With a surveillance video, identified individuals are assumed to move within a space-time continuum. Spatial and temporal interpolation methods can be used to estimate the possible trajectories between two known locations.
- **Stops**: locations on a track where a mobile object stays over a length of time beyond a specified duration. A car makes stops at traffic lights, a person stops by a convenience store, and a train stops at a city station. A specified duration may be 30 s, 10 min, an hour, or a week, etc. depending upon what meaningful information may be in the problem domain of interest.

- **Moves**: locations on a track where a mobile object stays shorter than the specified threshold duration. Locations where an object moves have instantaneous velocity, even though there may not be data available for velocity calculation. A track is composed of temporally ordered observed locations at which the mobile object of interest either stops or moves. In addition to the threshold duration, spatial and temporal granularities can influence the determination of stops and moves locations as well as locations along respective tracks.

Minimal spatial and temporal granularities are subject to observations and can influence decisions on points of stops and moves. At a fine space-time granule, for example sub-meter and second, detailed motion can be recognized, such as taking a step forward and the point will be determined as a move. However, if the space-time granule is at 10 m and 5 min, a step forward is likely to be within the granule and therefore will be considered as a stop. The final decision of stops and moves is subject to the defined space-time threshold, say 20 m in an hour, which will consider a collection of points within 20 m in an hour as stops, but for a point which next point is beyond 20 m in an hour will be consider a move point. Determination of a proper space-time granule depends on the instrument taking observations (such as GPS or twitter) as well as the application of interest.

There are two kinds of structures for the elements of paths, tracks, trajectories, stops, and moves: time-oriented or space-oriented structures. For time-oriented structures time is considered the first order principle of organization that connects points of stops and moves over space to form different kinds of movement lines. At a basic level there are locations of stops and moves that are recorded as observations. Estimated pathways between temporally consecutive locations of stops and moves are trajectories, while straight line connections between these locations form tracks. Connections between these locations based on the actual route travelled results in a path for the journey.

Alternatively, space is considered the first order principle of organization for space-oriented structures. Space connects points of stops and moves over time to form visit histories at locations. Some locations may be visited on a regular basis, and others may be visited only occasionally. Frequency and timing of visits may vary from time to time. Passing-by locations involves only points of moves, other locations may have only stop points or mixed stops and moves. Moves by a location may be at different speed or in different direction. Stops at a location may be long or short. Passing by a location that has in the past been a traditionally stopping may signal a change in spatial behavior (such as no longer eating at a given fast food restaurant) or pattern of life (such as no longer needing to use day care services for children). Changes in how an individual moves at locations can also be revealing, such that speeding out of a work parking lot may be indicative of urgency.

Chronological time (a.k.a. calendar time) can serve the temporal structure necessary for a time-oriented organization. Geographic space is two-dimensional and is commonly referenced by x and y coordinates. Regular tessellation of space is commonly used as the framework for discrete global grids, such as the Military Grid Reference System, regular hierarchical triangulation (Dutton 2000), hexagonal

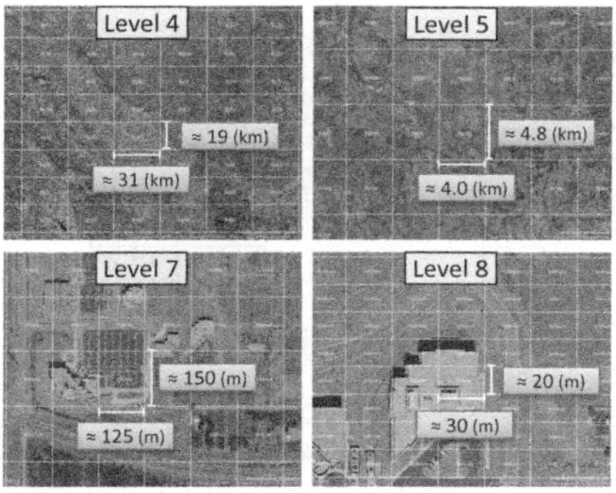

Geohash Length	Lat Bits	Lng Bits	Lat Error	Lng Error	Km Error
1	2	3	±23	±23	±2500
2	5	5	±2.8	±5.6	±630
3	7	8	±0.7	±0.7	±78
4	10	10	±0.087	±0.18	±20
5	12	13	±0.022	±0.022	±2.4
6	15	15	±0.0027	±0.0055	±0.61
7	17	18	±0.00068	±0.00068	±0.076
8	20	20	±0.000085	±0.00017	±0.019

Fig. 20.4 The hierarchical structure of Geohash grids progressively refines spatial granularity as the length of Geohash strings increases (Geohash). Geohash strings are highly efficient in coding spatial precision as compared to latitude and longitude pairs as well as other Cartesian coordinate systems

quad balanced structure (Tong et al. 2010), equal-areal cells (Beckers and Beckers 2012), and many other partitioning and subdivision systems (Lu et al. 2012). Any of the geographic coordinate systems or discrete global grid systems may serve the purpose of spatial ordering. One particular example is Geohash system (Geohash n.d.) that constitutes string-based prefixes as keys for coordinate sets in a hierarchical grid system (Fig. 20.4).

Geohash was invented by Gustavo Niemeyer to create a concise spatial indexing and data caching system for web services. Geohash is essentially a latitude and longitude geocode system that interleaves bits obtained from latitude-longitude pairs and converts the bits to a Geohash string using Base 32 character map. For example, the latitude-longitude pair (35.18, −97.44) falls in the box with Geohash code **9y68nuckb** with approximately 2 m precision at Level 9. The string length (e.g. level) of a Geohash code corresponds to the size of spatial granule. Generally speaking, points closer to each other are likely to have longer common prefixes in their Geohash codes. In most cases, all adjacent Geohash boxes only differ in the last digit of their Geohash codes, except for the edge boxes that adjacent boxes have different codes in the last two digits (Fig. 20.5). Exceptions also include Geohash boxes near the Equator and Greenwich Meridian.[1] The hierarchical organization of Geohash codes facilitates rapid data grouping and efficient proximity search to meet computational demands (Balkic et al. 2012). Geohash grids have been implemented in Google Map, MySQL, PostGIS and MongoDB as well as libraries or modules for Java, Javascript, Python, Perl, PHP, Pure Ruby, C, and many other programming languages.

[1]The issue is referred to as Geohash faultlines. Algorithms have been developed to address the faultline issues in spatial search. One algorithm with its source code is available at http://code.google.com/p/geohash-fcdemo/

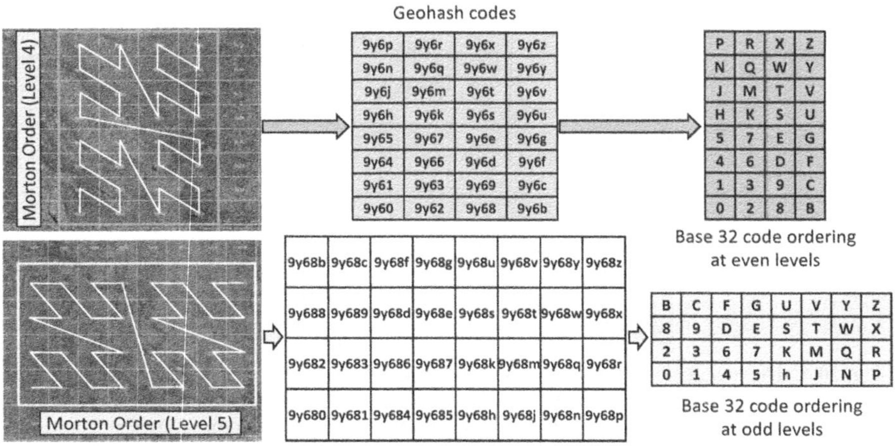

Fig. 20.5 Examples of Geohash codes at Levels 4 and 5 and the alternate ordering schemes for codes at even and odd levels. All codes at one level are of the same length. Codes in every 8 × 4 or 4 × 8 unit are only differentiated by their last digits

Both chronological time and Geohash space provide structures in which space-time elements of stops, moves, and tracks can be referenced to develop analytical methods and distill higher level information from primitive observations. Below are methods of space-time analytics based on the elements of tracks, stops, and moves and associated time-oriented and space-oriented structures to discern patterns of life with a case study to illustrate the procedures and potential findings.

20.5 Space-Time Analytics of Tracks to Elicit Patterns of Life

Here, patterns of life are space-time summaries of movements and from the summaries to project activities. Since trajectories and paths cannot be directly derived from location observations, additional assumptions, information and modeling projections are required to determine the most likely trajectories and paths taken by individuals. Methods to construct space-time summaries of movements are based on stops, moves, and tracks from observed or recorded locations. To facilitate the discussion, we use GPS observations for several individuals in Oklahoma, USA to step through Space-Time Analytics procedures.

20.5.1 Stop Analysis

Location-aware devices record GPS points at regular intervals. Unless the devices are equipped with a motion sensor, GPS location data files consist of streams of time-stamped spatial coordinates. Where an individual frequently visits and how the person usually travels to these places are fundamental space-time elements to understanding the person's pattern of life. Therefore, the first step of space-time analytics is to distinguish stops and moves on tracks that the person took.

A convenient way to differentiate which GPS points correspond to stops from those GPS points that were taken when one was in motion is based on how long one stayed at these locations, which is reflected in differences of time-stamps or by spatial clusters of GPS points (Palma et al. 2008). For tracks taken by individuals, clusters may not be apparent due to the number of sample points taken in periods of short duration. Therefore, we propose two scale parameters, a duration threshold (τ) and a distance threshold (δ) to determine stops on a track. Duration thresholds, e.g. greater than 5 min, may be used to include short duration stops such as the use of drive-through windows at business establishments. Longer duration thresholds may reflect locations for major activities, such as $\tau = 30$ min or longer for shopping or dining. A distance threshold is used to determine a geographical size of stop events. A smaller distance threshold may filter GPS sampling noise and determine stop events consisting of small locational variations, e.g. waiting at a bus stop in urban space, whereas a larger distance threshold can determine stay events that potentially include movement activities such as playing at a park.

Figure 20.6 illustrates the algorithm for differentiating stops and moves from collected GPS points based on two threshold values. A GPS track *TR* is composed of a sequence of 3- or 4-dimensional GPS points, $Tr = \{p_1, p_2, p_3, \ldots, p_k\}$ where k denotes the number of points in the track, $p_k = \{x, y, (z), t\}$ where (x, y) are longitude and latitude, z is altitude, and t is timestamp. Based on the distance threshold, the algorithm finds anchor points (*pa*), which differentiate between stops and moves, and each stop or move is composed of a set of GPS points between two consecutive anchor points. To find anchor points, we begin with the first and the last points of a track as anchor points (pa_1, pa_n). Then, we calculate a segment distance between two GPS points starting from the first point and check if it exceeds the distance threshold. In Fig. 20.6, the distance between p_1 and p_2 exceeds the distance

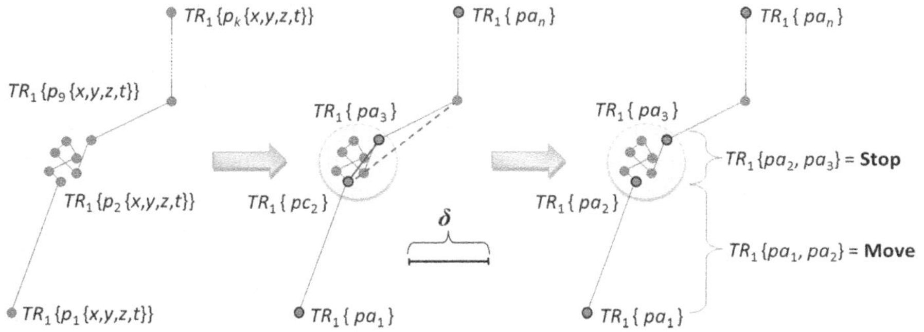

Fig. 20.6 A stop detection algorithm based on duration and distance thresholds

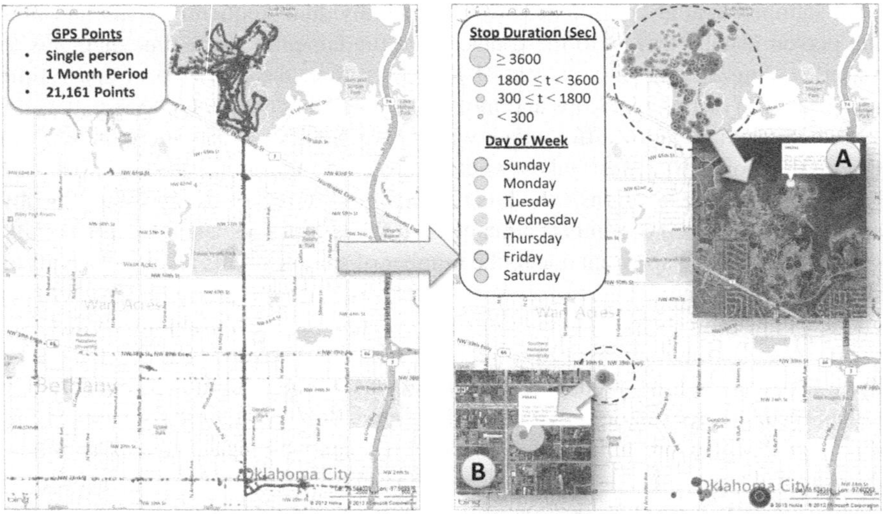

Fig. 20.7 GPS points and extracted stop-event distribution for a month (9/1/2009 to 9/30/2009)

threshold; therefore, the segment cannot formulate a stop. Moving to the next point, the distance between p_2 and p_3 is less than the threshold value, and thus the segment is a candidate for formulating a stop. When such a segment is identified, the first point becomes an anchor point (pa_2). Then we calculate the segment distance between the anchor point and its following points until the distance exceeds the threshold value. When the threshold is exceeded (a red dashed-line between p_2 and p_9), the last point (p_8) becomes an anchor point (pa_3). Lastly, if the duration between two anchor points (pa_2 and pa_3) is larger than the duration threshold, GPS points enclosed by those anchor points formulate a stop event; otherwise, anchor points will be released.

In our case study for stop analysis, we used GPS points of one individual collected over a 6-month period between 5/1/2009 and 10/31/2009. Figure 20.7 illustrates GPS points (left) and the stop-events distribution identified by the duration-distance thresholds algorithm, where $\tau = 1$ s and $\delta = 20$ m. In this case, the duration threshold is set to 1 s in order to extract all possible stops. The algorithm extracted stops with short-duration near the street intersection suggesting stops due to traffic lights. It also identified several long stops such as one in a residential area possibly for one's dwelling place, one in a commercial district located at south-east corner of an intersection (B), and one at a golf course nearby a lake (A).

After the stop locations are extracted from the GPS point data, the similarity between 2 days of stop points can be calculated for temporal analysis. For example, over the time range of a dataset, one day of stops can be compared to its previous day of stops to explore if any weekly stop pattern emerges, or one day of stops can be compared to the same day the week before to investigate if a pattern of stops is congruent or incongruent for a specific day of week. To measure similarity, various functions can be applied such as Sum-of-Pairs Distance (Li and Goodchild 2012), Dynamic Time Warping (Li et al. 2011), and Longest Common Subsequence (Böhm et al. 2011).

In our case study, we have employed a simple similarity measure, Sum-of-Closest-Pairs Distance (SCPD). In this measurement, we find the closest pair of points from two stop point datasets and calculate the sum of the distance between them. Given two sets of stop points A and B, SCPD is calculated as follows:

$$SCPD(A, B) = \sum_{i=1}^{n} \min_{a_i \in A, b_j \in B} d\left(a_i, b_j\right)$$

Figure 20.8 shows similarity distance comparisons for one day of stops to the previous day over a 6-month period (5/1/2009–10/31/2009). In this analysis, we use a duration threshold of 5 min in order to ignore very short stop events such as stops at traffic lights. The unit of similarity cost is measured in kilometer and a lower distance value denotes similar stop-events between 2 days. The color of data points represents a day of week, and it reveals weekly pattern of extracted stop-events. For example, Saturday's stop events are quite different from Friday's throughout the time period. In addition, Thursday's are also different from Wednesday. Moreover, stops between Saturday and Wednesday are relatively similar (low distance in the box **a** in Fig. 20.8). This regularity corresponds to the 5-day and 2-day weekly pattern suggesting that the individual may have a regular weekly work pattern. In fact, stop points for those 5 days (Saturday to Wednesday) are found in the area A in Fig. 20.7 and not the other 2 days (Thursday and Friday), which implies that the location could be his/her work place.

Alternatively, space can serve as the first order structure for analysis of patterns of life, investigating location-focused frequency and duration of stops distribution over time is a useful exploratory analysis. The bar chart in Fig. 20.9 visualizes stop-events frequency and duration distribution for two locations (A, B) in Fig. 20.7.

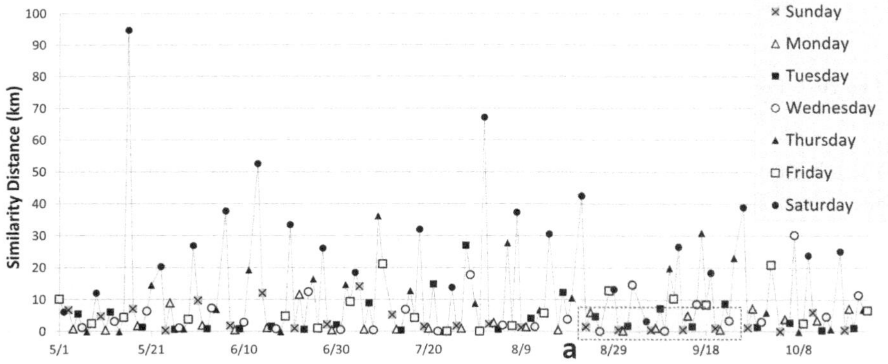

Fig. 20.8 Individual stop similarity over a 6-month period

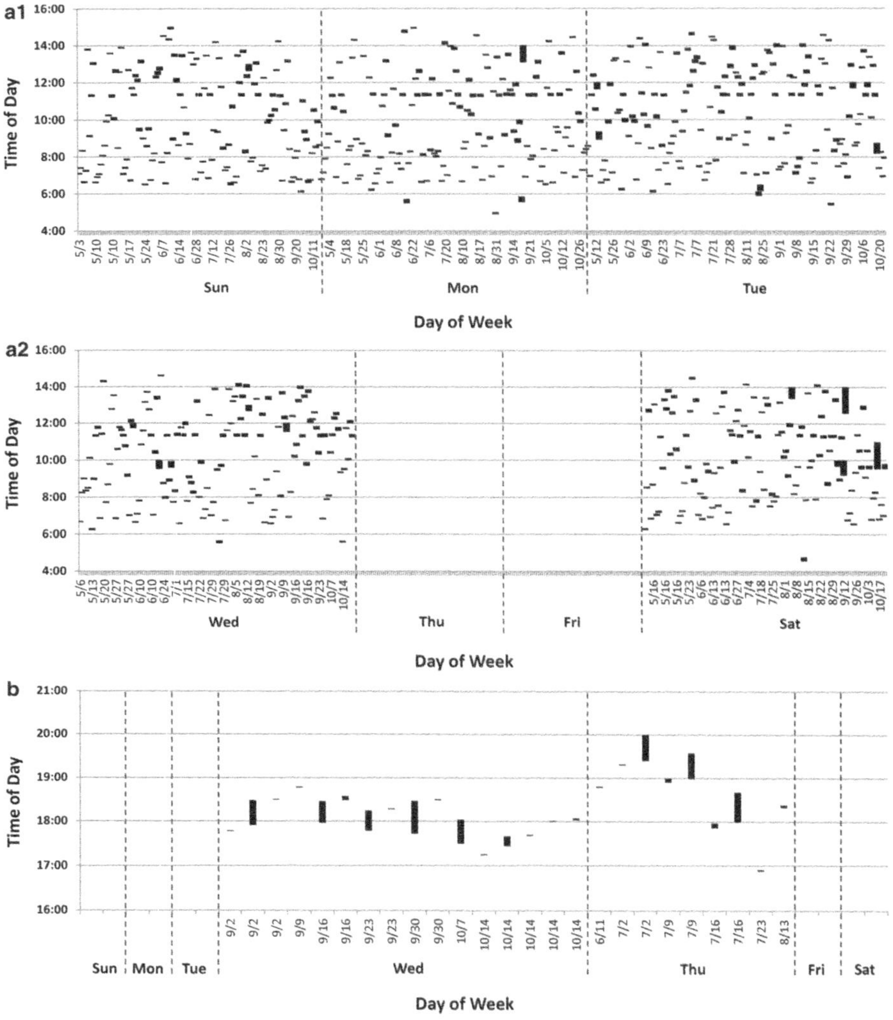

Fig. 20.9 Frequency and duration of stops over days of the week and over time of the day

Each bar on the graph represents a single stop-event and its vertical length represents the duration. The x-axis orders stops by day of the week. Two graphs from the top in Fig. 20.9 show stop-events frequency and duration for location A, while the bottom graph is for location B. This chart provides basic understandings of how an individual utilizes a particular place along with his/her life schedule and reveals, for example, a daily/weekly stop pattern regularity and pattern shift over time.

For the location A, first, the chart reconfirms that there is no stop-event on Thursday or Friday during a 6 month period. In addition, the stop-events observed in the location A are within a fixed time interval mostly between 6 a.m. and 2 p.m. These two patterns, 5 days in a week and 8 h in a day, further reinforce our previous indication to infer that the location A is likely to be his/her work place. On the other hand, the stop-events in location B are found only on Wednesday and Thursday between 5 p.m. and 8 p.m. The location B is a commercial area including several shops and restaurants, and the individual's activity may be shopping or eating out on Wednesday after work or on Thursday, which appears to be a day off work. In addition to these daily and weekly patterns, we can identify a pattern shift over time. Thursday's stop-events at the location B are only found in June, July, and August. The individual activity has been shifted from Thursday to Wednesday in September and October.

20.5.2 Track Analysis

A track consists of stops and moves. In the context of patterns of life, track analysis aims to reveal one's routines or regularity in space and time. The consideration includes regularity in time (e.g. leave home at 7 am for work), in space (e.g. take the same route to work), and in activities (e.g. do grocery shopping on the way home every Wednesday). Therefore, track analysis examines the degree to which points of stops and moves from track to track vary in space and time. One extreme case is a perfect match of stops and moves in space and time across all tracks for an individual, who follows a rigid daily routine and hence exhibits an unambiguously regular pattern of life. In such a unvarying daily routine, there is no deviance of stops and moves between tracks, and the degree of space-time deviance should be minimized or zero if all points of stops and moves across tracks are perfectly matched.

The other extreme case is a total random walk, in which every location in space and time enjoys the same probability being visited or passed by. The individual would have no routine resulting in an accidental pattern of life. The degree of space-time deviance would be maximized or approaching infinity depending on the numbers of stops and moves on these tracks. Most of our patterns of life fall in between the two extremes. Demographic characteristics (e.g. age, gender), social-economic factors (e.g. kinds of employment, means of transportation), local knowledge (e.g. new residents) and neighborhood effects (e.g. public safety, restricted zones) all contribute to the development of one's pattern of life in space and time. Nevertheless, space-time analytics of tracks to elicit patterns of life assumes that everyone will develop regular spatial routines over time, and the spatial routines may follow chronological cycles, e.g. hourly daily, weekly, monthly, seasonally, or other longer cycles. Space-time deviance is then expected to fluctuate according to these cycles.

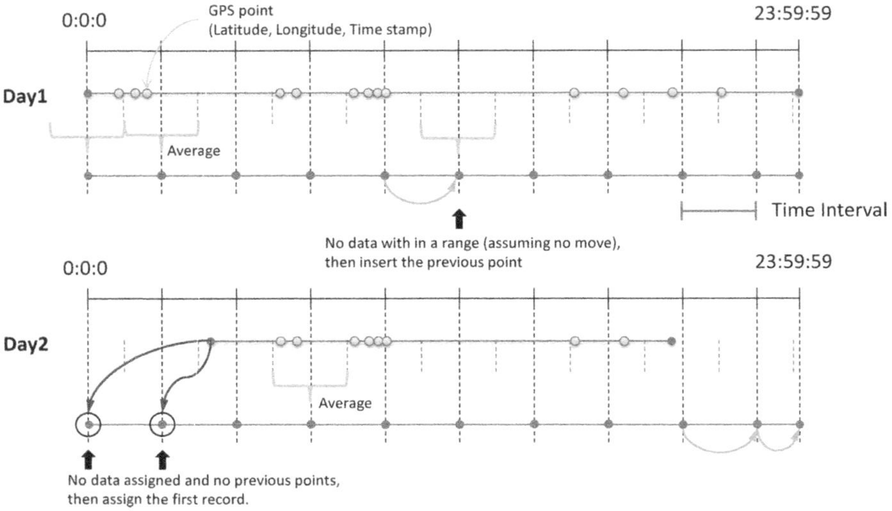

Fig. 20.10 Resampling tracks to match locations and time intervals to facilitate calculation of space-time deviance between two tracks

Determination of space-time differences between any two tracks is not trivial. GPS observations were likely taken at different time intervals and at different times. Even if a person leaves home at the same time and takes the same route for a 2 day period, GPS points of stops and moves along the track for the first day will not match exactly the track for the second day. Resampling is one technique that can be used to adjust the shifts in location and time of GPS observations (Fig. 20.10). Resampling along tracks allows all tracks in consideration to have the same number of GPS locations during a period of interest for comparison. The resampled time interval should be determined by spatial and temporal variability in the application domain of interest. For daily patterns, identification of hourly changes in space and time may be sufficient. For seasonal patterns, daily or weekly changes may be more meaningful than hourly variations.

Assuming that the time resampling interval is T (say an hour) in Fig. 20.10, the mean center of all GPS points taken $T \pm \Delta T/2$ (e.g. 30 min before and after an hour) will be assigned to the location to time T_i (e.g. on the hour). When there are no GPS points taken within $T \pm \Delta T/2$ (e.g. 30 min before and after the hour), the mean center calculated for T_{i-1} (e.g. the previous hour) is assumed to be the representative location at T_i (e.g. on the hour). In addition, when there is no GPS location in the initial hours of the day, the first known GPS location is assumed to be the locations for these initial hours. An example would be if an individual left home at 9 am, comparably later than 7 am on the previous day. The approach assumes that the individual was at home at 7 a.m. to 9 a.m. The resampling procedure results in the same number of locations and the same number of time intervals on all tracks in consideration for comparative analysis of space-time difference.

Furthermore, GPS locations carry spatial uncertainty. GPS technology at a survey grade can obtain centimeter accuracy, but consumer grade GPS receivers may have spatial uncertainty to tens of meters. Precise latitude and longitude readings of GPS may be misleading in terms of location accuracy. Therefore, the use of Geohash codes, instead of latitude-longitude readings, will better preserve the innate spatial uncertainty in track analysis.

After resampling and converting coordinates to Geohash codes, track analysis can compare locations at each time interval to determine if similar tracks were taken on these days. Geohash codes condense two dimensional latitude-longitude pairs to one text string, in which the distance comparison can be based on the length of common codes and the distance represented by the differences of digits at the end of the Geohash strings. Figure 20.11a shows an example of an individual that was

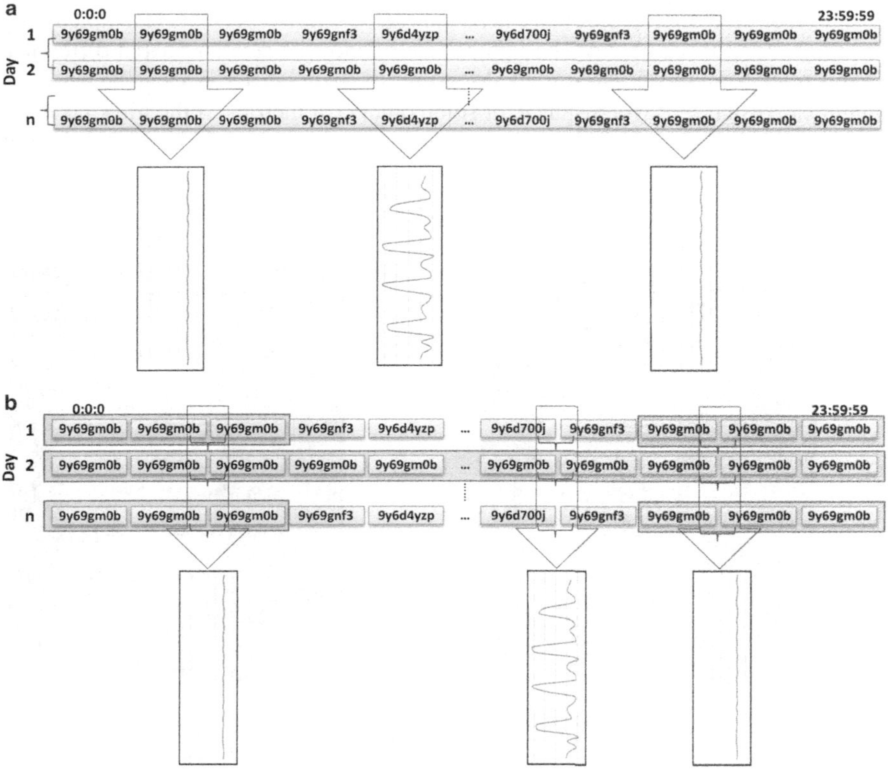

Fig. 20.11 Comparison of tracks to determine space-time deviance: (**a**) comparing locations across days at the same time to reveal changes in tracks; (**b**) comparing distances in adjacent locations to reveal changes in travel speed. The *vertical line graphs* show space-time deviance in locations between two tracks (**a**) or in adjacent locations on the same track (**b**) over the days of interest (Day1, Day2, ... Dayn)

at different locations in the fifth time interval across 2 days. This could be due to detours or traffic conditions that caused the individual to be ahead or behind the locations at this time on the previous days. Comparison can also be made using distances between adjacent locations on the same track on different days but in the same time intervals to determine changes in travel speeds from day to day (Fig. 20.11b). Furthermore, daily groupings of strings of consecutive Geohash codes (like the horizontal groupings in the shaded boxes in Fig. 20.11b) reveal the daily patterns of stops in common locations. In this illustrative example, the individual stayed (long stops) at **9y69gmob** in the morning and returned to the same location at night on Day_1 and Day_n, but the person stayed at the location all day on Day_2, which suggested that the location is likely to be the individual's residence.

In addition, differences in space and time across tracks as shown in separate line graphs in Fig. 20.11a can be converted to a space-time-track image with a brighter shade indicating a greater difference along tracks (Fig. 20.12). In a space-time-track image, each row corresponds to a track, and each column represents resampled time intervals along these tracks. The lighter a cell denotes a smaller spatial difference of this track from the previous track. Measures of spatial difference can be based on Euclidean distance, travel distance, or some other cost metrics. If the track taken on Day_i is identical to the track taken on Day_{i-1}, the row for Day_i in the image will

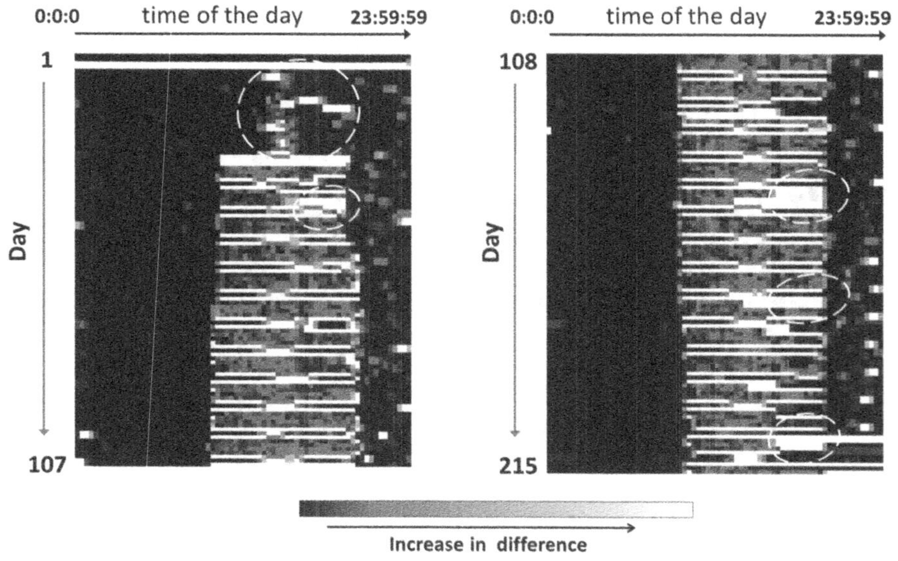

Fig. 20.12 A space-time-track image showing spatial differences in daily track comparison. The difference is determined for each time interval by the spatial difference of a track to that of the previous day. The *circles* mark examples that may suggest incongruent patterns or new activities

be black throughout. Otherwise, the spatial difference from the track on Day_{i-1} at $Time_j$ will be stored in the cell (Day_i, $Time_j$).

Figure 20.12 shows a space-time-track image based on an individual's GPS tracks for 215 days. There are two key re-occurring features in the image. One feature consists of repetitive sets of two bright lines starting around Day 35 to Day 215. The bright lines correspond to spatial differences between Fridays and Saturdays as well as Sundays and Mondays. The dark lines in-between reflect the greater similarity between tracks recorded for Saturdays and Sundays. The second re-occurring feature is the progression of the scattering bright spots developed over these days in the late night hours. These late-night outings could be for social events or a night job, which can be further explored by mapping stops and moves on these days and times to determine places that the individual visited. The space-time-track image also includes incongruent features. Examples are marked by circles in Fig. 20.12. These incongruent features could be indicative of new or occasional activities.

Furthermore, the daily total of spatial difference (e.g. the row sum in the space-time-track image) reflects the cumulative difference of a track compared to the track taken on the previous day. If an individual more or less took the same route to work, the person's daily track would be quite similar and therefore resulted in a small cumulative difference along a track. On the other hand, if one was wandering around town, the person's daily tracks could be very different and the cumulative difference would be large. Figure 20.13 shows a time series of daily cumulative differences of tracks based on the values in the space-time-track image in Fig. 20.12. The time

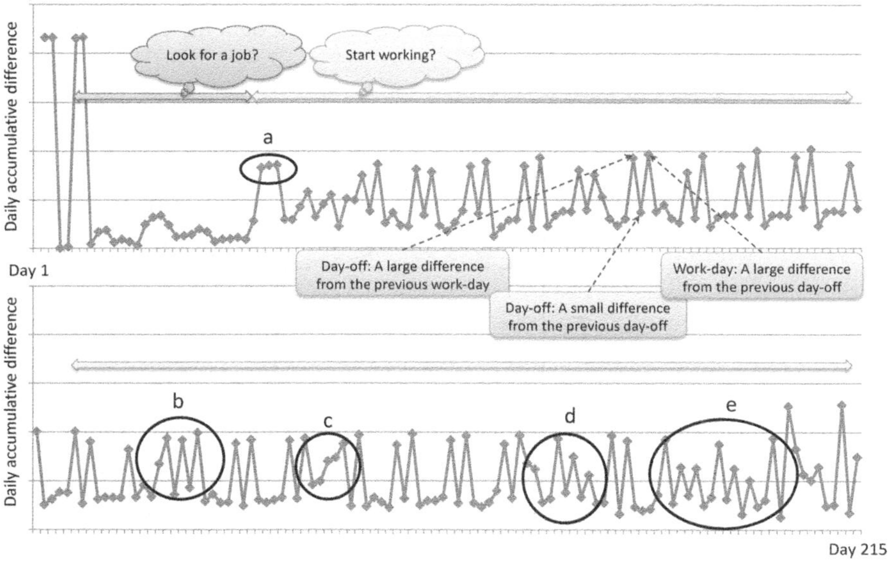

Fig. 20.13 A time series of daily accumulative differences among tracks

series shows comparably large fluctuations in the early days and then some small fluctuations before settling into a rather regular pattern.

A possible interpretation of the pattern is that the individual was new to the area, wandered around and narrowed his/her job search to a specific region. Once the individual found a job, the person tried different routes to work (circle a in Fig. 20.13) and then settled with a regular route to commute. The regular fluctuations correspond to track differences in weekdays and weekends. After the person settled down to a regular routine for a while, the person's pattern varied again, which might be due to exploration of one's neighborhoods or participation in social activities (circles b, c, and d in Fig. 20.13), and eventually grew to a new pattern (circle e in Fig. 20.13) which can be related to the late-night outings in Fig. 20.12. While the interpretation is speculative and requires validation with interviews or auxiliary data, the example shows that daily cumulative differences in track analysis can reveal the development of and changes in patterns of life implied by an individual's movements in space and time.

20.5.3 Identify Opportunities for Interactions in Space and Time

In the framework of time geography (Fig. 20.1), paths in proximity in space and time can be considered to coexist and therefore offer opportunities for interactions (Fig. 20.14). Similar ideas can be used to identify individuals with opportunities to interact based on their proximity in space and time. Studies may focus on opportunities for sustained interactions and therefore discount the cases when two

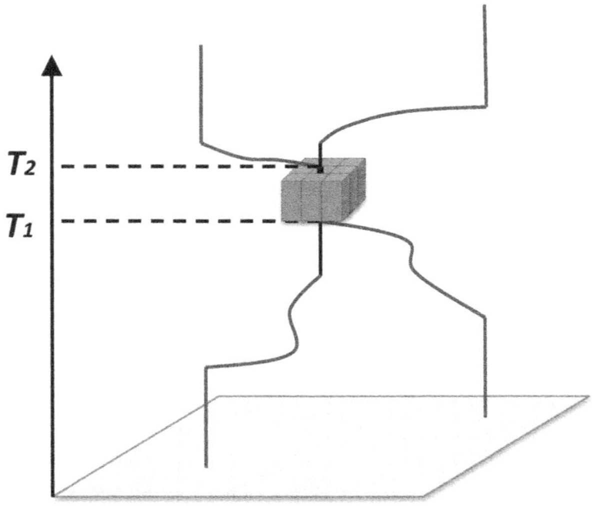

Fig. 20.14 Space-time coincidence indicative of opportunity for interactions

people pass by each other in transit, and therefore only locations of stops are considered for analysis. Other studies may include instantaneous interactions during passing.

The Geohash system provides a convenient spatial structure to determine proximity for its grouping efficiency. For a Level 9 Geohash, the default spatial proximity is set at 2 m, and for Level 8 Geohash, 30 m. All stops for all individuals of interest are included in the analysis. However, like all discrete grid systems, locations along the boundary of adjacent grid boxes can be within the proximity threshold even though these locations are not in the same grid box (i.e. the fault line issue). Therefore, proximity examination needs to include locations in all Geohash boxes adjacent to the box of interest.[2] Figure 20.15 provides an example of the adjacent Geohash boxes needed for analysis of nearby points. The spatial proximity needs to be adjusted to about 5 m for Level 9 and 45 m for Level 8 Geohashes.

Once the Geohash Level is determined, all stops falling in the same Geohash box as well as in adjacent Geohash boxes are grouped. The analysis then proceeds one group at a time to determine stops within temporal proximity. Both spatial proximity and temporal proximity should be defined based on application domains. Individuals can sit next to each other in a basketball game or run far apart in a marathon race. One can drop a package at a friend's house, and another person might pick up the package an hour later. There are four types of space-time constraints on communication (Janelle 1995). Spatially, communications can take place among individuals who are in presence or telepresence, and temporally, it can be synchronous or asynchronous. The four types of space-time constraints on communication can also guide for the conceptualization of interactions.

Once the proper thresholds for spatial proximity and temporal proximity are determined, Euclidean distance (or travel distance) and time differences can be calculated among locations in a group to determine stops that are within the desired spatial proximity and temporal proximity. Individuals within the spatial and temporal proximity can then be identified and connected for a social network of potential interactions. In a case study of 50 individuals in a suburban town (Moore, Oklahoma, USA), we define the space-time proximity as within one level 8 Geohash grid ($\sim$45 m) and 30 min. Eleven of the 50 individuals are within space-time proximity of synchronous presence over 553 days of the study period (3/4/2009–9/8/2010). The width of an edge corresponds to the relative frequency of the potential to meet. The connection between person ID 16 and person ID 30 is an example of a strong connection with a relatively high frequency for potential meetings in places where both individuals visited at about the same time (Fig. 20.16). Individuals with IDs not listed in the social network have no records of presence within the defined space-time proximity to other individuals in the study.

[2]The issue can be solved with the fault line algorithm. Its source code is available at http://code.google.com/p/geohash-fcdemo/

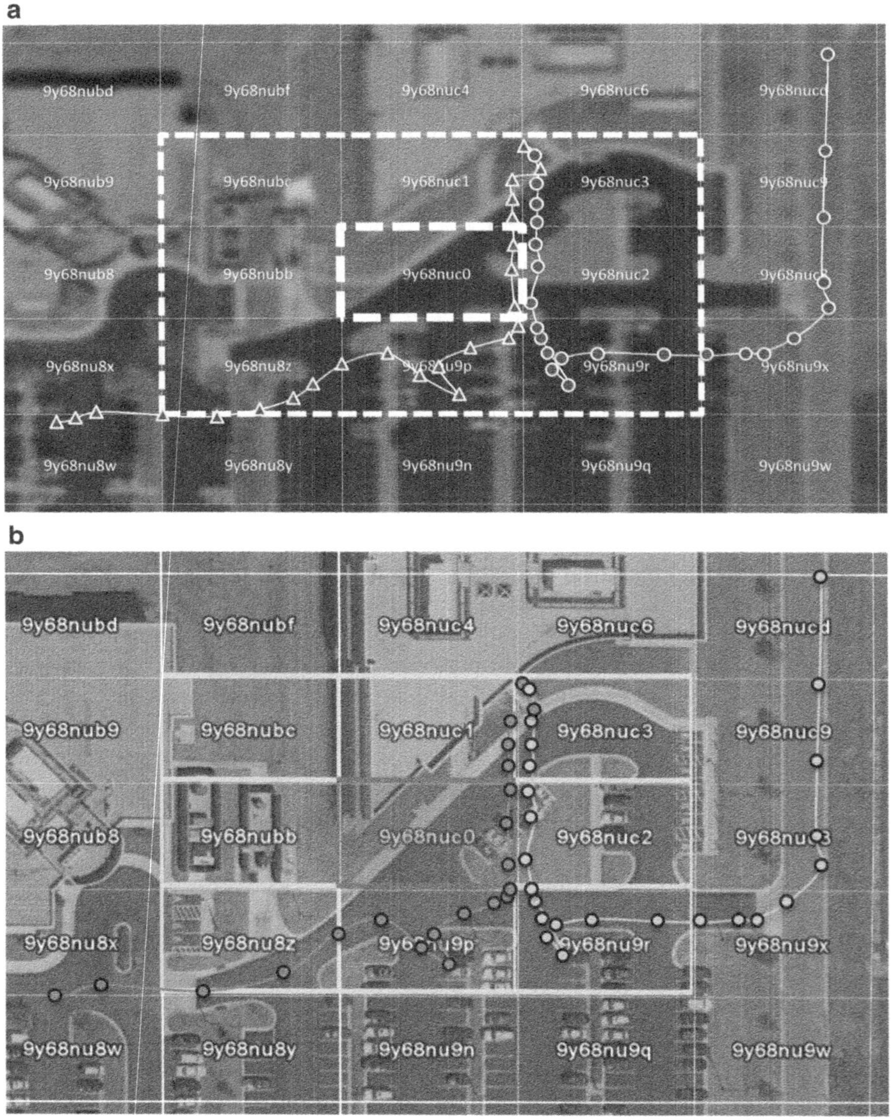

Fig. 20.15 Consideration of proximity should include locations in adjacent Geohash boxes

20.6 Concluding Remarks

Data Science and Big Data challenges are prominent research issues with the fast growing body of spatio-temporal data being generated across many domains and applications. With the popularity of GPS devices, location-aware devices (including

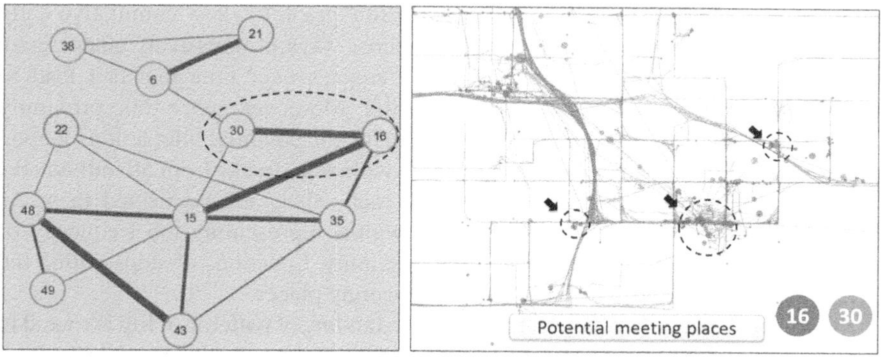

Fig. 20.16 An example of a social network based on space-time proximity that implies opportunities for interactions

smart phones), geotagged tweets and photographs, and numerous other means, we can attach location information and record movements for animals, people, goods, and services. While time geography offers a rich conceptual framework to examine spatial behavior and interactions among individuals, research on trajectory analysis, trajectory mining, and more generally, mobile object modeling has developed many statistical or computational methods to identify clusters or detect outliners in movement data. Many of these studies considered only geometric or semantic properties of trajectories. Instead, this chapter advocates for a space-time analytics approach to examine reoccurring movements of individuals and from the reoccurring movements to identify patterns of life and their opportunities for interactions based on proximity in space and time.

The chapter defines the terms: paths, tracks, trajectories, stops and moves, while the meanings of these terms may vary in the literature. Paths are actual routes taken by individuals, tracks are composed of observed locations, trajectories are interpolated paths between observed locations, and stops and moves are observed locations with durations of non-movement in transit. Space-time analytics of tracks emphasize elements of tracks and how the elements may be structured for analysis. Stops, moves, and track segments are considered in this chapter as elements of tracks, and Geohash boxes and chronological time are used to structure space and time for georeferencing and sequencing stops and moves to form tracks for analysis. Spatial clusters and time spent at locations are criteria used to identify stops along tracks. Analysis of stops reveals the place where an individual frequently visits and visiting patterns. Observed locations are resampled to form tracks with stops and moves in Geohash codes at common time intervals.

Comparison of corresponding Geohash strings at time intervals reveals how an individual might have taken a different route or changed travel behavior (left early/late or at a higher speed). In this chapter, we show three ways to compare tracks: (1) line graphs that show variations across multiple days; (2) space-time-track images to highlight high and low differences in tracks throughout a day and across multiple days; and (3) summative line graph to show cumulative daily track differences across multiple days. The three ways progressively characterize individual's movements, recurring track differences over time that can lead to inferences about the individual's pattern of life. Moreover, space-time proximity of stops along tracks is used as a proxy indicator for potential interactions among individuals. We briefly described a case study with daily tracks from 50 individuals, identified 11 in a space-time proximity to each other, and connected these 11 individuals in a social network with edge widths representing the frequency of the two individuals within the space-time proximity. Locations of stops within the proximity can be mapped to show potential meeting places.

Space-time analytics of tracks for the understanding of patterns of life is rooted in the premise that space-time research in GIScience must connect what we learn from the data to the underlying geographic dynamics that generate the data. Space-time data are observations about activities, events and processes. Space-time analytics aims to discern elements of these geographic dynamics and recognize structures among these elements to capture the spatiotemporal characteristics of activities, events and processes through geographic contextualization for interpretation and understanding. It is a divide-and-conquer as well as disaggregate-and-aggregate approach to first zoom in to disassembling the composition and then finding ways to organizing them to expose hidden meanings. The main goal of this chapter is to communicate such a conceptual framework for space-time research in GIScience. The potential of Data Science, Big Data, or BigSpatial research is much enhanced with emphases on making the connection between data and what the data represent as elements or in a structure.

Acknowledgment The research was in part supported by Award #2010-DE-BX-K005, awarded by the National Institute of Justice, Office of Justice Programs, U.S. Department of Justice. The opinions, findings, and conclusions or recommendations expressed in this publication are those of the authors and do not necessarily reflect those of the Department of Justice. NIJ defines publications as any planned, written, visual or sound material substantively based on the project, formally prepared by the award recipient for dissemination to the public.

References

Andrienko, N., & Andrienko, G. (2011). Spatial generalization and aggregation of massive movement data. *IEEE Transactions on Visualization and Computer Graphics, 17*(2), 205–219.

Balkic, Z., Sostaric, D., & Horvat, G. (2012). GeoHash and UUID identifier for multi-agent systems. *Lecture Notes in Computer Science, 7327*, 290–298.

Beckers, B., & Beckers, P. (2012). A general rule for disk and hemisphere partition into equal-area cells. *Computational Geometry, 45*(7), 275–283.

Berry, M. J. A., & Linoff, G. (1997). *Data mining techniques – For marketing, sales and customer support*. New York: Wiley.

Böhm, C., Feng, J., He, X., Mai, S. T., Plant, C., & Shao, J. (2011). *A novel similarity measure for fiber clustering using longest common subsequence*. Paper presented at the proceedings of the 2011 workshop on data mining for medicine and healthcare, San Diego, CA, USA.

Dodge, S., Weibel, R., & Lautenschütz, A.-K. (2008). Towards a taxonomy of movement patterns. *Information Visualization, 7*, 240–252.
Dodge, S., Weibel, R., & Forootan, E. (2009). Revealing the physics of movement: Comparing the similarity of movement characteristics of different types of moving objects. *Computers, Environment and Urban Systems, 33*(6), 419–434.
Dodge, S., Laube, P., & Weibel, R. (2012). Movement similarity assessment using symbolic representation of trajectories. [Article]. *International Journal of Geographical Information Science, 26*(9), 1563–1588. doi:10.1080/13658816.2011.630003.
Dutton, G. (2000). *Universal geospatial data exchange via global hierarchical coordinates*. Paper presented at the international conference on discrete global grids, Santa Barbara, CA.
Eagle, N., & Pentland, A. (2006). Reality mining: Sensing complex social systems. *Personal and Ubiquitous Computing, 10*(4), 255–268.
Eagle, N., & Pentland, A. (2009). Eigenbehaviors: Identifying structure in routine. *Behavioral Ecology and Sociobiology, 63*(7), 1057–1066.
Gaffney, S., & Smyth, P. (1999). Trajectory clustering with mixtures of regression models. In *Proceedings of 5th ACM SIGKDD international conference on knowledge discovery and data mining, San Diego, CA, 1999* (pp. 63–72). New York, NY, USA: ACM.
Gaffney, S., Robertson, A., Smyth, P., Camargo, S., & Ghil, M. (2007). Probabilistic clustering of extratropical cyclones using regression mixture models. *Climate Dynamics, 29*(4), 423–440.
Geohash. (n.d.). In *Wikipedia*. http://en.wikipedia.org/wiki/Geohash. Accessed 17 Dec 2012.
Giannotti, F., & Pedreschi, D. (2007). *Mobility, data mining, and privacy: Geographic knowledge discovery*. Berlin/Heidelberg: Springer.
Guo, D., Liu, S., & Jin, H. (2010). A graph-based approach to vehicle trajectory analysis. *Journal of Location Based Services, 4*(3), 183–199.
Hägerstraand, T. (1970). What about people in regional science? *Papers in Regional Science, 24*(1), 7–24. doi:10.1111/j.1435-5597.1970.tb01464.x.
Janelle, D. G. (Ed.). (1995). *Metropolitan expansion, telecommuting, and transportation* (The geography of urban transportation 2nd ed.). New York: Guilford Press.
Kwan, M.-P. (2004). GIS methods in time-geographic research: Geocomputation and geovisualization of human activity patterns. [Article]. *Geografiska Annaler: Series B, Human Geography, 86*(4), 267–280. doi:10.1111/j.0435-3684.2004.00167.x.
Laube, P., Dennis, T., Forer, P., & Walker, M. (2007). Movement beyond the snapshot: Dynamic analysis of geospatial lifelines. *Computers, Environment and Urban Systems, 31*, 481–501.
Lavalle, S., Lesser, E., Shockley, R., Hopkins, M. S., & Kruschwitz, N. (2011). Big data, analytics and the path from insights to value. *MIT Sloan Management Review, 52*(2), 21–32.
Lee, J.-G., Han, J., & Whang, K.-Y. (2007). *Trajectory clustering: A partition-and-group framework*. Paper presented at the Proceedings of the 2007 ACM SIGMOD international conference on management of data, Beijing, China.
Li, L., & Goodchild, M. (2012). Automatically and accurately matching objects in geospatial datasets. *Advances in Geo-Spatial Information Science, 10*, 71.
Li, H., Guo, C., & Qiu, W. (2011). Similarity measure based on piecewise linear approximation and derivative dynamic time warping for time series mining. *Expert Systems with Applications, 38*(12), 14732–14743. http://dx.doi.org/10.1016/j.eswa.2011.05.007
Long, J. A., & Nelson, T. A. (2012). A review of quantitative methods for movement data. *International Journal of Geographical Information Science, 27*(2), 292–318. doi:10.1080/13658816.2012.682578.
Lu, N., Cheng, C., Ma, H., & Yang, Y. (2012). Global discrete grid systems analysis and comparison. In *Geoscience and remote sensing symposium (IGARSS), 2012 IEEE International, 22–27 July 2012* (pp. 2771–2774). New York, NY, USA: IEEE.
Marek, K. (2011). Chapter 1: Web analytics overview. *Library Technology Reports, 47*(5), 5–10.
Miller, H. J., & Bridwell, S. A. (2009). A field-based theory for time geography. *Annals of the Association of American Geographers, 99*(1), 49–75.
Morrill, R. (2005). Hägerstrand and the 'quantitative revolution': A personal appreciation. *Progress in Human Geography, 29*(3), 333–336.

Palma, A. T., Bogorny, V., Kuijpers, B., & Alvares, L. O. (2008). *A clustering-based approach for discovering interesting places in trajectories*. Paper presented at the Proceedings of the 2008 ACM symposium on applied computing, Fortaleza, Ceara, Brazil.

Panagiotakis, C., Pelekis, N., Kopanakis, I., Ramasso, E., & Theodoridis, Y. (2012). Segmentation and sampling of moving object trajectories based on representativeness. *IEEE Transactions on Knowledge and Data Engineering, 24*(7), 1328–1343.

Patterson, D. J., Liao, L., Fox, D., & Kautz, H. (2003, January). Inferring high-level behavior from low-level sensors. In *UbiComp 2003: Ubiquitous computing* (pp. 73–89). Berlin/Heidelberg: Springer.

Popa, M. C., Rothkrantz, L. J. M., Shan, C., Gritti, T., & Wiggers, P. (2012). Semantic assessment of shopping behavior using trajectories, shopping related actions, and context information. [Short Communication]. *Pattern Recognition Letters, Online Preprint*. doi:10.1016/j.patrec.2012.04.015.

Shaw, S.-L., & Yu, H. (2009). A GIS-based time-geographic approach of studying individual activities and interactions in a hybrid physical–virtual space. *Journal of Transport Geography, 17*(2), 141–149.

Shen, Z., & Ma, K.-L. (2008). MobiVis: A visualization system for exploring mobile data. In *Visualization symposium, 2008. PacificVIS'08. IEEE Pacific, 5–7 March 2008* (pp. 175–182). New York: IEEE Geoscience and Remote Sensing Society.

Shen, Z., Ma, K.-L., & Eliassi-Rad, T. (2006). Visual analysis of large heterogeneous social networks by semantic and structural abstraction. *IEEE Transactions on Visualization and Computer Graphics, 12*(6), 1427–1439.

Shmueli, G., & Koppius, O. R. (2011). Predictive analytics in information systems research [Article]. *MIS Quarterly, 35*(3), 553–572.

Spaccapietra, S., Parent, C., Damiani, M. L., de Macedo, J. A., Porto, F., & Vangenot, C. (2008). A conceptual view on trajectories. *Data & Knowledge Engineering, 65*(1), 126–146.

Tong, X., Ben, J., & Wang, Y. (2010). A new effective hexagonal discrete global grid system: Hexagonal quad balanced structure. In *2010 18th international conference on geoinformatics, 18–20 June 2010* (pp. 1–6). New York: IEEE Geoscience and Remote Sensing Society.

Wagner, E. P. (2012). Data changes everything: Delivering on the promise of learning analytics in higher education [Article]. *Educause Review, 47*(4), 32.

Willems, N., van de Wetering, H., & van Wijk, J. J. (2009). Visualization of vessel movements. *IEEE VGTC Symposium on Visualization, 28*(3), 959–966.

Yan, Z., Giatrakos, N., Katsikaros, V., Pelekis, N., & Theodoridis, Y. (2011, August). *SeTraStream: Semantic-aware trajectory construction over streaming movement data*. Paper presented at the 12th international symposium on spatial and temporal databases (SSTD), Minneapolis, MN, USA.

Yan, Z., Chakraborty, D., Parent, C., Spaccapietra, S., & Aberer, K. (2012). Semantic trajectories: Mobility data computation and annotation. *ACM Transactions on Intelligent Systems and Technology (TIST), 9*(4), 1–34, Article 39.

Chapter 21
Real-Time Space-Time Integration in GIScience and Geography

Douglas Richardson

Have you also learned that secret from the river... that the river is everywhere at the same time, at the source and at the mouth, at the waterfall, at the ferry, at the current, in the ocean and in the mountains, everywhere, and that the present only exists for it, not the shadow of the past, not the shadow of the future?

(Hermann Hesse, *Siddhartha (1951)*)

21.1 Real-Time Space-Time Integration

Time is a multi-faceted concept, with deeply personal and subjective meanings as well as formally constructed but still arbitrary meanings, and ontologies for time in GIScience can be as complex as they are for space in Geography. However, at its simplest and in the practice of Geographic Information Science research, it is still useful to think about space-time research challenges in three basic temporal domains: (1) *past time* – documenting and interpreting historical events, places, and processes; (2) *future time* – modeling, simulating, or projecting future spatial scenarios; and (3) *real-time* –understanding and interacting with events, people, places, and spatial processes as they are emerging and mutually evolving in a dynamic and mobile world.

While of course interrelated, each of these simple categories has its own special set of challenges in gathering, understanding, and representing space-time integrated data. Historical time and geographic data, especially from the more distant past, is highly constrained by pre-existing and frequently indefinite or inconsistent time and space categories of bygone eras. Historical GIS cannot

This chapter has been reprinted with the permission of the Association of American Geographers and Taylor & Francis, LLC. Originally published: Douglas B. Richardson (2013) Real-Time Space–Time Integration in GIScience and Geography, *Annals of the Association of American Geographers,* 103:5, 1062–1071.

D. Richardson (✉)
Association of American Geographers, 1710 16th Street NW, Washington, DC 20009, USA
e-mail: drichardson@aag.org

M.-P. Kwan et al. (eds.), *Space-Time Integration in Geography and GIScience*, DOI 10.1007/978-94-017-9205-9_21

ignore and must struggle to work within and reconcile the profusion of its received categories of time and space (Richardson 2011). Future time GIScience work is less restricted, in the sense that its temporal and spatial constructs are not necessarily constrained by past categories or practices and can more freely be structured as the researcher or modeler wishes; its delineations of space or time need not be confined to those of a world or a time that is, or already has, occurred.

The possibility of autonomous and continuous *real-time space-time* geographic data creation, integration, and use as one encounters and travels through space is a relatively new and thoroughly transformative development for geography, GIS, and GIScience, as well as for science and society more broadly. The real-time interactive GPS/GIS functionality at the core of this transformation is largely responsible for the explosion of spatiotemporal[1] data being generated today, and is the underlying breakthrough enabling many new geographic research initiatives and myriad real-time space-time integrated geospatial applications in governments, businesses, and society. As others in this forum on space-time integration will discuss past time in the context of historical GIS, and the construction of future time for spatial modeling and so forth, I will limit my observations in this essay to the concept of *real-time space-time* integration in GIScience and geography.

21.2 Origins of Real-Time Space-Time Integration in GIS and GIScience

By the mid 1980s, geographic information systems (GIS) had developed into very useful cartographic tools for the computerized storage, manipulation, and graphic display of multiple thematic layers of locational data and attribute information, as well as some analytical functions. It greatly improved the efficiency of making and updating maps, compared with non-digital processes. However, GIS alone was a *post facto* system: its locational, feature, and attribute data were laboriously gathered and then integrated within the system well after it was collected. GIS was not fused with or interactive in real-time with the dynamic real world around it.

Similarly, GPS alone, while a significant development of the early 1980s, provided location but not surrounding context. As Abler (1993) noted, "GPS equipment will tell me where I am with great precision … But knowing precisely where I am may not be very helpful. Location, no matter how precisely specified,

[1]In what follows, spatial and spatiotemporal should be taken as including references to location and also to time where appropriate. GPS refers also to new similar positioning systems providing equivalent autonomous and accurate location and time. GIS refers to a broad range of geographic database systems, including but not limited to commercial GIS products. GIScience refers to research on the scientific and technological challenges associated with working with spatial or spatiotemporal data, including philosophical, ethical and social considerations integral to its practice. "Real-time" refers, as always when dealing with the electronic transmission and processing of measurements, to "near real-time."

is sterile in and of itself. Context determines whether knowledge of location is invaluable... Contextual knowledge immensely enriches the value of locational information" (133–134).

The pivotal enabling development in real-time space-time GIScience and geography was the invention and development of Real-Time Interactive (RTI) GPS/GIS software and systems in the late 1980s and early 1990s. This innovation enabled the continuous and mobile creation of highly accurate real-time and tightly-fused geospatial location, attribute, and temporal data, and its simultaneous integration with the GIS for interactive real-time use as one moved through and encountered ones environment. This transformative development has since functioned in geography, GIScience, cartography, and related fields as a core change agent, profoundly realigning basic traditional relationships and changing the ways geographic data is collected, mapped, and used, as well as how people encounter and experience their world.[2] RTI GPS/GIS also has greatly expanded research frontiers in geography and other disciplines, including the social sciences, transportation, health, natural resources, genetics, ecology, logistics, and the environmental sciences, among many others.

The early conceptual and technical challenges to achieving the real-time space-time interactive functionality of RTI GPS/GIS were formidable, and years of intensive research were required to overcome the multiple scientific and ontological challenges and to develop the complex technical procedures and methods to achieve fully functional automated RTI GPS/GIS mapping systems. These early efforts are well documented in US Patent 5,214,757 (Mauney et al. 1993) and the reader is referred to this document for a thorough summary of the scope and technical details of the original invention and its range of new methods. Additional archival materials addressing the early development, functionality, and use of RTI GPS/GIS technologies are available online.[3]

With the fundamental conceptual, ontological, scientific, and technological research necessary for the development of the first RTI GPS/GIS systems largely accomplished during the mid and late 1980s, proven technologies embodying RTI GPS/GIS functionality for real-time space-time geographies were firmly demonstrated by the late 1980s, and initial commercial systems came into increasingly widespread use for research and a wide array of governmental, business, and educational applications by the early 1990s (Richardson 1990). Figure 21.1 summarizes the already mature functionality of an early RTI GPS/GIS system of this period (in dark grey), and also illustrates two of many research and application domains, typical of subsequent outcomes of these innovations.

---

[2]One could argue that RTI GPS/GIS has functioned within the disciplines of Geography, GIScience, Cartography, and related fields, as what is termed in computer science and related industries as a "Disruptive Technology," impacting past processes and procedures broadly throughout these disciplines and industries (Bower and Christensen 1995).

[3]Archival documentation of early real-time space-time integration systems is available at www.aag.org/RTIGPSGIS.

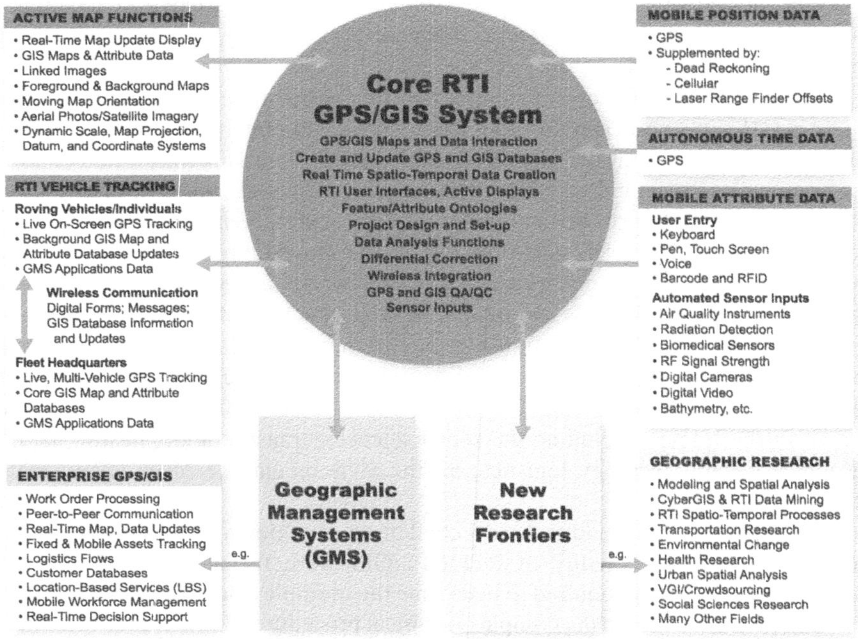

Fig. 21.1 Real-time interactive GPS/GIS system functionality (*in dark grey*) developed by the early 1990s, and two examples of the many subsequent outcomes of these technologies: (*1*) the generation of new research frontiers and capabilities in geography and many related fields (*in light grey*), and (2) the development and widespread implementation of "Geographic Management Systems" for real-time operations management by many governmental, business, and nongovernmental organizations (*in medium grey*)

At this early stage, RTI GPS/GIS provided mobile real-time generation of high-accuracy three-dimensional location (x, y, z coordinates), precise time, data on features and attributes via manual or automated electronic instrumentation input, and the real-time integration and interaction of these data within GIS or other spatial databases. While GPS was the preferred source of location, other supplemental methods such as dead reckoning, cellular signal, and laser rangefinders were also included in the invention and its early embodiments. Data from the GIS could be accessed in the field, and interactively created, updated, used, or acted upon in real-time. GIS vector maps, raster maps, and remotely sensed imagery could be manipulated and displayed as active or passive maps, and as foreground or background maps. Complex interactive data management, user interface, and QA/QC functions were developed to create tightly fused spatiotemporal data in real time and to interact and integrate that data continuously and in real time with GISs and other spatial databases. Special external data source ("XDS") software modules also were developed to enable the automated input of data from multiple electronic sensors (such as laser range finders, digital photo and video, radiation detectors, air quality and other environmental monitoring instruments, etc.) directly into the RTI GPS/GIS system (Richardson 1993).

The real-time fusion, analysis, visualization and interaction with dynamic geographic information and processes made possible by these early RTI GPS/GIS systems quickly engendered new research directions in geography and GIScience research as well as other scientific domains (Richardson et al. 2013). As Fig. 21.1 illustrates, transportation, health, environmental change, and modeling are just some of the examples. RTI GPS/GIS functionality also stimulated and produced dynamic new real-time operations management possibilities for businesses and governments, generating what I have termed *geographic management systems* (Richardson 2006). For example, these systems were developed early on for the real-time management of vehicle fleets to optimize package delivery operations or taxicab dispatch; electrical utility energy demand monitoring, distributed asset inventory and maintenance, and mobile work force work management; public safety activities by government agencies; United Nations international relief operations; automated supply chain and logistics operations for retail and manufacturing; and for tracking and responding to contagious disease outbreaks. For a more detailed discussion of the transformations generated in geography and GIS by the introduction of real-time geographic management systems, see Richardson and Solis (2004) and Abler and Richardson (2003).

In recent years, further diffusion and widespread adoption of RTI GPS/GIS has been facilitated by improvements in supporting technology infrastructure such as smaller and faster computers, better wireless systems, easier access to the internet and developments in web communications, increased resolution of remote sensing, and continued development of smaller and more portable environmental sensors. However, nearly all of the RTI GPS/GIS functionality now in use today was embodied in these early systems and their related patents.

Miniaturized RTI GPS/GIS functionality is now also embedded in consumer technologies such as cell phones, automobiles, and personal navigation devices and is spawning new geographic research and applications involving volunteered geographic information (VGI), crowd sourcing, and citizen science. The compelling experiential immediacy of these technologies is also increasingly mediating and changing how individuals interact with and experience the world, and each other (Monaghan 2013).

21.3 Real-Time Space-Time Fault Lines in Geography and GIScience

RTI GPS/GIS has created completely new possibilities within the realms of GIScience and GIS for interacting with, representing, making decisions about, and negotiating the world around us *as it is encountered and experienced in real time*.[4] At the individual level, this allows GIS to better approximate how individuals actually interact with their environment in daily life (Kwan 2009; McQuoid and Dijst 2012). Within larger GPS/GIS systems it also enables researchers to better understand and monitor complex real-time population dynamics at a larger scale, for example natural disaster evacuations during Hurricane Katrina or refugee migrations occurring from conflict zones in Africa or the Middle East.

The changes wrought by RTI GPS/GIS and its creation of new "real-time space-time" capabilities in GIScience and geography are extensive. One can discern its "fault lines," or discontinuities in multiple methods and practices in the disciplines of geography, cartography, and GIScience. Such fault lines signal discontinuities between what was possible prior to the introduction of transformative technologies, in this case RTI GPS/GIS functionality, and what then became possible following its development and introduction.

RTI GPS/GIS functionality has transformed mapping and cartographic methods worldwide, by enabling the collection and use of geographic information that is far more detailed, timely, accurate, immediate, and specific to a particular application than was possible before. It is largely responsible for the explosion in the amount of highly detailed, specific, and updated feature and attribute spatiotemporal data now available for GIS applications and for GIScience research. It has also enabled the integration into GIS of *mobile* electronic sensors, such as environmental pollutant monitors, digital cameras, bathymetric instruments, noise monitors, biometric health sensors, and so forth, resulting in vast amounts of new continuously georeferenced and time coded sensor data.

RTI GPS/GIS has spawned whole new industries, such as Location Based Services (LBS), and is a primary technology employed to create the navigational databases and street maps now used so widely by Google, Apple, MapQuest and other internet mapping providers. It is a core component of the management of

---

[4]As David Lowenthal (1961) pointed out, the focus of geographic enquiry is fundamentally congruous with the objects and structures and immediacy of common experience. More than any other field, he argues, "the subject matter of geography approximates the world of general discourse; the palpable present, the everyday life of man on earth, is seldom far from our professional concerns... geography observes and analyzes aspects of the milieu on the scale and in the categories that they are usually apprehended in everyday life" (241). Along similar lines, when referring to the experience of RTI GPS/GIS mapping, Ron Abler (1993) noted that,

> Few geographers will be able to resist the seduction of creating new maps interactively, in real time, amid the real phenomena the maps represent, while at the same time referring to and revising the background maps for the area contained in portable geographic information systems carried into the field. The capability for displaying to a person in the field background maps, the map being created, and the individuals changing position on both background and new maps will arouse the dormant field work virus lurking in even the most ardent armchair geographer. (135)

The immediacy and somewhat exhilarating experience of direct encounter and interaction with the real world in real time while using RTI GPS/GIS has been remarked on by many users of these technologies.

day-to-day real-time operations at most governmental agencies and large business organizations. It has, for better or worse, changed the ways wars are waged and military operations are carried out.

Another fault line created by the introduction of real-time space-time geographies is evident in the enormous broadening of participation in mapping and geographic knowledge production to include directly in the mapping process the actual subject matter experts themselves (e.g., the botanist, the environmental scientist, the lineman, the archeologist, etc.), or community activists (participatory GIS), human rights organizations, and most recently by ordinary citizens through volunteered geographic information (VGI) and crowd-sourced geographies.

21.4 Trends and Research Challenges

The pace of real-time geographic data creation and it usage in consumer and business applications is expected to accelerate in future years, resulting from three interrelated trends that together constitute a convergence of opportunities for new research (1) the explosion of real-time, spatiotemporal data from real-time interactive GPS/GIS-enabled devices, often integrated with distributed environmental sensor systems, satellite remote sensing, and crowd sourcing; (2) continuous advances in computing and mobile wireless technologies, Web service-oriented architectures, and geospatial cyberinfrastructure; and (3) the development of more sophisticated tools and methods for analyzing, modeling, and visualizing spatiotemporal data at scales ranging temporally from the everyday to the life course and spatially from the microscale to the global (Richardson et al. 2011).

As spatiotemporal data and applications have become ubiquitously available, and as the adoption of RTI GPS/GIS as a "geographic management system" has become more pervasive in society in recent years, research agendas for real-time space-time integration in geography and GIScience are also proliferating.[5] What follows below

[5]A special 3-day Symposium focused on the research status, recent advances, and research needs of space-time integration, modeling and analysis in geography and GIScience was organized by the Association of American Geographers (AAG) for its Annual Meeting in Seattle, April 12–16, 2011. This major research Symposium built on momentum from an earlier space-time analysis workshop co-sponsored by the AAG, Esri, the University of Redlands, and the University of Southern California in February of 2010. Research agendas presented here draw in part on collaborative activities undertaken by the author during these symposia and workshops. The space-time integration research presented at the AAG Symposium has generated, as intended, many ongoing outcomes. These include books, workshops, journal special issues, NSF and NIH research proposals, and this space-time forum in the *Annals of the Association of American Geographers*, which provides perspectives from each of the plenary speakers at the AAG Space-Time Symposium.

Space-time integration has of course long been the topic of study and speculation in geography and beyond. A great deal of productive research has been undertaken generally on space-time relationships in GIS and GIScience, with a few leading examples including Worboys (2005),

is a brief summary of key real-time space-time research and development agenda items that I believe warrant special attention and focus.

21.4.1 Temporal Scale

We should address temporal scale with the same degree of attention paid to spatial scale. Metadata frameworks for spatial-temporal data need to be developed to accommodate varying time systems. Efficient methods for integrating important legacy temporal databases that are at widely varying temporal scales (temporal data conversion) are needed.

21.4.2 Space-Time Models

We should inventory and assess existing space/time data structures and analysis models as found in different application domains, identify commonalities across those domains and models, and seek integration across domains. We should also identify the exceptions and outliers, and seek separate but linked solutions for these. Substantial funding from the National Science Foundation should be directed to the development of more sophisticated "now-casting" capabilities, as well as predictive and forecasting models which incorporate the growing wealth of RTI GPS/GIS spatiotemporal data.

21.4.3 Ontology-Based Frameworks

Ontological frameworks (or data dictionaries) to guide the development of real-time space-time research should link the properties of data and task characteristics to functional capabilities of visualization and analytical techniques to indicate fruitful chains/workflows of real-time spatial-temporal analysis operations. These frameworks will likely become key contributions to knowledge about the practice of spatial-temporal analysis. Workshops, interactive forums, and publications should be organized to develop shared frameworks.

Kwan (2004), Cressie and Wikle (2011), Miller (2004), Peuquet (2002) and Yuan (1996), and with broader treatments by Hawking (1988), Hägerstrand (1970), and Kennedy (2002).

21.4.4 Transforming Geographic Data into Meaningful Information

This is a central research challenge for all of GIScience. The key research question, fundamentally, is how do increasing quantities of geographic data become fused into meaningful information? Information is the rich relational context that dynamically binds data together, provides much of its utility, and allows people to choose among several otherwise indistinguishable courses of action (Abler and Richardson 2003). The integrated GIS component of RTI GPS/GIS offers some potential pathways forward, as the robust integrated data collected by these systems can be directly integrated in real time in the field into previously structured GIS databases, designed for specific tasks or for answering focused research questions. Essentially, the extended GIS itself can provide much of the necessary external relational context, not only about ones immediate surroundings, but also key related information necessary to answer specified questions in real time in the field based on the new real-time interactive GPS/GIS data being generated about events, places, and processes as they are emerging, evolving and interacting in real-time across space.

21.4.5 Ethics, Human Rights and Human Subject Protection

GIScience is an international and interdisciplinary enterprise. Examining the impacts of our research on others is a necessary component of research in our field. Concepts of human subject protection vary greatly from country to country, and also among academic disciplines. Examining fundamental human rights as a broader foundation for examining human subject protection in a transdisciplinary and transboundary field such as GIScience is a promising direction which demands our attention and further research (Albro and Richardson 2012).

21.4.6 User Interfaces and Communications

We should integrate new wireless communications bandwidth, high speed Internet options, and existing real-time GPS/GIS linkages to optimize dynamic background map displays and database use on the fly, in order to enable current real-time, real-world interactive systems and methods to incorporate change over very short time periods into GIS databases. We should enhance real-time user interfaces to match those available for computer games.

21.4.7 Analytical Tools

Existing real-time space-time interactive functions should be expanded to support additional visualization, query, and analysis specific to multiple application domains. Time-constrained decision making imposes less time for analytical processing. Broader utilization of current analytical tools such as interpolation, derived geometry, and dead reckoning to supplement GPS will develop better real-time decision support systems. Extensions of RTI GPS/GIS into indoor mapping and navigation also need additional research. Prior work from the 1990s on indoor extensions of GPS/GIS mapping using laser rangefinders and radio frequency ID (RFID) needs to be complemented by additional instrumentation to create seamless interfaces. We also should build on existing error budget assessment and QA/QC tools for identifying and measuring spatial-temporal uncertainty, and explore RTI GPS/GIS data intensive solutions to address the uncertain geographic context problem (Kwan 2012a, b).

21.4.8 Computing and Large Datasets

Although the explosion of real-time spatiotemporal data from RTI GPS/GIS functionality embedded in consumer devices and in vehicles (including piloted and unmanned aerial vehicles,[6] or UAVs) creates many new possibilities and opportunities, it also poses new challenges regarding data handling, analysis, and storage and raises new questions regarding data privacy and confidentiality. Ongoing research in distributed and cloud computing, CyberGIS (Wang 2010), and data mining of so-called big data (Li et al. 2010) represent current attempts to manage and find meaning and "caution" in large diffuse data sets. These research efforts, especially those which employ critical approaches, should be encouraged and funded. Related research also is needed to address the formidable challenges of archiving spatiotemporal data.

---

[6]RTI GPS/GIS technologies have long been used from helicopters and fixed-wing aircraft for rapid response to disasters such as earthquakes, forest fires, and floods (Mauney and Bottorff 1993). Continuous environmental sensor input to mobile RTI GPS/GIS systems was well developed by the early 1990s. Early automated sensor inputs included laser range finders, Geiger counters, air quality monitors, bathemetric sensors, radio frequency signal strength instruments, etc. Subsequent extensions of early RTI GPS/GIS laser range finder integration include the addition of multiple lasers and other features for current LIDAR (Light Detection and Ranging) systems. UAVs and other vehicles now integrate RTI GPS/GIS functionality with multiple sensor systems, including most of those available in the early 1990s.

21.4.9 Policy and Institutional Considerations

We must confront and address locational privacy and data confidentiality issues related to real-time space-time data.[7] Promising directions include the development of interactive Virtual Data Enclaves (VDE) for sharing confidential space-time data within the GIScience research community and beyond. Improved methods for spatiotemporal data access and sharing for sustainable development and environmental protection is also needed, especially in developing regions, and on terms as desired by those in the region. We should continue to foster cross-border coordination and international standards, and develop stronger public, private, and university research coordination. Finally, we must support existing and develop new advanced higher education programs to address the complex next generation research needs of real-time space-time interaction in GIScience and geography.

21.5 Transformational Research in Geography: Science, Technology, Innovation, and Discovery

Scientific and technological innovations in real-time space-time interactions have transformed geography, cartography and GIScience, and continue to drive research and applications within and well beyond these fields. Inherent within such transformative changes are interesting implications, and perhaps lessons, regarding how innovation in geography can occur and how it might be understood, nurtured and fostered in the future. In addition to the more specific research agenda suggestions related to real-time space-time integration above, I would like to close with some thoughts on the practice of GIScience itself and more broadly the practice of science in geography, and to pose a few provocative questions for consideration and discussion as we think about future directions of space-time integration research.

I believe we need to examine more closely how the actual processes, practices, and cultures of science in GIScience and geography currently function, and how they might function in the future. The process of innovation in our intertwined disciplines needs to be better understood, and documented. What is the actual role of

[7] Issues of privacy and dual use of RTI GPS/GIS technologies are also significant. I have addressed these questions in multiple venues (see for example, Richardson 2008), and as chair of the American Association for the Advancement of Science (AAAS) Science and Human Rights Coalition Steering Committee (Albro and Richardson 2012). The dilemma posed by the prospect of any scientific or technological innovation being used or abused for the wrong ends is a real and inevitable cause for concern. Norbert Weiner famously stated that, "I have seriously considered the possibility of giving up my scientific productive effort because I know of no way to publish without letting my inventions go into the wrong hands" (Roszak 1994). These issues have been the focus of a substantial body of scholarship during the past half century (see, for example, Pickles (1995). Harris and Weiner (1998), Onsrud (2003), and Chomsky (1969)).

theory and its interaction with practice and applications in GIScience? How is theory generated in GIScience? Are we over-privileging borrowed and brittle notions of theory from other disciplines in GIScience? And, more simply, what are our key theories, and in what ways do they serve or hinder innovation and discovery?

I would argue that the tendency to de-couple technologies and applications from theory in GIScience has not served us well. Research which takes a closer look at the ways in which science, technology, and applications-driven research interact in the actual practice of GIScience, as has been done in many other sciences (Pickering 1995; Kuhn 1996) is long overdue.

We also need to examine past and potential future linkages and synergies among public, private, and university real-time space time integration research. One can argue that research and innovation in the private sector in this area is as advanced as in the university, and has at times, especially earlier, outstripped research in GIScience in the universities. After all, the term GIScience was not coined until 1992 (Goodchild 1992, 2010), and a major focus of its early practice was to describe and attempt to understand what had already happened in the development of core GIS and geospatial technology functionality, and to begin to assemble it as a science. This is and has been a worthwhile and necessary undertaking, and those who engaged in this debate early on are to be commended. But it does raise the question of how all of this early development actually happened, prior to a formal GIScience that now aspires to guide it, and further, what drives what: the GIScience or the GITechnology? How did the unnamed science work then, and how does GIScience work now? What is its dynamic? What flows from what? Or is the reciprocal interaction between the technology and the science actually the real dynamic?

It is perhaps also time to examine current interactions and boundaries between GIScience and geography in more detail. In what ways are these boundaries fixed and rigid, or permeable and mobile? What does the intellectual trade across these boundaries look like? Is the flow balanced or dependent, and in which direction? Are the two fields diverging or converging?

Further, how does conceptual thinking as part of the iterative processes of overcoming research obstacles generate innovation in GIScience? What social structures of GIScience will best enable the next conceptual breakthroughs and the new methods, and what social values will shape the future of our coupled scientific and technological research program? And how do curiosity-driven research and applications-aware research actually differ in practice and intersect as drivers of GIScience, and how mutually exclusive are these categories? Current work in the philosophy and history of science has much to offer our examination of these questions. I suggest as a starting point the work of Andrew Pickering (e.g., 1995 *The Mangle of Practice)*, Helen Longino (e.g., 2002 *The Fate of Knowledge*), and Peter Galison (e.g., 2003 *Einstein's Clocks, Poincaré's Maps*).

In closing, I would like to thank all of my colleagues within the geography and GIScience communities who have contributed so constructively and critically to research at the intersection of these fields, and in advance for our continued discussions on these and related ideas in the future.

Acknowledgments I would like to thank Martin Dijst, the Editor of this special Forum on Space-Time Integration in Geography and GIScience, as well as the anonymous reviewers for their many helpful comments. I would also like to acknowledge Jean McKendry for her insightful thoughts on an earlier draft of this essay, and Martin Fox for his comments and assistance in compiling the archival materials relevant to this research. This research was supported in part by the National Cancer Institute of the NIH (R13CA162823) and the NSF (BCS-1244691). The content is solely the responsibility of the author and does not necessarily represent the official views of the NIH or the NSF.

References

Abler, R. F. (1993). Everything in its place: GPS, GIS, and geography in the 1990s. *The Professional Geographer, 45*(2), 131–139.

Abler, F., & Richardson, D. (2003). Geographic management systems for homeland security. In S. L. Cutter, D. B. Richardson, & T. J. Wilbanks (Eds.), *The geographical dimensions of terrorism* (pp. 117–124). New York: Routledge.

Albro, R., & Richardson, D. (Co-chairs). (2012). *Intersections of science, ethics and human rights: The question of human subjects protection*. Washington, DC: Science Ethics and Human Rights Working Group of the AAAS Science and Human Rights Coalition. http://srhrl.aaas.org/coalition/WG/2/Projects/HumanSubjects/Report.pdf. Accessed 1 Feb 2013.

Bower, J. L., & Christensen, C. M. (1995). Disruptive technologies: Catching the wave. *Harvard Business Review, 73*(1), 43–54.

Chomsky, N. (1969). *American power and the new mandarins*. New York: Routledge.

Cressie, N. A. C., & Wikle, C. K. (2011). *Statistics for spatio-temporal data*. New York: Wiley.

Galison, P. (2003). *Einstein's clocks, Poincaré's maps*. New York: W.W. Norton & Company Ltd.

Goodchild, M. F. (1992). Geographical information science. *International Journal of Geographical Information Systems, 6*(1), 31–45.

Goodchild, M. F. (2010). Twenty years of progress: GIScience in 2010. *Journal of Spatial Information Science, 1*, 3–20.

Hägerstrand, T. (1970). What about people in regional science? *Papers of the Regional Science Association, 24*, 7–21.

Harris, T., & Weiner, D. (1998). Empowerment, marginalization, and community-integrated GIS. *Cartography and Geographic Information Systems, 25*(2), 67–76.

Hawking, S. W. (1988). *A brief history of time: From the big bang to black holes*. Toronto: Bantam Books.

Hesse, H. (1951). *Siddhartha* (pp. 106–107). New York: New Directions Publishing and Bantam Books.

Kennedy, M. (2002). *The global positioning system and GIS* (2nd ed.). New York: Taylor & Francis.

Kuhn, T. S. (1996). *The structure of scientific revolutions* (3rd ed.). Chicago: University of Chicago Press.

Kwan, M.-P. (2004). GIS methods in time-geographic research: Geocomputation and geovisualization of human activity patterns. *Geografiska Annaler B, 86*(4), 267–280.

Kwan, M.-P. (2009). From place-based to people-based exposure measures. *Social Science & Medicine, 69*, 1311–1313.

Kwan, M.-P. (2012a). The uncertain geographic context problem. *Annals of the Association of American Geographers, 102*(5), 958–968.

Kwan, M.-P. (2012b). How GIS can help address the uncertain geographic context problem in social science research. *Annals of GIS, 18*, 245–255.

Li, Z., Ding, B., Han, J., & Kays, R. (2010). Swarm: Mining relaxed temporal moving object clusters. *Proceedings of the VLDB Endowment, 3*(1), 723–734.

Longino, H. E. (2002). *The fate of knowledge*. Princeton: Princeton University Press.
Lowenthal, D. (1961). Geography, experience, and imagination: Towards a geographical epistemology. *Annals of the Association of American Geographers, 51*, 241–260.
Mauney, T., & Bottorff, H. R. (1993, November/December). GIS and the Midwest flood. *ACSM Bulletin*, pp. 24–29.
Mauney, T., Kong, A., & Richardson, D. (1993). Interactive automated mapping system. U.S. *Patent No. 5,214,757* filed 29 September 1992 and issued 25 May 1993 (Continuation of Ser. No. 564,018, August 7, 1990).
McQuoid, J., & Dijst, M. (2012). Bringing emotions to time geography: The case of mobilities of poverty. *Journal of Transport Geography, 23*, 26–34.
Miller, H. J. (2004). What about people in geographic information science? In P. Fisher & D. Unwin (Eds.), *Re-presenting geographic information systems* (pp. 215–242). Hoboken: Wiley.
Monaghan, P. (2013, April 5). The human art of way-finding. *The Chronicle Review*, p. B17.
Onsrud, H. (2003). Openness versus security of geographic information. In S. Cutter, D. Richardson, & T. Wilbanks (Eds.), *The geographical dimensions of terrorism* (pp. 207–212). New York: Routledge.
Peuquet, D. J. (2002). *Representations of space and time*. New York: Guilford.
Pickering, A. (1995). *The mangle of practice*. Chicago: The University of Chicago Press.
Pickles, J. (1995). *Ground truth: The social implications of geographic information systems*. New York: Guilford.
Richardson, D. (1990). GeoResearch contracts for GEOLINK projects around the world. *ArcNews, 12*(2), 30.
Richardson, D. (1993). *GeoLink moves mapping into a whole new field—yours*. Washington, DC: GeoResearch, Inc.
Richardson, D. (2006). GIS&T: Transforming science and society. In D. DiBiase, M. DeMers, A. Johnson, K. Kemp, A. T. Luck, B. Plewe, E. Wentz, et al. (Eds.), *Geographic information science & technology body of knowledge* (pp. vii–x). Washington, DC: Association of American Geographers.
Richardson, D. (2008). Homeland security and context. In D. Z. Sui (Ed.), *Geospatial technologies and homeland security: Research frontiers and future challenges* (pp. vii–ix). New York: Springer.
Richardson, D. (2011). Geohistories. In M. Dear, J. Ketchum, S. Luria, & D. Richardson (Eds.), *GeoHumanities: Art, history, text at the edge of place* (pp. 209–214). New York: Routledge.
Richardson, D., McKendry, J., Goodchild, M., Kwan, M.-P., McLafferty, S., Tatalovich, Z., Stinchcomb, D., & Deeds, B. (2011). *Establishing an NIH-wide geospatial infrastructure for medical research: Opportunities, challenges, and next steps*. Washington, DC: Association of American Geographers. http://www.aag.org/AAG-NIH_Geospatial_Infrastructure. Accessed 30 Apr 2013.
Richardson, D., & Solis, P. (2004). Confronted by unsurmountable opportunities: Geography in society at the AAG's centennial. *The Professional Geographer, 56*(1), 4–11.
Richardson, D. B., Volkow, N. D., Kwan, M.-P., Kaplan, R. M., Goodchild, M. F., & Croyle, R. T. (2013). Spatial turn in health research. *Science, 339*, 1390–1392.
Roszak, T. (1994). *The cult of information*. Berkeley: University of California Press.
Wang, S. (2010). A cyberGIS framework for the synthesis of cyberinfrastructure, GIS, and spatial analysis. *Annals of the Association of American Geographers, 100*(3), 535–557.
Worboys, M. F. (2005). Event-oriented approaches to geographic phenomena. *International Journal of Geographical Information Systems, 19*(1), 1–28.
Yuan, M. (1996). Modeling semantical, temporal, and spatial information in geographic information systems. In M. Craglia & H. Couclelis (Eds.), *Geographic information research: Bridging the Atlantic* (pp. 334–347). London: Taylor & Francis.

The manufacturer's authorised representative in the EU is Springer Nature Customer Service Centre GmbH, Europaplatz 3, 69115 Heidelberg, Germany. If you have any concerns regarding our products, please contact ProductSafety@springernature.com

Printed and bound by CPI Group (UK) Ltd, Croydon, CR0 4YY
30/09/2026
02225436-0012